Handbook of AI-based Metaheuristics

Advances in Metaheuristics

Series Editors:

Anand J. Kulkarni, Symbiosis Center for Research and Innovation, Pune, India
Patrick Siarry, Universite Paris-Est Creteil, Laboratoire Images, Signaux et Systemes Intelligents, Vitry-sur-Seine, France

Handbook of AI-based Metaheuristics
Edited by Anand J. Kulkarni and Patrick Siarry

Metaheuristic Algorithms in Industry 4.0
Edited by Pritesh Shah, Ravi Sekhar, Anand J. Kulkarni, Patrick Siarry

For more information about this series please visit: https://www.routledge.com/Advances-in-Metaheuristics/book-series/AIM

Handbook of AI-based Metaheuristics

Edited by

Anand J. Kulkarni

Patrick Siarry

CRC Press is an imprint of the
Taylor & Francis Group, an **informa** business

First edition published 2022
by CRC Press
6000 Broken Sound Parkway NW, Suite 300, Boca Raton, FL 33487-2742

and by CRC Press
2 Park Square, Milton Park, Abingdon, Oxon, OX14 4RN

CRC Press is an imprint of Taylor & Francis Group, LLC

Library of Congress Cataloging-in-Publication Data

Names: Kulkarni, Anand Jayant, editor. | Siarry, Patrick, editor.
Title: Handbook of AI-based metaheuristics / edited by Anand Kulkarni, Patrick Siarry.
Description: First edition. | Boca Raton : CRC Press, 2022. | Series: Advances in metaheuristics | Includes works by Mohammed El-Abd and many others. | Includes bibliographical references and index.
Identifiers: LCCN 2021013497 | ISBN 9780367753030 (hbk) | ISBN 9780367755355 (pbk) | ISBN 9781003162841 (ebk)
Subjects: LCSH: Systems engineering--Data processing. | Artificial intelligence. | Metaheuristics. | Heuristic algorithms. | Mathematical optimization.
Classification: LCC TA168 .H3183 2022 | DDC 620/.0042028563--dc23
LC record available at https://lccn.loc.gov/2021013497

ISBN: 978-0-367-75303-0 (hbk)
ISBN: 978-0-367-75535-5 (pbk)
ISBN: 978-1-003-16284-1 (ebk)

Typeset in Times
by KnowledgeWorks Global Ltd.

Dedication

Anand J. Kulkarni would like to dedicate this handbook to his guru, Dr. S. B. Mujumdar, Honorable Chancellor of the Symbiosis International (Deemed University). It is a token of respect for his motivation, warmth, and unconditional support in all ways.

Contents

SECTION III Socio-inspired Methods

SECTION IV Swarm-Based Methods

Preface

The domain of optimization and computational intelligence is one of the inherent and essential parts of most of the disciplines associated with utilizing resources to their fullest. It is the need of today and the future. In the heart of the optimization domain are mathematical modeling of the problem and the solution methodologies. Today, the problems are becoming larger with growing complexity. Such problems are becoming cumbersome to be handled by traditional optimization methods. This motivated researchers to resort to artificial intelligence (AI)-based nature-inspired solution methodologies or algorithms. They are commonly referred to as metaheuristics. These algorithms are preferred as they are based on simple rules and can handle large complex real-world problems in a reasonable amount of time. They are classified into bio-inspired, physics-based, and socio-inspired algorithms and swarm-based algorithms.

The bio-inspired algorithms are motivated from genetic evolution of living animals and plants. The major bio-inspired optimization algorithms are evolution based, swarm based, and ecology based. The popular evolution-based algorithms are artificial neural network, genetic algorithms, differential evolution, genetic programming, evolution strategies, and so forth. The swarm-based methods are particle swarm optimization, ant colony optimization, artificial bee colony algorithm, fish swarm algorithm, intelligent water drops algorithm, bacterial foraging optimization algorithm, artificial immune system algorithm, firefly algorithm, group search optimizer, shuffled frog leaping algorithm, cat swarm optimization, gray wolf optimization algorithm, elephant search algorithm, chicken swarm optimization algorithm, moth flame optimization, and so forth. The ecology-based methods are invasive weed colony optimization, biogeography-based optimization, and so forth. The notable socio-inspired algorithms could be broadly classified into the methods inspired from socio-political ideologies, competitive behavior in sports, social and evolutionary cooperation/competition, cultural transition in humans, colonization and imperialistic competition, and memetic algorithms.

So far a wide variety of problems from almost every application domain has been successfully solved by the variations as well as by hybridized versions of these algorithms. The major domains are supply chain and transportation, packing, medicinal and healthcare, robotics and automation, image processing and steganography, machine learning, mechanical engineering design and manufacturing, real-time control systems, construction and planning, and so forth. Any development and improvement associated with the metaheuristics will lead to a huge impact on the above listed and allied areas as well as societal viewpoint.

Several researchers from industry and academia as well as Masters and PhD students around the globe are working in the metaheuristics and applications domain. *Handbook of AI-based Metaheuristics* intends to serve as a complete reference to the theoretical and mathematical formulations of the metaheuristics, their testing, and validation along with detailed illustrative solutions, solutions to certain classes of problems, and applications. In addition, this book may also serve as a platform to propose novel metaheuristic algorithms. Based on the overall domains, this book is divided into four sections: Bio-Inspired Methods, Physics and Chemistry-Based Methods, Socio-Inspired Methods, and Swarm-Based Methods.

SECTION I: SOCIO-INSPIRED METHODS

Chapter 1 by Mohammed El-Abd explains the theoretical and detailed mathematical formulation of brain storm optimization (BSO). It further presents a detailed literature review covering the contributions in the last 3 years, and it introduces the real-world application of power system reliability assessment for which the BSO is applied. The major limitations of the original BSO algorithm are also listed along with the rationale behind each one. Accordingly, the variations of the BSO developed for solving single-objective problems such as global-best BSO (GBSO), BSO with multi-information interaction (MIIBSO), BSO algorithm with flexible search length and memory-based selection (ASBSO), simplex search-based BSO (Simplex-BSO), vector grouping learning BSO (VGLBSO), multiple diversity driven BSO (MDBSO), and active learning BSO (ALBSO) are critically surveyed. Furthermore, a review of the many-objective BSO (MaOBSO), multi-objective brain storm optimization algorithm based on decomposition (MBSO/D), grid-based BSO algorithm (GMOBSO), *k*-means (KMOBSO), and group-based clustering (PMOBSO) is also provided. Importantly, the limitations and possible solutions opened the doors and avenues to explore the BSO algorithm further.

Chapter 2 by Filho et al. presents a critical review of the advances and developments of the Fish School Search algorithm (FSS) in the last decade. It includes theoretical and mathematical discussion of the variations such as improved binary FSS (IBFSS), simplified binary FSS (SBFSS), and mixed FSS (MFSS). Three approaches such as multi-objective FSS (MOFSS), multi-objective binary FSS (MOBFSS), and weight-based many objective FSS (wmoFSS) for multi- and many-objective optimizations are also discussed in detail. Moreover, two different multimodal optimization approaches such as density-based FSS (dFSS) and weight-based FSS (wFSS) are also described. Apart from these variations, two other methods for parallel processing referred to as parallel FSS (pFSS) and multi-threaded FSS (MTFSS) with an aim to accelerate the processing time are illustrated. A theoretical review of some of the important applications such as clustering, feature selection, image compression and thresholding, the knapsack problem, recommendation systems, smart grids, supply chain network planning, electrical impedance tomography, the mixed model workplace time-dependent assembly line balancing problem, and finite-element model updating is also provided in the chapter.

Chapter 3 by Liu et al. discusses an algorithm referred to as marriage in honey bees optimization (MBO). The chapter discusses in detail the honey bee colony structure and marriage process along with its artificial model representing the MBO algorithm. It is accompanied with a detailed illustration on the proposed MBO in continuous domain with five heuristics improving the solutions significantly. Experiments are conducted on several benchmark functions to evaluate the search ability of MBOs with each single heuristic and five incorporated heuristics. The benchmark functions with several dimensions are chosen. The performance of the algorithm is validated by comparing the results with contemporary optimization algorithms such as particle swarm optimization (PSO), ant colony optimization (ACO), and artificial bee colony (ABC). The computational complexity of the algorithm is also evaluated. Certain limitations of the MBO are listed, which can be of help for the readers to explore the avenues further.

The genetic algorithm (GA) architecture and its application scenario in real-world engineering optimization designs, especially in the domain of structural optimization, are of prime focus in Chapter 4 by R. Desai. Fundamental concepts and working principles with design representations available in GA are highlighted in the chapter. It also touches on several constraint handling approaches that have been implemented successfully in the optimization domain. Apart from these, the widespread applications of GA in various fields of structural optimization, deliberated with a great amount of details, are described. Most importantly, the working of the algorithm is demonstrated in those very details using a variety of examples, which underscored that the GA offers a powerful search technique with validated performance in almost all design optimizations.

SECTION II: PHYSICS AND CHEMISTRY-BASED METHODS

Chapter 5 by Hashemi et al. describes original as well as modified versions of the gravitational search algorithm (GSA) that is proposed for solving continuous problems. The chapter gives an overall survey of the exiting algorithms and provides a clear distinction between them and the

GSA. The chapter provides the GSA procedure along with the mathematical formulation. Most significantly, eight operators proposed by several researchers, which helped the algorithm evolve or modify, are illustrated in detail. The operators such as escape, Kepler, black hole, and disruption are inspired by physical theories and are related to astronomy, gravity, and relativity. The other operators include crossover, mutation, chaotic, and discrete local search. The detailed theoretical and mathematical descriptions of the variations of the GSA such as binary GSA, discrete GSA, mixed GSA, quantum GSA, constraint GSA, and multimodal GSA are also provided in the chapter. The review of several applications of the GSA and its modified versions is discussed. The applications are associated with power engineering, pattern recognition, communication engineering, control and electrical engineering, civil engineering, software engineering, mechanical engineering, and biology. It is important to mention that the chapter lists the limitations of the algorithm, which opens the door for the readers to work on.

Chapter 6 by Andrew O. Martin discusses the stochastic diffusion search (SDS) algorithm. It provides mathematical formulation involving the discrete dynamical systems model. The detailed analysis of the convergence time, convergence stability, convergence sensitivity, positive feedback, and so forth, are discussed. This chapter introduces a formalism that defines standard SDS as a set of interacting functions. It enabled variants of SDS to be described in terms of the behavior that is distinct from standard SDS. It demonstrates the modularity of SDS and the implementation of the variants for use in experiments. The formalism therefore facilitates the expression of variants of SDS in literature and in algorithmic implementation. In addition, a number of variants of SDS are also described. For each variant the individual level rules that define the variant are delineated. The advantages and disadvantages of the behavior of the variants are also discussed.

SECTION III: BIO-INSPIRED METHODS

Chapter 7 by Kashan et al. presents the league championship algorithm (LCA), which is based on the metaphor of sports competitions in leagues. The chapter describes the complete mathematical formulation of the algorithm and highlights its distinction with the other contemporary sport-based algorithms as well as nature-inspired methodologies. The underlying and important role of strengths, weaknesses, opportunities, threats (SWOT) analysis for generating new solutions is highlighted in the chapter. Further to the mathematical modeling and mechanism, the chapter reviews all the versions of the LCA developed so far by the researchers around the world since its first introduction in 2009. Along with these modified approaches, the chapter emphasizes several areas and applications to which they have been applied, viz. clustering problems, a variety of engineering problems, several types of scheduling problems, stock market predictions, transportation problems, image processing problems, and so forth. The chapter also underscores certain limitations of the algorithm and provides possible suggestive solutions and areas in which the modified versions could be tested.

A review of the use of cultural algorithms for both single- and multi-objective optimization is provided in Chapter 8 by Carlos Artemio Coello Coello. It provides a brief yet significant discussion on the theories of sociology and anthropology on which the phenomenon of cultural evolution is based. The description of the algorithm emphasizes the prominent components of the algorithm such as the population space, belief space, and communication protocol. A review of the modified versions of the cultural algorithm solving static and dynamic single-objective problems has been provided. The review emphasizes applications as well as mechanisms of modifications associated with the knowledge sources, such as situational knowledge, normative knowledge, topographical knowledge, history knowledge, and domain knowledge. The chapter also reviews a wide variety of prominent applications from electrical engineering, mechanical engineering, scheduling, image processing, finance, and so forth. Importantly, based on the review, several potential modifications and applications have been suggested, such as the use of a variety of cultural paradigms and reducing computational cost using parallelism. In addition, a need of the development of a generalized cultural algorithm-based multi-objective optimizer as well as many-objective optimizer is suggested with an emphasis on solving dynamic problems.

Chapter 9 by Toğan et al. discusses teaching-learning-based optimization (TLBO) with detailed elaboration on its theoretical development, while illustrating TLBO through numerical examples associated with minimization of time and cost of construction projects, under the category of a multi-objective optimization problem. The chapter describes in detail the original version of the TLBO as well as the detailed theoretical and mathematical discussion of modified versions such as the elitist TLBO (eTLBO) and the modified TLBO (mTLBO) approaches. The application of these TLBO versions solving the time-cost optimization (TCO) problem encountered in the field of construction management and execution is described involving mathematical formulations. The chapter illustrates two computational experiments demonstrating the application of TLBO and its variants on TCO type problems. This study lays the groundwork for future research works on applications of multi-objective optimization problems in various fields.

Chapter 10 by Yue-Jiao Gong introduces a novel optimization approach referred to as social learning optimization (SLO). The approach mimics the social learning process of humans. The motivation comes from the social learning behavior of humans that inherently exhibits the highest level of intelligence in nature. The chapter describes in detail the SLO method and associated mechanism with a focus on its major components such as attention, retention, reproduction, and motivation. In addition to the above four basic elements, human intelligence is addressed using aspects such as vicarious reinforcement and self-reinforcement. The algorithm of SLO is validated by solving several unimodal and multimodal functions of different dimensions and comparing the performance with particle swarm optimization, the genetic algorithm, and the differential evolution algorithm. In addition, the investigation on parameter setting is also carried out showing the effectiveness of certain suitable settings.

The real-world problems are inherently constrained in nature. The need of the development of a suitable constraint handling technique is underscored in Chapter 11 by Apoorva S. Shastri and Anand J. Kulkarni. In this chapter an emerging socio-inspired optimization metaheuristic referred to as multi-cohort intelligence (Multi-CI) is incorporated with a probability-based constraint handling approach. The constrained version of the Multi-CI algorithm is applied for solving four constrained test problems and two engineering design problems such as the welded beam design problem and the pressure vessel design problem. The solutions to these problems are compared with several well-known methods such as ant colony optimization, cultural algorithm, particle swarm optimization, genetic algorithm, differential evolution algorithm, and so forth. This contribution opens the door for the constrained version of the algorithm, which could be applied to real-world problems from the supply chain domain.

SECTION IV: SWARM-BASED METHODS

Chapter 12 by Teodorović et al. focuses on promoting the bee colony optimization (BCO). The chapter describes the biological background behind the BCO. Then a detailed theoretical and mathematical explanation of the BCO algorithm is provided. A significant part of this chapter is devoted to the review of recently developed versions of BCO and various recent applications including hard optimization problems as well as combinatorial optimization problems. In addition, the applications of these versions for practically important traffic, transportation and logistics, location, and scheduling domains are critically surveyed. The chapter also reviews a very recent variation of the BCO algorithm combined with fuzzy logic associated with the domain of continuous optimization, and stochastic optimization is discussed.

Chapter 13 by Magdalene Marinaki and Yannis Marinakis provides a new version of the bumble bees mating optimization (BBMO) algorithm referred to as the adaptive memory bumble bees mating optimization (ABBMO). The chapter provides a detailed theoretical and mathematical explanation of the BBMO and the ABBMO, clearly discussing the underlying differences. The practical usability of the algorithm is demonstrated by successfully solving the location routing problem with stochastic demands along with several benchmark instances. The performance is further successfully validated by comparing the computational results with other contemporary algorithms from the literature.

Chapter 14 by Rapanaki et al. proposes the parallel multi-start multi-objective glowworm swarm optimization (PMS-GSO) algorithm suitable for multi-objective routing problems. The proposed PMS-GSO is used for the solution of a complex vehicle routing problem, the multi-objective route-based energy reduction multi-depot vehicle routing problem (MERMDVRP). In all four different versions of the problem, from the one defined as the simplest, which had a problem without real-life constraints, and the one defined as the most difficult, which involves several real-world constraints based mainly to the behavior of the driver, the road, and the weather conditions. The computational performance of the proposed algorithm is validated by comparing the results with three other algorithms, such as the parallel multi-start non-dominated sorting genetic algorithm II (PMS-NSGA-II), the parallel multi-start multiobjective firefly (PMS-FIREFLY) algorithm, and the parallel multi-start multiobjective particle swarm optimization (PMS-NSPSO) algorithm by solving several benchmark instances.

Chapter 15 by Liwen Xie and Gai-Ge Wang theoretically and mathematically describes the monarch butterfly optimization (MBO) algorithm. A review of the existing algorithms is also carried out to underscore the distinct underlying principle of the MBO. The chapter critically reviews the variations of the MBO such as chaotic MBO (CMBO), opposition-based learning MBO (OMBO), opposition-based learning MBO (OPMBO), multi-strategy MBO (MMBO), self-adaptive population MBO (SPMBO), self-adaptive crossover operator MBO (GCMBO), F/T mutation, and so forth. Apart from this, hybridized versions of the MBO, such as differential evolution (LMBO-DE), differential evolution (DEMBO), differential evolution (DE-LSMBO), differential evolution (DEMBO), harmony search (HMBO), artificial bee colony (HAM), artificial bee colony (HAMBO-CHLD), firefly algorithm (MBO-FS), and simulated annealing (SAMBO) are also critically surveyed. Moreover the variants of the MBO such as binary MBO, discrete MBO, quantum MBO, and multi-objective MBO are explained in brief. The computational performance of the MBO algorithm is validated by solving several test problems as well as several large-sized knapsack test cases and further comparing them with several contemporary algorithms. Most significantly, in addition to several proven advantages of the MBO, the chapter also provided limitations of the algorithm, which may help the researchers in exploring new avenues.

Editors

Anand J. Kulkarni is an Associate Professor of Computational Intelligence Domain at the Symbiosis Center for Research and Innovation, Symbiosis International (Deemed University, Pune, India).

Patrick Siarry is a Professor of automatics and informatics at the University of Paris-Est Créteil, where he leads the Image and Signal Processing team in the Laboratoire Images, Signaux et Systèmes Intelligents (LiSSi).

List of Contributors

Hussein A. Abbass
School of Engineering and Information Technology
University of New South Wales Canberra
ACT Canberra, Australia

Isabela Maria Carneiro Albuquerque
Institut national de la recherche scientifique (INRS)
University of Quebec
Quebec City, Canada

Sreenatha Anavatti
School of Engineering and Information Technology
University of New South Wales Canberra
ACT Canberra, Australia

Alireza Balavand
Department of Industrial Engineering, Science and Research Branch
Islamic Azad University
Tehran, Iran

Hasan Basri Başağa
Civil Engineering Department Karadeniz Technical University
Trabzon, Turkey

Carmelo José Abanez Bastos-Filho
Computational Intelligence Research Group (CIRG)
University of Pernambuco (UPE)
Recife, Brazil

Carlos Artemio Coello Coello
Department of Computer Science
Researcher Cinvestav 3F
CINVESTAV-IPN
San Pedro Zacatenco
Mexico City, Mexico

Ma Guadalupe Castillo Tapia
Administration Department
UAM Azcapotzalco
Cabbage Reynosa Tamaulipas
Mexico City, Mexico

Anthony José da Cunha Carneiro Lins
Center for Research in Environmental Sciences and Biotechnology (NPCIAMB)
Catholic University of Pernambuco
Recife, Brazil

Mariana Gomes da Motta Macedo
BioComplex Laboratory
Computer Science Department
University of Exeter
Exeter, Devon, England

Rodrigo Cesar Lira da Silva
Computational Intelligence Research Group (CIRG)
University of Pernambuco (UPE)
Recife, Brazil

Tatjana Davidović
Mathematical Institute
Serbian Academy of Science and Arts Knez Mihailova
Belgrade, Serbia

Tayfun Dede
Civil Engineering Department Karadeniz Technical University
Trabzon, Turkey

Marcelo Gomes Pereira de Lacerda
Center for Informatics
Federal University of Pernambuco
Recife, Brazil

Fernando Buarque de Lima-Neto
Computational Intelligence Research Group (CIRG)
University of Pernambuco (UPE)
Recife, Brazil

Breno Augusto de Melo Menezes
Practical Computer Science Group
University of Münster
Münster, Germany

Ravindra Desai
Department of Civil Engineering
Walchand College of Engineering
Sangli, India

Clodomir Joaquim de Santana Junior
BioComplex Laboratory
Computer Science Department
University of Exeter
Exeter, Devon, England

João Luiz Vilar Dias
Computational Intelligence Research Group (CIRG)
University of Pernambuco (UPE)
Recife, Brazil

Mohammad Bagher Dowlatshahi
Department of Computer Engineering
Faculty of Engineering
Lorestan University
Khorramabad, Iran

Mohammed El-Abd
College of Engineering and Applied Sciences American University of Kuwait
Kuwait City, Kuwait

João Batista Monteiro Filho
Institut national de la recherche scientifique (INRS)
University of Quebec
Quebec City, Canada

Matthew Garratt
School of Engineering and Information Technology
University of New South Wales Canberra
ACT Canberra, Australia

Mohammadreza Gholami
ParSpion Pioneers Company
St. Shiraz, Iran
School of Computer Science and Engineering, South China University of Technology, Guangzhou, China

Yue-Jiao Gong
School of Computer Science and Engineering
South China University of Technology Guangzhou, China

Amin Hashemi
Department of Computer Engineering
Faculty of Engineering
Lorestan University
Khorramabad, Iran

Somayyeh Karimiyan
Department of Civil Engineering, Islamshahr Branch
Islamic Azad University
Islamshahr, Iran

Ali Husseinzadeh Kashan
Faculty of Industrial and System Engineering Tarbiat Modares University
Tehran, Iran

Anand J. Kulkarni
Symbiosis Center for Research and Innovation
Pune, India

Jing Liu
School of Engineering and Information Technology
University of New South Wales Canberra ACT Canberra, Australia

Magdalene Marinaki
Technical University of Crete School of Production Engineering and Management University Campus, Chania
Crete, Greece

Yannis Marinakis
Technical University of Crete School of Production Engineering and Management University Campus, Chania
Crete, Greece

Andrew Owen Martin
Department of Computing
Goldsmiths College
University of London
London, England

Hugo Amorim Neto
Computational Intelligence Research Group (CIRG)
University of Pernambuco (UPE)
Recife, Brazil

Hossein Nezamabadi-pour
Department of Electrical Engineering
Shahid Bahonar University of Kerman
Kerman, Iran

Miloš Nikolić
Faculty of Transport and Traffic Engineering
University of Belgrade, Vojvode Stepe
Belgrade, Serbia

Murilo Rebelo Pontes
Computational Intelligence Research Group (CIRG)
University of Pernambuco (UPE)
Recife, Brazil

Iraklis-Dimitrios Psychas
Technical University of Crete School of Production Engineering and Management
University Campus, Chania
Crete, Greece

Emmanouela Rapanaki
Technical University of Crete School of Production Engineering and Management
University Campus, Chania
Crete, Greece

Milica Šelmić
Faculty of Transport and Traffic Engineering
University of Belgrade, Vojvode Stepe
Belgrade, Serbia

Marwa Sharawi
College of Engineering and Applied Sciences
American University of Kuwait
Kuwait City, Kuwait

Apoorva S. Shastri
Symbiosis Institute of Technology
Symbiosis International (Deemed University)
Pune, India

Patrick Siarry
Universite Paris-Est Creteil
Laboratoire Images, Signaux et Systemes Intelligents
Vitry-sur-Seine, France

Hugo Valadares Siqueira
Department of Electrical Engineering
Federal university of Technology of Parana
Ponta Grossa-PR, Brazil

Fariba Soleimani
Razi Chemistry Research Center (RCRC), Shahreza Branch
Islamic Azad University
Isfahan, Iran

Dušan Teodorović
Faculty of Transport and Traffic Engineering
University of Belgrade, Vojvode Stepe
and
Serbian Academy of Sciences and Arts Knez Mihailova
Belgrade, Serbia

Vedat Toğan
Civil Engineering Department Karadeniz Technical University
Trabzon, Turkey

Gai-Ge Wang
Department of Information Science and Engineering
Ocean University of China
Qingdao, Shandong, China

Liwen Xie
Department of Information Science and Engineering
Ocean University of China
Qingdao, Shandong, China

Section I

Bio-Inspired Methods

1 Brain Storm Optimization Algorithm

Marwa Sharawi, Mohammadreza Gholami, and Mohammed El-Abd

CONTENTS

1.1 INTRODUCTION

The term *population-based metaheuristics* refers to one large class of algorithms that has been successfully applied to solve continuous and discrete optimization problems. *Population-based metaheuristics* maintain a population of individuals (i.e., solutions) that are updated, using some specified mechanism, over a number of iterations (i.e., generations) until some stopping criteria are met. Population-based metaheuristics are also referred to as *nature-inspired metaheuristics* or *bio-inspired metaheuristics* as their population update mechanism is usually biologically inspired. This class of algorithms could be further categorized based on the natural and/or biological inspiration behind their population update mechanism. The categories include [1] *breeding-based evolution*, *swarm intelligence*, *physics and chemistry*, *social human behavior algorithms*, *plants-based*, and *miscellaneous*.

According to the above categorization, the BSO algorithm [2, 3] is a social human behavior algorithm that mimics brainstorming processes. In a typical brainstorming process, a group of humans with different backgrounds, expertise, and abilities meet together to collect ideas or develop solutions for a problem at hand. The collected ideas could be further hybridized or enhanced to generate more ideas. Since its introduction in 2011, different researchers have studied BSO behavior, proposed improvements to BSO, hybridized BSO with other techniques, developed cooperative BSO frameworks, and applied BSO to real-world applications. Previous literature reviews on BSO can be found in [4, 5].

In this chapter, we explain the mathematical formulation of BSO, conduct a comprehensive literature survey for the last 3 years, detail how BSO is applied to one real-world application, and finally highlight promising future research directions.

The rest of this chapter is organized as follows: Section 1.2 gives details about the original BSO algorithm along with its limitations. Different improvements proposed in the literature, in the previous 3 years, to improve BSO are covered in Section 1.3. A real- world application for BSO is presented in Section 1.4. Section 1.5 concludes this chapter and proposes future research directions.

1.2 BRAIN STORM OPTIMIZATION

This section presents the mathematical formulation of BSO and highlights a number of limitations that were addressed in the literature.

1.2.1 Description of the Original BSO Algorithm

In BSO, a population is composed of a collection of *ideas*, where each idea represents a solution to the problem. In each iteration, the population is updated following three consecutive steps: *clustering*, *generation*, and *replacement*, as shown in Algorithm 1.1, where n is the population size (i.e., number of ideas) and m is the specified number of clusters.

ALGORITHM 1.1 THE BSO ALGORITHM

Require: n and m
1: Population initialization
2: Population evaluation
3: **while** stopping criteria not met **do**
4: **Cluster** population into m clusters using a clustering algorithm
5: **Generate** n new ideas
6: **Replace** the old population
7: **end while**
8: **return** best idea

The *clustering* step aims to combine similar ideas together and to help the population to converge into small regions. The clustering k-means algorithm was used in the original BSO and the best idea in each cluster was recorded as the *cluster center.*

The *generation* step is used to generate new ideas from existing ones. An intermediate individual X_n^i is generated taking information from:

1. A single cluster by setting it equal to one of the following:
   a. A cluster center of a probabilistically selected cluster cr:

$$X_n^i = center_{c_r} \tag{1.1}$$

b. A randomly selected idea j from a probabilistically selected cluster cr:

$$X_n^i = X_{cr}^j \tag{1.2}$$

2. Two clusters by setting it equal to one of the following:
   a. A random combination of the cluster centers of two probabilistically selected clusters cr_1 and cr_2:

$$X_n^i = r \times center_{cr_1} + (1-r) \times center_{cr_2} \tag{1.3}$$

   b. A random combination of two randomly selected ideas j and k from two probabilistically selected clusters cr_1 and cr_2:

$$X_n^i = r \times X_{cr_1}^j + (1-r) \times X_{cr_2}^k \tag{1.4}$$

One of the above four strategies is randomly selected based on the parameters $p_{one-cluster}$ (i.e., probability of considering a single cluster), $p_{one-center}$ (i.e., probability of considering a single center), and $p_{two-centers}$ (i.e., probability of considering two centers). A cluster is probabilistically selected based on its size (i.e., the number of ideas in the cluster).

Afterward, a perturbation step is applied to X_n^i to generate the new individual X_{new}^i using a step-size parameter ξ and Gaussian distribution as follows:

$$\xi = rand \times logsig\left(\frac{0.5 \times Max\_Iterations - Current\_Iteration}{k}\right) \tag{1.5}$$

$$X_{new}^i = X_n^i + \xi \times N(0,1) \tag{1.6}$$

where *rand* is a function that generates a uniformly distributed random number in the range [0,1), $N(0,1)$ represents a Gaussian distribution with mean 0 and a standard deviation of 1, and k is a coefficient to modify the slope of the *logsig*() function. The detailed BSO *generation* step is shown in Algorithm 1.2.

As the *logsig*() is a declining function, this results in less and less perturbation to be applied to newly generated ideas as the search progresses. Hence, BSO moves from an explorative behavior at the beginning, due to applying large perturbations, into an exploitative behavior toward the end due to having less perturbation.

Finally, in the *replacement* step, each newly generated idea X_{new}^i replaces the current X_{old}^i if it has a better fitness; otherwise, it will be discarded.

1.2.2 Limitations of the Original BSO Algorithm

The first version of the developed BSO algorithm explained above had a number of limitations that many works in the literature attempted to address.

First, the clustering stage introduced a computational burden and the adopted k-means algorithm required the number of clusters to be set in advance as a parameter of the algorithm. Earlier attempts to address this issue include the *simple grouping method* (SGM) [6], *random grouping* (RG) [7], *affinity propagation* (AP) [8], and *agglomerative hierarchical clustering* (AHC) [9].

Second, BSO lacked any re-initialization mechanism that could allow it to escape local minima. Such a mechanism exists in some other metaheuristics. For example, in the artificial bee colony (ABC) algorithm, if an individual does not improve for a predetermined number of iterations, it goes into a scout phase where it is randomly (or heuristically) re-initialized in the search space. Previous work introducing a re-initialization step was reported in [10–12].

ALGORITHM 1.2 THE BSO GENERATION STEP

Require: n, m, k, $p_{one-cluster}$, $p_{one-center}$, and $p_{two-centers}$
1: **for** each idea i **do**
2: **if** $rand < P_{one-cluster}$ **then**
3: Probabilistically select a cluster cr
4: **if** $rand < P_{one-center}$ **then**
5: $X_n^i = center_{cr}$
6: **else**
7: Randomly select an idea j in cluster cr
8: $X_n^i = X_{cr}^j$
9: **end if**
10: **else**
11: Probabilistically select two clusters cr_1 and cr_2
12: Randomly select two ideas $X_{cr_1}^j$ and $X_{cr_2}^k$
13: $r = rand$
14: **If** $rand < P_{two-centers}$ **then**
15: $X_n^i = r \times center_{cr_1} + (1-r) \times center_{cr_2}$
16: **else**
17: $X_n^i = r \times X_{cr_1}^j + (1-r) \times X_{cr_2}^k$
18: **end if**
19: **end if**
20: $\xi = rand \times logsig\left(\frac{0.5 \times Max\_Iterations - Current\_Iteration}{k}\right)$
21: $X_{new}^i = X_n^i + \xi \times N(0,1)$
22: **end for**

Third, having a fixed slope parameter k dictates that the *logsig*() function will have the same decay characteristics regardless of the function being optimized. This could result in undesired behaviors like having a sharp curve decrease, which leads to a quick switch from exploration to exploitation. This is not desirable for more complicated functions with many local optima and especially in high dimensions. Earlier attempts to propose dynamic scheduling include the works in [7, 12, 13].

Fourth, inspecting the *generation* step of the original BSO reveals that it actually involves two separate steps: *hopping* then *perturbation*. A new individual, X_{new}^i, in BSO is generated using step 21 in Algorithm 1.2. In this equation, the first term, X_n^i, is basically generated using one or more cluster centers, or one or more randomly selected individuals. Afterward, this term is perturbed and the generated individual, X_{new}^i, is compared with the old one X_{old}^i. This means that the newly generated individual X_{new}^i does not inherit any traits of the old individual X_{old}^i, with which it is compared. This is unlike other metaheuristics like genetic algorithms (GAs) and differential evolution (DE) through the crossover operator, or particle swarm optimization (PSO) through position updates in which the new individual is generated while inheriting from or updating old individuals. This renders BSO as a sort of memoryless algorithm in which there is no clear trajectory that an individual follows in the search space. Subsequently, this results in BSO being ineffective in exploring the search space. Moreover, in the case of using a single cluster center or random individual, all decision variables are passed to X^i_{new}, which again limits the explorative ability of BSO as it does not allow for cooperation or information sharing between different individuals.

1.3 LITERATURE REVIEW

Many works in the literature have been proposed to address the limitations highlighted above for the original BSO. This section presents a detailed literature review covering the improvements introduced into the BSO algorithm in the last 3 years.

1.3.1 Single-objective Optimization

A global-best BSO (GBSO) was proposed in [14]. In GBSOs, a fitness-based grouping strategy was adopted to equally allocate good individuals across different clusters. Moreover, the per-variable updates were introduced so that Eqs. (1.1) to (1.4) are repeated for every problem variable, and not only once per individual, allowing for more individuals sharing information. The same re-initialization and dynamic scheduling techniques as in [12] were incorporated. Finally, the newly generated idea was influenced by the global-best one with an increasing effect facilitated by a new parameter C as follows:

$$C = C_{min} + \frac{Current\_Iteration}{Max\_Iterations} \times (C_{max} - C_{min}) \tag{1.7}$$

$$X^i_{new} = X^i_{new} + rand(1, DimSize) \times C \times (Global\_Best - X^i_{new}) \tag{1.8}$$

Experiments proved the superior performance of the GBSO compared with the previous BSO improved algorithms. Further experiments demonstrated the highly competitive performance against other popular global-best heuristic algorithms. Moreover, GBSO exhibited a very desirable property by having an improved performance with higher dimensions on the hybrid and composition functions of the CEC14 benchmarks [15].

The authors in [16] proposed a simple BSO algorithm with a periodic quantum learning strategy (SBSO-PQLS). The clustering process was a simple individual clustering (SIU) strategy that sorts all individuals according to their fitness and then reasonably allocates them to different clusters. For the update strategy, when a new idea is generated from one cluster, it was either based on the cluster center or a combination of two randomly selected ideas in that cluster. However, when generating an idea from two clusters, this was based on a combination of two randomly selected ideas from the two clusters. In either case, when generating a combination of two ideas, this was independently done for each problem variable depending on a different random number, which is similar to the per-variable update proposed in [14]. Hence, X^i_n was defined as:

$$X^{id}_n = r_d \times X^{jd}_{cr_1} + (1 - r_d) \times X^{kd}_{cr_2} \tag{1.9}$$

where d represents the dth problem variables, cr_1 and cr_2 are the selected clusters (would be the same if the update is based on a single cluster), and j and k are the randomly selected ideas. Moreover, the work also adapted the step-size function, for each problem variable independently, according to the selected update strategy and the minimum and maximum boundaries of the search space. Finally, the work introduced a quantum-behaved individual update with periodic learning, which was applied after the above update strategy, to generate new momentum for similar ideas. SBSO-PQLS was compared with seven other BSO variants, PSO, and DE using the

CEC13 benchmark functions [17]. The results showed that SBSO-PQLS achieved a better global search performance and avoided premature convergence.

The work in [18] tested three different versions of BSO for locating multiple optima. The tested algorithms were the original BSO, BSO in objective space (BSO-OS) [19] with the Gaussian random variable, and BSO-OS with the Cauchy random variable. Algorithms were modified to preserve the found global/local optima until the end of the search. Experiments were conducted on eight benchmark functions and seven nonlinear equation system problems. Results showed that BSO-OS performed better than the fireworks algorithm (FWA) and PSO (with star topology) but worse than PSO (with the von Neumann topology) for several problems.

BSO with multi-information interaction (MIIBSO) was introduced in [20]. In MIIBSO, the clustering step was performed using the RG strategy in [7]. Moreover, MIIBSO adopted three patterns for their update strategy. For pattern I, when the update strategy is based on a single cluster, MIIBSO combined a cluster center and a random idea. This could be considered as a mixture of steps 5 and 8 in Algorithm 1.2 as follows:

$$X_n^i = r \times center_{cr} + (1-r) \times X_{cr}^j \tag{1.10}$$

For pattern II, and to remove redundancy, when updating an individual using information from two clusters, only a combination of two randomly selected ideas is considered. Finally for pattern III, which is repeated for every dimension j, Δ random ideas are selected from the population, where $2 \le \Delta \le N$, then problem variable j is extracted from the best idea among the selected Δ ideas and used to update dimension j in X_n^i. However, to reduce the computational complexity, pattern III is only adopted as a re-initialization mechanism if a certain individual has not been improved for a number of iterations. Finally, the step-size function is dynamically updated based on the search space size. MIIBSO was compared with 11 BSO algorithms and five other algorithms on the CEC13 benchmarks. MIIBSO was ranked number one when applied to problem sizes of 30 and 50.

In [21], the authors proposed a BSO algorithm with flexible search length and memory-based selection (ASBSO). To effectively explore the search space, ASBSO used different step lengths by providing M different values for the parameter k, referred to as different strategies. To adjust k, ASBSO employed two memory structures, namely success memory and failure memory. Both memories consist of M columns, one for each strategy, and L rows, for L consecutive iterations. An entry in the success(failure) memory is set to 1(0) when an individual is successfully improved using strategy m, where $1 \le m \le M$, at iteration l, where $1 \le l \le L$, and 0(1) otherwise. When an individual was updated, a strategy was probabilistically selected based on the success rates. When tested using the CEC13 benchmarks, ASBSO outperformed BSO-OS and was competitive with GBSO.

The authors in [22] introduced a simplex search-based BSO (simplex-BSO), which combined the explorative ability of BSO with the local search ability of the Nelder-Mead simplex (NMS) method. First, the BSO clustering and generation steps were modified. The clustering step only generated two clusters referred to as elites (the best 20% individuals) and normals. As for the generation step, an individual was only generated using elites (one or a combination of two) or normals (one or a combination of two) but not both. Second, the NMS method was applied in every iteration to exploit the area around the best idea. The NMS method only consumed $40 \times D$ function evaluations, where D is the problem size, before returning the best-found solution. Simplex-BSO was tested using the CEC17 benchmarks [23], with dimensionality 2 and 10. Experimental results showed that simplex-BSO provided a better performance than BSO and NMS as well as being the best algorithm at any given computational cost.

A vector grouping learning BSO (VGLBSO) was proposed in [24]. The clustering stage adopted the RG strategy. The update strategy considered the information exchange between subvectors of individuals. This could be considered as an in-between strategy with the two extremes as the entire individual update in the original BSO and the per-variable update in GBSO. Each individual is stochastically divided into τ subvectors with at least Δ problem variables each. Two new individual generating patterns, A and B, were devised to emphasize local search and global search, respectively. In pattern A, one cluster, cr, and one individual from that cluster, X_{cr}^{j}, were randomly selected. The new individual, X_{n}^{i}, was generated based on combining subvectors from the selected cluster center, $center_{cr}$, and the selected individual, X_{cr}^{j}, or by combining subvectors from the randomly selected individual, X_{cr}^{j}, and another randomly selected one, X_{cr}^{k} (k changes for every subvector). This is illustrated in the following equation:

$$S(X_{n}^{i},t)=\begin{cases} r\times S(X_{cr}^{k},t)+(1-r)\times S(X_{cr}^{j},t) & if \quad rand\leq 0.5 \\ r\times S(center_{cr},t)+(1-r)\times S(X_{cr}^{j},t) & otherwise \end{cases} \tag{1.11}$$

where $S(X_{n}^{i},t)$ denotes subvector t in X_{n}^{i}, $S(X_{cr}^{j},t)$ denotes subvector t in X_{cr}^{j}, $S(X_{cr}^{k},t)$ denotes subvector t in X_{cr}^{k}, and $S(center_{cr},t)$ denotes subvector t in the cluster center. On the other hand, pattern B considered information from different individuals that belong to two different clusters to the entire population (i.e., expressed as the mean of all cluster centers). It did not only rely on stochastic combinations as pattern A, but also included differential information. Hence, pattern B focused on the global exploration. Furthermore, to generate X_{new}^{i}, a perturbation is applied to X_{n}^{i} by either the step function combined with Gaussian distribution as in the original BSO or by adding a differential term that is similar to DE. The CEC13 benchmarks library was used to compare VGLBSO with another 12 BSO variants for a problem size of 50. Experimental results showed that VGLBSO ranked number 1 among the compared algorithms followed by GBSO.

To enhance the population diversity and make effective exploration/exploitation balance, the work in [25] combined BSO with the sine-cosine algorithm (SCA) [26]. In EBS-SCA, RG is used to cluster ideas. Meanwhile, an exploration phase was followed for the first 80% of iterations, while an exploitation phase was followed in the last 20%. In general, and to generate X_{n}^{i}, EBS-SCA combined updating the entire individual as in the original BSO with a per-variable update as in GBSO. For exploration, and after generating X_{n}^{i}, X_{new}^{i} was generated as follows:

$$X_{new}^{ij}=\begin{cases} X_{n}^{ij}+\lambda_1\times sin(\lambda_2)\times\lambda_3\times\left|X_{n}^{ij}-X_{old}^{ij}\right| & if \quad \lambda_4<0.5 \\ X_{n}^{ij}+\lambda_1\times cos(\lambda_2)\times\lambda_3\times\left|X_{n}^{ij}-X_{old}^{ij}\right| & otherwise \end{cases} \tag{1.12}$$

However, for exploitation, X_{new}^{i} was generated as follows:

$$X_{new}^{ij}=\begin{cases} X_{n}^{ij}+\lambda_1\times \sin(\lambda_2)\times\left|\lambda_3\times Global\_Best^{j}-X_{old}^{ij}\right| & if \quad \lambda_4<0.5 \\ X_{n}^{ij}+\lambda_1\times \cos(\lambda_2)\times\left|\lambda_3\times Global\_Best^{j}-X_{old}^{ij}\right| & otherwise \end{cases} \tag{1.13}$$

Note that these equations provided, to the best of our knowledge, the first attempt to generate new individuals based on old ones (using differential information). Experiments run using the CEC13 benchmarks [17] showed that EBS-SCA ranked first among six different algorithms followed closely by GBSO.

A multiple diversity driven BSO (MDBSO) was proposed in [27] to adapt the BSO parameters based on the population diversity. Two diversity measures were used, namely distance based and fitness based. Clustering was carried out using the k-means algorithm, while the update equations were completely replaced by two mutation operators, i.e., BLX-α and Gaussian. One of the two mutation operations was selected based on a probability that is adapted using the population diversity. MDBSO was tested using the CEC2017 benchmarks and had the best performance compared with other BSO variants followed by RGBSO and GBSO.

The work in [28] introduced an improved BSO (RMBSO) algorithm in which a slight relaxation selection and ensembles were used to improve the performance. In RMBSO, three update schemes were utilized. In the first two, generating X_n^i is as in the original BSO algorithm, then to generate X_{new}^i, either a *current-to-pbest/1* or *current-to-rand/1* perturbation approach was employed. In the third scheme, X_{new}^i was directly generated using a new *triangulation mutation* approach. In addition, the replacement step was slightly relaxed by accepting new ideas that are better than either one of the previous two individuals. RMBSO was evaluated using the CEC05 benchmarks [29] and compared with a number of BSO variants. Among the compared variants, RMBSO was ranked at number 1 followed by GBSO.

In [30], the authors proposed an active learning BSO (ALBSO) with a dynamically changed cluster cycle. For the clustering stage, the k-means algorithm was only executed after a random number of generations based on a new parameter d_c. For the update stage, when generating X_n^i from within a single cluster, ALBSO used the difference information between the cluster center and the mean individual of the population or the difference information between the cluster center and a random individual. However, when generating X_n^i using two clusters, the difference information between two cluster centers or the difference information between the cluster centers and a random individual was used. Although ALBSO showed a better performance than other metaheuristics when applied to the CEC05 benchmarks, it was not compared with other BSO variants.

Population interaction networks (PINs) were used in [31] to theoretically analyze the performance of BSO. The PIN was used to capture the relationships among individuals in BSO and was updated after every iteration. The work concluded that in lower dimensions, BSO produced good results as the population interaction of BSO follows a power law distribution. This is deemed effective for seeking an optimal solution. However, in higher dimension, the population interaction did not follow such a law; hence, the performance deteriorated. Through further experiments, the work proposed a combination of parameter settings that could be used to produce better results. Finally, the authors proposed that BSO should be modified to make sure that the population completely obeys a power law distribution to enhance the performance in higher dimensions.

1.3.2 Multi-objective Optimization

The authors in [32] developed an improved BSO algorithm for many-objective optimization. The many-objective BSO (MaOBSO) algorithm adopted new clustering methods with reference point allocation to increase the diversity and decision variables clustering to divide the variables into convergence-related and diversity-related variables. Convergence-related variables were necessary to find Pareto solutions that are close to the Pareto front. On the other hand, diversity-related variables makes the Pareto solutions more uniform. To optimize these variables, different approaches were taken. Roulette wheel selection and binary crossover were used to optimize convergence-related variables, while selection and mutation from the original BSO were used to optimize diversity-related variables. MaOBSO provided a superior performance when compared with NSGA-III [33] and MOEA/DD [34] when applied to the DTLZ [35] and MaF [36] benchmarks.

A multi-objective BSO algorithm based on decomposition (MBSO/D) was proposed in [37]. Using decomposition, every multi-objective problem is decomposed into a series of problems using weight vectors and aggregate functions. In MBSO/D, each given weight vector provided a cluster, and the best solution of each cluster was determined by the corresponding weight vector aggregation function. Moreover, the update equations were adopted from MOEA/D [38] while a new solution was generated using either one or three clusters. MBSO/D showed competitive performance when compared with other decomposition-based algorithms MOEA/D and MOHS/D [39] using the CEC09 [40] and DTLZ benchmarks.

Dynamic multi-objective optimization was tackled in [41] using a grid-based BSO algorithm (GMOBSO). Grid-based clustering was used to evenly divide the objective space across each objective and group the individuals located in the same grid into a cluster. A hybrid mutation strategy combining Gaussian-, Cauchy- and chaotic-based mutation operators was used to enhance the diversity and avoid premature convergence. The algorithm was tested using eight dynamic multi-objective benchmarks and compared with other BSO versions using different clustering techniques including *k*-means (KMOBSO) and group-based clustering (PMOBSO). Experimental results showed the improved performance of GMOBSO as it could find robust Pareto optimal solutions approximating the true Pareto-front and had longer survival times.

1.3.3 Constrained Optimization

To solve constrained optimization problems, the work in [42] combined a modified BSO (MBSO) algorithm with a simplified constraint consensus (CC) method (MBSO-R+V). The MBSO version used a simple grouping technique by randomly selecting cluster seeds from the population and assigning each idea in the same cluster with the closest seed. Moreover, the Gaussian perturbation term is replaced with a differential term while generating some problem variables randomly. MBSO was shown to be effective in constrained optimization [43]. To handle infeasible solutions, a new CC variant called restriction+violated (R+V) was proposed. In R+V, only the feasibility vector of the hardest constraint in turn is considered. Hence, only the gradient of that constraint is computed reducing the computational cost. MBSO-R+V was shown to be competitive with other state-of-the-art algorithms for constrained optimization problems. The work represented an extension of the same authors' previous work in [44].

1.4 REAL-WORLD APPLICATION

BSO and its variants have been successfully applied to many real-world applications including feature selection [45–47], fault detection [48], cloud computing [49], optimization of photovoltaic models [50], robotics and unmanned aerial vehicles (UAVs) [51–53], electromagnetic applications [54], vehicle routing [55], quality of service [56], renewable energy [57], energy optimization [58], matching ontologies [59], image classification [60], flowshop scheduling [61], graph planarization [62], virtual network mapping [63], wireless sensor network deployment [8], artificial neural networks [64], and emission prediction [65]. In this section, BSO is applied to one real-world problem: the power system reliability assessment problem.

1.4.1 Power System Reliability Assessment

Reliability of the power supply is defined as the ability of a power system to supply customers in an acceptable degree of adequacy and security. Reliability plays an important role in power system planning. Within this competitive environment, fast and accurate power system reliability assessment techniques are necessary power systems where reliability is considered as a constraint. As shown in Figure 1.1, assessment methods in power systems are mainly applied to three different hierarchical levels (HLs) [66]. At HLI, the total system generation is examined to

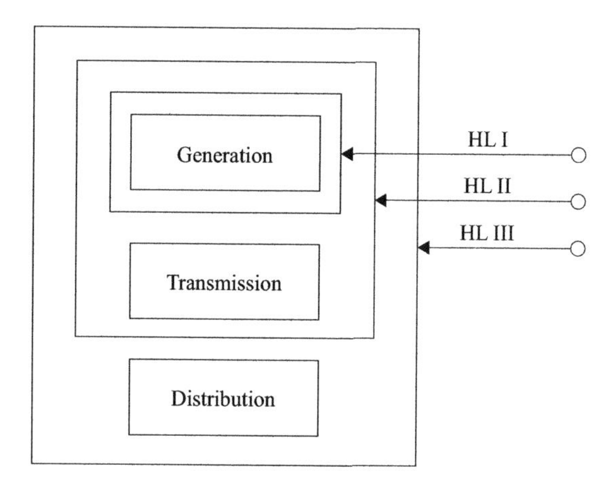

FIGURE 1.1 Power system hierarchical levels (HLs).

determine its adequacy to meet the total system load requirements. At this level, the transmission lines are ignored and considered as completely reliable elements. In HLII studies, the adequacy analysis is usually termed composite system or bulk transmission system evaluation. The HLIII studies include all the three functional zones of the power system, starting at generation points and terminating at the individual consumer load points. In this experimental study, the reliability indices of generation (HLI) and composite power system (HLII) are calculated.

Both analytical methods and simulation-based methods could be applied to this problem. However, analytical methods are not suitable for large systems, while simulation-based methods have an increased computational burden when increasing the reliability level. Using population-based metaheuristics offer the balance when applied to the reliability assessment of a power systems problem. Previous works showed that such methods can reduce the amount of sample states, reduce the computational time, and achieve better coverage [67, 68].

1.4.2 Problem Definition

In a power system reliability study, each element (generation unit of transmission line) has its own force outage rate (FOR), which is calculated using failure μ and repair rate λ:

$$FOR_i = \frac{\lambda_i}{\lambda_i + \mu_i} eq1 \tag{1.14}$$

Each power system state is defined according to the elements status (1 = on and 0 = off). A probability value PS is calculated for each state j as follows:

$$Availability_i = \begin{cases} FOR_i & if \quad status = 0 \\ 1 - FOR_i & otherwise \end{cases} \tag{1.15}$$

$$PS_j = \prod_{i=1}^{N_g - N_t} Availability_i \tag{1.16}$$

where N_g is the number of generation units and N_t is the number of transmission lines.

In addition, each system state j should be evaluated as a success or failure state. The state is evaluated as a success if the generated power is more than the hourly demand. Otherwise, if the generated power is less than the hourly demand, the system state is considered a failure. In case of a failure state, the difference of the generated power and the demand is referred to as the energy not supplied (ENS). But, in composite power system reliability studies, evaluation is more complicated and a linear programming minimization problem based on DC load flow equations are used to evaluate the sampled state (described in detail in next section). If it is a failure state the amount of load curtailment for the whole system and the share of each load bus in this amount will be determined.

Finally, after evaluation of all possible states of the system, reliability indices such as LOLE (loss of load expected) and EENS (expected energy not supply) can be calculated for the whole system.

1.4.3 Idea Encoding

GBSO is applied in this work to the power system reliability application. In GBSO, each idea represents a solution to the problem. Hence, each idea represents a system state in which all generation units/transmission lines are considered as either on (up state) or off (down state). In this case, if a generation unit status is set to 1, the generation unit is assumed to generate power equal to its capacity. If it is 0, then the generation unit is down and no power is generated. Regarding the transmission line, if the status is 0, this line is considered to be out of service. The fitness of an idea is the probability calculated according to the availability of units and transmission lines in Eq. 1.16. One idea representing the state of system is shown in Figure 1.2. In this state all of the units are up and only one line (number 8) is down.

To calculate reliability indices, only failure states are considered. Regarding the generation system (HLI) this is simple stage. The generated power of each system state (idea) is calculated using Eq. 1.14. If the generated power is more than load demand, then it is considered as a success state. If generated power is less than load demand, then it is considered a failure state and used to calculate reliability indices. To calculate the generated power, the following equation is utilized:

$$G_j = \sum_{i=1}^{N_g} g_i \times b_i \tag{1.17}$$

where G_j is the generated power of state j, N_g is the number of generation units, g_i is the capacity of unit i, and b_I is the status of unit i.

In composite power system reliability studies (HLII), the calculation is more complicated. To recognize that a state (idea) is a failure or not, a minimization linear programming problem is solved. Simply, an optimized DC load flow is done to determine the contribution of each bus in load curtailment with a focus on minimizing the total not supplied power. If it is a failure state,

Generation units at bus I				Generation units at bus II							Transmission Lines								
20 MW	10 MW	40 MW	40 MW	20 MW	20 MW	20 MW	20 MW	20 MW	5 MW	5 MW	L1	L2	L3	L4	L5	L6	L7	L8	L9
1	1	1	1	1	1	1	1	1	1	1	1	1	1	1	1	1	1	0	1

FIGURE 1.2 The idea encoding scheme.

the amount of load curtailment (LC) for the whole system and the share of each load bus in this amount are determined. These values are saved for each sampled state and then used to calculate adequacy indices for the whole system and for load buses. Each state is evaluated using a linear programming optimization model based on DC load flow equations. For the sampled state, based on network configuration and status of transmission lines, the susceptance matrix B is modified. The amount of available real power generation at each PV bus (generation bus) is also updated according to the status of generation units installed at such a bus. In this optimization problem, the power generated at each bus and status of transmission lines along with the configuration of system and transmission line parameters are known. Moreover, the minimum and maximum amounts of transmission line capacity and amount of peak load in each bus are given. The power generated and angle at each bus (except the reference bus) are variables of the problem. The objective of this optimization linear program is to minimize the total LC for the whole system:

$$min\sum_{i=1}^{n} LC_i \tag{1.18}$$

where LC_i is the load curtailment at bus i and n is the total number of system buses.

This optimization problem is subject to the following constraints:

- Real power balance at each system bus:

$$PD_i = PG_i - \left(\sum_{i=1}^{k} p_{ij} - LC_i\right) \tag{1.19}$$

where PD_i is the load demand at bus i at the peak load condition, PG_i is the real power generation at bus i, and P_{ij} is the power flow from bus i to j.
- Real power flow limits on each transmission line:

$$P_{ij}^{min} \leq p_{ij} \leq p_{ij}^{max} \quad (i = 1, 2, \ldots N_t) \tag{1.20}$$

where N_t is the number of transmission lines.
- Maximum amount of load curtailed at each load bus:

$$0 \leq LC_i \leq PD_i \quad (i = 1, 2, \ldots N_t) \tag{1.21}$$

where PD_i is the load demand at bus I at the peak load condition.
- Maximum and minimum available real power at each PV bus:

$$PG_i^{min} \leq PG_i \leq PG_i^{max} \quad (i = 1, 2, \ldots N_v) \tag{1.22}$$

where N_v is the number of buses that have installed generation.

Finally, the power flow p_{ij} is defined as:

$$p_{ij} = \frac{\theta_i - \theta_j}{X_{ij}} \tag{1.23}$$

where θ_i and θ_j are the angles at buses i and j (with bus 1 assumed to be the reference bus with $\theta_1 = 0$) and X_{ij} is the reactance value for line i to j (which is provided in the transmission line data).

TABLE 1.1
Reliability Results for the Compared Algorithms—RTS-79

Measure	Unit Addition	GA	GBSO
LOLE	9.355	9.324	**8.890**
EENS	1168	1163	**1085**

1.4.4 Experimental Results

In this section, GBSO, PSO, and GA are compared when applied to the problem at hand. All algorithms have a population size of 50. For GBSO, all parameters are set similar to [14]. For PSO, $w = 0.7$, $c_1 = 1.9$, and $c_2 = 2.1$. For GA, the mutation probability $P_m = 0.04$ and is applied by flipping one randomly selected chromosome. A one-point crossover operator is applied with a probability $P_c = 0.8$. All algorithms are run for 20 independent runs and 1000 iterations per a single run.

GBSO is used to decrease the amount of sampled states needed to calculate the reliability indices. In other words, instead of evaluating all possible states $2^{N_t+N_g}$, the most probable failure states are sampled and saved using BSO. Each system state is considered as an idea in BSO. First, a set of system states is generated randomly. The fitness of an idea is defined as state probability PS_j. In the adopted GBSO, fitness-based grouping is used to cluster system states with the best state (state with the highest probability) selected as the cluster center. In each iteration, new sample states are generated and failure states are saved and used to calculate the reliability indices.

GBSO is tested on RTS-79, which consists of 32 generating units. Table 1.1 shows a comparison between the unit addition method, GA [69] and GBSO.

GBSO has also been tested on the Roy Billinton Test System (RBTS), which has 11 generation units, installed at bus 1 and 2, and 9 transmission lines, as shown in Figure 1.3. The results of GBSO are given in Table 1.2. The results are compared with sequential sampling and state sampling methods in [70] and PSO in [71].

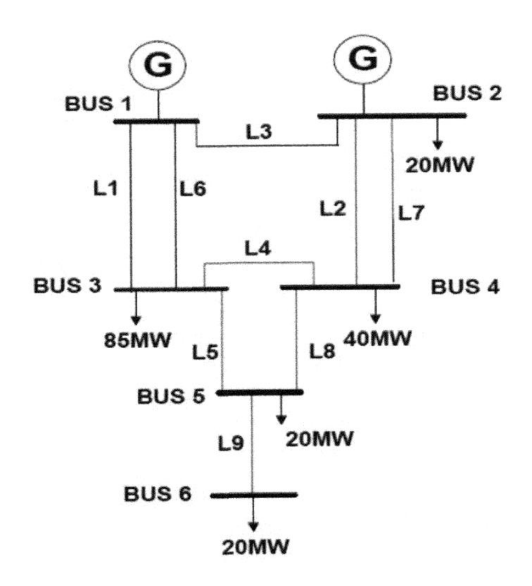

FIGURE 1.3 The RBTS system.

TABLE 1.2
Reliability Results for the Compared Algorithms—RBTS

Measure	Sequential Sampling	State Sampling	GA	PSO	GBSO
LOLE	86.399	88.58	85.02	85.79	**80.41**
EENS	1081.01	1082.63	1079.4	1081.02	**998.67**

1.5 CONCLUSIONS AND FUTURE RESEARCH DIRECTIONS

In this chapter, the BSO algorithm was presented and explained. A comprehensive literature survey covering the last 3 years was conducted. Moreover, GBSO was shown to be applicable to many real-world applications. A simple study was conducted to present the applications of GBSO to the power system reliability problem and compared with PSO and GA.

Based on the conducted literature surveys, a number of promising future research directions are identified as follows:

- Although BSO has many parameters (population size, cluster size, and probabilities), only one identified publication attempted to carry a full sensitivity study to identify the best combination of parameters settings for BSO. An important direction is to develop an adaptive mechanism for these parameters to alleviate the burden of having to set these parameters for different problems.
- Only one identified publication attempted to address the identified limitation of the update equation. More research work could be conducted on improving the update strategies to build on old information.
- The application of BSO algorithms has not been fully explored in large-scale optimization.
- Although many BSO variants emerged in the last 3 years, new variants do not often compare with all older variants but rather a small set. A good direction is to compare the most prominent BSO variants up to date with different benchmarks to identify the state-of-the-art algorithms among these variants.

REFERENCES

1. Javier Del Ser Salvador Garcia Amir Hussain Daniel Molina, Javier Poyatos and Francisco Herrera. Comprehensive taxonomies of nature- and bio-inspired optimization: Inspiration versus algorithmic behavior, critical analysis and recommendations. https://arxiv.org/abs/2002.08136, 2020.
2. Y. Shi. Brain storm optimization algorithm. In *Proc. International Conference on Swarm Intelligence*, pages 303–309, 2011.
3. Y. Shi. An optimization algorithm based on brainstorming process. *International Journal of Swarm Intelligence Research*, 2(4):35–62, 2011.
4. S. Cheng, Q. Qin, J. Chen, and Y. Shi. Brain storm optimization algorithm: A review. *Artificial Intelligence Review*, 46:1–14, 2016.
5. Shi Cheng, Yifei Sun, Junfeng Chen, Quande Qin, Xianghua Chu, Xiujuan Lei, and Yuhui Shi. A comprehensive survey of brain storm optimization algorithms. In *Proc. IEEE Congress on Evolutionary Computation*, pages 1637–1644, 2017.
6. Z. Zhan, J. Zhang, Y. Shi, and H. Liu. A modified brain storm optimization algorithm. In *Proc. of IEEE Congress on Evolutionary Computation*, pages 1969–1976, 2012.
7. Z. Cao, Y. Shi, X. Rong, B. Liu, Z. Du, and B. Yang. Random grouping brain storm optimization algorithm with a new dynamically changing step size. In *Proc. International Conference on Swarm Intelligence*, pages 364–387, 2015.

8. J. Chen, S. Cheng, Y. Chen, Y. Xie, and Y. Shi. Enhanced brain storm optimization algorithm for wireless sensor networks deployment. In *Proc. International Conference on Swarm Intelligence*, pages 373–381, 2015.
9. J. Chen, J. Wang, S. Cheng, and Y. Shi. Brain storm optimization with agglomerative hierarchical clustering analysis. In *Proc. 7th International Conference on Swarm Intelligence, ICSI*, pages 115–122, 2016.
10. S. Cheng, Y. Shi, Q. Qin, Q. Zhang, and R. Bai. Population diversity maintenance in brain storm optimization algorithm. *Journal of Artificial Intelligence and Soft Computing Research*, 4(2):83–97, 2014.
11. S. Cheng, Y. Shi, Q. Qin, T. O. Ting, and R. Bai. Maintaining population diversity in brain storm optimization algorithm. In *Proc. Of IEEE Congress on Evolutionary Computation*, pages 3230–3237, 2014.
12. Mohammed El-Abd. Brain storm optimization algorithm with re-initialized ideas and modified step size. In *Proc. IEEE Congress on Evolutionary Computation*, pages 2682–2686, 2016.
13. D. Zhou, Y. Shi, and S. Cheng. Brain storm optimization algorithm with modified step-size and individual generation. In *Proc. International Conference on Swarm Intelligence*, pages 243–252, 2012.
14. M. El-Abd. Global-best brain storm optimization algorithm. *Swarm and Evolutionary Computation*, 37:27–44, 2017.
15. J. J. Liang, B-Y. Qu, and P. N. Suganthan. Problem definitions and evaluation criteria for the CEC 2014 special session and competition on single objective real-parameter numerical optimization. Technical report, Computational Intelligence Laboratory, Zhengzhou University, Zhengzhou, China and Nanyang Technological University, Singapore, 2014.
16. Zhenshou Song, Jiaqi Peng, Chunquan Li, and Peter X. Liu. A simple brain storm optimization algorithm with a periodic quantum learning strategy. *IEEE Access*, 6:19968–19983, 2017.
17. J. J. Liang, B. Y. Qu, P. N. Suganthan, and A. G. Hernandez-Diaz. Problem definitions and evaluation criteria for the CEC2013 special session on real-parameter optimization. Technical Report 201212, Zhengzhou University, China, 2013.
18. Shi Cheng, Junfeng Chen, Xiujuan Lei, and Yuhui Shi. Locating multiple optima via brain storm optimization algorithms. *IEEE Access*, 6:17039–17049, 2018.
19. Yuhui Shi. Brain storm optimization algorithm in objective space. In *Proc. IEEE Congress on Evolutionary Computation*, pages 1227–1234, 2015.
20. Chunquan Li, Zhenshou Song, Jinghui Fan, Qiangqiang Cheng, and Peter X. Liu. A brain storm optimization with multi-information interactions for global optimization problems. *IEEE Access*, 6:19304–19323, 2018.
21. Yang Yu, Shangce Gao, Yirui Wang, Jiujun Cheng, and Yuki Todo. ASBSO: An improved brain storm optimization with flexible search length and memory-based selection. *IEEE Access*, 6:36977–36994, 2018.
22. Chunquan Li, Dujuan Hu, Zhenshou Song, Feng Yang, Zu Luo, Jinghui Fan, and Peter X. Liu. A vector grouping learning brain storm optimization algorithm for global optimization problems. *IEEE Access*, 6:78193–78213, 2018.
23. N. H. Awad, M. Z. Ali, J. J. Liang, B. Y. Qu, and P. N. Suganthan. Problem definitions and evaluation criteria for the CEC 2017 special session and competition on single objective bound constrained real-parameter numerical optimization. Technical report, Nanyang Technological University, Singapore, 2017.
24. Wei Chen, Yingying Cao, Shi Cheng, Yifei Sun, Qunfeng Liu, and Yun Li. Simplex search-based brain storm optimization. *IEEE Access*, 6:75997–76006, 2018.
25. Chunquan Li, Zu Luo, Zhenshou Song, Feng Yang, Jinghui Fan, and Peter X. Liu. An enhanced brain storm sine cosine algorithm for global optimization problems. *IEEE Access*, 7:28211–28229, 2019.
26. Seyedali Mirjalili. SCA: A sine cosine algorithm for solving optimization problems. *Knowledge-Based Systems*, 96:120–133, 2016.
27. Yang Yu, Shangce Gao, Yirui Wang, Zhenyu Lei, Jiujun Cheng, and Yuki Todo. A multiple diversity-driven brain storm optimization algorithm with adaptive parameters. *IEEE Access*, 7:126871–126888, 2019.

28. Yuehong Sun, Jianxiang Wei, Tingting Wu, Kelian Xiao, Jianyang Bao, and Ye Jin. Brain storm optimization using a slight relaxation selection and multi-population based creating ideas ensemble. *Applied Intelligence*, 50:3137–3161, 2020.
29. P. N. Suganthan, N. Hansen, J. J. Liang, K. Deb, Y.-P. Chen, A. Auger, and S. Tiwari. Problem definitions and evaluation criteria for the CEC 2005 special session on real-parameter optimization. Technical Report 2005005, ITT Kanpur, India, 2005.
30. Zijian Cao and Lei Wang. An active learning brain storm optimization algorithm with a dynamically changing cluster cycle for global optimization. *Cluster Computing*, 22:1413–1429, 2019.
31. Yirui Wang, Shangce Gao, Yang Yu, and Zhe Xu. The discovery of population interaction with a power law distribution in brain storm optimization. *Memetic Computing*, 11:65–87, 2019.
32. Yali Wu, Xinrui Wang, Yulong Fu, and Guoting Li. Many-objective brain storm optimization algorithm. *IEEE Access*, 7:186572–186586, 2019.
33. K. Deb and H. Jain. An evolutionary many-objective optimization algorithm using reference-point-based nondominated sorting approach, Part I: Solving problems with box constraints. *IEEE Transactions on Evolutionary Computation*, 18(4):577–601, 2014.
34. K. Li, K. Deb, Q. Zhang, and S. Kwong. An evolutionary many-objective optimization algorithm based on dominance and decomposition. *IEEE Transactions on Evolutionary Computation*, 19(5):694–716, 2015.
35. K. Deb. Multi-objective genetic algorithms: Problem difficulties and construction of test problems. *Evolutionary Computation*, 7(3):205–230, 1999.
36. R. Cheng, M. Li, Y. Tian, X. Zhang, S. Yang, Y. Jin, and X. Yao. A benchmark test suite for evolutionary many-objective optimization. *Complex Intelligent Systems*, 3(1):67–81, 2017.
37. Cai Dai and Xiujuan Lei. A multiobjective brain storm optimization algorithm based on decomposition. *Complexity*, 2019:5301284.
38. Q. Zhang and H. Li. MOEA/D: a multiobjective evolutionary algorithm based on decomposition. *IEEE Transactions on Evolutionary Computation*, 11(6):712–731, 2007.
39. I. A. Doush, M. Q. Bataineh, and M. El-Abd. The hybrid framework for multi-objective evolutionary optimization based on harmony search algorithm. In *Real-Time Intelligent Systems*, pages 134–142, 2017.
40. Q. F. Zhang and P. N. Suganthan. Final report on CEC'09 MOEA competition. Technical report, The School of CS and EE, University of Essex, UK and School of EEE, Nangyang Technological University, Singapore, 2009.
41. Yinan Guo, Huan Yang, Meirong Chen, Dunwei Gong, and Shi Cheng. Grid-based dynamic robust multi-objective brain storm optimization algorithm. *Soft Computing*, 24:7395–7415, 2020.
42. Adriana Cervantes-Castillo and Efren Mezura-Montes. A modified brain storm optimization algorithm with a special operator to solve constrained optimization problems. *Applied Intelligence*, 50:4145–4161, 2020.
43. Farhad Pourpanah, Ran Wang, Xizhao Wang, Yuhui Shi, and Danial Yazdani. MBSO: A multi-population brain storm optimization for multimodal dynamic optimization problems. In *Proc. IEEE Symposium Series on Computational Intelligence*, pages 673–679, 2019.
44. Adriana Cervantes-Castillo and Efren Mezura-Montes. A study of constraint-handling techniques in brain storm optimization. In *Proc. IEEE Congress on Evolutionary Computation*, pages 3740–3746, 2016.
45. Farhad Pourpanah, Chee Peng Lim, Xizhao Wang, Choo Jun Tan, Manjeevan Seera, and Yuhui Shi. A hybrid model of fuzzy min-max and brain storm optimization for feature selection and data classification. *Neurocomputing*, 333:440–451, 2019.
46. Farhad Pourpanah, Yuhui Shi, Chee Peng Lim, Qi Hao, and Choo Jun Tan. Feature selection based on brain storm optimization for data classification. *Applied Soft Computing*, 80:761–775, 2019.
47. Farhad Pourpanah, Ran Wang, and Xizhao Wang. Feature selection for data classification based on binary brain storm optimization. In *Proc. 6th International Conference on Cloud Computing and Intelligent Systems*, pages 108–113, 2019.
48. Noaya Otaka, Yoshikazu Fukuyama, Yu Kawamura, Kenya Murakami, Adamo Santana, Tatsuya Lizaka, and Tetsuro Matsui. Refrigerated showcase fault detection by a correntropy based artificial neural network using modified brain storm optimization. In *Proc. IEEE Symposium Series on Computational Intelligence*, pages 808–814, 2019.

49. S. Thanga Revathi, N. Ramaraj, and S. Chithra. Brain storm-based whale optimization algorithm for privacy protected data publishing in cloud computing. *Cluster Computing*, 22:3521–3530, 2019.
50. Zeyuan Yan, Chunquan Li, Zhenshou Song, Liwei Xiong, and Chaoyang Luo. An improved brain storming optimization algorithm for estimating parameters of photovoltaic models. *IEEE Access*, 7:77629–77641, 2019.
51. Qian Zhou and She sheng Gao. 3D UAV path planning using global-best brain storm optimization algorithm and artificial potential field. In *Proc. International Conference on Intelligent Robotics and Applications*, pages 765–775, 2019.
52. Eva Tuba, Ivana Strumberger, Dejan Zivkovic, Nebojsa Bacanin, and Milan Tuba. Mobile robot path planning by improved brain storm optimization algorithm. In *Proc. IEEE Congress on Evolutionary Computation*, pages 1–8, 2018.
53. Cong Zhang, Xiaobin Xu, Yuhui Shi, Yimin Deng, Cong Li, and Haibin Duan. Binocular pose estimation for UAV autonomous aerial refueling via brain storm optimization. In *Proc. IEEE Congress on Evolutionary Computation*, pages 254–261, 2019.
54. Alwaleed Aldhafeeri and Yahya Rahmat-Samii. Brain storm optimization for electromagnetic applications: Continuous and discrete. *IEEE Transactions on Antennas and Propagation on Antennas and Propagation*, 67(4):2710–2722, 2019.
55. Yang Shen, Mingde Liu, Jian Yang, Yuhui Shi, and Martin Middendorf. A hybrid swarm intelligence algorithm for vehicle routing problem with time windows. *IEEE Access*, 8:93882–93895, 2020.
56. Shunshun Peng, Hongbing Wang, and Qi Yu. Multi-clusters adaptive brain storm optimization algorithm for QoS-aware service composition. *IEEE Access*, 8:48822–48835, 2020.
57. Xing-Rui Chen, Jun-Qing Li, Yuyan Han, Ben Niu, Lili Liu, and Biao Zhang. An improved brain storm optimization for a hybrid renewable energy system. *IEEE Access*, 7:49513–49526, 2019.
58. Mayuko Sato, Yoshikazu Fukuyama, Mohammed El-Abd, Tatsuya Iizaka, and Tetsuro Matsui. Total optimization of energy networks in smart city by cooperative coevolution using global-best brain storm optimization. In *Proc. IEEE Congress on Evolutionary Computation*, pages 681–688, 2019.
59. Xingsi Xue and Jiawei Lu. A compact brain storm algorithm for matching ontologies. *IEEE Access*, 8:43898–43907, 2020.
60. C. Narmatha, Sarah Mustafa Eljack, Afaf Abdul Rahman Mohammed Tuka, S. Manimurugan, and Mohammed Mustafa. A hybrid fuzzy brain-storm optimization algorithm for the classification of brain tumor MRI images. *Journal of Ambient Intelligence and Humanized Computing*, 2020.
61. Jian-Hua Hao, Jun-Qing Li, Yu Du, Mei-Xian Song, Peng Duan, and Ying-Yu Zhang. Solving distributed hybrid flowshop scheduling problems by a hybrid brain storm optimization algorithm. *IEEE Access*, 7:66879–66894, 2019.
62. Zhe Xu, Xiaofang Li, Xianglian Meng, and Yanting Liu. A distributed brain storm optimization for numerical optimization and graph planarization. *IEEE Access*, 7:39770–39781, 2019.
63. D. Palanikkumar and S. Priya. Brain storm optimization graph theory (BSOGT) and energy resource aware virtual network mapping (ERVNM) for medical image system in cloud. *Journal of Medical Systems*, 43:37, 2019.
64. Z. Cao, X. Hei, L. Wang, Y. Shi, and X. Rong. An improved brain storm optimization with differential evolution strategy for applications of ANNs. *Mathematical Problems in Engineering*, 2015:1–18, 2015.
65. Kofi Baah Boamah, Jianguo Du, Daniel Adu, Claudia Nyarko Mensah, Lamini Dauda, and Muhammad Aamir ShafiqueKhan. Predicting the carbon dioxide emission of china using a novel augmented hypo-variance brain storm optimisation and the impulse response function. *Environmental Technology*, 2020.
66. R. Billinton and R. N. Allan. *Reliability Assessment of Large Electric Power System.* Kluwer Academic Publishers, Boston, 1988.
67. Mohammed Benidris, Salem Elsaiah, and Joydeep Mitra. Power system reliability evaluation using a state space classification technique and particle swarm optimisation search method. *IET Generation, Transmission and Distribution*, 9(14):1865–1873, 2015.
68. Mohamadreza Gholami, Reza Sharifi, and Hamid Radmanesh. Development of composite power system effective load duration curves by using a new optimization method for assessment composite generation and transmission reliability. *International Journal of Power and Energy Research*, 1(1):41–52, 2017.

69. N. Samaan and C. Singh. Adequacy assessment of power system generation using a modified simple genetic algorithm. *IEEE Transactions on Power Systems*, 17(4):974–981, 2002.
70. R Billinton and A. Sankarakrishnan. A comparison of Monte Carlo simulation techniques for composite power system reliability assessment. In *Proc. IEEE Communications, Power, and Computing*, pages 145–150, 1995.
71. Mohammadreza Gholami, Seied Hadi Hoseini, and Meisam Mohamad Taheri. Assessment of power composite system annualized reliability indices based on improved particle swarm optimization and comparative study between the behaviour of GA and PSO. In *Proc. IEEE International Power and Energy Conference*, 2008.

2 Fish School Search

Account for the First Decade

Carmelo José Abanez Bastos-Filho, Fernando Buarque de Lima-Neto, Anthony José da Cunha Carneiro Lins, Marcelo Gomes Pereira de Lacerda, Mariana Gomes da Motta Macedo, Clodomir Joaquim de Santana Junior, Hugo Valadares Siqueira, Rodrigo Cesar Lira da Silva, Hugo Amorim Neto, Breno Augusto de Melo Menezes, Isabela Maria Carneiro Albuquerque, João Batista Monteiro Filho, Murilo Rebelo Pontes, and João Luiz Vilar Dias

CONTENTS

2.1 INTRODUCTION

Evolutionary and swarm-based algorithms are optimization methods greatly inspired in the theory of evolution of species and gregarious animal behavior, respectively. Both comprise two versatile families of algorithms within the computational intelligence discipline. The bio-inspired aspect of both families' population-based techniques brings vast possibilities and capabilities for solving hard search and optimization tasks in highly nonlinear complex contexts. Applications excel in discrete and in continuous search spaces and are very popular in an ample range of challenging domains such as scheduling, routing, complex networks, and others [1, 2].

In particular, swarm-based algorithms profit significantly from the cooperation and interaction of their processing units, the simple reactive particles of the swarms. These properties endow the swarm with fast and effective proactive adaptive responses. Moreover, collectively, the particles can seamlessly carry out an unusual combination of local and global search capabilities. The evoked behavior of the swarm results in a powerful abstraction with a fascinating practical emergence of self-organization and smart-communication strategies. The swarm dynamics mimics the group behavior observed in ant colonies, schools of fish, flocks of birds, pack of wolves, and others [3].

Each algorithm of the comprising families of the swarm clique has parameters that control the evoked behavior during the search or optimization according to a given fitness function. Tuning these parameters is not an easy task for practitioners because it is not simple to define parameter values that evoke optimal behavior and because of how each parameter relates to the desired results.

Since its inception, the swarm-based approach became an unprecedented selection of tools capable of tackling large and challenging problems. This is mostly because of their inner functioning mechanisms, which allow smart internal synergistic gains so that the known curse of dimensionality is somehow tamed. Consequently, scalability is no longer the Achilles heel. From a research standpoint, fortunately, many other aspects related to solving real-world problems pressed for new developments, namely (1) varied data representation (e.g., binary, discrete, continuous, and mixed), (2) multimodal output, (3) multi-objective capabilities, and (4) more effective use of the computational resources (e.g., external parallelizations).

Fish School Search (FSS) is a swarm-intelligence subfamily of algorithms proposed by Bastos Filho and Lima Neto in 2008 and first published in 2009 [2]. In this algorithm, each fish weight represents the success obtained during the search and directly influences the individual and school movements. The embedded mechanisms of feeding and coordinated movement make all fish "swim" toward the positive gradient to gain weight (and hence find local and global better positions). By design, the heavier the fish, the more influential it is for guiding the search.

Moreover, this one definition is seminal for a swift functioning of FSS, which uses the barycenter as a frame of reference to decide overall directions toward better search space regions. Another unique feature of FSS, also based on the barycenter, is its ability to switch between exploration and exploitation automatically. The barycenter is also very handy for parallelizing FSS, as it is one of the few global variables that need to be shared in parallel implementations. This chapter accounts for several variations of FSS produced over the last decade, which address the pressing aspects of real-world optimization.

Many bio-inspired algorithms have been proposed in the last decade, and some papers have shown that most of the recent proposals use similar computational mechanisms [4]. Nevertheless, we can only observe some basic mechanisms to represent and aggregate the useful information obtained by the simple reactive agents along the search process. Among these proposals, we can cite positional information (deployed in particle swarm-based algorithms), markers in the environment (employed in artificial bee colony-based and ant colony optimization-based algorithms), and indicators that accumulate the degree of success of the simple agents (proposed in the FSS and also deployed in the Glowworm Swarm Optimization).

The idea of accumulating success along the search indicates that a specific simple reactive agent is worth influencing others. FSS works well for continuous optimization problems with a single optimization, and it has influenced other researchers to propose variations for other problems, such as optimization in binary problems, multi-objective optimization, many-objective optimization, and multimodal optimization. This chapter presents some proposals for binary optimization, three approaches for multi- and many-objective optimizations, and two different multimodal optimization proposals. We also show two different approaches for parallel processing, which aim to accelerate the processing time. We finalize this chapter giving some examples of applications of those recent approaches in real-world problems.

2.2 ORIGINAL FISH SCHOOL SEARCH

The original version of FSS can be described as a continuous and uni-modal metaheuristic inspired by the collective behavior of fish schools. The proposal is based on the idea that each agent (fish) has an innate memory of their success in search space, such as their weights. Besides that, the proposal is composed by two new operators based in Fish School behavior.

The operators proposed by the authors are grouped in two classes: feeding and swimming. The feeding operator is based on the idea that to find food, the fish in the school can move independently. In its search, a fish can grow or diminish its weight, depending of its result in this task. As a result, the equation proposed considers that the new fish's weight is proportional to the fitness variation in the new position, as shown in Eq. 2.1.

$$W_i(t+1) = W_i(t) + \frac{f[\vec{x}_i(t+1)] + f[\vec{x}_i(t)]}{max\left\{\left|f[\vec{x}_i(t+1)] - f[\vec{x}_i(t)]\right|\right\}} \tag{2.1}$$

where $W_i\ (t)$ is the weight of the fish i in the iteration t, $x_i\ (t)$ is the position of the fish i, and $f[x_i\ (t)]$ is its fitness.

Regarding the feeding operator, it is important to notice that the weight evaluation occurs once at every iteration. Also, there is a parameter in the proposal called the weight scale (W_{scale}), which limits the maximum weight of a fish. Lastly, each fish is created with its weight equal to $\frac{W_{scale}}{2}$.

In FSS, the swimming operator is a combination of three different actions: individual, collective-instinctive, and collective-volitive movements. Individual movement performs a displacement in each fish using the individual step parameter ($step_{ind}$), as described in Eq. 2.2.

$$\Delta x_i(t+1) = step_{ind}(t) \cdot 2 \cdot rand \cdot direction \tag{2.2}$$

where $\Delta x_i(t+1)$ is the movement displacement; $step_{ind}(t)$ is the individual step parameter in iteration t; and *rand* and *direction* are two random values, where *rand* is generated by a uniform distribution in the interval [0,1] and *direction* is randomly chosen between –1 or 1. In the equation, $step_{ind}$ can be decreased linearly to provide better exploitation in the later iterations.

The individual movement is only properly performed if the new temporary position (Eq. 2.3) is better than the current fish position. In this case, the current fish position is updated.

$$\vec{temp} = \vec{x}_i(t) + \Delta x_i(t+1) \quad (2.3)$$

Collective-instinctive is a movement based on the instantaneous success of all fish of the school. In this movement, the fish that were successful in the individual movement influence the fish school in a direction. Each fish is repositioned using the Eq. 2.4.

$$\vec{x}_i(t+1) = \vec{x}_i(t) + \frac{\sum_{i=1}^{N} \Delta x_{indi} f[\vec{x}_i(t+1)] + f[\vec{x}_i(t)]}{\sum_{i=1}^{N} \{f[\vec{x}_i(t+1)] - f[\vec{x}_i(t)]\}} \quad (2.4)$$

where Δx_{indi} is the displacement of the fish i due to the individual movement the current iteration.

Collective-volitive is the last movement in FSS iteration, and it is important to improve the exploration capability of algorithm. It is performed as a small step drift to every fish position with regard to the school's barycenter. In summary, if the fish school is increasing its weight, the radius of the school should contract; otherwise, it should dilate. To do that, the barycenter is calculated using Eq. 2.5.

$$Bari(t) = \frac{\sum_{i=1}^{N} \vec{x}_i(t) W_i(t)}{\sum_{i=1}^{N} W_i(t)} \quad (2.5)$$

If the overall weight of the school increases, the new position is calculated using Eq. 2.6; otherwise, Eq. 2.7 is used.

$$\vec{x}_i(t+1) = \vec{x}_i(t) - step_{vol} \times rand \times [\vec{x}_i(t) - Bari(t)], \quad (2.6)$$

$$\vec{x}_i(t+1) = \vec{x}_i(t) + step_{vol} \times rand \times [\vec{x}_i(t) - Bari(t)], \quad (2.7)$$

where $rand$ is a random number uniformly generated in the interval [0,1] and the $step_{vol}$ is an FSS parameter that is decreased linearly along the iterations.

Finally, the pseudocode for FSS is described in Algorithm 2.1

ALGORITHM 2.1 ORIGINAL FISH SCHOOL SEARCH

1: Initialize fish in the swarm
2: **while** stop criteria is not attained **do**
3:   **for each** fish in Fish School **do**
4:     Update position applying the individual operator
5:     Apply feeding operator
6:     Apply collective-instinctive movement
7:     Apply collective-volitive movement
8:   **end for**
9:   update the individual and volitive steps
10: **end while**

Although the FSS has shown promising results for mono-objective optimization problems in continuous spaces, the original version presents a high dependency with the parametric values. Besides, there are some other challenges related to different types of optimization problems from the perspective of the decision space and objective space.

We have performed a search using the terms FSS, Fish School Search, and Fish School in different databases, such as Google Scholar, IEEE Xplore, and ScienceDirect.

2.3 VARIANTS OF FISH SCHOOL SEARCH

In the previous section the original FSS was covered showing its operators and parameters. In this section, we briefly discuss others variants of FSS proposed in the literature.

2.3.1 Mono-Objective FSS

This section presents some approaches based on the FSS for different types of mono-objective optimization problems.

2.3.1.1 Binary Fish School Search (BFSS)

In 2014, Sargo et al. [5] modified the FSS internal mechanisms to map the continuous domain to the binary domain. They proposed the Binary Fish School Search (BFSS), where each fish is encoded as a binary vector with N bits representing a solution. The algorithm was applied in feature selection (FS), and the results showed that the BFSS improved sensitivity and obtained a smaller number of features compared with other algorithms in the same problem.

In BFSS, for each fish, the bits initialization follows Eq. 2.8.

$$x_{ij} = \begin{cases} 1, & \text{if } u > 0.5. \\ 0, & \text{otherwise.} \end{cases}, \mathrm{i} = 1,2 \ldots N, \mathrm{j} = 1,2 \ldots Nb \tag{2.8}$$

where u is a random number uniformly generated between 0 and 1, N is the number of fish, and Nb is the number of bits in each fish position related to the number of the dimensions of the problem.

In the individual movement, the individual step ($S_{ind}(t)$) parameter is used as a threshold that regulates the flip probability of a bit position (Eq. 2.9).

$$x_{ij} = \begin{cases} \bar{x}_{ij}, & \text{if } v < S_{ind}(t). \\ x_{ij}, & \text{otherwise.} \end{cases} \tag{2.9}$$

where v is a random number uniformly generated in the interval [0,1]. $S_{ind}(t)$ is the same parameter as the FSS continuous proposal that decreases linearly along with the iterations.

Collective-instinctive movement is performed using a resultant vector ($\vec{I}$) calculated using the relationship between the position and the fitness of the school, as described in Eq. 2.10.

$$\vec{I}(t) = \frac{\sum_{i=1}^{N} x_{indi} \left\{ f_x\left[x_i(t+1)\right]\right\} - f_x\left[x_i(t)\right]}{\sum_{i=1}^{N} \left\{ f\left[x_i(t+1)\right]\right\} - f_x\left[x_i(t)\right]} \tag{2.10}$$

Then, the $thres_c$ parameter is used to create a threshold. It is created by the multiplication of the max value of $\vec{I}$. Next, a new binary vector $\vec{I}_b$ is calculated as follows: if the values of the position of $\vec{I}$ are below the threshold, the current vector position is considered 0, otherwise, it is 1. Next, for each fish, one dimension that does not have the same value of $\vec{I}_b$ is flipped, making the new position closer to success fish.

The collective-volitive movement follows a similar approach. The barycenter ($\vec{B}$) is calculated using the barycenter equation proposed in FSS. After that, the $thres_v$ parameter is used to calculate the binary barycenter ($\vec{B}_b$) using the same steps described previously in the collective-instinctive movement. If the school's overall weight increases, each fish has one dimension flipped that does differ regarding $\vec{B}_b$. Otherwise, it flips one dimension that has the same value, moving the fish to the opposite of barycenter ($anti\text{-}\vec{B}_b$).

BFSS presented slightly better results when compared with other FS methods in three benchmark data sets. It was also used for predicting intensive care unit readmission. The proposal achieved superior results in this problem due to the improvement in its sensitivity.

2.3.1.2 Improved Binary Fish School Search (IBFSS)

In 2016, Carneiro and Bastos-Filho [6] proposed the Improved Binary Fish School Search (IBFSS). The authors carried out a detailed parametric analysis in the BFSS, and they demonstrated that some adaptations could improve the algorithm results in FS.

In the new proposal, the initialization strategy was only active in 25% of bits in the fish position using Eq. 2.11. They explained that it allows a better FS exploration.

$$x_{ij} = \begin{cases} 1, & \text{if } u < 0.25. \\ 0, & \text{otherwise.} \end{cases}, \mathrm{i} = 1,2\ldots N, \mathrm{j} = 1,2\ldots Nb \tag{2.11}$$

Carneiro and Bastos-Filho also changed the individual movement. As described previously, the BFSS uses $S_{ind}(t)$ as a threshold to flip the bits (Eq. 2.9). In IBFSS, the same parameter is used as the number of flips performed in each fish position.

The modifications in collective-instinctive movement improved the algorithm convergence. $\vec{I}$ is obtained using an equation similar to Eq. 2.10, but in the IBFSS equation, the fish position (x_{ij}) is replaced by the fish position variation ($\Delta x_i j$) as typically used in FSS. Besides that, more than one bit could be flipped in each fish position in the improved version, accordingly Algorithm 2.2.

ALGORITHM 2.2 PSEUDOCODE OF IMPROVED COLLECTIVE-INSTINCTIVE MOVEMENT

1: $j = 1$
2: numberOfFlips = % of Instinctive Step
3: **for each** fish in Fish School **do**
4:   **while** j ≤ numberOfFlips **do**
5:     $r = Random(1,\ldots,M)$
6:     $x_{ir} = \overline{x_{ir}}$
7:     j = j + 1
8:   **end while**
9: **end for**

where *numberofFlips* is a percentage of the number of features that are marked to flip, i.e., the dimensions of vector $\vec{I}(t)$ that are different from fish position; r is a random number in interval [0, M]; x_{ir} is the rth dimension of the position of fish i; and M is the number of features that are marked to be flipped.

The collective-volitive movement was changed to use the volitive step, which is an FSS parameter. This parameter is used to determine the number of bits that could be flipped to dilate or contract toward the barycenter. This new approach is shown in Algorithm 2.3.

ALGORITHM 2.3 PSEUDOCODE OF IMPROVED COLLECTIVE-VOLITIVE MOVEMENT

1: j = 1
2: numberOfFlips = % of Volitive Step
3: **for each** fish in Fish School **do**
4:   **while** j ≤ numberOfFlips **do**
5:     $r = Random(1,...,K)$
6:     $x_{ir} = \overline{x_{ir}}$
7:     j = j + 1
8:   **end while**
9: **end for**

where *numberOfFlips* is a volitive step percentage of the fish *i*, *r* is a random number in interval [1, *K*], and *K* is the number of features that are marked to be flipped.

The results showed that IBFSS overcame the original binary version, BFSS. The new proposal reduced the number of features for classification, and it increased the accuracy of the classifier in FS.

2.3.1.3 Simplified Binary Fish School Search (SBFSS)

Santana et al. proposed, in 2019 [7], a new binary version of the FSS named simplified BFSS (SBFSS). The main idea behind the SBFSS is to be a more straightforward yet robust binary optimizer. Different from the previous mono-objective BFSS, the SBFSS does not use any transfer functions to transform the position of the fish from the continuous to the binary domain.

In the individual displacement, each fish selects two random dimensions ($D = d1, d2$) to change the value and only moves if the new position is better than the previous one. Eq. 2.12 describes this modification.

$$x_{i,d}(t+1) = \overline{x_{i,d}(t)}, \tag{2.12}$$

where d is in D. Next, in the collective, instinctive displacement, for each successful fish and in every dimension d, they sum the fitness variation of the fish that have $X_{i,d} = 0$ and compare with the sum of the fitness variation of fish that had $X_{i,d} = 1$, as illustrated by Algorithm 2.4.

ALGORITHM 2.4 INSTINCTIVE VECTOR

1: **for** each selected fish i **do**
2:   **for** each selected dimension d **do**
3:     **If** $x_{i,d} == 1$ **then**
4:       $ones\_cont_d = ones\_cont_d + \Delta f_i$
5:     **else**
6:       $zeros\_cont_d = zeros\_cont_d + \Delta f_i$
7:     **end if**
8:   **end for**
9: **end for**

where $zeros\_cont_d$ and $ones\_cont_d$ represent, respectively, the sum of the fitness variation of the successful fish that have $X_{i,d}$ equal to 0 and 1. The displacement will be toward the dimensions that presented the greatest variation, as indicated by Eq. 2.13.

$$x_{i,d}(t+1) = \begin{cases} 1, & \text{if } zeros\_cont_d < ones\_cont_d \\ 0, & \text{otherwise.} \end{cases} \tag{2.13}$$

Lastly, the volitive movement adopts a similar strategy, but there are three fish selected with a binary tournament. These three fish are used to represent the barycenter of the swarm, and this procedure is explained in Algorithm 2.5.

ALGORITHM 2.5 BARYCENTER CALCULATION

1: Apply a binary tournament to select three fish from the school
2: **for** each selected dimension d **do**
3: **for** each selected fish i **do**
4: **If** $x_{i,d} = 1$ **then**
5: $ones\_weight_d = ones\_weight_d + W_i$
6: **else**
7: $zeros\_weight_d = zeros\_weight_d + W_i$
8: **end if**
9: **end for**
10: **If** $zeros\_weight_d < ones\_weight_d$ **then**
11: $B_d(t+1) = 1$
12: **else**
13: $B_d(t+1) = 0$
14: **end if**
15: **end for**

where B is the binary vector that represents the barycenter. Similarly to the standard FSS, the swarm will move toward the position of the barycenter, but in the SBFSS, each fish will update at most two dimensions randomly selected.

With the modifications proposed, the SBFSS overcame the previous binary versions of the FSS in the OneMax and 0/1 knapsack problems. Moreover, in the experiments performed, the new BFSS was around 50% faster and less computationally expensive than Binary Particle Swarm Optimization (BPSO) and IBFSS. Compared with other algorithms, such as BPSO, Binary Artificial Bee Colony and Binary Genetic Algorithm (BABC), and BGA, it produced superior results in high-dimensional problems.

2.3.1.4 Mixed (MFSS)

A mixed version of FSS (MFSS) was implemented to cope with the necessity of tackling problems where the domain contains continuous and discrete variables. The mechanisms work in the same way as the vanilla FSS for the continuous variables, but new procedures have been added to keep the same movement characteristics for discrete domains.

The fish school's initialization usually occurs for the continuous variables, and each fish assumes a possible position in the search space with proper values. The values are randomly set following a uniform distribution that covers the whole domain. For the discrete variables, a non-uniform distribution can also be used. For example, for a certain discrete variable x, where

$\{x \mid -1 \le x \le 1\}$, the probabilities can follow a different distribution, and a value could have a higher chance of being selected compared with the other. The distribution can be set according to the necessities of the problem.

Each movement operator has its function inside FSS. In each of them, handling the discrete variables occurs in the same way. For example, during the individual movement phase, each fish will try a new random position. For a discrete variable, like the one mentioned above, a fish will try to take a random step inside the region [−1, 1]. To do it so, each fish generates a random value inside this interval. The biggest and the smallest value are used to generate thresholds, as shown in Eq. 2.14. For these calculations, the parameter $thresh\_multiplier$ can be set to increase or decrease the probability of taking a step or not.

$$\begin{cases} positive\_threshold = max(random\_steps[dim_x]) * thresh\_multiplier; \\ negative\_threshold = min(random\_steps[dim_x]) * thresh\_multiplier; \end{cases} \tag{2.14}$$

If the random value generated by a fish i is more significant than the positive threshold, this fish shall take a +1 step for this variable. If it is smaller than the negative threshold, it shall take a −1 step for this variable. Otherwise, it shall remain in the same position for the corresponding variable. This process also has to respect the boundaries of the discrete variable (Eq. 2.15). The same process is repeated for each of the discrete variables.

$$rand\_step_i = \begin{cases} +1 \text{ if } random\_step_i[dim_x]) > positive\_threshold; \\ -1 \text{ if } random\_step_i[dim_x]) < negative\_threshold; \\ 0 \end{cases} \tag{2.15}$$

The process is the same for all other FSS movement operators. The threshold and discrete step calculations are performed using the displacement vector for the instinctive movement and the volitive movement. By doing so, MFSS can cope with the main algorithm's main concepts (i.e., expansion and contraction) for discrete/continuous domains.

Tan et al. applied MFSS to the inference of gene regulatory network (IGRN) problem with very satisfactory results [8]. For this problem, each fish represented a solution that had continuous variables and discrete variables with different domains. Some discrete variables could assume values among −1, 0, and 1, while other discrete variables were restricted to −0.5 and 0.5. This setup shows the flexibility of this approach and the capability of tackling very complex problems such as the IGRN problem.

2.3.2 Multi/Many-Objective

In this section, we present the FSS-based algorithms for multi- and many-objective optimization.

2.3.2.1 MOFSS

In 2015, Bastos-Filho and Guimarães proposed the multi-objective FSS (MOFSS) algorithm that adapts the original FSS algorithm to multi-objective problems [9]. The main steps of the MOFSS algorithm are similar to FSS, but two steps are added: update the external archive (EA) and execute turbulence on EA.

MOFSS uses an EA to store reasonable solutions and to guide the school. Because of the cost of storing and updating the EA, the size is limited by the crowding distance (CD) method [10, 11]. The addition of EA prevents the school from entirely losing a set of already found promising solutions [10, 12]. The turbulence operator on EA randomly changes a subset of solutions, avoiding school stagnation.

In summary, MOFSS executes the following steps on a loop: the individual movement, the EA update process, the feeding operator, the collective-instinctive, the collective-volitive, the EA update process again, the turbulence on EA, and the EA update process. On the individual movement, MOFSS chooses a leader from the EA to guide each fish (the movement is performed even if the new solution is worse than the previous one). The feeding operator incorporates the dominance of solutions in its calculation. The dominance of a fish i, $D_i(t)$, is defined as $1-\frac{R_i(t)}{max(R_i(t))}$ where $R_i(t)=\sum_{j\in N, j\succ i} S_j(t)$ and $S_j(t)=\sum_{i\in N}[1 \mid j \succ i]$. The symbol $\succ$ represents dominance. In this way, each fish updates the weight as $w_i(t+1)=w_i(t)+\Delta w_i(t)D_i(t)$. The $\Delta w_i(t)$ is updated based on the parameters a and b (defined a priori). The value of a should be higher than the value of b. If the current position dominates the previous one, the $\Delta w_i(t)$ is increased by a. If the new position is dominated by the previous one, the $\Delta w_i(t)$ is decreased by a. When comparing current and previous positions is indifferent, the value of a CD of EA solutions is considered. The closest EA solutions to the current and previous positions are compared, and if the closest EA solution to the current position presents the highest value in terms of CD, the $\Delta w_i(t)$ is increased by b. If the opposite is true, the $\Delta w_i(t)$ is decreased by b. Instead of using the fitness variation, the instinctive movement uses the variation of weight for updating the position. On the collective movement, MOFSS picks a leader from the EA to control the magnitude of the school's exploitation or exploration, so the barycenter is represented here by the leader.

2.3.2.2 MOBFSS

The multi-objective BFSS (MOBFSS) algorithm [13] adapts the main operators of the three FSS versions: the multi-objective approach used by MOFSS [9], and the binary approach applied in BFSS [5] and IBFSS [6]. Each fish in MOBFSS is represented by its fitness, weight, and position (binary vector). The calculation of weight depends on the dominance concept, the use of the EA to perpetuate good solutions, and the application of a truncation method named crowding distance [9–11, 14].

The main steps on MOBFSS are similar to MOFSS: individual movement, calculate dominance, feeding operator, update EA, instinctive movement, volitive movement, update EA, and turbulence operator. The individual movement randomly chooses positions to flip for each fish. In the instinctive movement, the fish only flip the dimensions in which the majority of the fish had successfully flipped in the same direction (1 to 0 or 0 to 1). As the collective-instinctive movement differs from the other versions of FSS, we show the pseudocode in Algorithm 2.6. The parameter $flip_{inst}$ controls the threshold of how many fish flipped their dimension in the same direction to allow the swarm to flip in the same direction and dimension. The parameters r_d and $step_{inst}$ are a random value and the step instinctive. MOBFSS selects leaders from the EA to guide the school to contract and expand on the volitive movement. If a random leader from the EA has a better weight than a fish i, a random dimension d of fish i that is not in common with the leader is flipped. Otherwise, a random dimension that the leader and the fish i has in common will be flipped. The turbulence operator chooses a random leader from the EA and randomly flips some dimensions, and if the new position is better than the current one, the leader moves to the new position.

Several versions of MOBFSS were studied to minimize the computational cost of the MOBFSS [15]. MOBFSS-1-LS was the version with the highest efficiency on the FS problem, and MOBFSS-3-LS was the version with the smallest computational cost for the same problem. MOBFSS-1-LS and MOBFSS-3-LS substitute the turbulence operator with a local search mechanism proposed on Binary Multi-objective PSO with crowd distance roulette wheel (BMOPSO-CDR) [16]. Moreover, MOBFSS-3-LS concentrates the search on the collective movements, excluding the individual movement.

2.3.2.3 Weight-Based Many-Objective Fish School Search (wmoFSS)

The main goal behind weight-based many-objective FSS (wmoFSS) is to adapt a multimodal version of FSS, the weight-based FSS (wFSS) [3], to solve problems with many objectives without the need to store the best solutions found throughout the search in an EA. The proposed algorithm inherits the same operators as the original version of FSS, and a new clustering operator is included to promote diversity during the search by splitting the swarm into subgroups. Moreover, in wmoFSS, the optimization problem is decomposed into single-objective subproblems, and each one is assigned to one sub-swarm that is specialized in solving it. Compared with the original version of FSS, the most crucial difference in wmoFSS is that each fish has now a vector **w** of weights in which a single component represents the corresponding weight of a fish considering each objective individually. Moreover, an augmented version of wmoFSS, including simulated binary crossover (SBX) [17], was also proposed to improve its performance in multimodal many-objective problems.

ALGORITHM 2.6 COLLECTIVE-INSTINCTIVE MOVEMENT OF MOBFSS

1: for each fish **do**
2: **If** fish moved in the individual movement **then**
3: **for each** d as dimension **do**
4: $I_d = I_d + (x_d(t+1) - x_d(t))$
5: **end for**
6: **end if**
7: **end for**
8: **for each** first **do**
9: **for each** d as dimension **do**
10: **if** $I_d > 0$ **and** $|I_d| >= flip_{inst}$ **and** $r_d > step_{inst}$ **then**
11: $x_d(t+1) = 1$
12: **else if** $I_d < 0$ **and** $|I_d| <= flip_{inst}$ **and** $r_d > S_{inst}$ **then**
13: $x_d(t+1) = 0$
14: **end if**
15: **end for**
16: **end for**

2.3.3 Multimodal (Niching Algorithms)

In this section, we present the proposals for multimodal optimization.

2.3.3.1 Density-Based Fish School Search (dFSS)

In 2011, Madeiro et al. proposed the first niching version of the FSS algorithm, the density-based FSS (dFSS) [18]. In this algorithm, the fish looks for multiple local optima in the search space using the concept of food sharing, rather than of single fish foraging. The nearby individuals' influence and group up around each other along the search process, which splits the big school into smaller ones. The number of individuals around a given fish (density) dictates how influential the given fish can be. The operators of the dFSS have several changes, as mentioned below.

In the feeding operator the Δf_i is shared among all fish following Eq. 2.16.

$$\Delta f_i = P_i \sum_{j=1}^{N} \frac{1}{\left(d_{R_{ij}}\right)^{q_{ij}}} = \frac{\Delta f_i}{\sum_{j=1}^{N} \frac{1}{\left(d_{R_{ij}}\right)^{q_{ij}}}} = P_i \tag{2.16}$$

where $d_{R_{ij}}$ is the normalized distance $\frac{d_{ij}}{min_{d_{ik}}}$, q_{ij} is the number of fish k for which $d_{ik} < d_{ij}$, and P_i is the amount of food the fish i will receive after sharing. In Eq. 2.17 $C(i,j)$ is the amount of food fish j will receive from fish i.

$$C(i,j) = \frac{P_i}{\left(d_{R_{ij}}\right)^{q_{ij}}} = \frac{\Delta f_i}{\left(d_{R_{ij}}\right)^{q_{ij}} \sum_{k=1}^{N} \frac{1}{\left(d_{R_{ik}}\right)^{q_{ik}}}} \tag{2.17}$$

After each iteration all fish update their weight according to Eq. 2.18. Q is the number of fish that found food at that iteration.

$$\vec{w}_i(t+1) = \vec{w}_i(t) + \sum_{j=1}^{Q} \frac{\Delta f_j}{\left(d_{R_{ij}}\right)^{q_{ij}} \sum_{k=1}^{N} \frac{1}{\left(d_{R_{jk}}\right)^{q_{jk}}}} \tag{2.18}$$

This version also has a memory operator (Eq. 2.19), where each fish i has a memory of each other fish j represented by M_{ij}, and it determines the influence of fish j over fish i. There is a parameter ρ that controls that influence.

$$M_{ij}(t+1) = (1-\rho)M_{ij}(t) + \frac{\Delta f_j}{\left(d_{R_{ij}}\right)^{q_{ij}} \sum_{k=1}^{N} \frac{1}{\left(d_{R_{jk}}\right)^{q_{jk}}}} = (1-\rho)M_{ij}(t) + C(j,i) \tag{2.19}$$

The collective-instinctive movement follows Eq. 2.20.

$$\vec{I}_i(t) = \frac{\sum_{j=1}^{N} \Delta x_j M_{ij}}{\sum_{k=1}^{N} M_{ik}} \tag{2.20}$$

A partitioning operator was added to this approach. At each iteration, the primary school is partitioned into subgroups. A fish i will be in the same subgroup of fish j if (Eq. 2.21):

$$M_{ij} = maxM_{ik} \vee M_{ji} = maxM_{jk}, k = 1,2,\ldots,N \tag{2.21}$$

The partitioning operator follows Algorithm 2.7.

ALGORITHM 2.7 PARTITION OPERATOR

1: **while** There is fish in the main school **do**
2: Choose a fish i randomly in the main school
3: Create a new subgroup S_i
4: Put fish i in subgroup S_i
5: Remove fish i from the main school
6: Find other fish j in the main school that satisfies $(x+5)$
7: **while** there exists fish j in the main school **do**
8: Put fish j in subgroup S_i
9: Remove fish j from the main school
10: Set $i = j$
11: Find other fish j in the main school that satisfies $(x+5)$
12: **end while**
13: **end while**

The individual movement follows Eqs. 2.22–2.25. In which $decay_{min}$, $decay_{max_{ini}}$, and $decay_{max_{end}}$ are parameters whose values are in the interval [0,1]

$$step_{ind_i}(t+1) = decay_i * step_{ind_i}(t) \tag{2.22}$$

$$decay_i = decay_{min} - \left(\frac{R_i(t) - min(R_j(t))}{max(R_j(t)) - min(R_j(t))} \right) (decay_{min} - decay_{max}(t)) \tag{2.23}$$

$$decay_{max}(t) = decay_{max_{ini}} \left(\frac{decay_{max_{end}}}{decay_{max_{ini}}} \right)^{\frac{t}{T_{max}}} \tag{2.24}$$

$$R_i(t) = \sum_{j=1}^{Q} \frac{\Delta f_j}{\left(d_{R_{ij}}\right)^{q_{ij}} \sum_{k=1}^{N} \frac{1}{\left(d_{R_{jk}}\right)^{q_{jk}}}} \tag{2.25}$$

The collective-volitive movement is performed for each subgroup. According to Eq. 2.26.

$$\vec{x}(t+1) = \vec{x}(t) + (1 - decay_{max}(t))(\vec{B}(t) - \vec{x}(t)) \tag{2.26}$$

In most of the experiments, the dFSS outperformed some of the state-of-the-art niching algorithms at the time it was proposed. However, the dFSS algorithm presents two drawbacks: it adds new parameters to the original FSS' set parameters, which are intensely sensitive to the characteristics of the search space, and it presents a computational complexity of $O(n^4)$, where n is the number of fish.

2.3.3.2 Weight-Based Fish School Search (wFSS)

To overcome the dFSS' drawbacks, Lacerda and de Lima-Neto proposed the second niching algorithm of the FSS family, wFSS [19]. Such an algorithm is intended to use the weight of the fish, which is already present in the original algorithm as a success indicator, to guide the big schools' splitting mechanism into smaller schools.

In the wFSS algorithm, before starting a new iteration, each fish must decide which individual will be its leader during the current iteration. The decision can be to keep swimming by itself, follow a new leader, or keep following the same leader. Such a decision is guided by the weight of the candidate leader, which is selected randomly. The heavier the fish, the more likely to be chosen as a leader by a lighter fish. It means that the heaviest fish create sub-schools that swarm around multiple local optima, refining the search process in these regions and return them as multiple solutions.

The rationale of such a linking definition mechanism is as follows: for each fish a, another fish b is chosen. If b is heavier than a, then a now has a link between them and follows b (i.e., b leads a). This rule will be referred to as the first linking rule of the wFSS algorithm. However, if a already has a leader c, a first checks if c is lighter than itself. If this is true, the link is broken. Then, it checks if the potential new leader b is heavier than itself. In case this is true, a new link is created from a to b (i.e., b leads a). This is the second linking rule of the wFSS algorithm. In case such a condition is false, the sum of the weights of the followers of a is compared with the sum of the weight of the followers of the new potential leader b. If the sum of a is higher than the sum of b, a link with b is created. This is the third linking rule of the wFSS algorithm.

Besides the linking mechanism, as mentioned earlier, the wFSS presented modified collective movements. In the collective-instinctive movement, each fish calculates its displacement vector using its information combined with its leader information. Such a vector is calculated as defined in Eq. 2.27, where t and T are the current iteration and the maximum number of iterations, respectively; i and l are the current fish and its leader's indexes, respectively; and L is a binary variable that is set to 1 if the fish i has a leader and zero otherwise.

$$\vec{x}_i(t+1) = \vec{x}_i(t) + \frac{t}{T}\left(\frac{\Delta\vec{x}_i(t)\Delta f(\vec{x}_i(t)) + L\Delta\vec{x}_l(t)\Delta f(\vec{x}_l(t))}{\Delta f(\vec{x}_i(t)) + Lf(\vec{x}_l(t))}\right) \tag{2.27}$$

In the volitive movement, the barycenter is also calculated individually for each fish, considering its information and its leader information. Eq. 2.28 shows how its calculated, where w_i and w_l are the weights of the fish i and its leader l.

$$\vec{B}_i(t+1) = \frac{\vec{x}_i(t)w_i(t) + L\vec{x}_l(t)w_l(t)}{w_i(t) + Lw_l(t)} \tag{2.28}$$

The experiments showed that the wFSS overcame its predecessor dFSS with considerably less computational effort (i.e., $O(n^2)$) and without adding new parameters to the ones already inherited from the original FSS. It has also shown superior performance against a few state-of-the-art algorithms in most experimental scenarios.

2.3.4 Parallelization

This section aims to present two strategies for parallelization.

2.3.4.1 Parallel Fish School Search (pFSS)

This proposal was based on the original FSS algorithm, implemented with parallel processing in CPU and GPU, which was the main objective of the version called pFSS (Parallel FSS) [20]. Because the original FSS is structured in several functions representing its operators, a parallelizable approach was designed according to the existing approaches in the parallel development architecture known as CUDA from NVIDIA. In this version, an analysis was made considering the dynamic use of resources, the organization of data memory within the GPU, and the application of different processing optimization resources. The authors presented a scalability analysis according to the variation in the number of dimensions in the models described by hypercomplex functions. The study also verified the applicability of techniques to reduce latency and improve the maximum bandwidth in the communication between CPU and GPU for different hardware configurations on CUDA. Other approaches were tested for the hierarchy of threads, memory allocation modeling, and CPU-GPU communication, and using warps, streams, and loop unrolling techniques. Some approaches were applied to multiple GPUs, processing to speed up the performance, creating direct communication between different GPUs, and being a device host. Another proposal with subsets of schools was processing in different GPUs at the same time. The experiments were executed in different GPU configurations, exploring multiple dimensionalities for benchmark functions. The best analysis for multiple GPUs showed speedups of 127 and 93 times in 1000 dimensions in the test functions.

2.3.4.2 Multi-Threaded Fish School Search (MTFSS)

To allow cheaper computers (i.e., not equipped with GPUs) to benefit from multiple CPU cores, Lacerda and Lima Neto proposed a multi-threaded FSS (MTFSS) [21]. In this approach, each fish has its behavior executed within an individual thread of which creation, execution, and death are managed by the runtime environment and the operating system. It is essential to mention that the rationale of the original algorithm was kept. Well-known benchmark functions were used to evaluate the speedup of the MTFSS compared with the standard FSS and check if there are statistically significant changes in the ability of the new algorithm to find reasonable solutions. The experiments were carried out on a regular personal computer instead of expensive setups. The results showed that the new version of the algorithm could achieve impressive growing speedups for increasingly higher problem dimensionality than the standard FSS, without losing the original algorithm's ability to find suitable solutions without any need for more powerful hardware (e.g., parallel computers).

2.4 APPLICATIONS

In this section, we list some examples of applications of the FSS into several types of optimization problems. There are two groups of examples: application of FSS on classes of problems and applications on flagship problems of different fields. This is just a sample and much more can be found in the FSS web page at https://fbln.me/fss.

2.4.1 Clustering

Clustering methods are a well-known class of techniques developed to coherently aggregate data in groups. The process is guided by separating the samples based on their similarities and dissimilarities. The models split the objects into groups in an unsupervised way using as inputs their common features [22].

FSS was also applied in the clustering field. For instance, Serapião et al. proposed an adaptation on the structure of the FSS [2] to deal with clustering problems [23]. In this study, the authors suggest that each fish is a candidate solution containing the centroids' spatial coordinates. The computational results considering 13 benchmark data sets, revealed that the FSS overcame the *k*-means and the particle swarm optimization (PSO).

Another example of application is the work from Ferreira et al. where they developed clustering versions of the PSO, artificial bee colony algorithm (ABC), and FSS to group crude (petroleum) oil based on its quality, which is measured considering physicochemical properties [24].

Lastly, Menezes and Coelho addressed the FSS to deal with two real-world data sets: (1) identifying sets of genes of *Saccharomyces cerevisiae* sharing compatible expression patterns across subsets of samples and (2) performing automated suggestions for users based on the opinions of other users with similar interests [25]. The computational results revealed the capability of the FSS to overcome the genetic algorithms (GA) and PSO.

2.4.2 Feature Selection

Three binary versions, BFSS, IBFSS, and MOBFSS, have been tested in the FS problem. In this problem, the algorithm needs to find the subset of features to represent the data and discard the non-relevant ones. BFSS was tested using three benchmark data sets from the UCI machine learning repository, German (credit card), Sonar, and WDBC [5]. In contrast, IBFSS has been tested with Sonar, German, Ionosphere, Vehicle, Colon, and WDBC data sets from the same repository [6]. Finally, MOBFSS was tested by Wine, Ionosphere, and Sonar [13, 15].

FS is also applied to improve the diagnosis of diseases such as breast cancer [26]. Breast cancer is the deadliest cancer for women, and detecting cancer earlier is the highest chance for a cure. To allow breast cancer diagnosis to be fast and more effective, Macedo et al. [15] also applied the MOBFSS to select the best features extracted from thermography images. Thus, FSS algorithms can also be applied to improve other diagnoses, such as prostate cancer, diabetes, and tuberculosis.

2.4.3 Image Compression and Thresholding

We can also find in the literature examples of applications of the FSS into the image processing field. For instance, Mishra et al. applied the FSS to the task of finding the optimal threshold value to segment the gray-scale images [27]. The authors reported that the proposed approach achieved results superior to traditional image processing techniques.

Ferreira and Madeiro [28] have applied the algorithm to image compression. The authors used the FSS in as a clustering algorithm as part of a channel-optimized vector quantization (COVQ) process. The goal was to use the FSS to generate dictionaries that could be used to compress the image. They tested this approach's robustness by simulations of the transmission of the compressed image via a channel with noise. Then this image was reconstructed, and they measured the error compared with the original image. The results obtained showed the superiority of the FSS-based approach in terms of the quality of the images reconstructed compared with the more traditional COVQ techniques.

2.4.4 Knapsack Problem

The knapsack problem maximizes the sum of values of items composed of weight that fits into a knapsack with a limited weight. In the 0/1 knapsack problem, the items can be selected (1) or not (0) to be inside the knapsack, and the weight of all the items can never be higher than a threshold.

SBFSS was applied for the 0/1 knapsack problem [7], overcoming FSS-based algorithms, BGA, and BPSO in the majority of the experiments. For 50 dimensions, SBFSS had a better performance than the other FSS-based algorithms and the BGA. For 100 dimensions, SBFSS could overcome BABC, BPSO, BGA, and FSS-based algorithms.

2.4.5 Recommendation Systems

Recommendation systems are models that, based on previous information, are trained to learn users' preferences and recommend similar products, experiences, or services. These systems have a wide range of applications, from e-commerce to educational platforms, and in the literature, we can find a variety of successful applications.

In 2019, Bova et al. proposed using the FSS as part of a recommendation system [29]. The idea was to employ the FSS to reduce the search space's size composed by the users' preferences. This reduction improved the system's performance in large data sets and helped to enhance the quality of the recommendations. The results obtained indicate that the proposed approach increases the accuracy of the preference prediction.

2.4.6 Smart Grids

A smart grid is a name given to an electric grid that incorporates sustainable technologies that help improve the system; for example, smart electricity meters, smart appliances, and energy-efficient resources/devices are components of smart grids.

In 2017, the FSS and the wFSS were applied to optimize a system's performance to improve the energy consumption in lighting equipment. The results obtained indicate that both the FSS and wFSS reduced the consumption of the system producing solutions comparable to other deterministic algorithms such as the Lpsolve and GLPK [30].

2.4.7 Supply Chain Network Planning

A supply chain is defined as a network of organizations connected by the flow of materials, information, and financial resources in different activities that produce products and services driven by the demands of the customers. The management of such networks is a crucial capability for modern companies. In this context, the supply chain network planning problem (SCNP) aims to determine an optimal allocation of demands and capacities in inventory, transportation, and production facilities for a tactical time horizon, consisting of several planning periods, aiming to fulfill customer demands at minimal costs.

Hellingrath et al. have applied the FSS algorithm to solve a few benchmark scenarios [31]. In their paper, its performance was compared with differential evolution (DE) and PSO. The FSS demonstrated the ability to deal with the high number of local optima in this problem domain, overcoming the PSO and DE algorithms because they got stuck in local optima quickly and were found to be unable to find good solutions.

2.4.8 Electrical Impedance Tomography

Electrical impedance tomography (EIT) is a low-cost, non-invasive, and radiation-free technology with several applications in industry, geophysics, and health. Dos Santos et al. in 2018, employed the FSS to reconstruct the EIT images based on the electrical conductivity signature of the object scanned with the EIT machine [32].

They performed a set of experiments with different images, and the qualitative analysis of the results reveals that the images reconstructed with the FSS were consistent with the anatomy of the object and had a satisfactory resolution. Furthermore, the reconstruction error rate of the images when compared with the original was inferior to 5%.

2.4.9 Mixed Model Workplace Time-dependent Assembly Line Balancing Problem

Balancing assembly lines (in large industrial plants) comprises a set of optimization problems of enormous practical interest to manufacturing industry due to the relevant frequency of this type of production paradigm, and of its sheer complexity.

In 2017, Monteiro Filho et al. used the basic FSS and the FSS with stagnation avoidance routine (FSS-SAR) [33] to solve a suggested problem of balancing assembly lines, the mixed model workplace time-dependent assembly line balancing problem (MMWALBP) [34]. This problem was proposed with the intention of including pressing issues of real assembly lines in the optimization problem to make it utilizable mainly for assembly lines producing big-sized products. The MMWALBP considers mixed model nature, many workplaces per workstation, and an additional time between tasks that requires displacement of operators to work in more than one task.

Monteiro Filho et al. [34] have compared vanilla FSS, FSS-SAR, and PSO to minimize the number of open workplaces (operators) for a given cycle time and their workload for the execution. For that an external procedure during the fitness calculation was added to solve issues regarding the balancing problem. The authors noticed that regarding the number of active workplaces all algorithms achieved the equivalent outputs. Although FSS versions presented solutions with lower workload, which is desirable, a clear performance distinction pointed toward FSS-SAR.

2.4.10 Finite-Element Model Updating

Finite-element models (FEMs) are widely used in engineering, usually as the numerical models of real mechanical structures. However, results obtained by the FEM approach are different when compared with the experimental measurements. The difference between both is caused by the not-so-easy tuning of the model parameters and some mathematical simplifications assumed during modeling. The correcting procedure (of the numerical models) is referred to as model updating. Basically, the parameters are adjusted to minimize the error between measurements and model.

In 2015, Boulkaibet et al. have used FSS twice as the optimizer for the model-updating procedure. First, a customized version of the Fish School Search (FSSb) algorithm and second, an innovative use of the volitive operator that was incorporated into the PSO algorithm were applied to the FEM updating problem [35]. As a proof of concept, both algorithms were tested on the updating of real structures, namely the unsymmetrical H-shaped beam and a GARTEUR SM-AG19 structure. The results produced by both algorithm versions were more accurate results than the GA, which is an encouraging result for many possible applications in engineering.

2.5 CONCLUSION

This chapter accounts for over a decade of research work related to FSS. The aim was to depict a representative but not complete set of examples of variations and applications of FSS. By far it is not a complete account. Research labs working with FSS, a complete list of papers published, and more details on the FSS family can be found in the FSS web page at https://fbln.me/fss.

Since its introduction, many versions of FSS were produced for dealing with distinct classes of problems and addressing challenging applications. However, some features of FSS were kept intact and still are unique: (1) the ability to self-control exploration and exploitation with no need for re-starting, (2) compact representation of knowledge regarding the history of the search, (3) reduced number of global variables, (4) a modular set of operators, and (5) preparedness for multimodal computations. These combined features allow (1) less need for expert interaction, as the algorithm by itself decides when to change the search mode; (2) ready utilization of success for more thoughtful inferences, as hidden correlations can be drawn directly from the search history; (3) swift parallelizations, as few global variables are easier to keep up in many parallel platforms; (4) hybridizations with other swarm techniques, such as operators like the volitive, which greatly improve the host autonomy (e.g., PSO); and (5) application in real-world problems, as some optimal solutions may not be feasible. The sheer number of versions of FSS developed so far is an indisputable flexibility characteristic of the initial proposal. The competitive results obtained time and again are also indisputable proof of the ability of the FSS family to tackle large and challenging problems. Above all, the large number of research groups still working with FSS, 10 years past its inception, is an encouraging sign that many more research has to come. Scalability, flexibility, economy of resources, and modularity are the most striking observable characteristics of the computations with FSS so far.

In this chapter the reader could have perceived how data representation, operators, and new features were experimented with and played with. The overall result is that FSS now possesses the striking ability to optimize binary, discrete, continuous, and mixed types of variables.

The now available combination of multimodal and multi-objective optimization capability is innovative, unique, and very handy. At this moment a parallel, multimodal, many-objective, and mixed variable version is under testing. This is just one of the newest versions being tested. Not many other algorithms are simultaneously equipped with this multitude of functionalities.

All those highlights encourage and convince the authors that the future of FSS is long and with plenty of possibilities. On top of the commonplace, reducing the number of operators, the authors suggest the following research avenues for the interest researchers: (1) try out new internal data representation schema, (2) conceive new updating and communication methods, (3) think of effective policy transfer to cut down training times, and (4) perhaps persist on hybridizing FSS with other similar savvy algorithms to foster the emerging adaptive behavior.

This chapter does not intend to be a survey considering all fish-inspired algorithms; rather it intends to show how the FSS family has grown in the last decade.

ACKNOWLEDGMENTS

Along with this decade of hard research work on FSS, the dozens of publications produced, and hundreds of conferences attended, several theses presented were only possible by the steady public funding of basic research in Brazil. The authors of this decade's account would like to sincerely thank the many supporting funding agencies. We would like to thank: CNPq (Brazilian National Council for Scientific and Technological Development), CAPES (Brazilian Coordination for the Improvement of Higher Education Personnel), FACEPE (Science and Technology Foundation of the State of Pernambuco-Brazil), and AvH (Alexander von Humboldt Foundation-Germany). A special thanks goes to the alma mater of most of the authors of this chapter, the UPE (University of Pernambuco), the centenary POLI (Polytechnic School of Engineering), and the exciting EComp/PPGEC (Undergraduate and Postgraduate Computing Engineering Programs). Without the support of all the above, none of the results reported here would have been possible. In conclusion, the heads of the FSS Project, Prof. Fernando Buarque and Prof. Carmelo Bastos-Filho, stress the importance that public as well as private funds are kept flowing to support science and technology, and hope that the trust in education and scientific research is not obliterated by the growing negationism of science.

REFERENCES

1. Xin-She Yang, editor. *Nature-Inspired Optimization Algorithms*, page i. Elsevier, Oxford, 2014.
2. Carmelo J. A. Bastos-Filho, Fernando B. de Lima-Neto, Anthony J. C. C. Lins, Antônio I. S. Nascimento, and Marília P. Lima. Fish school search. In *Nature-Inspired Algorithms for Optimisation*, pages 261–277. Springer, 2009.
3. Fernando Buarque De Lima-Neto and Marcelo Gomes Pereira de Lacerda. Weight based fish school search. In *2014 IEEE International Conference on Systems, Man and Cybernetics (SMC)*, pages 270–277. IEEE, 2014.
4. Kenneth Sörensen. Metaheuristics—the metaphor exposed. *International Transactions in Operational Research*, 22(1):3–18, 2015.
5. Joao A. G. Sargo, Susana M. Vieira, João M. C. Sousa, and Carmelo J. A. Bastos Filho. Binary Fish School Search applied to feature selection: Application to ICU readmissions. *IEEE International Conference on Fuzzy Systems*, pages 1366–1373, 2014.
6. Raphael F. Carneiro and Carmelo J. A. Bastos-Filho. Improving the Binary Fish School Search Algorithm for feature selection. *2016 IEEE Latin American Conference on Computational Intelligence, LA-CCI 2016 – Proceedings*, 2017.
7. Clodomir J. Santana, Carmelo J. A. Bastos-Filho, Mariana Macedo, and Hugo Siqueira. SBFSS: Simplified Binary Fish School Search. In *2019 IEEE Congress on Evolutionary Computation (CEC)*, pages 2595–2602. IEEE, 2019.
8. Yukun Tan, Fernando Buarque de Lima-Neto, and Ulisses Braga Neto. Pallas: Penalized maximum likelihood and particle swarms for inference of gene regulatory networks from time series data. *bioRxiv*, 2020.

9. Carmelo J. A. Bastos-Filho and Augusto C. S. Guimarães. Multi-objective fish school search. *International Journal of Swarm Intelligence Research (IJSIR)*, 6(1):23–40, 2015.
10. Kalyanmoy Deb, Amrit Pratap, Sameer Agarwal, and T. Meyarivan. A fast and elitist multiobjective genetic algorithm: NSGA-II. *IEEE Transactions on Evolutionary Computation*, 6(2):182–197, 2002.
11. Eckart Zitzler, Marco Laumanns, and Lothar Thiele. Spea2: Improving the strength pareto evolutionary algorithm. *TIK-Report*, 103, 2001.
12. Antonio J. Nebro, Juan José Durillo, Jose Garcia-Nieto, Carlos A. Coello Coello, Francisco Luna, and Enrique Alba. SMPSO: A new PSO-based metaheuristic for multi-objective optimization. In *2009 IEEE Symposium on Computational Intelligence in Multi-Criteria Decision-Making (MCDM)*, pages 66–73. IEEE, 2009.
13. Mariana Macedo, Carmelo J. A. Bastos-Filho, Susana Vieira, and João Sousa. Multi-objective binary fish school search. In *Critical Developments and Applications of Swarm Intelligence*, pages 53–72. IGI Global, 2018.
14. Robson A. Santana, Murilo Rebelo Pontes, and Carmelo J. A. Bastos-Filho. A multiple objective particle swarm optimization approach using crowding distance and roulette wheel. In *2009 Ninth International Conference on Intelligent Systems Design and Applications*, pages 237–242. IEEE, 2009.
15. Mariana Macedo, Carmelo J. A. Bastos-Filho, and Ronaldo Menezes. Improved multi-objective binary fish school for feature selection. In *The Thirty-First International Flairs Conference*, 2018.
16. Luciano S. de Souza, Ricardo B. C. Prudêncio, and Flávia de A Barros. A hybrid binary multi-objective particle swarm optimization with local search for test case selection. In *2014 Brazilian Conference on Intelligent Systems*, pages 414–419. IEEE, 2014.
17. Kalyanmoy Deb and Ram Bhushan Agrawal. Simulated binary crossover for continuous search space. *Complex Systems*, 9(2):115–148, 1995.
18. Salomão Sampaio Madeiro, Fernando Buarque de Lima-Neto, Carmelo José Albanez Bastos-Filho, and Elliackin Messias do Nascimento Figueiredo. Density as the segregation mechanism in fish school search for multimodal optimization problems. In *Proceedings of the Second International Conference on Advances in Swarm Intelligence – Volume Part II*, ICSI'11, pages 563–572. Springer-Verlag, Berlin, Heidelberg, 2011.
19. Fernando B. de Lima-Neto and Marcelo Gomes P. de Lacerda. Weight based fish school search. In *2014 IEEE International Conference on Systems, Man, and Cybernetics (SMC)*, pages 270–277, 2014.
20. Anthony J. C. C. Lins, Fernando B. de Lima-Neto, and Carmelo J. A. Bastos-Filho. Paralelização de algoritmos de busca baseados em cardumes para plataformas de processamento gráfico. *Learning and Nonlinear Models*, 13:3–20, 2015.
21. Marcelo Gomes Pereira de Lacerda and Fernando Buarque de Lima-Neto. A multithreaded implementation of the fish school search algorithm. In Clara Pizzuti and Giandomenico Spezzano, editors, *Advances in Artificial Life and Evolutionary Computation*, pages 86–98. Springer International Publishing, Cham, 2014.
22. Elliackin Figueiredo, Mariana Macedo, Hugo Valadares Siqueira, Clodomir J. Santana Jr, Anu Gokhale, and Carmelo J. A. Bastos-Filho. Swarm intelligence for clustering—a systematic review with new perspectives on data mining. *Engineering Applications of Artificial Intelligence*, 82:313–329, 2019.
23. Adriane B. S. Serapião, Guilherme S. Corrêa, Felipe B. Gonçalves, and Veronica O. Carvalho. Combining k-means and k-harmonic with fish school search algorithm for data clustering task on graphics processing units. *Applied Soft Computing*, 41:290–304, 2016.
24. F. Ferreira, T. Ciodaro, J. M. de Seixas, G. Xavier, and A. Torres. Clustering crude oil samples using swarm intelligence. In *Anais do 13 Congresso Brasileiro de Inteligência Computacional*, pages 1–6. SBIC, 2017.
25. Lara Menezes and André L. V. Coelho. Mining coherent biclusters with fish school search. In *International Conference in Swarm Intelligence*, pages 573–582. Springer, 2011.
26. Maíra Araújo de Santana, Jessiane Mônica Silva Pereira, Fabrício Lucimar da Silva, Nigel Mendes de Lima, Felipe Nunes de Sousa, Guilherme Max Silva de Arruda, Rita de Cássia Fernandes de Lima, Washington Wagner Azevedo da Silva, and Wellington Pinheiro dos Santos. Breast cancer diagnosis based on mammary thermography and extreme learning machines. *Research on Biomedical Engineering*, 34(1):45–53, 2018.

27. Debashis Mishra, Utpal Chandra De, Isita Bose, and Bishwojyoti Pradhan. Fish school search approach to find optimized thresholds in gray-scale image. In *Fifth International Conference on Computing, Communications and Networking Technologies (ICCCNT)*, pages 1–4. IEEE, 2014.
28. Felipe A. B. S. Ferreira and Francisco Madeiro. A fish school search based algorithm for image channel-optimized vector quantization. In *2016 IEEE International Conference on Systems, Man, and Cybernetics (SMC)*, pages 001680–001685. IEEE, 2016.
29. V. Bova, Yu Kravchenko, S. Rodzin, and E. Kuliev. Hybrid method for prediction of users' information behavior in the internet based on bioinspired search. *Journal of Physics: Conference Series*, 1333(3): 032008, 2019.
30. Pedro Faria, Ângelo Pinto, Zita Vale, Mahsa Khorram, Fernando B. de Lima-Neto, and Tiago Pinto. Lighting consumption optimization using fish school search algorithm. In *2017 IEEE Symposium Series on Computational Intelligence (SSCI)*, pages 1–5. IEEE, 2017.
31. Bernd Hellingrath, Dennis Horstkemper, Diego de S. Braga, Luis Filipe de A. Pessoa, Fernando Buarque de Lima-Neto, and Marcelo Gomes Pereira de Lacerda. Application of the fish school search for the supply chain network planning problem. In A. de P. Braga and C. J. A. Bastos Filho, editors, *Anais do 11 Congresso Brasileiro de Inteligência Computacional*, pages 1–6. SBIC, Porto de Galinhas, PE, 2013.
32. Wellington Pinheiro dos Santos, Ricardo Emmanuel de Souza, Reiga Ramalho Ribeiro, Allan Rivalles Souza Feitosa, Valter Augusto de Freitas Barbosa, Victor Luiz Bezerra Arajo da Silva, David Edson Ribeiro, and Rafaela Covello de Freitas. Electrical impedance tomography using evolutionary computing: A review. In D. P. Acharjya and V. Santhi, editors, *Bio-Inspired Computing for Image and Video Processing*, pages 93–128. CRC Press, Boca Raton, 2018.
33. João Batista Monteiro-Filho, De Albuquerque Isabela, Fernando Lima-Neto, and Filipe V. S. Ferreira. Optimizing multi-plateau functions with FSS-SAR (stagnation avoidance routine). In *2016 IEEE Symposium Series on Computational Intelligence (SSCI)*, pages 1–6. IEEE-SSCI, 2016.
34. João Batista Monteiro-Filho, IMC Albuquerque, and Fernando B. de Lima-Neto. Solving mixed model workplace time-dependent assembly line balancing problem with FSS algorithm. *arXiv preprint arXiv:1707*.06132, 2017.
35. I. Boulkaibet, L. Mthembu, Fernando B. De Lima-Neto, and T. Marwala. Finite element model updating using fish school search and volitive particle swarm optimization. *Integrated Computer-Aided Engineering*, 22(4):361–376, 2015.

3 Marriage in Honey Bees Optimization in Continuous Domains

Jing Liu, Sreenatha Anavatti, Matthew Garratt, and Hussein A. Abbass

CONTENTS

Marriage in honey bees optimization (MBO) is a swarm-based algorithm that was originally designed to solve discrete problems by getting inspiration from the marriage process of honey bees. In this chapter, MBO is adapted to solve problems in continuous domains by using the adapted representation way and four proposed heuristics. Experiments are conducted on benchmark functions to evaluate the search ability of MBOs with each single heuristic and four incorporated heuristics.

3.1 INTRODUCTION

The collective behaviors of swarms have been observed in a number of natural species and inspired the concept of swarm intelligence (SI), which was first proposed in 1980s by Beni [1, 2]. The individuals in the swarm may be relatively not intelligent, but they can show high intelligence as a whole swarm that is highly self-organized and self-adaptive via the interaction among them [3, 4]. SI has increasingly attracted the attention from researchers and has inspired a number of swarm optimization algorithms, such as ant colony optimization (ACO) [5] based on ant foraging behavior and particle swarm optimization (PSO) [6] based on birds flocking.

Honey bees are typical social insects with some behaviors such as division of the labor, communications, and foraging, which have inspired some algorithms. The first bees optimization algorithm was marriage in honey bees optimization (MBO) [7, 8] in 2001. In 2005, the artificial bee colony (ABC) algorithm [9] was introduced, which seems to follow design principles similar to differential evolution.

MBO was proposed to solve the propositional satisfiability problem. Its distinguishing feature from all other swarm optimization algorithms is that it works with partial solutions. MBO mimics the marriage process of bees, which evolve from a solitary colony to an eusocial colony. The bee colony usually consists of queen(s), drones, workers, and broods. Queens start a mating flight with a dance performance to attract drones to mate with them. The sperm of drones are partial solutions, and they accumulate in the spermatheca and will be selected randomly by queens to generate broods, which will be raised by workers.

After the emergence of MBO, some efforts have been made to improve the performance of MBO. For example, a conventional annealing approach is used to determine the pool of drones by Teo and Abbass [10]. Based on the honey bee mating model in MBO, Afshar et al. [11] present an improved honey bee mating optimization (HBMO) algorithm for the optimization of the reservoir operation. A new algorithm called wolf pack search (WPS) was presented and incorporated into the local search process of MBO by Yang et al. [12]. The new WPS-MBO has better convergence characteristics than classical MBO and generic algorithm (GA) when solving some popular functions and TSP. Celik and Ulker [13] propose an improved MBO by adding a levy flight algorithm for queen mating flight and considering the neighborhood during the improvement process by worker bees. Based on the advantages of MBO, it has been successfully applied to some other real-world problems, such as integrated circuit industry output forecasting [14], hydro systems design and operation [15], and materialized view selection [16].

As MBO was initially designed for discrete problems, it cannot be applied to solve problems in continuous domains directly. In this chapter, MBO is adapted to continuous domains without major changes to the core concept of MBO. The representation of MBO for continuous problems is designed, and four heuristics employed for the worker to improve the population are proposed. Experiments are conducted on two different sets of benchmark functions to compare the MBO with other swarm optimization algorithms. The search ability of MBO with each single heuristic and four incorporated heuristics as well as the computational complexity of MBO are discussed.

The rest of this chapter is structured as follows. Section 3.2 introduces the colony structure and marriage process of honey bees and then presents the artificial model of MBO. The section is then followed by the description of proposed MBO in continuous domains in Section 3.3. Section 3.4 presents the experimental setup, comparison algorithms, and results and analysis, followed by the conclusion in Section 3.5.

3.2 MARRIAGE IN HONEY BEES

3.2.1 Colony Structure

There are queen(s), drones, workers, and broods in a honey bee colony generally. The queens are the sexually developed females whose function is reproduction and the drones are the male bees whose primary function is to fertilize the queen during her mating flight. The queens lay both fertilized and unfertilized eggs, which are the broods of the colony. The fertilized eggs are potential queens or workers, whereas the unfertilized eggs will grow into drones. Workers are sexually undeveloped females who take responsibility of brood care.

3.2.2 The Mating Flight

The mating flight is for the queen to mate with drones. During the flight, the queen is found by the drones and generally mates with 7 to 20 drones [17]. Sperm of the drones accumulate in the spermatheca, forming the genetic pool of the honey bee colony for fertilization. At each time of fertilization, the sperm are retrieved randomly by the queen to fertilize the eggs.

3.2.3 The Artificial Analogue Model

By mimicking the above colony structure and the honey bees' mating process, the artificial analogue model of MBO was proposed to address optimization problems [7]. In MBO, the queen is initialized with an amount of energy and some speed, both of which reduce during the mating flight. When the energy is consumed or the spermathaca is full, the queen will stop the mating flight and return to the nest. In the flight, the queen mates with the drone encountered with the following probability:

$$p(Q,D) = exp\left(\frac{-difference}{speed}\right) \tag{3.1}$$

where $p(Q,Dr)$ is the probability of mating between the queen Q and the drone Dr, which means adding the sperm of drone Dr to the spermathaca of queen Q; *speed* is the speed of Q; and *difference* is the absolute difference between the fitness of Dr and Q. The energy and speed of the queen are reducing during the flight based on following equations:

$$energy(t+1) = energy(t) - step \tag{3.2}$$

$$speed(t+1) = \lambda \cdot speed(t) \tag{3.3}$$

$$step = \frac{0.5 \cdot energy}{M} \tag{3.4}$$

where t is the index for the queen's transition in the environment where the queen might mate with the drone; λ is a coefficient ranging from 0 to 1; *step* is the size of energy reduction in each transition; and M is the size of the spermatheca.

As described before, there are queens, drones, workers, and broods in the honey bee colony. In the MBO analogue model, queens mate with drones and generate the broods, which are raised by workers. Then the best bees among the queens and broods will be the queens for the next generation. The pseudocode of the artificial analogue model of the MBO is presented in Algorithm 3.1. The queens are initialized randomly at the beginning, followed by the mating flights. During the mating flights, each queen constructs its spermatheca by adding drone sperm to it according to each drone's probability p until the *energy* is exhausted or the spermatheca is full. Subsequently, a queen selected randomly and sperm selected from the queen's spermatheca based on $p(Q,D)$ generate brood genotypes by crossover and mutation. Then workers, which represent different heuristic methods, are used to improve the broods' genotypes. After the generation of broods, queens will be updated if the best brood is better than the worst queen. The bee colony keeps generating and raising broods until the termination conditions are met and the best queen will be outputted as the best solution.

ALGORITHM 3.1 THE PSEUDOCODE OF THE GENERIC MARRIAGE IN HONEY BEES OPTIMIZATION MODEL

```
1:  initialise workers
2:  randomly generate the queens
3:  apply local search to get good queens
4:  while termination conditions not met do
5:      for each queen do
6:         initialise the speed, energy, position and step
7:         while energy > 0 do
8:           the queen moves in some speed and chooses drones probabilistically
9:           if a drone is selected and the spermmatheca is not full then
10:             add the selected drone's sperm to the queen's spermatheca
11:          end if
12:          update the energy and speed of the queen based on Eqs. 3.2 and 3.3
13:        end while
14:     end for
15:     generate broods by crossover and mutation
16:     improve broods using workers
17:     update the fitness value of broods
18:     while the best brood is better than the worst queen do
19:        replace the worst queen with the best brood
20:        remove the best brood form the brood list
21: end while
22: end while
```

3.3 MARRIAGE IN HONEY BEE OPTIMIZATION IN CONTINUOUS DOMAINS

Based on the generic MBO discussed above, the MBO is adapted and applied to solve problems in continuous domains in this section.

3.3.1 Representation

The candidate solution of the optimization problem is represented by the genotype where the length equals the dimension of the problem and each cell corresponds to a variable. A genotype usually refers to a pair of alleles, each of which is inherited from one parent. Each allele consists of a vector of real numbers and the solution is represented by the average of the pair of alleles. The fitness of the genotype is the objective value of the optimization problem.

A drone is represented by one allele as the drone is haploid, whereas the queen is represented by the genotype. Each queen has speed, energy, and spermatheca along with the genotype. The speed and energy are initialized randomly ranging from 0.5 to 1 at the beginning of each mate flight. If a drone is selected to mate with the queen, its sperm will be stored in the spermatheca. The broods are generated by inheriting an allele from one sperm and another allele from the queen.

3.3.2 MBO in Continuous Domains

The pseudocode of the MBO algorithm in continuous domain is presented in Algorithm 3.2. The parameters are initialized at the beginning of the algorithm, including the number of queens N_q, the number of the workers N_w, the number of the broods N_b, the size of the queen's spermatheca M, and the constant λ. N_w also refers to the number of the heuristics employed in the algorithm, M represents the maximum number of successful matings per queen in a single flight, and λ is the reduction ratio of the queen's speed.

Then the mating flight for each queen starts. In the mating flight, each queen has its energy and speed, which are initialized randomly before the commencement of each mating flight. The

ALGORITHM 3.2 THE PSEUDOCODE OF MARRIAGE IN HONEY BEES OPTIMIZATION ALGORITHM IN CONTINUOUS DOMAIN

Input: a set of variables $X = \{X^d, d = 1,...,D\}$, an objective function $f(X)$

1: Initialise parameters: number of queens, workers, and broods N_q, N_w, N_b, size of the spermatheca M, a constant $\lambda \in (0,1)$
2: **while** termination conditions not met **do**
3: **for** $i = 1,...,N_q$ **do**
4: initialise the speed *speed*$_i$, energy *energy*$_i$ and genotype X_i of the queen, size of energy reduction *step*
5: generate drone X_{drone} randomly
6: **while** *energy*$_i > 0$ **do**
7: calculate probability p based on Eq. 3.1
8: **If** *rand* $< p$ and the number of sperm $< M$ **then**
9: add the selected drone's sperm to the queen's spermatheca
10: **end if**
11: update the energy and speed of the queen based on Eqs. 3.2 and 3.3
12: change the drone's genotype in each dimension with the probability of *speed* based on Eq. 3.5
13: **end while**
14: **end for**
15: **for** $j = 1,...,N_b$ **do**
16: select a queen in proportion to her fitness
17: select a sperm from the spermatheca of the queen randomly
18: generate broods by crossover and mutation based on Algorithm 3.3
19: select a worker in proportion to its fitness
20: improve broods using the select work
21: update the fitness value of workers
22: **end for**
23: **while** the best brood is better than the worst queen **do**
24: replace the worst queen with the best brood
25: remove the best brood form the brood list
26: **end while**
27: **end while**

Output: the best solution X_*, the best objective value $f(X_*)$

energy decides the duration of the flight while the speed represents the probability of flipping the drone's genotype. The drone is generated randomly in the feasible area at first and mated with the queen probabilistically. If the drone is selected, its sperm will be added to the queen's spermatheca. During the mating flight, the queen's energy reduces to nearly zero when the mating flight stops and the queen returns to the hive. The queen's speed reduces as well during the flight, which means that the neighborhood searched by the queen decreases. The drone's genotype is changed during the mating flight with the probability of *speed* based on the following equation:

$$\begin{aligned} Dr^d &= Dr^d \cdot r_2, \textit{if } r_1 < speed; \\ Dr^d &= Dr^d, otherwise; \, d = 1,...D \end{aligned} \tag{3.5}$$

where r_1, r_2 are random values in (0,1).

The breeding starts after the mating flights of all queens. To construct the brood *Brood*, a queen Q is selected in proportion to the fitness and a sperm *Sperm* is selected from the selected queen's spermatheca. The brood is generated by crossover and mutation in discrete domains. To adapt it to continuous domains, we merge the queen and the sperm and then mutate it as shown in Algorithm 3.3. *LB* and *UB* are the lower bound and upper bound of the variables respectively.

ALGORITHM 3.3 THE PSEUDOCODE OF CROSSOVER AND MUTATION IN CONTINUOUS MBO

1: $a = rand(D,1)$
2: $b = (a < 0.5)$
3: $Brood = b.*Q + (1-b).*Sperm$
4: $c = ceil(rand.*D)$
5: $Brood(c,1) == LB + (UB - LB).*rand;$

After that, a worker is selected in proportion to its fitness and employs the corresponding heuristic to improve the brood. When all broods are generated, they will be compared with the queens and the best broods will replace the worst queens to make sure that all the queens are better than the broods. Then the rest of the broods are killed and the new mating flight starts. The above process is continuous until the termination conditions are met.

3.3.3 The Heuristics Employed for Workers

Workers take the responsibility of raising broods. This means that the broods are growing to better solutions by employing the workers. A number of heuristics are applied for workers to shape broods. Each worker/heuristic has a fitness value that records the amount of improvement achieved in the broods' genotype by applying this heuristic to raise them. In each generation, one worker is selected for one brood in proportion to the worker's fitness value. A different number and type of heuristics can be selected by users to adapt the MBO to address specific problems. In this chapter, we present four relatively simple heuristics to give some examples of employing workers to improve the brood. These heuristics can be altered as required. The maximum number of iteration times for each heuristic is set at 1000. The process for each heuristic to improve a brood is briefly described as follows:

H1: A random direction is generated in this method. A new brood is constructed by moving the old one in this direction using a step size until no further improvements could be made.

H2: Copy the value of one variable to another and keep doing that until the objective does not improve.

H3: For each variable, round the value to a random precision. If the solution improves, round the value to a random precision again, or else move to the next variable.

H4: A queen Q is generated randomly in the feasible space and the direction from the old brood B_o to the queen is calculated using $De = Q - B_o$. If Q is better than B_o, another new brood is generated by $B_n = Q + rand * De$; otherwise, $B_n = B_o - rand * De$. *Rand* is a random value in $(0,1)$. If B_n is better than B_o, B_n is saved as the old brood for the next iteration.

3.4 EXPERIMENTS

3.4.1 Experimental Setting

The experiments are conducted on an Intel i7-6700 CPU (3.40 GHz) and 16 GB RAM using MATLAB R2017b. MBO is compared with the basic versions of three well-known SI algorithms, including PSO [18], continuous ACO [19], and ABC [9] as the proposed MBO is the basic version. We adapt MBO to solve continuous problems in this chapter without changing the design principles of the original discrete MBO. Similar to MBO, ACO was originally proposed for addressing discrete problems and then adapted to solve continuous problems, which makes the continuous ACO a perfect comparison. We tested them on two suits of benchmark functions with different complexity to evaluate its performance. The first suit included nine unimodal functions $f_1 - f_9$ and nine multimodal functions $f_{10} - f_{18}$, whose function names and search range are shown in Table 3.1. The second suit is adopted from CEC17 competition on single objective real-parameter numerical optimization [20], which is much more complex than the first test suit. Please note that the second problem of the CEC17 test suit is excluded by the proposers due to its unstable behavior. We optimized these functions of four dimensions D (10D, 30D, 50D, and 100D). The initial populations are generated uniformly random in the search space. All experiments were run 30 times. The swarm size is 50 and the maximum fitness evaluations (MaxFES) is set to $10000 * D$. Based on the parameters setting in the discrete MBO [7] and the preliminary experimental testing, the parameters of continuous MBO are set as $N_q = 2$; $N_w = 5$; $N_b = 48$; $M = 21$; $\lambda = 0.9$; the maximum number of times to improve a brood by a worker is set to 1000. The parameters of the comparison algorithm are set based on the suggestions of the corresponding literature as follows. In PSO, the two coefficients $c_1 = 2, c_2 = 2$, the maximum velocity is set to 2 and the inertia weight

TABLE 3.1
The Functions and Their Search Range

No.	Function Name	Search Range	No.	Function Name	Search Range
f_1	Sphere	$[-100,100]^D$	f_{10}	Rastrigin	$[-5.12,5.12]^D$
f_2	Schwefel1.2	$[-100,100]^D$	f_{11}	Schwefel	$[-500,500]^D$
f_3	Rosenbrock	$[-2.048,2.048]^D$	f_{12}	Griewank	$[-600,600]^D$
f_4	Sumpowers	$[-10,10]^D$	f_{13}	Ackley	$[-32.768,32.768]^D$
f_5	Step	$[-100,100]^D$	f_{14}	Michalewicz10	$[-10\pi,10\pi]^D$
f_6	Schewfel2.22	$[-500,500]^D$	f_{15}	Michalewicz20	$[-10\pi,10\pi]^D$
f_7	Quartic	$[-50,50]^D$	f_{16}	Weierstrass	$[-0.5,0.5]^D$
f_8	Elliptic	$[-100,100]^D$	f_{17}	Penalized1	$[-50,50]^D$
f_9	Levy	$[-100,100]^D$	f_{18}	Penalized2	$[-50,50]^D$

$w \in [0.4, 0.9]$. In ABC, the number of iterations allowed for no improvement of the solution before it is abandoned is set to 1000. In ACO, the pheromone evaporation rate is set to 1 and the parameter influencing the distribution of solution weights is set to 0.5.

3.4.2 Experimental Results

3.4.2.1 The Comparison of Different Heuristics

In this chapter, four heuristics employed by workers in MBO are presented in Section 3.3.3, including H1, H2, H3, and H4. To evaluate the performance of each heuristic, we implemented the experiments to compare the performance of MBOs employing only one heuristic. The average and standard deviations (in parentheses) of fitness values obtained by 30 runs using MBO with single heuristic are presented in Table 3.2. The best results for each function are shown in

TABLE 3.2
Comparative Results of Different Heuristics on 30D Benchmark Function

Function No.	H1	H2	H3	H4
f_1	1.66e+03*	1.13e-04*	7.46e+03*	**6.78e-05**
	(2.04e+03)	(6.21e-04)	(7.82e+03)	(7.91e-05)
f_2	4.43e+02*	**4.96e+00**	3.46e+02	1.77e+01*
	(2.40e+02)	(1.34e+01)	(1.01e+03)	(7.37e+00)
f_3	1.82e+02*	**1.47e+01**	2.90e+01*	2.80e+01*
	(1.85e+02)	(1.40e+01)	(5.85e-02)	(8.49e-01)
f_4	4.66e+01*	3.82e-12*	**0.00e+00**	4.81e-13*
	(1.22e+02)	(1.62e-11)	(0.00e+00)	(9.05e-13)
f_5	1.79e+03*	**0.00e+00**	5.39e+03*	2.93e+00*
	(2.40e+03)	(0.00e+00)	(6.59e+03)	(2.13e+00)
f_6	8.54e+36*	1.47e-03*	**0.00e+00**	1.12e+02*
	(4.68e+37)	(4.26e-03)	(0.00e+00)	(1.63e+02)
f_7	2.03e+04*	3.68e-09*	**0.00e+00**	3.82e-06*
	(4.67e+04)	(1.89e-08)	(0.00e+00)	(8.32e-06)
f_8	8.45e+05*	8.11e-04*	**0.00e+00**	2.36e+04*
	(7.71e+05)	(2.15e-03)	(0.00e+00)	(3.57e+04)
f_9	1.86e+03*	**4.41e-02**	2.19e+03*	5.10e+00*
	(5.00e+03)	(7.19e-02)	(3.34e+03)	(7.08e+00)
f_{10}	2.81e+01*	1.73e-04*	**0.00e+00**	2.51e+01*
	(3.91e+01)	(7.94e-04)	(0.00e+00)	(4.05e+01)
f_{11}	−6.57e+03*	**−1.26e+04**	−3.95e+03*	−7.15e+03*
	(4.34e+02)	(0.00e+00)	(4.33e+02)	(7.16e+02)
f_{12}	1.11e+02*	**8.15e-05**	7.88e-01	1.27e-02*
	(9.16e+01)	(3.36e-04)	(1.62e+00)	(1.38e-02)
f_{13}	1.04e+01*	3.33e-03*	**8.88e-16**	3.40e+00*
	(3.31e+00)	(5.25e-03)	(4.01e-31)	(9.58e-01)
f_{14}	−1.04e+01*	**−2.67e+01**	−8.30e+00*	−1.66e+01*
	(7.76e-01)	(1.08e+00)	(7.51e-01)	(1.10e+00)
f_{15}	−9.51e+00*	**−2.58e+01**	−7.44e+00*	−1.57e+01*
	(8.27e-01)	(1.47e+00)	(8.26e-01)	(1.12e+00)

TABLE 3.2 (*Continued*)
Comparative Results of Different Heuristics on 30D Benchmark Function

Function No.	H1	H2	H3	H4
f_{16}	3.06e+00*	9.49e-02*	**0.00e+00**	4.59e+00*
	(3.21e+00)	(1.70e-01)	(0.00e+00)	(4.76e+00)
f_{17}	1.62e+01*	**6.92e-07**	1.10e+00*	5.26e+02*
	(8.53e+01)	(1.07e-06)	(1.96e-01)	(7.04e+01)
f_{18}	1.82e+03*	**4.33e-06**	2.57e+00*	2.34e+03*
	(1.04e+03)	(6.72e-06)	(9.02e-02)	(2.64e+02)

Note: Bold indicates the best result and * means statistical difference.

bold. The Wilcoxon rank sum tests are implemented to test the statistical significance between the best result and the other results. The * in the table means that the result of this algorithm is statistically different from the best result with 95% certainty. This table shows that H2 and H3 perform better on these test functions. Specifically, MBO with H2 achieved the best results of 10 functions and MBO with H3 achieved the best results on 7 functions. However, MBO with H4 outperforms H2 and H3. This is consistent with the "no free lunch" theorem [21], which claims that no optimization method can be the generic super solver.

Table 3.3 presents the comparative results between MBO with single heuristic and MBO with all four heuristics. The "b" and "w" in the table means that MBO with all four heuristics is significantly better (worse) than MBO with this single heuristic, while "–" means there is no significant difference. We can see from the table that MBO with multiple heuristics performs

TABLE 3.3
Comparative Results of Different Heuristics on 30D Benchmark Function

Function No.	H1	H2	H3	H4
f_1	b	b	b	b
f_2	b	w	–	–
f_3	b	–	b	b
f_4	b	b	w	b
f_5	b	–	b	b
f_6	b	b	–	b
f_7	b	b	–	b
f_8	b	b	w	b
f_9	b	b	b	b
f_{10}	b	b	–	b
f_{11}	b	–	b	b
f_{12}	b	b	b	b
f_{13}	b	b	–	b
f_{14}	b	b	b	b
f_{15}	b	b	b	b
f_{16}	b	b	–	b
f_{17}	b	b	b	b
f_{18}	b	b	b	b

better than MBO with a simple heuristic. For example, after the incorporation of other heuristics, MBO obtained better results on f_1, f_3, f_5, and so forth, compared with the algorithm with only H3 employed. This means that the employment of multiple heuristics in MBO can utilize the advantages of different heuristics and then improve the performance of the algorithm. However, MBO with four heuristics cannot outperform MBO with single H3 on some functions, such as f_4, f_8 and so forth. This indicates that the incorporation of multiple heuristics might also weaken the search capability of a single heuristic. In the following experiments for comparing MBO with other algorithms, we employed the MBO with all four heuristics.

3.4.2.2 The Comparison of MBO with other Algorithms on the Suit of Simple Functions

The average and standard deviations (in parentheses) of fitness values obtained by 30 runs using the four algorithms for 18 relatively simple benchmark functions on 10D, 30D, 50D, and 100D are presented in Tables 3.4–3.7. The best results for each function are shown in bold and the * stands for statistical significance between this result and the best result with 95% certainty.

TABLE 3.4
Comparative Results of Algorithms on 10D Benchmark Function

Function No.	PSO	ACO	ABC	MBO
f_1	9.25e-58*	**3.13e-86**	1.32e-16*	2.02e-36
	(3.19e-57)	(8.30e-86)	(5.33e-17)	(1.11e-35)
f_2	**1.11e-20**	2.98e-14*	1.95e+01*	5.62e-10*
	(2.23e-20)	(4.76e-14)	(2.29e+01)	(3.08e-09)
f_3	2.89e+00*	1.15e+00*	**3.39e-01**	6.63e-01*
	(5.95e-01)	(9.25e-02)	(5.42e-01)	(2.02e+00)
f_4	3.80e-95*	**1.97e-148**	3.08e-17*	4.26e-77*
	(2.02e-94)	(7.37e-148)	(1.66e-17)	(2.33e-76)
f_5	**0.00e+00**	**0.00e+00**	**0.00e+00**	**0.00e+00**
	(0.00e+00)	(0.00e+00)	(0.00e+00)	(0.00e+00)
f_6	1.85e-24*	2.04e-45*	3.35e-16*	**0.00e+00**
	(1.01e-23)	(3.38e-45)	(8.61e-17)	(0.00e+00)
f_7	2.48e-94*	6.77e-132*	4.14e-17*	**0.00e+00**
	(1.16e-93)	(1.88e-131)	(1.75e-17)	(0.00e+00)
f_8	4.95e-53*	5.31e-83*	1.00e-16*	**0.00e+00**
	(2.57e-52)	(2.49e-82)	(3.59e-17)	(0.00e+00)
f_9	**1.35e-31**	**1.35e-31**	3.66e-15*	**1.35e-31**
	(6.68e-47)	(6.68e-47)	(9.56e-15)	(6.68e-47)
f_{10}	1.79e+00*	2.69e+01*	**0.00e+00**	**0.00e+00**
	(1.15e+00)	(4.86e+00)	(0.00e+00)	(0.00e+00)
f_{11}	–2.25e+03*	–2.35e+03*	**–4.19e+03**	**–4.19e+03**
	(3.85e+02)	(1.07e+02)	(1.85e-12)	(1.85e-12)
f_{12}	4.55e-01*	2.97e-01*	5.92e-03*	**0.00e+00**
	(2.88e-01)	(1.00e-01)	(5.75e-03)	(0.00e+00)
f_{13}	4.44e-15*	4.32e-15*	8.70e-15*	**8.88e-16**
	(4.01e-30)	(6.49e-16)	(2.54e-15)	(4.01e-31)
f_{14}	–9.47e+00*	–4.20e+00*	**–9.70e+00**	–9.44e+00*
	(2.69e-01)	(2.96e-01)	(9.12e-02)	(3.43e-01)

TABLE 3.4 (*Continued*)
Comparative Results of Algorithms on 10D Benchmark Function

Function No.	PSO	ACO	ABC	MBO
f_{15}	–9.18e+00*	–3.78e+00*	**–9.54e+00**	–9.24e+00*
	(4.37e-01)	(2.49e-01)	(9.29e-02)	(4.49e-01)
f_{16}	**0.00e+00**	**0.00e+00**	**0.00e+00**	**0.00e+00**
	(0.00e+00)	(0.00e+00)	(0.00e+00)	(0.00e+00)
f_{17}	6.22e-02*	**4.71e-32**	2.89e+02*	1.24e-31
	(1.51e-01)	(0.00e+00)	(6.11e+01)	(4.19e-31)
f_{18}	**1.50e-33**	**1.50e-33**	5.10e+02*	3.53e-24*
	(0.00e+00)	(0.00e+00)	(9.59e+01)	(1.93e-23)

Note: Bold indicates the best result and * means statistical difference.

TABLE 3.5
Comparative Results of Algorithms on 30D Benchmark Function

Function No.	PSO	ACO	ABC	MBO
f_1	**9.49e-44**	4.66e-17*	7.97e-16*	1.19e-38*
	(2.73e-43)	(5.09e-17)	(1.28e-16)	(6.50e-38)
f_2	**1.46e-03**	3.20e+04*	5.47e+03*	1.80e+01*
	(1.57e-03)	(4.82e+03)	(1.63e+03)	(2.31e+01)
f_3	2.44e+01*	2.32e+01*	**9.49e+00**	1.93e+01*
	(1.09e+01)	(1.30e-01)	(6.45e+00)	(1.28e+01)
f_4	**1.01e-64**	9.29e-13*	1.92e-09*	1.01e-42*
	(5.45e-64)	(2.09e-12)	(4.27e-09)	(5.52e-42)
f_5	**0.00e+00**	**0.00e+00**	1.00e-01	**0.00e+00**
	(0.00e+00)	(0.00e+00)	(3.05e-01)	(0.00e+00)
f_6	1.76e-17*	8.97e+17*	1.82e-15*	**1.23e-39**
	(6.30e-17)	(4.91e+18)	(2.45e-16)	(6.57e-39)
f_7	6.44e-55*	1.11e-11*	4.10e-16*	**1.11e-82**
	(2.44e-54)	(2.88e-11)	(1.22e-16)	(6.08e-82)
f_8	1.79e-40*	2.94e-15*	7.89e-16*	**1.19e-41**
	(4.45e-40)	(2.57e-15)	(1.74e-16)	(6.53e-41)
f_9	1.65e-31*	4.37e-16*	2.67e-11*	**1.35e-31**
	(5.03e-32)	(4.59e-16)	(4.05e-11)	(6.68e-47)
f_{10}	2.43e+01*	2.23e+02*	2.42e-14*	**0.00e+00**
	(6.05e+00)	(1.11e+01)	(6.55e-14)	(0.00e+00)
f_{11}	–5.60e+03*	–4.95e+03*	–1.26e+04	**–1.28e+04**
	(1.80e+03)	(2.06e+02)	(0.00e+00)	(1.01e+03)
f_{12}	1.25e-02*	2.66e-02*	2.47e-04*	**0.00e+00**
	(1.79e-02)	(8.80e-02)	(1.35e-03)	(0.00e+00)
f_{13}	1.19e-14*	1.68e-09*	5.70e-14*	**8.88e-16**
	(3.53e-15)	(8.28e-10)	(8.67e-15)	(4.01e-31)
f_{14}	–2.39e+01*	–6.82e+00*	–2.77e+01	**–2.78e+01**
	(2.85e+00)	(4.99e-01)	(2.69e-01)	(1.06e+00)
f_{15}	–2.31e+01*	–6.07e+00*	–2.64e+01*	**–2.71e+01**
	(2.10e+00)	(3.84e-01)	(3.86e-01)	(1.25e+00)

(*Continued*)

TABLE 3.5 (*Continued*)
Comparative Results of Algorithms on 30D Benchmark Function

Function No.	PSO	ACO	ABC	MBO
f_{16}	9.11e-04*	5.34e-09*	5.92e-15*	**0.00e+00**
	(3.07e-03)	(7.01e-09)	(4.60e-15)	(0.00e+00)
f_{17}	1.07e-01*	9.02e-16*	5.21e+02*	**1.57e-32**
	(1.90e-01)	(1.23e-15)	(6.31e+01)	(8.35e-48)
f_{18}	**4.99e-33**	4.86e-18*	2.31e+03*	3.01e-25*
	(8.73e-33)	(5.88e-18)	(2.57e+02)	(1.65e-24)

Note: Bold indicates the best result and * means statistical difference.

TABLE 3.6
Comparative Results of Algorithms on 50D Benchmark Function

Function No.	PSO	ACO	ABC	MBO
f_1	**5.19e-35**	2.33e+00*	1.94e-15*	3.01e-33*
	(1.28e-34)	(1.03e+00)	(6.61e-16)	(1.52e-32)
f_2	**5.73e-01**	1.21e+05*	2.19e+04*	3.03e+02*
	(3.02e-01)	(1.23e+04)	(3.09e+03)	(2.28e+02)
f_3	5.58e+01*	1.52e+02*	**3.02e+01**	3.31e+01*
	(2.77e+01)	(7.13e+01)	(1.30e+01)	(2.21e+01)
f_4	**7.45e-37**	3.35e+08*	1.55e-05*	4.98e-31
	(4.07e-36)	(7.46e+08)	(4.61e-05)	(1.77e-30)
f_5	**0.00e+00**	1.01e+01*	1.43e+00*	**0.00e+00**
	(0.00e+00)	(3.19e+00)	(1.04e+00)	(0.00e+00)
f_6	2.87e+01*	2.43e+75*	3.50e-15*	**2.78e-41**
	(1.15e+02)	(1.27e+76)	(3.77e-16)	(1.52e-40)
f_7	1.99e-37*	3.72e+05*	1.04e-15*	**7.64e-77**
	(9.25e-37)	(2.35e+05)	(3.36e-16)	(4.18e-76)
f_8	3.05e-31*	5.18e+01*	2.35e-15*	**1.41e-38**
	(5.15e-31)	(2.00e+01)	(2.16e-15)	(7.63e-38)
f_9	2.20e-02*	1.27e+01*	2.41e-11*	**1.35e-31**
	(4.47e-02)	(6.90e+00)	(3.77e-11)	(6.68e-47)
f_{10}	5.89e+01*	4.94e+02*	7.64e-12*	**0.00e+00**
	(1.49e+01)	(1.41e+01)	(3.97e-11)	(0.00e+00)
f_{11}	−1.07e+04*	−6.76e+03*	−2.09e+04*	**−2.09e+04**
	(1.65e+03)	(2.35e+02)	(4.75e+01)	(0.00e+00)
f_{12}	7.06e-03*	9.44e-01*	1.61e-11*	**0.00e+00**
	(8.46e-03)	(7.31e-02)	(6.04e-11)	(0.00e+00)
f_{13}	2.41e-14*	5.34e-01*	1.26e-13*	**8.88e-16**
	(4.04e-15)	(2.29e-01)	(1.31e-14)	(4.01e-31)
f_{14}	−3.62e+01*	−8.89e+00*	−4.45e+01*	**−4.61e+01**
	(5.21e+00)	(5.24e-01)	(4.74e-01)	(1.40e+00)
f_{15}	−3.38e+01*	−7.58e+00*	−4.22e+01*	**−4.47e+01**
	(4.29e+00)	(4.59e-01)	(5.93e-01)	(2.09e+00)
f_{16}	7.43e-02*	2.31e+00*	6.02e-14*	**0.00e+00**
	(2.74e-01)	(3.05e-01)	(2.20e-14)	(0.00e+00)

TABLE 3.6 (*Continued*)
Comparative Results of Algorithms on 50D Benchmark Function

Function No.	PSO	ACO	ABC	MBO
f_{17}	1.13e+00*	2.55e+01*	6.31e+02*	**1.03e-32**
	(4.42e+00)	(6.33e+00)	(5.32e+01)	(4.67e-33)
f_{18}	7.90e+00*	3.77e-01*	4.43e+03*	**2.06e-29**
	(3.02e+01)	(1.72e-01)	(3.94e+02)	(1.06e-28)

Note: Bold indicates the best result and * means statistical difference.

TABLE 3.7
Comparative Results of Algorithms on 100D Benchmark Function

Function No.	PSO	ACO	ABC	MBO
f_1	1.48e-23*	8.33e+04*	8.93e-15*	**1.48e-23**
	(2.76e-23)	(5.56e+03)	(5.86e-15)	(6.14e-23)
f_2	**7.80e+01**	4.91e+05*	1.04e+05*	2.82e+03*
	(1.46e+01)	(4.63e+04)	(8.23e+03)	(1.36e+03)
f_3	1.50e+02*	1.37e+04*	1.10e+02	**9.37e+01**
	(4.22e+01)	(1.22e+03)	(2.40e+01)	(1.77e+01)
f_4	**4.21e-17**	1.41e+51*	2.44e+72*	1.67e-10
	(1.51e-16)	(4.34e+51)	(1.33e+73)	(9.11e-10)
f_5	**0.00e+00**	8.42e+04*	8.10e+00*	**0.00e+00**
	(0.00e+00)	(5.71e+03)	(4.44e+00)	(0.00e+00)
f_6	2.92e+03*	2.28e+189*	9.83e-15*	**2.04e-35**
	(1.44e+03)	(Inf)	(1.43e-15)	(1.12e-34)
f_7	**5.38e-22**	1.06e+09*	8.44e-14*	1.19e-12*
	(1.47e-21)	(1.39e+08)	(2.19e-13)	(6.52e-12)
f_8	5.35e-20*	3.92e+06*	1.86e-14*	**1.75e-33**
	(1.49e-19)	(3.86e+05)	(1.87e-14)	(9.59e-33)
f_9	2.34e+00*	1.33e+05*	1.63e-10*	**1.35e-31**
	(5.86e+00)	(9.88e+03)	(5.32e-10)	(6.68e-47)
f_{10}	1.77e+02*	1.30e+03*	3.33e-02*	**0.00e+00**
	(2.63e+01)	(2.33e+01)	(1.82e-01)	(0.00e+00)
f_{11}	−1.93e+04*	−1.02e+04*	−4.16e+04*	**−4.25e+04**
	(4.42e+03)	(4.30e+02)	(1.15e+02)	(2.55e+03)
f_{12}	2.46e-03*	7.51e+02*	3.20e-10*	**0.00e+00**
	(7.56e-03)	(5.00e+01)	(1.74e-09)	(0.00e+00)
f_{13}	5.46e-13*	1.90e+01*	4.59e-13*	**8.88e-16**
	(5.13e-13)	(1.57e-01)	(1.41e-13)	(4.01e-31)
f_{14}	−6.30e+01*	−1.24e+01*	−8.55e+01*	**−9.10e+01**
	(5.67e+00)	(7.16e-01)	(7.80e-01)	(1.61e+00)
f_{15}	−5.59e+01*	−1.07e+01*	−8.00e+01*	**−8.64e+01**
	(1.10e+01)	(5.55e-01)	(1.36e+00)	(2.77e+00)
f_{16}	3.65e+00*	1.27e+02*	8.39e-13*	**0.00e+00**
	(2.37e+00)	(3.16e+00)	(7.79e-13)	(0.00e+00)
f_{17}	3.86e+00*	3.90e+02*	7.29e+02*	**4.71e-33**
	(7.40e+00)	(2.29e+01)	(5.32e+01)	(2.09e-48)
f_{18}	2.56e+02*	3.42e+03*	1.00e+04*	**7.27e-23**
	(1.22e+02)	(2.38e+02)	(5.17e+02)	(3.92e-22)

Note: Bold indicates the best result and * means statistical difference.

The convergence graphs of these four algorithms on 30D benchmark functions are presented in Figures 3.1 and 3.2. It should be noted that the fitness values are scaled by the logarithm ($\log_{10}$) in the convergence graphs. When the algorithms converge fast to the optima with the fitness value 0, $\log_{10} 0$ does not show in the graphs. This makes some convergence curves in the graphs (e.g., Figure 3.1e) incomplete. In other words, the incomplete convergence curves mean fast convergence to the optima.

We can find from Table 3.4 that when optimizing 10D functions, MBO obtained the best results on 10 functions (5 unimodal functions and 5 multimodal functions), and the difference from the results obtained by other algorithms (PSO, ACO, and ABC) are mostly statistically significant. Among these 10 functions, all the four algorithms achieved the optima of f_5 *and* f_{16}, whereas MBO and ABC both achieved the optima of f_{10}. PSO, ACO, and MBO obtained the same fitness values of f_9, whereas PSO and ACO obtained the same fitness values of $f_1 8$. For the rest of the functions, PSO, ACO, and ABC performed best on 2, 3, and 3 problems, respectively.

When addressing 30D functions, the number of best results obtained by MBO increases to 13 as shown in Table 3.5. It is worth noting that MBO can still achieve the optima of 30D f_5, f_{10}, and f_{16} while some comparative algorithms cannot, even though they are able to find the optima when the dimension is 10D. It also shows the same feature when dimension increases to 50D and 100D. For example, ACO missed the optima of 50D and 100D f_5 although it can find the optima when the dimension is 10D and 30D. The number of best results achieved by MBO increases to 14 for 50D functions as shown in Table 3.6 and increases to 15 for 100D functions as shown in Table 3.7. This means that when the dimension of the optimization problem increases, MBO can keep its competitive search ability while other algorithms might be trapped in local optima.

Looking at the convergence graphs of algorithms presented in Figs. 3.1 and 3.2, it is obvious that MBO has a faster convergence speed compared with PSO, ACO, and ABC in most of the cases. For example, when optimizing f_2, MBO achieves the optima within 5000 fitness evaluations *FEs* while PSO, ACO, and ABC are still stuck in local optima with 300000 *FEs*. Based on the above, we can conclude that MBOs are competitive in most of cases and the increasing problem dimension has less impact on the performance of MBO compared with other algorithms.

3.4.2.3 The Comparison of MBO with other Algorithms on CEC17 Test Suit

To further evaluate the capability of MBO, we compared MBO with PSO, ACO, and ABC on the CEC17 test suit, which is much more complex. The average error values and standard deviations (in parentheses) of error values obtained by 30 runs using the four algorithms for the CEC17 test suit on 10D, 30D, 50D, and 100D are presented in Tables 3.8–3.11. To evaluate the overall performance of algorithms, we ranked the error values of each algorithm on each problem first, then averaged each algorithm's ranks on 29 problems. After that, we ranked the average ranks again and got the overall ranks (OR) of four algorithms on the test suit. The ranks are presented in Table 3.12.

We can observe from these tables that it is difficult for these algorithms to find the optima of these problems due to the problems' high complexity. The four algorithms show diverse performance on different problems. Specifically, when addressing 10D problems, PSO, ACO, ABC, and MBO perform best on 3, 6, 14, and 7 problems, respectively. For 30D problems, PSO, ACO, ABC, and MBO achieved the best results on 10, 3, 9, and 7 problems, respectively. When the dimension increases to 50D, PSO, ACO, ABC, and MBO get the best results on 9, 6, 5, and 9 problems, respectively. For 100D problems, the number of best results that PSO, ACO, ABC, and MBO achieved becomes 15, 1, 5, and 8. We observe that MBO is not competitive enough on the CEC17 test suit, although it performs quite good on the first test suit. It is also worth noting that MBO's performance is less impacted by the increase of the dimension, which is consistent to what we noted in the experiments on the first test suit.

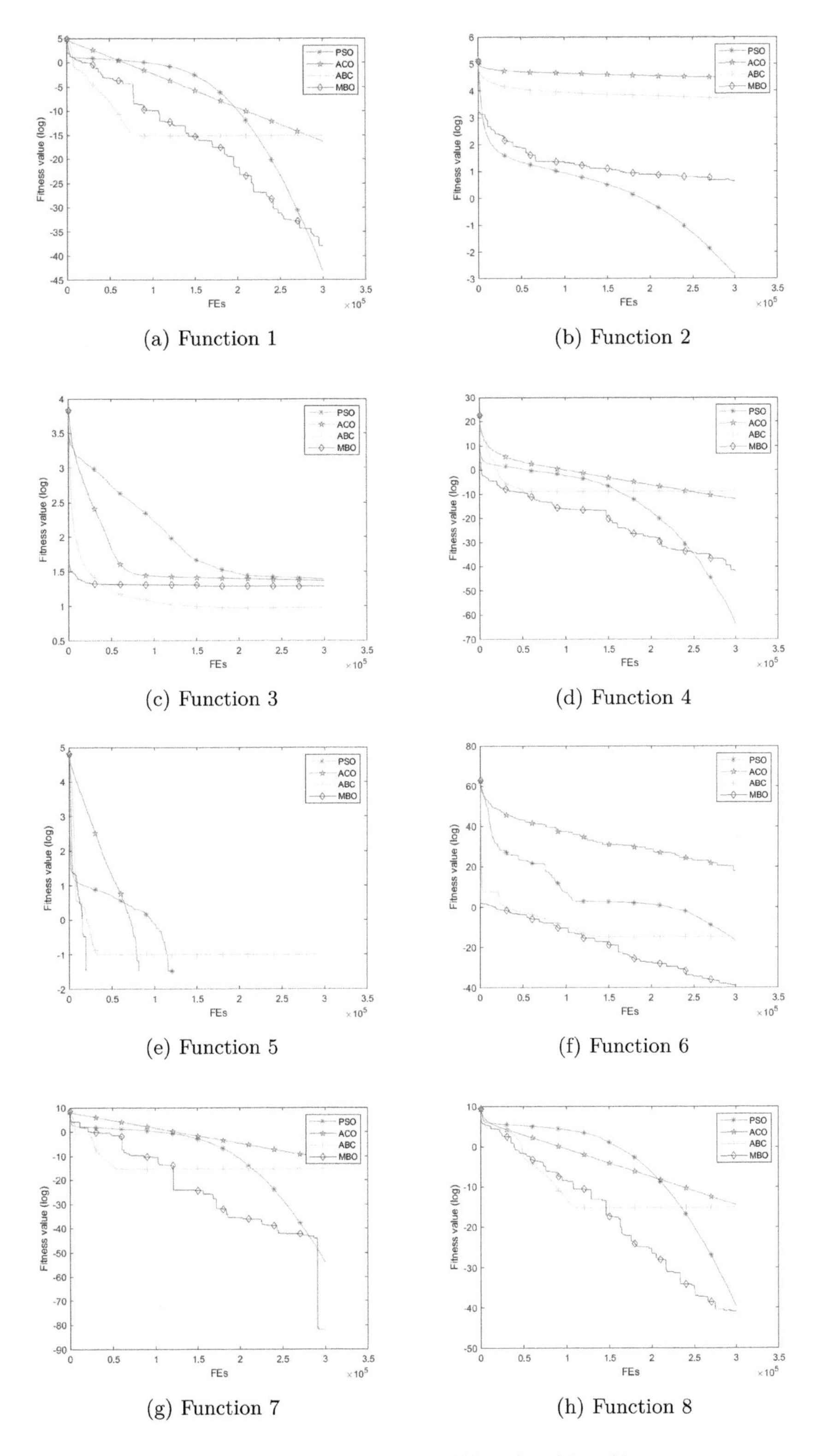

FIGURE 3.1 The convergence graphs of algorithms on 30D unimodal problems.

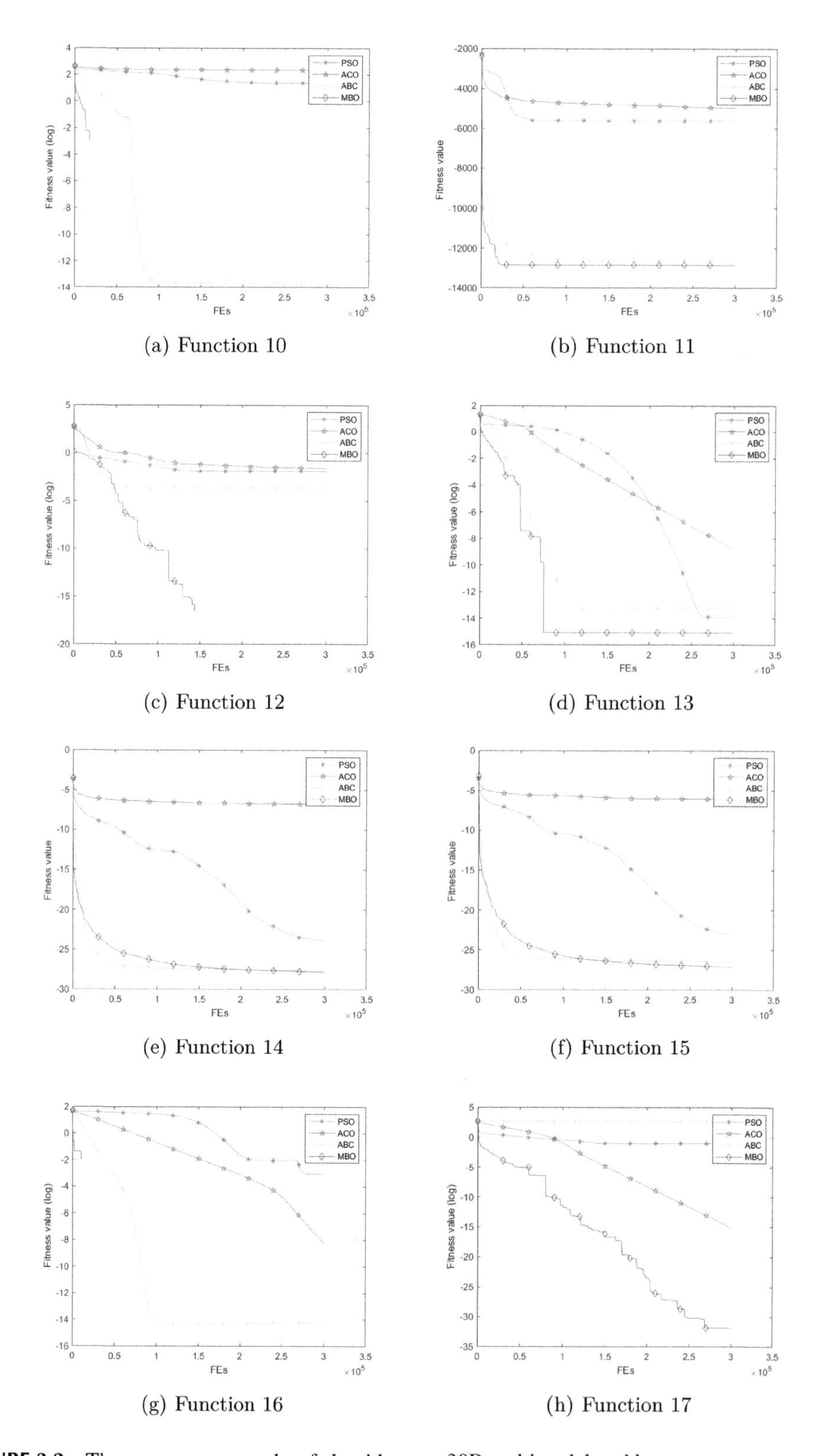

(a) Function 10 (b) Function 11 (c) Function 12 (d) Function 13 (e) Function 14 (f) Function 15 (g) Function 16 (h) Function 17

FIGURE 3.2 The convergence graphs of algorithms on 30D multimodal problems.

TABLE 3.8
Comparative Results of Algorithms on CEC17 10D Test Suit

Function No.	PSO	ACO	ABC	MBO
p_1	1.02e+03	4.02e+03*	**5.50e+02**	1.89e+03*
	(2.80e+03)	(3.85e+03)	(6.10e+02)	(1.93e+03)
p_3	**0.00e+00**	1.75e-09	1.10e+04*	2.05e+01*
	(0.00e+00)	(7.12e-09)	(4.33e+03)	(2.05e+01)
p_4	**2.91e-01**	1.30e+00*	7.12e-01*	5.87e+00*
	(1.62e-01)	(6.81e-02)	(5.67e-01)	(2.58e+00)
p_5	4.76e+01*	2.77e+01*	**1.14e+01**	2.40e+01*
	(1.35e+01)	(3.99e+00)	(3.88e+00)	(6.01e+00)
p_6	1.84e+01*	**0.00e+00**	**0.00e+00**	6.87e+00*
	(7.24e+00)	(0.00e+00)	(0.00e+00)	(3.74e+00)
p_7	3.62e+01*	3.84e+01*	**2.21e+01**	4.04e+01*
	(1.29e+01)	(3.45e+00)	(3.70e+00)	(9.06e+00)
p_8	2.91e+01*	2.79e+01*	**1.27e+01**	1.54e+01*
	(1.18e+01)	(4.43e+00)	(3.91e+00)	(3.35e+00)
p_9	1.14e+02*	**0.00e+00**	6.50e-01*	4.69e+01*
	(8.69e+01)	(0.00e+00)	(1.05e+00)	(4.24e+01)
p_{10}	9.34e+02*	8.92e+02*	**4.06e+02**	5.28e+02*
	(2.93e+02)	(3.51e+02)	(1.51e+02)	(1.87e+02)
p_{11}	2.64e+01*	**6.26e+00**	1.89e+01*	1.37e+01*
	(1.30e+01)	(1.06e+00)	(1.41e+01)	(6.16e+00)
p_{12}	1.29e+04	1.32e+04	1.51e+05*	**1.28e+04**
	(9.79e+03)	(9.89e+03)	(9.80e+04)	(1.17e+04)
p_{13}	6.91e+03*	6.80e+03*	4.11e+03*	**3.51e+02**
	(6.56e+03)	(7.40e+03)	(3.21e+03)	(5.63e+02)
p_{14}	2.27e+02*	1.53e+03*	1.07e+03*	**2.98e+01**
	(2.30e+02)	(3.84e+03)	(1.01e+03)	(2.13e+01)
p_{15}	5.05e+02*	2.00e+03*	5.39e+02*	**4.51e+01**
	(7.45e+02)	(3.63e+03)	(6.02e+02)	(7.82e+01)
p_{16}	2.20e+02*	**1.20e+01**	3.73e+01	9.11e+01
	(1.31e+02)	(5.74e+00)	(4.36e+01)	(8.17e+01)
p_{17}	5.68e+01*	4.59e+01*	**5.62e+00**	3.07e+01*
	(4.24e+01)	(1.50e+00)	(3.71e+00)	(1.05e+01)
p_{18}	6.80e+03*	2.26e+04*	4.42e+03*	**1.00e+03**
	(7.87e+03)	(1.36e+04)	(2.06e+03)	(1.66e+03)
p_{19}	1.41e+03*	5.35e+03*	4.21e+02*	**1.02e+01**
	(2.09e+03)	(8.55e+03)	(1.02e+03)	(1.75e+01)
p_{20}	8.77e+01*	1.23e+01*	**1.98e+00**	3.46e+01*
	(5.61e+01)	(3.66e+01)	(3.84e+00)	(1.14e+01)
p_{21}	1.99e+02*	2.25e+02*	1.16e+02*	**1.11e+02**
	(6.74e+01)	(1.83e+01)	(2.51e+01)	(3.95e+01)
p_{22}	1.37e+02	1.04e+02	**8.80e+01**	1.09e+02*
	(1.85e+02)	(1.12e+00)	(2.67e+01)	(4.84e+00)
p_{23}	4.15e+02*	3.22e+02*	**3.20e+02**	3.27e+02*
	(4.89e+01)	(3.20e+00)	(5.05e+00)	(1.03e+01)

(Continued)

TABLE 3.8 (*Continued*)
Comparative Results of Algorithms on CEC17 10D Test Suit

Function No.	PSO	ACO	ABC	MBO
p_{24}	3.58e+02*	3.57e+02*	**1.13e+02**	1.63e+02
	(1.59e+02)	(3.30e+00)	(2.85e+01)	(1.02e+02)
p_{25}	4.24e+02*	4.40e+02*	**2.99e+02**	4.15e+02*
	(2.28e+01)	(1.69e+01)	(1.02e+02)	(2.24e+01)
p_{26}	5.18e+02*	7.21e+02*	**1.81e+02**	3.04e+02*
	(3.10e+02)	(5.16e+02)	(7.61e+01)	(1.49e+02)
p_{27}	4.98e+02*	**3.90e+02**	3.97e+02*	3.91e+02
	(5.46e+01)	(8.55e-01)	(4.10e+00)	(7.33e+00)
p_{28}	3.54e+02*	6.11e+02*	**3.13e+02**	3.66e+02
	(4.15e+01)	(4.89e+00)	(7.53e+01)	(6.06e+01)
p_{29}	3.53e+02*	**2.65e+02**	2.92e+02*	2.83e+02*
	(6.79e+01)	(1.11e+01)	(2.53e+01)	(2.06e+01)
p_{30}	**3.87e+03**	4.82e+05*	8.61e+04*	1.14e+04*
	(2.40e+03)	(4.84e+05)	(1.35e+05)	(1.38e+04)

Note: Bold indicates the best result and * means statistical difference.

TABLE 3.9
Comparative Results of Algorithms on CEC17 30D Test Suit

Function No.	PSO	ACO	ABC	MBO
p_1	**8.82e+01**	5.01e+03*	6.64e+02*	6.57e+06*
	(8.40e+00)	(5.72e+03)	(5.85e+02)	(7.61e+06)
p_3	**1.71e-04**	4.80e+04*	1.53e+05*	2.24e+04*
	(1.59e-04)	(7.15e+03)	(2.06e+04)	(5.67e+03)
p_4	5.29e+01	8.57e+01*	**5.00e+01**	1.43e+02*
	(2.91e+01)	(9.73e-02)	(2.90e+01)	(1.57e+01)
p_5w	1.83e+02*	2.06e+02*	**1.05e+02**	1.34e+02*
	(3.34e+01)	(1.19e+01)	(1.81e+01)	(2.15e+01)
p_6	4.82e+01*	2.24e-04*	**0.00e+00**	1.24e+01*
	(7.61e+00)	(7.08e-05)	(0.00e+00)	(2.49e+00)
p_7	3.40e+02*	2.50e+02*	**1.33e+02**	1.59e+02*
	(6.68e+01)	(1.04e+01)	(1.65e+01)	(2.56e+01)
p_8	1.34e+02*	2.14e+02*	1.19e+02*	**1.05e+02**
	(2.73e+01)	(1.00e+01)	(1.57e+01)	(2.40e+01)
p_9	2.91e+03*	**2.24e-06**	3.12e+03*	8.81e+02*
	(7.97e+02)	(5.91e-06)	(1.10e+03)	(4.75e+02)
p_{10}	4.13e+03*	6.89e+03*	**2.87e+03**	3.16e+03*
	(7.53e+02)	(2.15e+02)	(3.35e+02)	(3.43e+02)
p_{11}	**1.01e+02**	1.55e+02*	1.49e+03*	1.93e+02*
	(2.46e+01)	(1.18e+01)	(9.96e+02)	(3.38e+01)

TABLE 3.9 (*Continued*)
Comparative Results of Algorithms on CEC17 30D Test Suit

Function No.	PSO	ACO	ABC	MBO
p_{12}	**3.13e+04**	4.39e+05*	1.96e+06*	2.11e+07*
	(1.98e+04)	(4.11e+05)	(1.06e+06)	(1.65e+07)
p_{13}	**1.05e+04**	1.43e+04	1.40e+05*	3.58e+04*
	(9.20e+03)	(1.47e+04)	(1.06e+05)	(2.88e+04)
p_{14}	**3.72e+03**	2.65e+04*	6.19e+05*	3.90e+03
	(2.31e+03)	(2.71e+04)	(5.53e+05)	(3.52e+03)
p_{15}	6.41e+03	**3.58e+03**	3.89e+04*	3.89e+03
	(8.43e+03)	(4.26e+03)	(3.29e+04)	(4.38e+03)
p_{16}	1.22e+03*	1.28e+03*	**8.54e+02**	1.06e+03*
	(3.68e+02)	(1.64e+02)	(1.63e+02)	(2.73e+02)
p_{17}	6.19e+02*	5.29e+02*	4.05e+02*	**2.06e+02**
	(2.44e+02)	(7.18e+01)	(1.44e+02)	(8.78e+01)
p_{18}	1.70e+05*	3.24e+06*	1.01e+06*	**8.09e+04**
	(1.44e+05)	(2.35e+06)	(7.42e+05)	(4.05e+04)
p_{19}	**7.78e+03**	8.42e+03	9.14e+04*	8.06e+03
	(1.01e+04)	(1.21e+04)	(7.81e+04)	(1.21e+04)
p_{20}	6.74e+02*	3.54e+02*	4.42e+02*	**2.90e+02**
	(1.69e+02)	(1.30e+02)	(1.26e+02)	(6.78e+01)
p_{21}	3.69e+02*	3.95e+02*	2.93e+02	**2.73e+02**
	(3.82e+01)	(1.07e+01)	(6.67e+01)	(8.87e+01)
p_{22}	3.37e+03*	7.00e+03*	1.24e+03	**1.34e+02**
	(2.14e+03)	(2.84e+02)	(1.61e+03)	(1.70e+01)
p_{23}	9.73e+02*	5.45e+02*	**4.43e+02**	4.60e+02
	(1.20e+02)	(1.18e+01)	(3.34e+01)	(2.89e+01)
p_{24}	9.94e+02*	6.08e+02	**5.26e+02**	5.67e+02*
	(8.05e+01)	(9.32e+00)	(2.02e+02)	(3.94e+01)
p_{25}	**3.78e+02**	3.87e+02*	3.85e+02*	4.32e+02*
	(1.98e+00)	(1.43e-01)	(1.37e+00)	(1.81e+01)
p_{26}	3.84e+03*	2.99e+03*	9.08e+02	**7.51e+02**
	(2.26e+03)	(1.21e+02)	(1.00e+03)	(6.03e+02)
p_{27}	7.49e+02	**5.08e+02**	5.24e+02*	5.24e+02*
	(4.08e+02)	(6.15e+00)	(5.35e+00)	(1.83e+01)
p_{28}	**3.51e+02**	4.09e+02*	3.99e+02	4.97e+02*
	(6.03e+01)	(8.81e+00)	(6.20e+00)	(2.72e+01)
p_{29}	1.11e+03*	8.90e+02	**8.10e+02**	8.90e+02
	(2.94e+02)	(1.68e+02)	(1.22e+02)	(1.57e+02)
p_{30}	**2.57e+03**	4.33e+03*	3.12e+04*	2.78e+05*
	(2.68e+03)	(2.32e+03)	(1.37e+04)	(2.06e+05)

Note: Bold indicates the best result and * means statistical difference.

TABLE 3.10
Comparative Results of Algorithms on CEC17 50D Test Suit

Function No.	PSO	ACO	ABC	MBO
p_1	**3.54e+03**	1.01e+07*	7.76e+03*	2.55e+08*
	(4.27e+03)	(4.17e+06)	(5.14e+03)	(1.50e+08)
p_3	**5.35e-01**	1.44e+05*	3.04e+05*	7.22e+04*
	(2.24e-01)	(1.70e+04)	(3.58e+04)	(1.03e+04)
p_4	8.63e+01*	2.01e+02*	**6.46e+01**	3.49e+02*
	(3.82e+01)	(3.08e+01)	(2.95e+01)	(4.28e+01)
p_5	3.02e+02*	4.48e+02*	2.58e+02*	**2.24e+02**
	(3.57e+01)	(1.15e+01)	(3.09e+01)	(6.13e+01)
p_6	5.49e+01*	3.32e+00*	**0.00e+00**	1.17e+01*
	(5.17e+00)	(4.78e-01)	(0.00e+00)	(1.15e+00)
p_7	6.87e+02*	5.53e+02*	2.85e+02	**2.84e+02**
	(7.75e+01)	(1.75e+01)	(2.80e+01)	(2.82e+01)
p_8	3.10e+02*	4.47e+02*	2.54e+02*	**2.10e+02**
	(3.84e+01)	(1.41e+01)	(2.52e+01)	(5.12e+01)
p_9	9.13e+03*	**1.18e+03**	1.70e+04*	3.26e+03*
	(1.33e+03)	(2.82e+02)	(3.18e+03)	(1.05e+03)
p_{10}	7.04e+03*	1.32e+04*	5.52e+03	**5.49e+03**
	(1.12e+03)	(3.44e+02)	(4.47e+02)	(5.09e+02)
p_{11}	**1.50e+02**	2.64e+02*	4.38e+03*	7.52e+02*
	(3.50e+01)	(2.13e+01)	(2.21e+03)	(2.09e+02)
p_{12}	**6.41e+05**	4.41e+07*	1.16e+07*	8.94e+07*
	(3.68e+05)	(3.54e+07)	(4.41e+06)	(6.09e+07)
p_{13}	6.48e+03*	**5.26e+03**	2.14e+05*	1.63e+05*
	(4.14e+03)	(6.20e+03)	(1.19e+05)	(1.70e+05)
p_{14}	**3.26e+04**	1.90e+05*	4.43e+06*	7.93e+04*
	(2.42e+04)	(1.37e+05)	(2.76e+06)	(5.77e+04)
p_{15}	8.67e+03	**6.03e+03**	5.61e+04*	1.11e+05*
	(1.03e+04)	(5.47e+03)	(6.10e+04)	(1.70e+05)
p_{16}	1.81e+03*	3.05e+03*	1.57e+03	**1.42e+03**
	(3.97e+02)	(1.62e+02)	(2.75e+02)	(2.95e+02)
p_{17}	1.58e+03*	1.91e+03*	1.33e+03*	**1.13e+03**
	(4.25e+02)	(1.45e+02)	(2.13e+02)	(2.22e+02)
p_{18}	**1.73e+05**	1.00e+07*	5.65e+06*	6.01e+05*
	(8.18e+04)	(4.98e+06)	(3.39e+06)	(3.20e+05)
p_{19}	1.49e+04	**1.44e+04**	1.30e+05*	7.62e+04*
	(1.01e+04)	(1.28e+04)	(7.35e+04)	(7.63e+04)
p_{20}	1.53e+03*	1.67e+03*	1.25e+03*	**7.24e+02**
	(2.91e+02)	(1.76e+02)	(2.04e+02)	(1.78e+02)
p_{21}	6.57e+02*	6.42e+02*	**4.61e+02**	4.88e+02*
	(6.18e+01)	(1.23e+01)	(5.14e+01)	(3.66e+01)
p_{22}	8.12e+03*	1.34e+04*	6.48e+03	**4.28e+03**
	(9.78e+02)	(4.80e+02)	(3.87e+02)	(3.27e+03)
p_{23}	1.65e+03*	8.51e+02*	7.20e+02*	**6.93e+02**
	(2.10e+02)	(1.49e+01)	(2.70e+01)	(2.85e+01)

TABLE 3.10 (*Continued*)
Comparative Results of Algorithms on CEC17 50D Test Suit

Function No.	PSO	ACO	ABC	MBO
p_{24}	1.49e+03*	**8.90e+02**	1.14e+03*	9.02e+02
	(1.76e+02)	(1.09e+01)	(6.68e+01)	(4.78e+01)
p_{25}	**4.45e+02**	5.37e+02*	5.21e+02*	6.94e+02*
	(1.94e+01)	(4.97e+00)	(1.60e+01)	(4.55e+01)
p_{26}	8.15e+03*	5.39e+03*	**2.84e+03**	2.88e+03
	(2.84e+03)	(1.19e+02)	(1.66e+03)	(9.80e+02)
p_{27}	2.45e+03*	**5.89e+02**	6.98e+02*	7.42e+02*
	(6.21e+02)	(3.03e+01)	(4.64e+01)	(4.20e+01)
p_{28}	**4.60e+02**	4.76e+02*	4.86e+02*	6.99e+02*
	(1.47e+01)	(9.87e+00)	(1.39e+01)	(5.49e+01)
p_{29}	1.83e+03*	2.08e+03*	**1.39e+03**	1.41e+03
	(3.36e+02)	(2.07e+02)	(1.73e+02)	(2.61e+02)
p_{30}	**1.12e+03**	1.04e+06*	1.11e+06*	1.42e+07*
	(2.03e+02)	(3.08e+05)	(1.77e+05)	(1.46e+07)

Note: Bold indicates the best result and * means statistical difference.

TABLE 3.11
Comparative Results of Algorithms on CEC17 100D Test Suit

Function No.	PSO	ACO	ABC	MBO
p_1	**7.73e+03**	3.90e+10*	9.44e+03	1.86e+09*
	(9.48e+03)	(3.70e+09)	(7.27e+03)	(4.61e+08)
p_3	**1.50e+03**	5.21e+05*	6.96e+05*	2.26e+05*
	(8.58e+02)	(3.94e+04)	(5.43e+04)	(1.63e+04)
p_4	**2.18e+02**	2.74e+03*	2.41e+02	8.71e+02*
	(5.41e+01)	(3.16e+02)	(2.35e+01)	(1.44e+02)
p_5	7.58e+02*	1.13e+03*	8.11e+02*	**5.89e+02**
	(5.58e+01)	(3.58e+01)	(5.52e+01)	(7.56e+01)
p_6	5.72e+01*	4.86e+01*	**0.00e+00**	1.27e+01*
	(3.15e+00)	(2.05e+00)	(0.00e+00)	(7.93e-01)
p_7	1.94e+03*	3.38e+03*	8.74e+02*	**8.46e+02**
	(1.83e+02)	(1.83e+02)	(5.89e+01)	(5.04e+01)
p_8	8.95e+02*	1.14e+03*	8.31e+02*	**5.89e+02**
	(6.09e+01)	(2.92e+01)	(5.51e+01)	(7.08e+01)
p_9	1.93e+04*	2.79e+04*	6.46e+04*	**1.46e+04**
	(1.58e+03)	(2.52e+03)	(8.04e+03)	(3.42e+03)
p_{10}	1.48e+04*	3.00e+04*	**1.39e+04**	1.41e+04
	(1.29e+03)	(4.81e+02)	(6.75e+02)	(6.51e+02)

(*Continued*)

TABLE 3.11 (*Continued*)
Comparative Results of Algorithms on CEC17 100D Test Suit

Function No.	PSO	ACO	ABC	MBO
p_{11}	**1.01e+03**	2.92e+04*	9.26e+04*	1.80e+04*
	(1.66e+02)	(3.73e+03)	(1.75e+04)	(3.67e+03)
p_{12}	**1.50e+06**	2.51e+09*	5.08e+07*	7.42e+08*
	(5.88e+05)	(8.51e+08)	(1.39e+07)	(3.73e+08)
p_{13}	**9.82e+03**	1.16e+04	9.78e+04*	9.35e+06*
	(5.74e+03)	(1.00e+04)	(6.78e+04)	(1.16e+07)
p_{14}	**1.51e+05**	1.15e+07*	2.07e+07*	6.65e+05*
	(3.79e+04)	(7.33e+06)	(8.30e+06)	(1.94e+05)
p_{15}	**2.67e+03**	5.21e+03*	1.29e+05*	3.04e+05*
	(2.58e+03)	(5.70e+03)	(7.81e+04)	(1.35e+06)
p_{16}	4.05e+03	8.41e+03*	**3.88e+03**	4.07e+03
	(6.21e+02)	(4.36e+02)	(4.73e+02)	(5.67e+02)
p_{17}	**3.33e+03**	5.51e+03*	3.39e+03	3.33e+03
	(5.58e+02)	(2.54e+02)	(3.23e+02)	(4.01e+02)
p_{18}	**4.64e+05**	6.38e+07*	1.43e+07*	9.27e+05*
	(1.86e+05)	(2.15e+07)	(4.26e+06)	(3.95e+05)
p_{19}	**2.70e+03**	5.45e+03	9.63e+05*	1.53e+06*
	(2.90e+03)	(7.62e+03)	(5.61e+05)	(1.54e+06)
p_{20}	3.57e+03*	4.89e+03*	3.65e+03*	**2.39e+03**
	(6.41e+02)	(1.98e+02)	(2.96e+02)	(2.43e+02)
p_{21}	1.80e+03*	1.39e+03*	1.02e+03*	**8.93e+02**
	(1.77e+02)	(3.74e+01)	(6.90e+01)	(7.00e+01)
p_{22}	1.65e+04	3.05e+04*	**1.58e+04**	1.62e+04*
	(1.35e+03)	(3.99e+02)	(7.46e+02)	(2.95e+03)
p_{23}	3.16e+03*	1.56e+03*	**9.39e+02**	1.03e+03*
	(3.32e+02)	(1.99e+01)	(5.07e+01)	(3.62e+01)
p_{24}	3.19e+03*	1.89e+03*	1.59e+03*	**1.53e+03**
	(2.80e+02)	(1.88e+01)	(5.11e+01)	(4.69e+01)
p_{25}	**7.26e+02**	7.40e+03*	7.56e+02*	1.46e+03*
	(6.03e+01)	(9.14e+02)	(3.40e+01)	(1.37e+02)
p_{26}	1.97e+04*	1.42e+04*	1.05e+04*	**9.07e+03**
	(8.30e+03)	(2.07e+02)	(2.55e+03)	(1.74e+03)
p_{27}	3.43e+03*	**6.96e+02**	8.18e+02*	9.18e+02*
	(1.18e+03)	(2.55e+01)	(3.40e+01)	(5.91e+01)
p_{28}	**5.21e+02**	1.08e+04*	5.79e+02*	1.22e+03*
	(2.28e+01)	(3.00e+03)	(1.43e+01)	(2.31e+02)
p_{29}	**3.90e+03**	6.38e+03*	4.45e+03*	4.32e+03*
	(5.58e+02)	(2.77e+02)	(2.58e+02)	(6.98e+02)
p_{30}	**4.47e+03**	4.01e+04*	2.14e+04*	5.90e+07*
	(4.82e+03)	(1.11e+04)	(6.70e+03)	(2.68e+07)

Note: Bold indicates the best result and * means statistical difference.

TABLE 3.12
Ranks of Algorithms on CE'17 Benchmark Function (10D, 30D, 50D, and 100D) Based on the Fitness

10D	P_1	P_3	P_4	P_5	P_6	P_7	P_8	P_9	P_{10}	P_{11}	P_{12}	P_{13}	P_{14}	P_{15}	P_{16}	P_{17}
PSO	2	1	1	4	4	2	4	4	4	4	2	4	2	2	4	4
ACO	4	2	3	3	1	3	3	1	3	1	3	3	4	4	1	3
ABC	1	4	2	1	1	1	1	2	1	3	4	2	3	3	2	1
MBO	3	3	4	2	3	4	2	3	2	2	1	1	1	1	3	2
10D	P_{18}	P_{19}	P_{20}	P_{21}	P_{22}	P_{23}	P_{24}	P_{25}	P_{26}	P_{27}	P_{28}	P_{29}	P_{30}	Mean	Std	OR
PSO	2	2	3	3	4	4	4	2	3	4	2	4	1	2.83	1.07	3
ACO	4	4	2	4	2	2	3	4	4	1	4	1	4	2.69	1.20	2
ABC	1	1	1	2	1	1	1	1	1	2	1	2	2	1.52	0.78	1
MBO	3	3	4	1	3	3	2	3	2	3	3	3	3	2.93	0.84	4
30D	P_1	P_3	P_4	P_5	P_6	P_7	P_8	P_9	P_{10}	P_{11}	P_{12}	P_{13}	P_{14}	P_{15}	P_{16}	P_{17}
PSO	1	1	2	3	4	4	3	3	3	1	1	1	1	3	3	4
ACO	3	3	3	4	2	3	4	1	4	2	2	2	3	1	4	3
ABC	2	4	1	1	1	1	2	4	1	4	3	4	4	4	1	2
MBO	4	2	4	2	3	2	1	2	2	3	4	3	2	2	2	1
30D	P_{18}	P_{19}	P_{20}	P_{21}	P_{22}	P_{23}	P_{24}	P_{25}	P_{26}	P_{27}	P_{28}	P_{29}	P_{30}	Mean	Std	OR
PSO	2	1	4	3	3	4	4	1	4	4	1	4	1	2.55	1.27	3
ACO	4	3	2	4	4	3	3	3	3	1	3	3	2	2.83	0.93	4
ABC	3	4	3	2	2	1	1	2	2	2	2	1	3	2.31	1.17	1
MBO	1	2	1	1	1	2	2	4	1	3	4	2	4	2.31	1.07	1
50D	P_1	P_3	P_4	P_5	P_6	P_7	P_8	P_9	P_{10}	P_{11}	P_{12}	P_{13}	P_{14}	P_{15}	P_{16}	P_{17}
PSO	1	1	2	3	4	4	3	3	3	1	1	2	1	2	3	3
ACO	3	3	3	4	2	3	4	1	4	2	3	1	3	1	4	4
ABC	2	4	1	2	1	2	2	4	2	4	2	4	4	3	2	2
MBO	4	2	4	1	3	1	1	2	1	3	4	3	2	4	1	1
50D	P_{18}	P_{19}	P_{20}	P_{21}	P_{22}	P_{23}	P_{24}	P_{25}	P_{26}	P_{27}	P_{28}	P_{29}	P_{30}	Mean	Std	OR
PSO	1	2	3	4	3	4	4	1	4	4	1	3	1	2.48	1.18	3
ACO	4	1	4	3	4	3	1	3	3	1	2	4	2	2.76	1.12	4
ABC	3	4	2	1	2	2	3	2	1	2	3	1	3	2.41	1.02	2
MBO	2	3	1	2	1	1	2	4	2	3	4	2	4	2.34	1.17	1
100D	P_1	P_3	P_4	P_5	P_6	P_7	P_8	P_9	P_{10}	P_{11}	P_{12}	P_{13}	P_{14}	P_{15}	P_{16}	P_{17}
PSO	1	1	1	2	4	3	3	2	3	1	1	1	1	1	2	1
ACO	4	3	4	4	3	4	4	3	4	3	4	2	3	2	4	4
ABC	2	4	2	3	1	2	2	4	1	4	2	3	4	3	1	3
MBO	3	2	3	1	2	1	1	1	2	2	3	4	2	4	3	2
100D	P_{18}	P_{19}	P_{20}	P_{21}	P_{22}	P_{23}	P_{24}	P_{25}	P_{26}	P_{27}	P_{28}	P_{29}	P_{30}	Mean	Std	OR
PSO	1	1	2	4	3	4	4	1	4	4	1	1	1	2.03	1.24	1
ACO	4	2	4	3	4	3	3	4	3	1	4	4	3	3.34	0.81	4
ABC	3	3	3	2	1	1	2	2	2	2	2	3	2	2.38	0.94	3
MBO	2	4	1	1	2	2	1	3	1	3	3	2	4	2.24	1.02	2

Abbreviation: OR, Overall rank.

TABLE 3.13
Computational Complexity (T_0 = 0.04)

	D = 10 (T_1 = 0.12)			D = 30 (T_1 = 0.55)			D = 50 (T_1 = 1.29)			D = 100 (T_1 = 5.18)		
	T_2	$\frac{(T_2-T_1)}{T_0}$	Rank	T_2	$\frac{(T_2-T_1)}{T_0}$	Rank	T_2	$\frac{(T_2-T_1)}{T_0}$	Rank	T_2	$\frac{(T_2-T_1)}{T_0}$	Rank
PSO	0.23	2.61	1	0.73	4.48	1	1.51	5.46	1	5.89	17.72	1
ACO	1.13	25.15	2	2.18	40.69	4	3.49	54.71	4	10.09	122.13	4
ABC	1.64	37.70	4	2.10	38.74	3	2.87	39.23	3	7.09	47.39	3
MBO	1.38	31.50	3	1.61	26.50	2	2.31	25.50	2	5.95	19.25	2

3.4.2.4 Computational Complexity

The complexity of the algorithm is evaluated using the method described in [20]. First, compute the time T_0 for running a test program. Then record the time T_1 just for computing p_{18} in CEC17 test suit with 200,000 evaluations on each dimension and evaluate the time T_2 for optimizing Problem 18 using the algorithm with 200,000 evaluations on the same dimension. Calculate T_2 for five times and the average of five T_2 is T_2. $\frac{(T_2-T_1)}{T_0}$ reflects the computational complexity of the algorithm. The results of T_1, T_2, and $\frac{(T_2-T_1)}{T_0}$ and the rank of all algorithms based on $\frac{(T_2-T_1)}{T_0}$ for each problem dimension are presented in Table 3.13. As shown in the table, PSO has the lowest computational complexity, followed by ACO for 10D and MBO for 30D, 50D, and 100D problems. ACO and ABC are more time-consuming than PSO and MBO, especially for high dimensional problems. It can also be observed that MBO's value of $\frac{(T_2-T_1)}{T_0}$, which reflects the computational complexity, decreases with the increase of dimension.

3.5 CONCLUSION

In this chapter, MBO in continuous domains is presented and applied to address continuous optimization problems by adapting the original discrete MBO. The representation and four heuristics employed for workers in MBO are presented to solve problems in continuous domains. Experiments are conducted to compare the performance of MBO with some well-known SI algorithms (PSO, ACO, and ABC) on two suits of well-known benchmark functions. The first test suit consists of 18 relatively simple functions, whereas the second test suit is from the CEC17 competition, which is much more complex. We also evaluated the performance of each single heuristic and compared them to MBO with four heuristics employed. The experimental results proved that the incorporation of multiple heuristics improves the performance of MBO and demonstrated the superior search ability and scalability of MBO. However, MBO has some difficulties when dealing with highly complex problems. More efficient methods should be designed and incorporated into MBO to improve its search capability. There are several parameters in MBO whose sensitivity can be further analyzed to figure out the optimal setting for them. The design of the adaptive parameters setting is a promising way to improve the performance of MBO. How to better design, select, and utilize multiple heuristics for addressing specific problems also should be further investigated.

APPENDIX 3.1 MBO CODE

```
% Initialise:
SwarmSize=50; % the size of the swarm
D=10; % the dimensions of problem
MaximumFEs=D*10000; % the number of maximum Fitness Evaluations
CurrentFEs=1; % the number of current Fitness Evaluations
UpperBound=-100; % the upper bound of search space
LowerBound=100; % the lower bound of search space
NumberOfQueens=2; % number of queens
NumberOfWorkers=5; % number of workers
NumberOfBroods=48; % number of broods
SpermathecaSize=21; % spermatheca size
ImproveTimes=D*10; % The maximum number of times to improve a brood by
a worker
Lambda=0.9;
StepSize=2;
Population= LowerBound+rand(D,SwarmSize)*(UpperBound-LowerBound);
FitnessValue=feval(Problem, Population, ProblemIndex);
[~, SortIndex]=sort(FitnessValue);
Population=Population(:,SortIndex);
FitnessValue=FitnessValue(SortIndex);
QueensPheno=Population(:,1:NumberOfQueens);
QueensFitnessValue=FitnessValue(1:NumberOfQueens);
Gbest=Population (:,1);
GbestValue=FitnessValue(1);
ConvergenceData=zeros(1,MaximumIterations+1);
ConvergenceData(1)=GbestValue-Optimum;
CurrentIterations=1;
Spermatheca=zeros(Dimension, SpermathecaSize,NumberOfQueens);
SpermathecaSizeOfQueen=zeros(1,NumberOfQueens);
BroodsPheno=repmat(Gbest,1,NumberOfBroods);
BroodsFitnessValue=repmat(GbestValue,1,NumberOfBroods);
WorkersFitness=ones(1,NumberOfWorkers);
while CurrentIterations<=MaximumIterations
      for i=1:NumberOfQueens
         Energy=0.5*rand+0.5;
         Speed=0.5*rand+0.5;
         Step=0.5*Energy/SpermathecaSize;
         Drone=LowerBound+(UpperBound-LowerBound).*rand(Dimension,1);
         CountSpermathecaSize=0;
         while Energy>0
               DroneFitnessValue=feval(Problem, Drone, ProblemIndex);
               CurrentIterations=CurrentIterations+1;
               ConvergenceData(CurrentIterations)=DroneFitnessVa
                 lue- Optimum;
               DroneProbability=exp(-(FitnessValue(i)-
                 DroneFitnessValue)./Speed);
               if rand<DroneProbability &&
                 CountSpermathecaSize<SpermathecaSize
                      CountSpermathecaSize=CountSpermathecaSize+1;
                      Spermatheca(:,CountSpermathecaSize,i)=Drone;
               end
               Energy=Energy-Step;
```

```
            Speed=lambda*Speed;
            ChangeIndex=rand(Dimension,1)<Speed;
            if ~ isempty(ChangeIndex)
                 Drone(ChangeIndex,1)=Drone(ChangeIndex,1).
                   *rand(size  (Drone(ChangeIndex,1)));
            end
      end
       if CountSpermathecaSize==0
          CountSpermathecaSize=CountSpermathecaSize+1;
          Spermatheca(:,CountSpermathecaSize,i)=Drone;
       end
       SpermathecaSizeOfQueen(i)=CountSpermathecaSize;
end
for i=1:NumberOfBroods
       % select a queen j
       QueenIndex=ceil(rand*NumberOfQueens);
       % select a sperm
       SpermIndex=ceil(rand*SpermathecaSizeOfQueen(QueenIndex));
       % crossover the queen's genome with the selected sperm
       a=rand(Dimension,1);
       b=a<0.5;
       BroodsPheno(:,i)=b.*QueensPheno(:,QueenIndex)+(1-b).
         *Spermatheca(:,SpermIndex, QueenIndex);
       % mutate the brood's genotype
       b=ceil(rand.*Dimension);
       BroodsPheno(b,i)=LowerBound+(UpperBound-LowerBound).*rand;
       BroodsFitnessValue(i)=feval(Problem, BroodsPheno(:,i),
         ProblemIndex);
       CurrentIterations=CurrentIterations+1;
       ConvergenceData(CurrentIterations)=BroodsFitnessValu
         e(i)- Optimum;
       OriginalFit=BroodsFitnessValue(i);
       CountsToNotChanged=0;
       if mod(i,10)==1
          ProbabilityOfWorker=WorkersFitness/
            (sum(WorkersFitness)+realmin);
       end
       WorkerIndex=RouletteWheelSelection(ProbabilityOfWorker);
       % use a worker to improve the drone's genetype
       switch WorkerIndex
          case 1
                Random=2*rand(Dimension,1)-1;
                RandomDirection=Random/sqrt(Random'*Random);
                NewBroodPheno=BroodsPheno(:,i)+RandomDirection*
                  StepSize;
                NewBroodFitnessValue=feval(Problem, NewBroodPheno,
                  ProblemIndex);
                CurrentIterations=CurrentIterations+1;
                ConvergenceData(CurrentIterations)=NewBroodFitnessValue-
                  Optimum;
                for k=1:ImproveTimes
                   if NewBroodFitnessValue< BroodsFitnessValue(i)
                      BroodsFitnessValue(i)=NewBroodFitnessValue;
                      BroodsPheno(:,i)=NewBroodPheno;
```

```
            NewBroodPheno=NewBroodPheno+RandomDirection*
              StepSize;
            NewBroodFitnessValue=feval(Prob
            lem,   NewBroodPheno, ProblemIndex);
            CountsToNotChanged=0;
         else
         Random=2*rand(Dimension,1)-1;
         RandomDirection=Random/sqrt(Random'*Random);
         NewBroodPheno=BroodsPheno(:,i)+RandomDirection*
           StepSize;
         NewBroodFitnessValue=feval(Problem, NewBroodPheno,
           ProblemIndex);
         CountsToNotChanged=CountsToNotChanged+1;
      end
      CurrentIterations=CurrentIterations+1;
      ConvergenceData(CurrentIterations)=
        BroodsFitnessValue(i)- Optimum;
      if CurrentIterations>=MaximumIterations+1
         break
      end
      if CountsToNotChanged> Dimension
         break
      end
      end
case 2
      for k=1:ImproveTimes
         NewBroodPheno=BroodsPheno(:,i);
         a=ceil(Dimension*rand); b=ceil(Dimension*rand);
         NewBroodPheno(a,1)=NewBroodPheno(b,1);
         NewBroodFitnessValue=feval(Problem, NewBroodPheno,
           ProblemIndex);
         if NewBroodFitnessValue< BroodsFitnessValue(i)
            BroodsFitnessValue(i)=NewBroodFitnessValue;
            BroodsPheno(:,i)=NewBroodPheno;
            CountsToNotChanged=0;
         else
            CountsToNotChanged=CountsToNotChanged+1;
         end
         CurrentIterations=CurrentIterations+1;
         ConvergenceData(CurrentIterations)=
           BroodsFitnessValue(i)- Optimum;
         if CurrentIterations==MaximumIterations+1
            break
         end
         if CountsToNotChanged> Dimension
            break
         end
      end
case 3
      k=1;
      while k<=ImproveTimes
         NewBroodPheno=BroodsPheno(:,i);
         Dx=1;
         while Dx<= Dimension
```

```
                temp=NewBroodPheno(Dx,1);
                digits=num2str(abs(temp));
                leftdigits=find(digits=='.');
                if isempty(leftdigits)
                   leftdigits=length(digits);
                   Precision=randi([0,leftdigits]);
                else
                   rightdigits=length(digits)-leftdigits;
                   Precision=randi([-(rightdigits-1),
                     leftdigits-1]);
                end
                NewBroodPheno(Dx,1)=roundn(NewBroodPheno
                  (Dx,1), Precision);
                NewBroodFitnessValue=feval(Problem,
                 NewBroodPheno, ProblemIndex);
                if NewBroodFitnessValue< BroodsFitnessValue(i)
                   BroodsFitnessValue(i)=NewBroodFitnessValue;
                   BroodsPheno(:,i)=NewBroodPheno;
                   CountsToNotChanged=0;
                else
                   CountsToNotChanged=CountsToNotChanged+1;
                   Dx=Dx+1;
             end
             k=k+1;
             CurrentIterations=CurrentIterations+1;
             ConvergenceData(CurrentIterations)=
               BroodsFitnessValue(i)- Optimum;
             if CurrentIterations>=MaximumIterations+1 ||
             CountsToNotChanged> Dimension || k> ImproveTimes
                 break
               end
             end
             if CurrentIterations>=MaximumIterations+1
                 break
               end
          end
   case 4
          for k=1:ImproveTimes
             QueenIndex=ceil(rand*NumberOfQueens);
             Direction=QueensPheno
               (:,QueenIndex)-BroodsPheno(:,i);
             if QueensFitnessValue(QueenIndex)<
               BroodsFitnessValue(i)
                 NewBroodPheno=QueensPheno(:,QueenIndex)+rand*
                   Direction;
                 NewBroodPheno=BoundLimits(NewBroodPheno,
                   LowerBound,UpperBound);
                 NewBroodFitnessValue=feval
                   (Problem, NewBroodPheno, ProblemIndex);
                 if NewBroodFitnessValue<BroodsFitnessValue(i)
                    BroodsFitnessValue(i)=NewBroodFitnessValue;
                    BroodsPheno(:,i)=NewBroodPheno;
                    CountsToNotChanged=0;
                 else
                    CountsToNotChanged=CountsToNotChanged+1;
```

```
                    end
                else
                    NewBroodPheno=BroodsPheno(:,i)-rand*Direction;
                    NewBroodPheno=BoundLimits(NewBroodPheno,
                      LowerBound,UpperBound);
                    NewBroodFitnessValue=feval
                      (Problem, NewBroodPheno, ProblemIndex);
                    if NewBroodFitnessValue<BroodsFitnessValue(i)
                        BroodsFitnessValue(i)=NewBroodFitnessValue;
                        BroodsPheno(:,i)=NewBroodPheno;
                        CountsToNotChanged=0;
                    else
                        CountsToNotChanged=CountsToNotChanged+1;
                    end
                end
                CurrentIterations=CurrentIterations+1;
                ConvergenceData(CurrentIterations)
                  =BroodsFitnessValue(i);
                if CurrentIterations>=MaximumIterations+1
                    break
                end
                if CountsToNotChanged> Dimension
                    break
                end
            end
        end
        WorkersFitness(WorkerIndex)=WorkersFitness(WorkerIndex)+
          (abs(OriginalFit-BroodsFitnessValue(i)))/
          (abs(OriginalFit)+realmin);
        if mod(i,10000)==1
            WorkersFitness=ones(1,NumberOfWorkers);
        end
        if CurrentIterations>= MaximumIterations+1
            break
        end
    end
    NewFitnessValue=[QueensFitnessValue BroodsFitnessValue];
    NewPopulation=[QueensPheno BroodsPheno];
    [~, SortIndex]=sort(NewFitnessValue);
    NewPopulation=NewPopulation(:,SortIndex);
    NewFitnessValue=NewFitnessValue(SortIndex);
    QueensPheno=NewPopulation(:,1:NumberOfQueens);
    QueensFitnessValue=NewFitnessValue(1:NumberOfQueens);
end
    Gbest=NewPopulation (:,1);
    GbestValue=NewFitnessValue(1);
end
```

REFERENCES

1. Gerardo Beni. The concept of cellular robotic system. In *Proceedings of the IEEE International Symposium on Intelligent Control, 1988*, pages 57–62. IEEE, 1988.
2. Gerardo Beni and Jing Wang. Swarm intelligence in cellular robotic systems. In *Robots and Biological Systems: Towards a New Bionics?*, pages 703–712. Springer, 1993.

3. James Kennedy. Swarm intelligence. In *Handbook of Nature-Inspired and Innovative Computing*, pages 187–219. Springer, 2006.
4. Ying Tan and Zhong-Yang Zheng. Research advance in swarm robotics. *Defence Technology*, 9(1):18–39, 2013.
5. Marco Dorigo and Thomas Stützle. The ant colony optimization metaheuristic: algorithms, applications, and advances. In *Handbook of metaheuristics, pages* 250–285. Springer, 2003.
6. J. Kennedy and R. Eberhart. Particle swarm optimization. In *IEEE International Conference on Neural Networks, Proceedings of ICNN'95*, volume 4, pages 1942–1948. IEEE, 1995.
7. Hussein A. Abbass. MBO: Marriage in honey bees optimization-a haplometrosis polygynous swarming approach. In *Proceedings of the 2001 Congress on Evolutionary Computation CEC2001*, pages 207–214, 2001.
8. Hussein A. Abbass. A single queen single worker honey-bees approach to 3-SAT. In *Proceedings of the 3rd Annual Conference on Genetic and Evolutionary Computation*, pages 807–814, 2001.
9. Dervis Karaboga. An idea based on honey bee swarm for numerical optimization. Technical report, Technical report-TR06, Erciyes University, Engineering Faculty, Computer Engineering Department, 2005.
10. Jason Teo and Hussein A. Abbass. A true annealing approach to the marriage in honey-bees optimization algorithm. *International Journal of Computational Intelligence and Applications*, 3(2):199–211, 2003.
11. Abbass Afshar, O. Bozorg Haddad, Miguel A. Marino, and B. J. Adams. Honey-bee mating optimization (HBMO) algorithm for optimal reservoir operation. *Journal of the Franklin Institute*, 344(5):452–462, 2007.
12. Chenguang Yang, Xuyan Tu, and Jie Chen. Algorithm of marriage in honey bees optimization based on the wolf pack search. In *The 2007 International Conference on Intelligent Pervasive Computing (IPC 2007)*, pages 462–467. IEEE, 2007.
13. Yuksel Celik and Erkan Ulker. An improved marriage in honey bees optimization algorithm for single objective unconstrained optimization. *The Scientific World Journal*, 2013:370172, 2013.
14. Ping-Feng Pai, Shun-Ling Yang, and Ping-Teng Chang. Forecasting output of integrated circuit industry by support vector regression models with marriage honey-bees optimization algorithms. *Expert Systems with Applications*, 36(7):10746–10751, 2009.
15. O. Bozorg Haddad and A. M. B. O. Afshar. MBO (marriage bees optimization), a new heuristic approach in hydrosystems design and operation. In *Proceedings of the 1st International Conference On Managing Rivers in the 21st Century: Issues and Challenges. Penang, Malaysia*, pages 21–23, 2004.
16. Biri Arun and T. V. Vijay Kumar. Materialized view selection using marriage in honey bees optimization. *International Journal of Natural Computing Research (IJNCR)*, 5(3):1–25, 2015.
17. Julian Adams, Edward D. Rothman, Warwick E. Kerr, and Zila L. Paulino. Estimation of the number of sex alleles and queen matings from diploid male frequencies in a population of *Apis mellifera*. *Genetics*, 86(3):583–596, 1977.
18. Yuhui Shi and Russell Eberhart. A modified particle swarm optimizer. In *Evolutionary Computation Proceedings, 1998. IEEE World Congress on Computational Intelligence*, pages 69–73. IEEE, 1998.
19. Krzysztof Socha and Marco Dorigo. Ant colony optimization for continuous domains. *European Journal of Operational Research*, 185(3):1155–1173, 2008.
20. N. H. Awad, M. Z. Ali, J. J. Liang, B. Y. Qu, andP. N. Suganthan. Problem definitions and evaluation criteria for the CEC 2017 special session and competition on single objective bound constrained real-parameter numerical optimization. *Technical Report.* Nanyang Technological University Singapore, 2016.
21. David H. Wolpert and William G. Macready. No free lunch theorems for optimization. *IEEE Transactions on Evolutionary Computation*, 1(1):67–82, 1997.

4 Structural Optimization Using Genetic Algorithm

Ravindra Desai

CONTENTS

of genetic material is carried out at a randomly chosen physical section of parent chromosomes. In single crossover only one portion of a parent chromosome is exchanged with the identically located portion of a second parent, in double crossover two identically located portions of parent chromosomes are exchanged, and in uniform crossover the entire chromosome is scanned for possible exchange. The locations on the parent chromosome, at which sections are made, are defined as crossover sites. Mutation is the third and last stochastic operator that brings heuristic change in newly created offspring by merely altering its single gene or a few genes, which is decided based on mutation probability. These three operations are fundamental to any GA and, indeed, they introduce advancement of generations to have better and better evolved populations. Technically, GA explores the entire search space and exploits potential regions for possible global optima with the help of these transformative operators.

4.2.1 Fitness: The Gateway of Evolution

The principle of survival of the fittest is key for evolution in genetics. In GA, the current population is called parent population and what is created in the next step is known as offspring population. Fitness is a measure of competence, which also serves as selection criteria, to choose eligible parents to create a new evolved generation of population. The fitness of a chromosome depends on how well it solves the problem at hand. To specify the quality of a chromosome accurately, a well-structured fitness function is required to be designed, which can be used as simple and effective index. In general, such a function is a composite of objective and constraint functions. So, in the sense of computation, fitness is a measure of the quality of an individual assessed by its soundness in objective function(s) and compliance with constraint functions. While constraint functions indicate the weakness of the candidate solution in terms of degree of violation, the objective function explains its relative merit in the population. These two important functions serve as effective tools to accurately define the fitness of any individual, and they are often used together to grade every individual of population on a uniform scale of fitness. Several methods have been devised to quantify fitness of an individual in this way, and they are generally included under constraint handling schemes, which will be dealt with separately in detail under literature review section of this chapter.

In general, the design of the fitness function is problem dependent and an adaptive design that progressively tunes itself to generation advancement will be highly useful. The common issues to be addressed while defining and quantifying fitness functions are as stated below.

a. Dynamic variation in fitness function value in successive generations.
b. Noisy fitness functions in uncertain environments.
c. Most of the computational time in GA is consumed by fitness function evaluations and efficient ways are necessary to tackle it.
d. Proper fitness variance must be provided by the specified fitness function in both minimization and maximization problems.

4.2.2 Design Representations: Encoding Schemes

Based on the context of the optimization problem, appropriate representation must be chosen to designate the structure of the chromosome template (genotype) for an individual candidate of population, which allows efficient and fast processing of GA. Such a representation is referred to as the *encoding scheme*, and binary, real, integer, or order type of encoding schemes have been implemented in classical GA to solve optimization problems in various domains. Typically, binary coding for Boolean, integer coding for combinatorial or discrete problems, real coding for

y1	y2	y3
x1	x2	x3

(a)

10001000	11100011	00000101
5.25	225.45	11.5

(b)

FIGURE 4.1 Binary and real coded chromosome in GA. (a) Genotype with three genes. (b) Phenotype with three instances (alleles).

n dimensional continuous space, and order-oriented coding for logistic optimization problems have been used. Figure 4.1 illustrates concepts associated with GA terminology: the gene and its instance, the chromosome genotype and phenotype, and the two most commonly used encoding schemes in structural optimization problems. Figure 4.2 shows the single-point crossover operation, and Figure 4.3 depicts the mutation operation in binary coded chromosomes.

Binary and real coded encoding schemes are the two most widely used representations in structural optimization problems. In case of truss topology optimization problems binary representation has been used widely and falls under the classification of combinatorial optimization. However, real coded representation has been found suitable in search space characterized by multidimensional continuous field such as sizing, or optimal control problems in structural optimization.

The issue of selecting an appropriate representation has always been a crucial decision in any optimization problem. Binary coding may demand intensive computational efforts in case of high-precision floating point design space. Hamming distance, i.e., total non-matching bits at identical locations of two binary strings, results in a Hamming cliff (i.e., large difference in real and binary representation of two numbers). For instance, the decimal distance in 3 and 4 is one; however, the hamming distance between their respective binary representation [0 1 1] and [1 0 0] is 3. Therefore, the position of genes in the chromosome becomes extremely important in binary coding unless uniform crossover has been adopted. Mapping from binary space to domain space was used successfully in several large-scale problems; however, computational efficiency has been a major challenge. To overcome such issues the binary equivalent of crossover and mutation operators for real coded representations have been developed, such as blended crossover (BLX-α) [3] and simulated binary crossover (SBX) [4] for crossover operators, and random mutation [5], polynomial mutation [6], and normally distributed mutation and so forth for mutation operators. Historically, fine-tuned complex engineering design problems and large-scale real-world structural optimization problems have been successfully solved with such representations, where binary representation was rendered ineffective. Appendix 4.3 gives the important encoding formulations discussed above.

4.2.3 Variation Operators

As noted, GA is a population-based evolution algorithm that seeks a solution to an optimization problem through exploration and exploitation by generation advancement. While

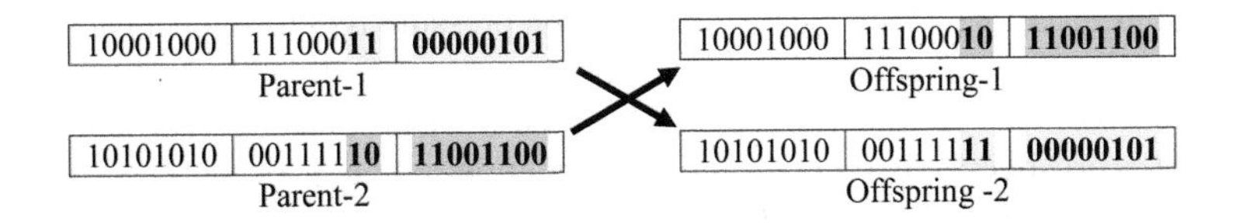

FIGURE 4.2 Single-point crossover operation in the binary chromosome. Crossover site is randomly chosen at the seventh bit of the second gene.

FIGURE 4.3 Mutation operation in binary chromosome. Randomly chosen last bit of the first gene is flipped in offspring-2.

exploration deals with searching of maximum solution space to avoid premature convergence, exploitation aims at reaching the global optima with high accuracy. Numerous modifications in the design of canonical mechanisms have been implemented efficiently to attain the goal of global optima for a given problem representation, especially in the selection, encoding, and constraint handling schemes. In the following discussion, we describe the important basic features of GA architecture and their significance in the evolution process, and we review attempts to implement various modified versions available in the scientific literature.

4.2.3.1 Selection Schemes: Sampling Mechanisms

Reproduction implies selecting most prospective parents (chromosomes) from the current population and putting them into a mating pool to create a more fit next generation. To collect a good quality, well-balanced, and diversified population of parents a sound selection strategy is essential. Success in terms of convergence speed and achievement of global optima critically depends on characteristics of the selection method. Several selection methods are devised, which are usually based on theory of probability, for the purpose of creating the most fit and diversified successive generations.

The two most important facets of genetic evolution are population diversity and selective pressure [7]. Population diversity is an ability of GA to represent and explore the entire search space, and selective pressure is the degree to which the best candidate solution is favored; although it indicates the quality of solution regarding global optima. These two factors have strong interdependence: an increase in the selective pressure reduces the population diversity, and vice versa. If selective pressure is too high in the successive generations then the GA search tends to premature convergence, on the other hand, a weak selective pressure can make the search ineffective. The selection schemes therefore must provide a right selection pressure and maintain the population diversity in successive generations. Selection intensity is another parameter in GA used to measure selection pressure, which is defined as anticipated change in mean fitness of population after selection, normalized by population fitness standard deviation before selection. It is preferred that, for a given selection intensity, the selection method should provide higher standard deviation of parent fitness to achieve the goal of trouble free convergence.

Several schemes have been investigated [8–11] in detail to refine the GA search and make it more efficient. In general, selection methods are all probability based and fall under two major classifications: proportionate based and ordinal based. The former method chooses parents directly based on fitness values, whereas the later method uses a ranking mechanism to select parents. The ordinal-based methods have a general advantage in that they are insensitive to fitness scaling and translation, which means the selection pressure remains unchanged. We discuss and explain in Appendix 4.2 some commonly implemented selection methods with their relative merits and demerits. The selection operator is also referred to as the reproduction operator because it chooses prospective parents from the current population, which is a mix of parents and their offspring together. To reduce sampling variance procedures of fitness, scaling has been suggested in [2]. It helps to keep the selective pressure and population diversity in the right balance and improve the performance of GA.

4.2.3.2 Crossover: Recombination

Creating a minimum of two offspring from two parents randomly chosen from the mating pool requires a crossover operator. A newly born child inherits genetic information from its parents; similarly, the crossover operator exchanges the information in chromosomes at random sections. The simple method of exchange is to clip the two chromosomes at a random location and interchange either the right or left cut portion of the chromosome with an identical portion of the other, which is known as the single-point crossover method. The clipping location, i.e., crossover point in chromosomes, is known as a site. Single-point, two-point, multipoint, and uniform crossover are the most common crossover techniques. The crossover operator roughly mimics the biological recombination between two single-chromosome organisms. Figure 4.2 demonstrates a general crossover operation in binary coded chromosomes. A crossover rate is a uniform probability ($Pc \in [0\ 1]$) indicating the percentage of recombination from the total number of parent population in the mating pool. Usually a crossover rate of 0.8–0.9 is used, which also means 80%–90% probability exists for each prospective parent to pair and participate in offspring generation. In this way, crossover has been considered as the most disruptive operation in the sense that it imparts powerful exploring capacity to GA by way of generating quite different candidate solutions in decision space for each generation. As GA advances, the exploring tends to exploiting due to reduced population variance and approaching convergence to optima because of several closely matched individuals.

4.2.3.3 Mutation: Randomization

Mutation operation imparts a diverse exploration ability to GA that allows searching in decision space that is unexplored by past generations. Mutation randomly alters a gene in the offspring chromosome by which a complete change occurs in the constitution of an individual offspring. The characteristic of such a genetically modified chromosome is so unique that it no longer belongs to the pool of current offspring. In other words, the mutated chromosome represents a completely new decision space that is unexplored. In case of binary representation, for example, this operator randomly flips some of the bits in a chromosome from 0 to 1, and vice versa. By such an operation few individuals of the new population are modified (mutated) to mimic the diversification of population in nature. Mutation operation is performed according to a certain mutation rate or probability for each bit of chromosome. In general, a 1%–2% mutation rate is accepted, which means that only 1%–2% of offspring chromosome have a chance of being altered. At the individual offspring level this probability translates to modifying the bit/gene with a prescribed mutation rate, i.e., every bit/gene has only a 1%–2% chance of alteration. The mutation operation is illustrated in Figure 4.3.

4.2.4 Stopping Criterion

Formal convergence criterion in metaheuristic optimization techniques like GA is unavailable. For recursive GA it is essential to define a stopping criterion in some way. Several such criterion have been proposed and implemented effectively in practice. The widely used stopping criteria implemented in GA applications are as below.

a. *Fitness limit:* Once GA reaches a user-specified value of fitness the termination is set to occur. The average and best fitness of the population are key to such criterion.
b. *Stall generations:* Once the weighted average change of the population fitness attains value below the user-specified tolerance value the GA is terminated.
c. *Number of generations (iterations):* The GA is terminated after a specified large number of generation cycles are executed. This criterion is required to ensure that GA stops

if the other convergence criteria are not met. In general, proper parameter tuning is essential for appropriate convergence of any metaheuristic optimization algorithms. In case of GA the population size, crossover rate, mutation rate, penalty factors and so on are crucial parameters.

d. *Time limit:* Maximum execution time can be imposed to stop the optimization process.
e. *Stall time limit:* The GA is terminated when the best fitness value in the population does not improve in a time interval.

4.2.5 Steps in GA Optimization

Before proceeding to a discussion of examples of GAs, it is worth noting the algorithmic steps of optimization using GA in general. The stepwise procedure for general optimization problems using GA is illustrated in a flowchart in Figure 4.4.

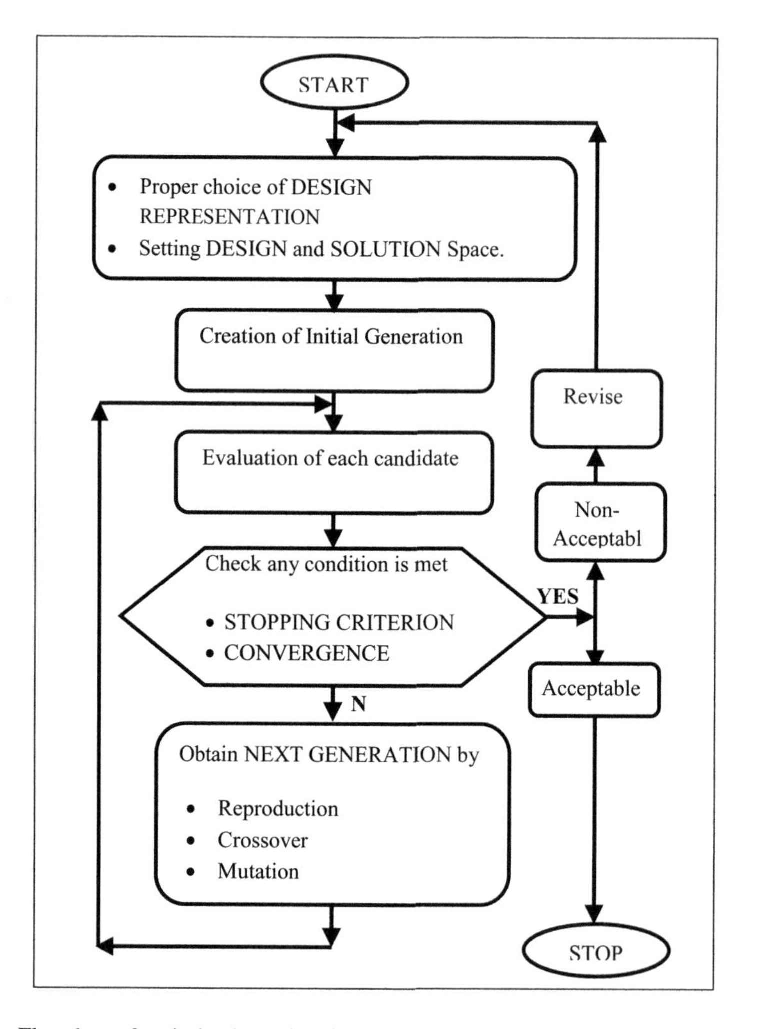

FIGURE 4.4 Flowchart of optimization using GA.

4.3 CONSTRAINT HANDLING IN GA OPTIMIZATION

GA does not have any direct approach to deal with constraints of an optimization problem like it has for an objective function using variation operators. During solution stages infeasible offspring appear as part of the evolution process. These are dealt with by constraint handling methods in all metaheuristic algorithms like GA. It is completely detrimental to the solution process, and contrary to the evolution principle of providing equal chance to every individual, to eliminate all such infeasible offspring from the population. In fact, many such individuals possess rich genetic information and it is a well-known fact that optimal solutions usually lie on boundaries of feasible regions. To allow uniform exploration of search space for preserving diversity, and to nurture the evolution process, such individuals are admitted in the next generation with techniques of constraint handling. Modified fitness function design is necessary for infeasible individuals in such techniques, which changes the degree of fitness of these individuals based on the amount and number of constraint violations. In general, the nature of constraint handling fitness functions largely depend on type and context of real-world optimization problems such as structural optimization. It is important to note that constraint handling fitness function design plays a crucial role in convergence of GA in all large-scale optimization problems.

There are several approaches of constraint handling techniques in the following discussion.

The penalty function approach has been widely used in structural and other domain optimization problems [12]. In this approach the fitness function $\Phi(X)$ is defined for the individual by modifying the objective function f(X) with penalty function ψ. The general form of such a fitness function is

$$\Phi(\mathbf{x}) = f(\mathbf{x}) \pm \psi \tag{4.4}$$

Ψ may be either a penalty for any general infeasible individual i, or a cost of repairing such an individual for making it feasible. The penalty term in Eq. 4.4 is set to zero for feasible individuals.

Following guidelines by [13], several variants [14–17] of the penalty function were constructed and implemented successfully with their own merits and demerits. The main task in usage of any such penalty function is fine-tuning of their parameters for the optimization problem under consideration. The dynamic penalty method was built and implemented by [14], and [15] recognized the significance of adaptive penalties and introduced the approach as a promising research direction. Then [16] and [17] used the augmented Lagrange method for large structural optimization with GA. GA integrated structural optimization with constraint handling has been presented in [18]. The self-adaptive method was proposed and shown to be more effective in quality of solution and convergence in [19], which has presented a comparative study on important benchmark problems used in structural optimization with GA. However, it admits that parallel implementation is necessary for expected performance speed. In the same work, several good attempts made by others for design and implementation of penalty function have been discussed in detail with their constructs and behavior. The reader is advised to get acquainted with the same. Other adaptive techniques [20, 21] find the relative normalized values of penalty coefficient for each violated constraint in each cycle adaptively, using the same cycle magnitudes of all violated constraints. These coefficients can then be used to obtain the fitness function from objective function and each of the violated constraint functions. Recently, new ranking-based constraint handling techniques [22–27] have evolved. Tournament selection method (TSM) [23] takes into account the relative merit of candidate solutions in terms of objective function or constraint violation and ranks them accordingly under each category to decide and evaluate their fitness. Pairwise comparison is made based on three criteria for selection: (1) a feasible solution (FS) is always preferred over an infeasible one; (2) between two FSs, the one having the best objective function value is preferred; and (3) between two infeasible solutions (ISs), the one having the lowest constraint violation value is preferred. In stochastic ranking (SR) [22] whether the ranking is to be made by objective function comparison or sum of constraint violation is decided based on a user-supplied probability parameter p_f. Objective functions are compared if $p_f = 1$, which is true when adjacent solutions to be compared are both feasible. The sets of other p_f values are used for different such comparison scenarios. The method compares adjacent neighbors only, which is also sensitive to values of parameter p_f, and authors have suggested a range of p_f values based on few set of experiments. Global competitive ranking (GCR) [24] ranks the individual by comparing it with the remaining individuals in the population. User-defined parameter p_f stipulates the amount of penalization for an IS, which is conceived as the balance factor to derive the fitness. In the Ho and Shimizu ranking (HSR) [25] method three distinct ranks are assigned to specify the fitness of an individual: the first rank compares the objective function, the second compares squared sum of constraint violations, and last one compares the total number of constraints violated. In the balanced ranking method (BRM) [26] dynamic manipulation of ranking of feasible and infeasible individuals with the help of adaptive penalty expression has been incorporated when any adjustable/user-specified parameters are not essential. Multiple constraint ranking (MCR) (MCR) technique [27] sets up queues of ranking for objective function and each of the constraint values. The individual with the best objective function value and lowest constraint violation receives first rank under each head; that individual receives highest fitness, which has the lowest sum total of all ranks. Most of these rank-based methods have been tested on benchmark problems of literature and can lead to more research in that direction. The reader will also find a useful review on constraint handling methods, implemented before this decade, in [28].

4.4 STRUCTURAL OPTIMIZATION USING GENETIC ALGORITHMS: AN OVERVIEW

4.4.1 Space Trusses Design Optimization

It appears that [29] applied GA for the first time to optimize the overall weight of a 10-bar steel plane truss. The simple genetic algorithm (SGA) was applied in [30] to optimize a trussed-beam roof structure and three-bar truss structure. Using GAs as function optimizers, key distinction between genetic search and random search has been presented by [31] analyzing optimal population size, encoding, selection, crossover, and mutation over binary alphabets. A two-beam grillage structure, a two-element thin-walled cantilever torsional rod subjected to sinusoidal excitation, and the dynamic response of a 10-bar plane truss have been illustrated and represent the non-convex search spaces. Discrete optimization of generalized trusses using GA has been performed in [32].

A system of genetic search (EVOLVE) developed by [33] handled a mixed integer, discrete and continuous design variables approach that was used to solve various design optimization problems. Fitness sharing and automatic constraint handling were incorporated in binary encoding/decoding schemes. A specially designed crossover operator could find the positions of crucial bit locations in a chromosome for crossover operation. A general method of handling constraints based on the behavioral memory paradigm was developed by [34] using a series of constraints complying with different fitness functions. A 10-bar (two-dimensional [2D]) and a 25-bar (three-dimensional [3D]) truss optimization were implemented using the procedure developed. It determined that the success of the entire process was greatly dependent on population diversity.

Early attempts at multi-objective optimization [35] designed a GA that generated the Pareto set in single run by implementing a utility function and sharing, which demonstrated a 10-bar truss optimization.

Hybrid methods combining classical approaches and GA, parallel computing paradigms, and novel design representations were also implemented successfully in early research efforts on GA. A minimization of overall weight of a space truss subjected to multiple constraints of displacements, stresses, and discrete standard cross-sections (c/s) was employed by [36]. Devising a quadratic penalty function to obtain the unconstrained problem of the design optimization of a 12-bar truss, a 25-bar truss, and a 72-bar truss has been illustrated. A new approach of a hybrid GA integrating the penalty function method and primal-dual method was developed by [37]. To minimize the unpredictable behavior of penalty function coefficient sequential minimization similar to the Lagrangian method was used. To solve a large-scale real-world optimization problem, a parallel computing procedure was adopted using GA by [38]. The strategy of parallel computing was successfully implemented for faster convergence to optimize a 17-member plane truss and 848-element space truss, and the unconstrained problem was obtained using a methodology of penalty function and augmented Lagrangian method. In further development the same authors [39] successfully employed GA for large-scale structural optimization using massively parallel supercomputers. In an attempt to employ simultaneous optimization of member sizes, shape, and topology of space trusses, a methodology that adopted a completely heterogeneous representation of decision space was implemented by [40]. Element connectivity and boundary conditions were represented by Boolean design variables for topology considerations, node locations were treated as continuous design variables, and both discrete and continuous variable representation was used to define member cross-sectional properties. Absence of deformation, members with zero force, and instability issues were tackled by the exception handling concept. Fitness recomputing was avoided by identifying and storing information of duplicate chromosomes. A discrete search space model was employed in [41] with a view to make GA computationally efficient for optimization of truss structures and was shown to have better performance.

The composite solution method, which combines Min-Max criterion and GA, was developed by [42] and claimed to be computationally more efficient than GA methods implementing Pareto-optimal or other techniques for multi-objective optimization of the structural design of trusses. Its only limitation, as mentioned by the authors, is that it requires the provision of an initial ideal vector of chromosomes by defining appropriate weights, which is still in the development stage. Their work also summarizes different methods available for multi-objective optimization and highlights several other approaches devised by researchers in GA methodology for multi-objective optimization. A brief history of multi-objective optimization in structural engineering is mentioned in good detail.

A single objective of weight minimization of structure was achieved by employing dynamic constraints [43] of natural frequencies and frequency responses including stresses. A different method called binary linear zero-one programming is proposed that handles the combinatorial optimization problem with any number of discrete design variables. Plane trusses of 10- and 25-bar configurations were demonstrated for optimization results and compared with branch and bound and GA method. However, influence of any real-life dynamic excitation is not included in any of the test cases to signify the role of dynamic constraints.

A great number of other studies implemented GA [44–50] either with different design representations or combining GA with other metaheuristics, such as fuzzy logic, for optimization of space structures.

4.4.2 Optimization in Reinforced Cement and Pre-stressed Concrete Structures

A great majority of papers on structural optimization deal with weight minimization or cost optimization of reinforced cement concrete (RCC) multi-story buildings or components thereof. Weight minimization by optimum detailing and overall cost optimization of reinforced cement concrete (RCC) buildings has been presented by [51] and [52], respectively. Design optimization of RCC continuous beam by [53] and RCC frame structures using GA have been proposed by [54, 55], and [56].

Optimized design of RCC flat slabs using discrete design variables and finite-element (FE) analysis has been dealt with by [57, 58], and [59]. To optimize shape and structural characteristics of hollow core slabs the 3D FE analysis was combined [60] with control on specified constraints of local and global stresses.

Cost optimization to determine the optimal number of spans and c/s of pre-stressed concrete bridges has been investigated [61]; they used uniform crossover and the controlled mutation in GA. Using linear and quadratic penalty functions to handle constraint violations, an integrated GA and FE analysis were performed by [62] to account for labor and formwork cost in addition to material costs for optimizing low-cost RCC frames.

4.4.3 Optimization in Steel Structures

Cost and/or weight optimization and efficient performance in a seismic event have been found to be the main focus for a large number of research studies on framed and other steel structures. Simultaneous shape, topology, and response optimization satisfying multiple constraints has pushed several investigations into exploration and implementation of different ideas bringing good versatility to SGA. The following discussion focuses on such concepts in addition to the usual optimization goals.

Elitism and other problem-specific schemes were also incorporated by some works successfully using the GA optimization method. For instance, in search of the minimum volume of welded I-section frames subjected to extremely nonlinear constraints, two distinctly different operators have been introduced, the making-clones operator and laboratory operator, by [63]. The first operator resembles the elitist operator that deliberately copies the fittest member to the next population. The usual crossover and mutation operators were also not a common feature of their algorithm. With analogous efforts, the concept of the "mirroring function" was illustrated [64] to optimize symmetrical structures, which reduced the computational time to half the expected time. Likewise, two different mating pools were used [65], one for structural members and one for their beam-column connection types, in optimum design of steel moment frames. Along similar lines, a eugenics evolutionary theory has been integrated with GA [66] for cost optimization. Their method uses a new selection operator, which leads each member of the population to have a descendant without loss of genetic material. Optimum cost design of an underground cylindrical liquid storage tank [67] and a welded plate girder bridge [68] implemented the elitism procedure to speedup convergence. The second study used discrete variable formulation in GA.

There are instances that reveal traditional engineering judgment has been reinforced/enhanced by GA. In other words, GA offers handy assistance to decision makers to identify near optimal solutions for problems where simultaneous consideration of all design parameters is restricted. For example, it was demonstrated [69] that a lighter design could be obtained for a large, complex real-life structure with heuristic two-phase GA. Analogous heuristic was used [70] in satisfying the architectural constraints and exploring the simultaneous sizing, topology, and shape optimization.

Extensive review has been presented on structural optimization of steel structures using GA by [71] and a similar review on life cycle cost optimization of steel structures using GA has been considered by [72].

Nonlinear behavior of steel frames with provision of semi-rigid connections for optimization of steel frames has been dealt with by [73]. Optimization of the degree of rigidity of semi-rigid connections was focused on by [74] to target the minimum weight of nine different steel frames under nonlinear and linear analysis.

The study [75] showed that GA can be one of the better choices in large-scale optimization problems of high-rise building structures subjected to discontinuous constraints of commonly used design codes. The performance of several approaches, such as classical, optimality criterion, and various metaheuristic optimization algorithms, have been compared with cost optimization of high-rise and super-high-rise RCC and steel structures. It has been reported that along with GA the hybrid algorithm developed by the study has delivered consistent and effective results for up to 15% cost optimization of large-scale civil engineering structures, especially buildings.

A near optimal trade-off solution between structural weight and story drifts was obtained [76] using a cluster-based non-dominated sorting genetic algorithm (NSGA)-II in investigating rehabilitation of 2D steel X-braced frames. The weak-beam–strong-column design principle has been complied with [77] by maximizing the story energy dissipation ratios of buckling restrained braces (BRB), using NSGA-II in midrise steel frames satisfying strength, stability, and story drift constraints. Optimum angular orientation and height-wise distribution of mega bracings in tall steel structures were obtained [78]; moreover, convergence time was reported to be reduced by applying a different selection and crossover operators simultaneously.

Design optimization methodology called EGAwMP has been proposed [79] for real-life steel skeletal structures using generational GA. Steel skeletal structural optimization using the real-life steel profiles and design codes was performed for cost optimization. To enhance the exploration capability of GA it was augmented with artificial neural network and the methodology was named as EGAwMP. The proposed neural network implementation was also utilized for prediction of more accurate design variables, in each population run, associating with design strategy adopted. The work is practically important as it uses the standard sections which have predefined properties. Such an approach is very useful in optimization as continuous design variables strategy suffers manufacturing inadequacy.

A multi-objective Pareto archived genetic algorithm (PAGA) was used [80] to optimize topology shape and size of elliptical and spherical steel dome structures and claimed to be more efficient. Optimal design of geometrically nonlinear steel dome structures of elliptical and spherical shapes has been performed. PAGA, which has the ability to archive the Pareto solutions throughout the search history, allows us to pick the requested optimal design from the archive for a better decision by the designer. The proposed algorithm has a property of simultaneously minimizing the weight of dome and nodal deflections while maximizing member forces. Normalized values of available strength and nodal displacements were used as a constraint checking mechanism to overcome the difficulty of integrating optimization algorithm with FE analysis software. It was shown that PAGA can be effective for dome optimization.

The work [81] examined the optimized design of a bridge subjected to earthquake loading. To reduce the time of the optimization procedure, two phase optimization and parallel computation procedures were adopted. In the first phase the quantity of two major materials of construction was optimized using trade-off analysis of multi-objective optimization. The second phase was employed to obtain the trade-off between bridge cost and its performance under seismic loading.

Nonlinear dynamic time history analysis was employed using several ground motions scaled to the response spectrum of the European code. GA was used to perform optimization using the methodology of NSGA-II for Pareto front analysis. A statistical procedure was followed, wherein the design variables were analyzed to find out the sensitivity of objective function to them and identify critical variables by examining populations of all runs of GA.

4.4.4 Optimal Control of Structures

Earthquakes have put major environmental loads on all types of civil engineering structures. Every attempt is made to protect the structures from damage by following the appropriate design philosophy. One of the significant developments in this field is the energy dissipation mechanism introduced by auxiliary devices in the building structures. Such mechanisms achieve earthquake response control in a desirable manner, keeping the structures at minimal or no damage. The effective and optimal usage of these systems has been at the core of interest for innumerable works in the last three decades and still continues. Several applications implemented GAs as a powerful tool for the optimal control of vibrating systems and structures. In general, the goal of such procedures is to optimize the structural performance expending minimal control effort while satisfying the specified constraint conditions. Passive and active optimal control strategies have been developed using GA in a mono-objective or multi-objective context.

4.4.4.1 Viscous, Viscoelastic, and other Dampers

Damping devices such as fluid viscous dampers (FVD), viscoelastic dampers (VE), and others have shown promising performance in dissipating vibration energy efficiently. When buildings or other structures are subjected to earthquake forces, these devices quickly respond to dissipate the undesired energy of vibration while protecting the main structure from structural and non-structural damage. The objective of achieving optimal performance of structures in a seismic environment at minimal usage of these devices has been formulated with GA in several works.

The investigations [82–84] employed GA approaches as an alternative to a gradient-based search to find the optimal placement of dampers. A linear behavior model of frame and state space approach was used for the analysis. It was determined that GA is especially suitable where the performance index is not a continuous function of the design variables.

Hybrid optimization method was proposed [85] where the property of GA to approach near global optimum region is combined with sequential quadratic programming (SQP) characteristics to capture the right path and converge to optimum very quickly and accurately. The transition from GA to SQP is suggested at a point when the difference between the best designs of two successive generations is within 10%. The study believes it to be more rational in the general case when there are more complex and non-convex design problems exhibiting many local optima. Two test cases from the literature were evaluated for weight optimization, such as a six-story space frame with 63 elements and a 20-storey space frame with 1020 members, to emphasize the efficiency of the hybrid method proposed.

Formulation of a discrete and nonlinear optimization problem was introduced [86] to determine the optimal position of actuators in tall buildings using GA. A 16-story building under 18 different earthquake excitations was analyzed. It has been reported that the control system and structural properties of a building are the main factors influencing the optimal placement of control devices rather than characteristics of earthquake excitations. The study also highlights points of comparison between conventional and GA-based optimization methods.

Optimized distribution of viscous devices by GA using frequency domain stochastic analysis was found out by [87] for the multi-story benchmark control building. H_2 and $H\infty$ norms of the transfer function were utilized as objective functions to consider both the average and peak response of the structure and to evaluate performance of the viscous damping system. It concluded that each objective needs different distribution and the priority of objectives governs a particular choice.

Control strategy incorporating norms of root-mean square (RMS) of absolute acceleration and interstory drift has been employed [88] with magnetorheological (MR) dampers for building systems. Optimal placement of MR dampers was achieved using a GA.

For buildings subjected to severe earthquake excitations, a GA-based multi-objective procedure was developed [89] to minimize the number of plastic hinge occurrences with optimal properties of FVD.

In their optimal control strategy [90] and [91] provided a framework for multi-objective optimization methodology in evolutionary computing by GA to optimize performance of a building to given seismic excitation by means of energy dissipation devices. It described how to incorporate multiple devices in chromosome modeling including weakening characteristics as an additional part of it to represent degraded structural stiffness for retrofitting objectives. Optimization of interstory drifts and floor accelerations was demonstrated with the Pareto front concept using the novel fitness index. Simultaneous layout of BRBs, viscous fluid (VF) dampers, and VE dampers was considered with four ground motions input to framed example buildings. Though this work was able to set certain direction to multi-objective optimization with passive devices, it also admitted the shortfalls in the study. Using the friction principle for damping [92] presented slip load and topology optimization of pall friction dampers in a multi-objective context. Controlled response was compared with that of uncontrolled response of the original frame. Peak values of top-story displacement and acceleration and peak base shear were used as objectives, and their minimum norm considered for assessment of performance, with uniform and optimized distribution of devices. Series of analyses were performed on a 10-story example building with 10 different earthquake time histories implementing the NSGA-II [93] optimization algorithm to demonstrate the results.

4.4.4.2 Tuned Mass Dampers

Robust control procedure was proposed by [94] using a tuned mass damper (TMD) device. Initially a multi-objective optimization was applied by using GA to minimize peak values of floor displacement, velocity, and acceleration. NSGA-II was employed to find Pareto optimal solutions.

Adaptive tuned mass damper (ATMD) controllers were used for reduction of wind-excited vibrations in tall buildings [95], which implemented rank-based fitness scaling to prevent the premature convergence. Likewise, the optimal design parameters of the TMDs have been found [96] by using a GA as a means to reduce the seismic accelerations of high-rise cross-laminated timber buildings.

Single ATMD applied to structures under low-moderate earthquakes has been investigated by [97]. To control the torsional responses of asymmetric buildings the concept of coupled TMD has been implemented by [98]. A smart tuned mass damper (STMD) has been devised [99] for integrated optimal design and control of a 60-story diagrid building structure subjected to wind and seismic excitations. A multi-objective GA was applied to optimize six different objectives and good performance has been reported.

4.4.4.3 Active and Hybrid Controls

Active control poses a complex optimization scenario due to various parameters and components to be considered simultaneously for seismic response control. The most challenging among such expectations consider the number, placement, controller gain, and type of control system for optimal structural control. GAs have been found to be equally useful tools in active control strategies.

Minimization of linear regulator quadratic control cost and maximization of modal control ability with robustness was employed by [100] in the multi-objective optimal control problem. A 45-bar plane truss layout optimization with efficient actuator placement was solved using weighted sum methodology with multiple constraints of eigenvalues, asymptotical stability, and total truss weight. This appears to be one of the few early attempts on evolutionary multi-objective optimization of trusses available in the literature.

Straightforward formulation of GA has been used [101] to guide placement of actuators within regular multi-story buildings. The 16-story building drift control by a predetermined actuator control mechanism was demonstrated as the objective of the study.

To predict and control response in a structure using a limited number of feedback measurements is a non-trivial task. Design of controller gains to control response of building systems using binary encoded GA has been demonstrated in [102]. Generation of optimal control gains in the presence of sensor noise was demonstrated for a wind-excited 76-story building installed with an ATMD at roof level. Peak and RMS values of displacement and accelerations were considered for response control. In addition, robustness criteria for the controller was also formulated in GA.

Optimization of the hybrid system control gains was applied [103] for 2D buildings using binary encoding of GAs. A comparative study of active, passive, and hybrid control systems for optimal control using real coded GA was presented in [104], with optimal control gain for ATMD devices using limited sensor arrangements. Similar comparisons for 3D structures were carried out [105] for optimal response control by passive and active control devices using GA.

The robust controller needs to address uncertainties arising in structure in realistic environments so that a truly stable and expected controlled performance is achieved. Two such uncertainties are system properties (e.g., mass and stiffness) and sensor output noise. The fuzzy logic controller (FLC) design algorithm was found to be effective for such matters by several studies. Peak responses were controlled to a desired level, and controllers were optimized combining FLC algorithms with GA by several studies [106–110], which considered 2D and 3D building structures under single and multi-objective scenarios.

A multi-level GA has been discussed and illustrated in [111], and fuzzy logic methodology has been combined with GA considering nonlinear structural response in [112]. Placement, sizing, and feedback control gains of a novel piezoelectric sensor-actuator to minimize the vibration of shell structures has also been demonstrated in [113]. Active control strategies for high-rise building structures using GA were also demonstrated in [114, 115].

It can be noted that all these works emphasize selection of the optimal number and placement of actuators and/or sensor and type of control algorithms (e.g., linear quadratic regulator, acceleration feedback control, or FLC), which was addressed effectively by GA.

4.4.5 Multi-Objective Optimization Using Genetic Algorithms

Several real-world problems require multiple objectives to be optimized simultaneously in addition to multiple constraints. In general, such objectives are conflicting in nature, which increases the complexity of the problem, and need robust strategies to find an amicable solution. Very often domain-specific knowledge and intervention is essential to arrive at an efficient decision that is helpful to protect all important priorities in a balanced manner.

A large number of function evaluations and multiple constraint handling is a major challenge in multi-objective optimization problems. Several approaches have been proposed and evaluated in the history of GA to address them and achieve the goals as discussed above.

Some earlier notable works on multi-objective optimization by GA in structural optimization include [116, 117]. Extensive contributions in multi-objective optimization of structures using GA and advancement of GA in multi-objective optimization have been presented in several significant works [118–122]. Many more works related to multi-objective criterion and GA that offer good insight include [123, 124] and [125].

Multi-objective optimization of truss structure design has been implemented in [41, 43, 49]. Structural control with multi-objective optimization has been dealt with in [90–94]. Active control strategies with multi-objective optimization have been implemented in [107, 108, 113]. Other works that solved multi-objective problems using GA include steel structures [77], FE analysis [126], dissipative connection designs [127], and adaptive constraint strategy [128].

A comprehensive state-of-the-art survey on multi-objective strategies by evolutionary algorithms and GA has been presented by [129] and [130]. In a recent investigation [131] compared 45 different multi-objective algorithms for their efficiency in many criteria such as metamodeling techniques, problem size, constraint handling strategies, and evolution model. Selection criteria of a particular algorithm has been proposed in their work to choose from among different algorithms so that computational efficiency is accomplished.

4.5 GA APPLICATIONS IN STRUCTURAL OPTIMIZATION

We illustrate the applications of GA in structural optimization through several examples. In the following sections the test functions, benchmark problem, and real-world applications have been explained.

4.5.1 Example 1: Optimization of a Cylindrical Can

For the purpose of simplifying the discussion, we explain a short problem of cost minimization in this example and illustrate the working of a simple GA in binary encoding. Consider an example of a *cylindrical can* whose diameter is D and height h, which has to hold a maximum volume V_0 of certain liquid. It is desired in this problem to optimize the total surface area of the can, closed from all sides, so that the cost of material that goes into making the can is minimized. We express the optimization problem as follows.

Find optimum design variables

$$\mathbf{x}^* = (D^*, h^*) \tag{4.5}$$

To

$$\text{Minmize } f(\mathbf{x}) = 2\pi D^2 + \pi Dh \tag{4.6}$$

That satisfy,

Constraints:

$$V = V_0 \tag{4.7}$$

$$D_{min} \le D \le D_{max} \tag{4.8}$$

$$h_{min} \le h \le h_{max} \tag{4.9}$$

There are two design variables (DVs), i.e., x1 = D and x2 = h in this single-objective optimization problem, which is a nonlinear function of the DVs. In terms of GA the individual of a population, i.e., chromosome, can be expressed as a pack of two genes that represent x1 and x2, the two DVs. In other words, genotype is a two-cell structure that holds position of x1 and x2, and the phenotype will take different values of these DVs for each individual of population. We choose a binary encoding scheme for representation of these real continuous variables. In binary coding 4-, 8-, and 16-bit precision (p) can be followed; this is decided based on accuracy required in the decision variables and problem solution. The set of equations as shown here can be followed for mapping from binary to real space or vice versa.

$$X_{real} = X_L + p_f . X_{integer} \tag{4.10}$$

$$X_{integer} = (X_{real} - X_L) / p_f \tag{4.11}$$

$$p_f = (X_U - X_L)/(2^p - 1) \tag{4.12}$$

where, p is bit precision; X_U and X_L are maximum and minimum values of decision variables, respectively; X_{real} is the required value of decision variable in real space; and $X_{integer}$ is the decoded value (Eq. 4.11) of the binary counterpart of X_{real}. Using these transformation equations the initial population of candidate solutions can be set up that consist of a set of decision variables involved in the problem. The variation operators then guide the population toward optimization. The binary decoded integer, i.e., $X_{integer}$, decimal value for 8-bit representation is obtained as follows:

$$01011101 \rightarrow X_{integer} = 0x2^7 + 1x2^6 + 0x2^5 + 1x2^4 + 1x2^3 + 1x2^2 + 0x2^1 + 1x2^0 = 93$$

Using the Eqs. 4.10–4.12 the values of *can* diameter D and height h can be obtained once the maximum and minimum limits on their values are available. For example, for $D_{min} = 100$ mm, $D_{max} = 150$ mm, $h_{min} = 150$ mm, and $h_{max} = 250$ mm; any values for D and h can be interpolated based on the above equations. Procedure of initialization of population, its fitness evaluation and selection, crossover, and mutation are illustrated for this problem in Appendix 4.1 for a single cycle of generation using the population size of six candidates. The constraint violation of the upper limit on the volume of cylinder expressed by Eq. 4.7 can be included in this procedure using a suitable penalty factor for objective function and accordingly fitness can be reduced. In this way generation cycles can be advanced until convergence criteria are met, as explained elsewhere in this paper.

4.5.2 Example 2: Optimization Performance of GA for Test Functions

We examine the performance of real coded GA on a few test problems in this example. Several test functions have been designed to check the performance of metaheuristic optimizations in scientific literature. The real coded GA developed for the purpose of the present study uses a uniformly distributed crossover operator on continuous decision space and a Gaussian-normal distributed mutation operator, both of which are explained in Appendix 4.3. The population size of 20 and maximum generation cycles of 50 and convergence and mutation rates of 0.9 and 0.01, respectively, are used to illustrate the performance of GA for the three test functions, namely sphere function, Rastrigin function, and Ackley function. The formal equations of these test functions are as given in Appendix 4.3. The number of decision variables used in case of sphere function was five. Figs. 4.5–4.7 show the best cost versus number of GA cycles, respectively, for

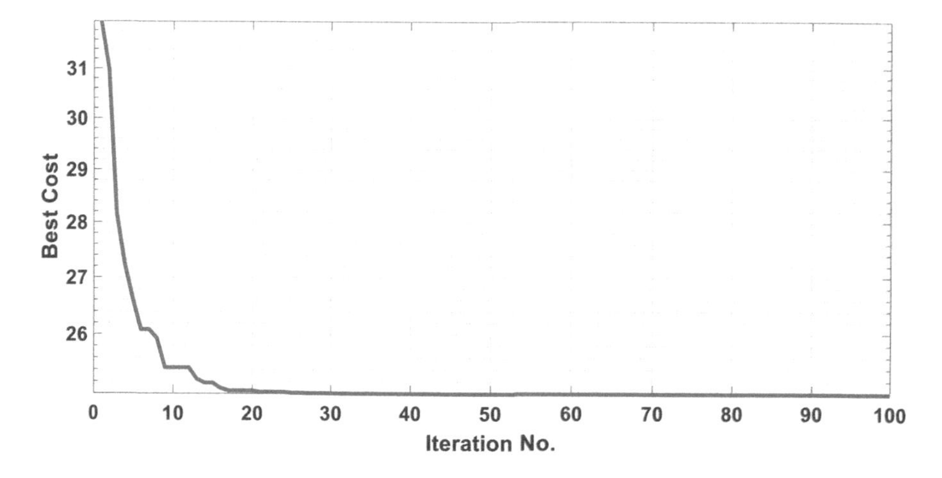

FIGURE 4.5 Sphere function convergence history. True Best = 0, True Solution=0.

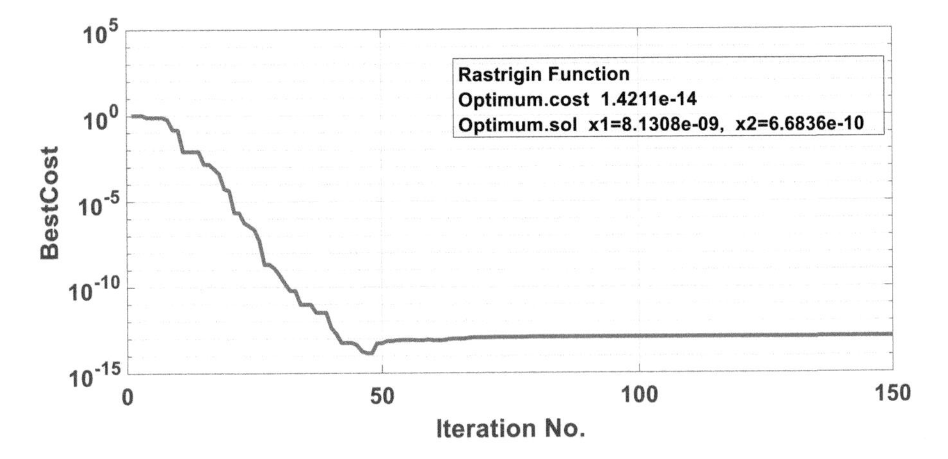

FIGURE 4.6 Rastrigin function convergence history with best cost and best solution. True Best = 0, True Solution x1=x2=0.

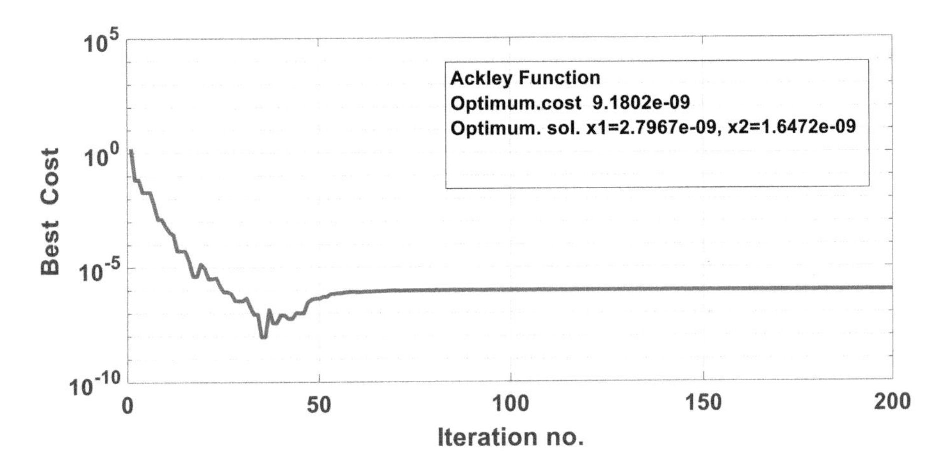

FIGURE 4.7 Ackley function convergence history with best cost and best solution. True Best = 0, True Solution x1=x2=0.

these functions. It can be observed that the GA could reach to true optimum for sphere function but not so in the case of the other two functions. Such behavior of metaheuristics is common, and performance depends on tuning parameters and robust design of metaheuristic algorithms.

4.5.3 Example 3: Weight Minimization of Two Different Cantilever Structures

Next we consider two more design examples of nonlinear structural optimization as shown in Figures 4.8 and 4.9. Two cantilever beams with rectangular and hollow circular c/s are subjected to single vertical point load at the free end of the cantilever. The objective here is to minimize weight, i.e., material quantity with specified constraints on flexural stress induced at the fixed end of cantilever beam and maximum deflection occurring at its tip. Both of these problems are identical in context with difference in the details. The binary or real coded representation can be employed to solve the problem with GA.

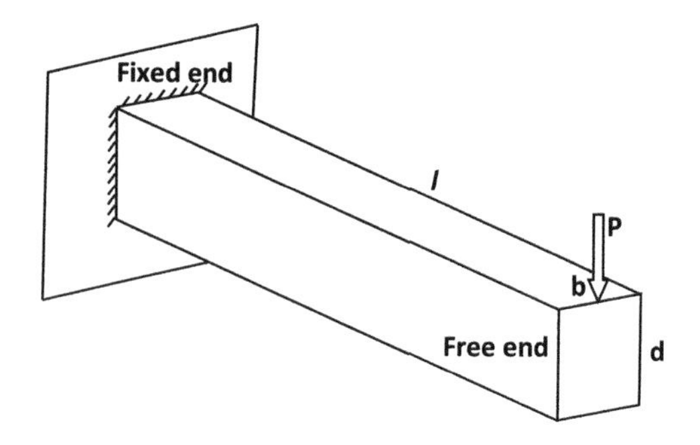

FIGURE 4.8 Cantilever beam with rectangular c/s.

Optimization Problem Statement of rectangular c/s

Find optimum design variables

$$\mathbf{x}^* = \left(D^*, b^*\right) \tag{4.13}$$

To *Minimize*

$$\left(\mathrm{f}(\mathbf{x}) = W\right) = \rho l b D \tag{4.14}$$

Constraints:

$$\sigma \leq \sigma_{lim} \tag{4.15}$$

$$\delta \leq \delta_{lim} \tag{4.16}$$

Design Variable limits

$$D_{min} \leq D \leq D_{max} \tag{4.17}$$

$$b_{min} \leq b \leq b_{max} \tag{4.18}$$

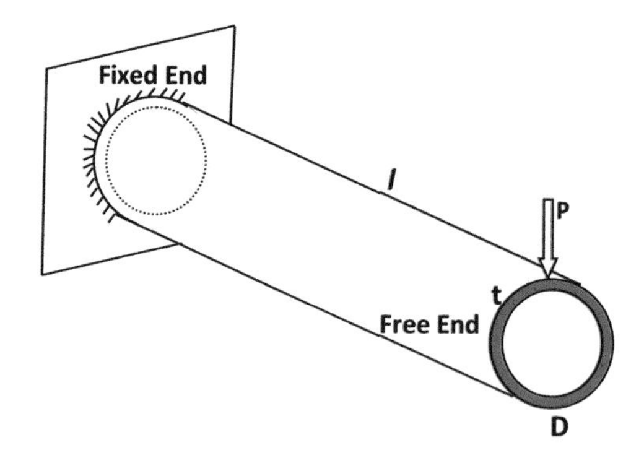

FIGURE 4.9 Cantilever with hollow circular c/s.

where σ is maximum bending stress at the root of the cantilever due to maximum bending moment $M = Pl$ and δ is the maximum deflection at the free end of the cantilever considering bending strain energy alone.

$$.\sigma = \frac{M}{Z} = 6M / bD^2 \tag{4.19}$$

$$\delta = \frac{Pl^3}{3EI} \tag{4.20}$$

$$\mathrm{I} = \frac{bD^3}{12} \tag{4.21}$$

Optimization Problem Statement of circular c/s

Find optimum design variables

$$\mathbf{x}^* = \left(D^*, t^*\right) \tag{4.22}$$

To *Minimize*

$$\left(\mathrm{f}(\mathbf{x}) = W\right) = \rho l \pi (Dt - t^2) \tag{4.23}$$

Constraints:

$$\sigma \le \sigma_{lim} \tag{4.24}$$

$$\delta \le \delta_{lim} \tag{4.25}$$

Design Variable limits

$$D_{min} \le D \le D_{max} \tag{4.26}$$

$$t_{min} \le t \le t_{max} \tag{4.27}$$

where σ is the maximum bending stress at root of the cantilever due to maximum bending moment $M = Pl$ and δ is the maximum deflection at the free end of the cantilever considering bending strain energy alone.

$$.\sigma = \frac{M}{Z} = \mathrm{MD} / 2\mathrm{I} \tag{4.28}$$

$$\delta = \frac{Pl^3}{3EI} \tag{4.29}$$

$$\mathrm{I} = \frac{\pi}{64}\left(8D^3t - 24D^2t^2 + 32D^2t^3 - 16t^4\right) \tag{4.30}$$

PSEUDOCODE FOR BENCHMARK TEST PROBLEMS: EXAMPLE 1, 2, 3

begin procedure

1. Initialize genetic algorithm parameters
2. Initial population created randomly between the set limits of [DV]
3. Initial evaluation of objective/fitness function
4. *while* stopping criterion is not fulfilled [Max. Iterations limit posed]
   i. Apply fitness proportional *selection for reproduction*
   ii. Apply *crossover* with crossover rate
   iii. Apply *mutation* with mutation probability
   iv. Check individuals for DV limits violation
   v. Evaluate Constraint function violation of each individual of Population for each constraint.
   vi. Evaluate objective/fitness function
   vii. Complete the pool of new parents using recent last and new population
   viii. The population is updated

end while

5. PLOT the convergence trend i.e. f_{obj} v/s No. of iterations
6. Report best f_{obj} and corresponding best DV

end procedure

4.5.4 Example 4: Weight Minimization of 10-Bar Truss Benchmark Problem

In this example we implement GA for optimization of a benchmark problem tested several times by different metaheuristic algorithms in research literature. It is required to optimize the weight of the 10-bar truss shown in Figure 4.10 subjected to two joint loads of 100 kips each. Under the action of these loads the two prescribed constraints for design optimization of the truss are the stress in any bar due to compression or tension is limited to 25 ksi (172 MPa) and joint deflection limit in any direction is ±2 inches (50.8 mm). The properties of structure are Young's modulus 10^4 ksi (68,950 MPa) and material density 0.1 lb/in^3 (2770 kg/m^3). The design variables are c/s area of members that are restricted to have a c/s area not less than 0.1 in^2 (64.5 mm^2). [Unit convergence

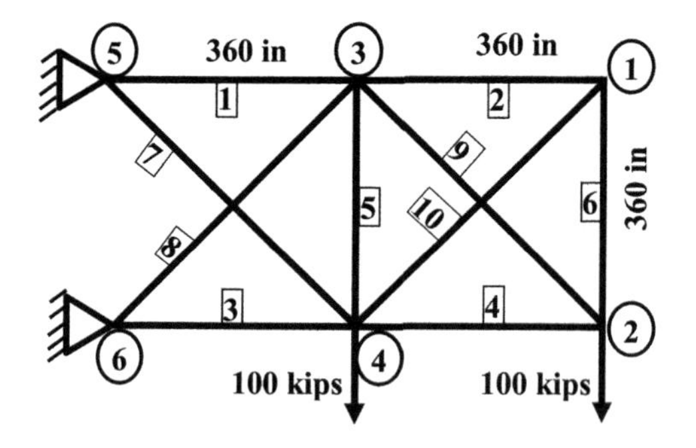

FIGURE 4.10 The 10-bar plane truss structure.

factors used are 1 lb = 0.454 kg, 1 in = 25.4 mm, 1 ksi = 6.895 MPa, 1 lb/in^3 = 27,705 kg/m^3, and 1 kip = 4.448 kN].

With population size of 30 the maximum generation cycles of 5000 was imposed as stall limit; convergence and mutation rates were kept to 0.9 and 0.01 respectively. The formulation of optimization problem is stated as below.

Find optimum design variables (c/s Area A of members).

$$\mathbf{x}^* = (A^*) \tag{4.31}$$

To Minimize (the weight of the truss)

$$(\mathrm{f}(\mathbf{x}) = W) = \rho.\sum_{i=1}^{10} l_i A_i \tag{4.32}$$

Constraints:

$$\sigma \leq \sigma_{lim} \tag{4.33}$$

$$\delta \leq \delta_{lim} \tag{4.34}$$

Design Variable limits

$$A_{min} \leq A_i \leq A_{max} \tag{4.35}$$

where σ is maximum axial stress in any member and δ is the maximum deflection of a joint in the *x*- or *y*-direction in plane of truss structure; σ_{lim} and δ_{lim} are, respectively, the maximum allowed values of these quantities as specified in the optimization problem; ρ is material density; and l_i, A_i are member length and c/s area of general member *i*. The upper limit A_{max} on c/s area of members is trivial in a finite space problem like this one.

PSEUDOCODE FOR TRUSS PROBLEM: EXAMPLE 4

begin procedure

1. Initialize genetic algorithm parameters [Pop_Size=30,Cross_rate=0.9,mut_prob=0.1]
2. Initial population created randomly between the set limits of *c/s area of members*, the DV.
3. Initial evaluation of objective/fitness function [i.e *Weight of Truss*]
4. *while* stopping criterion is not fulfilled [*Max. iterations = 5000*]
   i. Apply fitness proportional *selection for reproduction*
   ii. Apply *crossover* with crossover rate
   iii. Apply *mutation* with mutation probability
   iv. Check individuals for c/s area limits violation
   v. Evaluate Constraint function violation by Structural Analysis of each individual (i.e.Truss).
   vi. Evaluate objective/fitness function [i.e *Weight of Truss*]
   vii. Complete the pool of new parents using last and new population based on fitness ranking
   viii. The population is updated

end while

5. PLOT the convergence trend i.e. f_{obj} v/s No. of iterations
6. Report best f_{obj} and corresponding best DVs

end procedure

TABLE 4.1
Comparison of Various Metaheuristic Algorithm Results for 10 Bar Truss Problem (Figure 4.7)

Area of c/s of Member (in^2)	HS [132]	PSO [133]	PSO [134]	IHS [135]	TLBO [136]	GA (Present study)
A1	30.15	33.5	33.469	30.5222	30.4286	30.448
A2	0.102	0.100	0.11		0.100	0.102
A3	22.71	22.766	23.177	23.2005	23.2436	23.1
A4	15.27	14.417	15.475	15.2232	15.3677	15.399
A5	0.102	0.100	3.649	0.100	0.100	0.102
A6	0.544	0.100	0.116	0.5513	0.5751	0.5445
A7	7.541	7.534	8.328	7.4572	7.4404	7.6707
A8	21.56	20.467	23.34	21.0367	20.9665	21.548
A9	21.45	20.392	23.014	21.5288	21.533	21.426
A10	0.100	0.100	0.19	0.100	0.100	0.100
weight (lb)	5057.88	5024.21	5529.5	5060.82	5060.96	5092.536

Abbreviations: IHS, Improved harmony search; HS, Harmony search; GA, Genetic algorithm; PSO, Particle swarm optimization; TLBO, Teaching learning-based algorithm.

The performance of GA has been compared with the results of several other metaheuristic optimization techniques applied on the same problem. Table 4.1 shows such a comparison. It can be noted form the table that GA shows its satisfactory performance in all aspects compared with other metaheuristic techniques. The convergence history of GA for this problem is shown in Figure 4.11. (Note: Better performance of GA is possible with a different parameter tuning and designs of GA other than that are employed by this study. For instance, the SBX operator and polynomial mutation operator can be implemented as variation operators given in Appendix 4.3 to solve the optimization problem for better performance than shown here.)

4.5.5 Example 5: Optimal Passive Control of Seismic Vibrations of Buildings Using GA

During a seismic event, a certain amount of energy is input into a structure and by virtue of structural geometry this input energy is transformed into both kinetic (oscillations) and potential (strain) energy. Introducing supplemental damping into the structural system has been an effective method to attenuate the seismic shock before it reaches the structural system. The devices acting as supplemental damping either absorb the earthquake energy or dissipate it. Examples of such devices are fluid viscous devices, VE devices, friction devices, and so forth. We consider one such system of damping wherein FVDs provide energy dissipation in the building frame. The fundamental equation of vibratory motion of a building frame fitted with such devices in different stories is as follows.

$$[M]\{\ddot{u}\}+[C]\{\ddot{u}\}+[K]\{u\}+\{F_d\}=-[M]\{I\}\ddot{u}_g \quad (4.36)$$

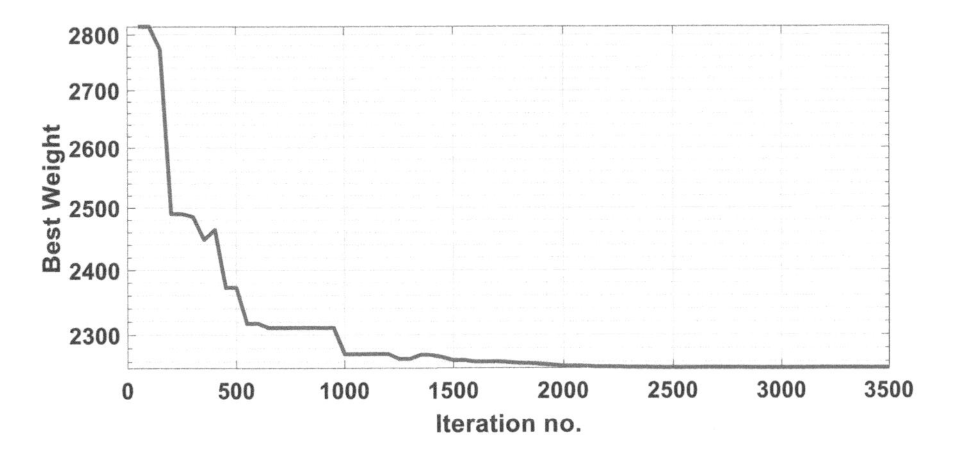

FIGURE 4.11 Convergence history of weight minimization of the 10-bar truss problem. (Weight in kilograms.)

where [M], [C], and [K] are system mass matrix, inherent damping matrix of system, and stiffness matrix of the lateral force resisting system, respectively; $\{\ddot{u}\}$, $\{\dot{u}\}$, and $\{u\}$ are relative floor acceleration, velocity, and displacement vectors, respectively; $\{F_d\}$ indicates damping forces in the devices; {I} is the influence vector associated with floor masses; and $\ddot{u}_g$ represents strong ground motion acceleration. In general, devices exhibit a nonlinear forcing function and any suitable numerical scheme of solution from the family of time-stepping techniques can be implemented. The Newmark-β stepping method has been followed in this study because of its accuracy and stability. The generalized expression of forces in the FVD, which are proportional to relative velocity $\dot{x}$ across its ends, is given as

$$F_d(t) = c_d |\dot{x}|^{\infty} \operatorname{sgn}(\dot{x}) \tag{4.37}$$

where c_d is a generalized damping coefficient and α may take values in the range of about 0.25–2. In case of seismic vibrations $\alpha = 1$ has been recommended and used by this study. The area under hysteresis loop shown in Figure 4.12 gives energy dissipated by these devices in one cycle of oscillations of building fitted with these devices (Figure 4.13).

The 5- and 10-story building frames have been considered that are designed for a live load of 3 kN/m^2 and actual dead load of columns, beams, slabs, and walls as per IS-456 (2000). Framing of each building in one of the principal directions (X) has been designed to the satisfy code permitted drift of 0.4% at design basis earthquake (DBE) defined by IS-1893 (2016); whereas the framing system in the other principal direction (Y) is deliberately designed to violate drift criteria for the purpose of evaluating the amount of damping and subsequent optimization. Such a situation typically represents a distressed building requiring some retrofitting strategy such as the damping technology considered in this study. To verify the responses to actual earthquake ground motion, the building performance has been evaluated for the frequently preferred El Centro strong ground motion. The El Centro simple grounding method (SGM) has been scaled to DBE as given by IS 1893 (2016). The optimal allocation of damping devices has been obtained by augmenting the GA optimizer developed in this study with the dynamic analysis of building structure. The optimization problem statement is formulated as given by Eq. 4.38-4.41.

Find a vector of optimum damping coefficients (n stories of building)

$$\mathbf{Cd}^* = \left(Cd_1^*, Cd_2^*, Cd_3^* \ldots\ldots, Cd_n^*\right) \tag{4.38}$$

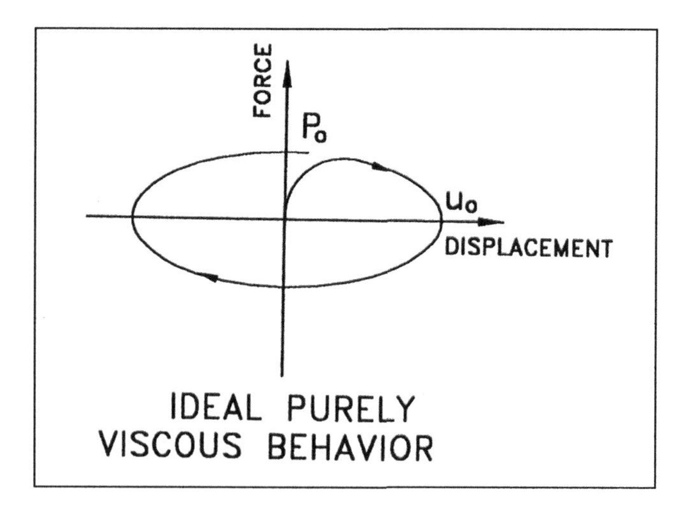

FIGURE 4.12 Fluid viscous damper (FVD) ideal hysteresis loop.

That satisfy,

The m inequality constraints:

$$g_r(\mathbf{Cd}^*) \leq 0 \qquad (\mathrm{r} = 1, 2, 3\ldots\ldots \mathrm{m}) \tag{4.39}$$

$$\left(g_{max} = \frac{\Delta \mathrm{u}_{max}}{\Delta \mathrm{u}_{lim}} - 1\right) \leq 0 \tag{4.40}$$

which gives the optimized aggregate weighted objective function,

$$\overline{\mathrm{f}}(\mathbf{Cd}^*) = \gamma_1 \frac{\Delta u_{max}}{\Delta u_{lim}} + \gamma_2 \frac{VB_{optimum}}{W_{seismic}} \tag{4.41}$$

where cd_i story damping coefficient is he design variable; $\Delta \mathrm{u}_{max}$ is the maximum interstory drift; $\Delta \mathrm{u}_{lim}$ is the limiting interstory drift as per IS-1893:2016 (0.4% of story height);

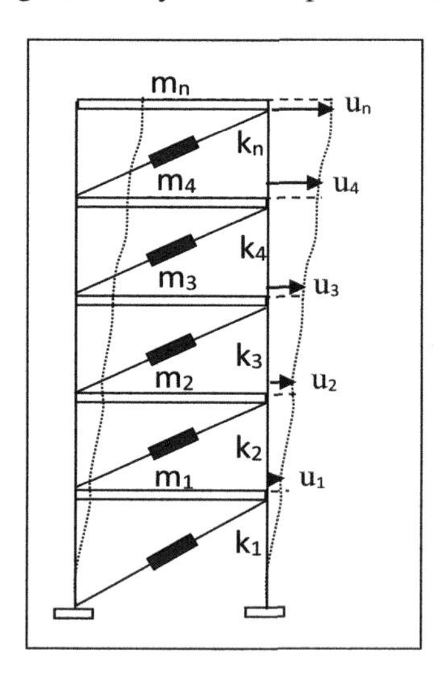

FIGURE 4.13 Building frame with damping devices.

$VB_{optimum}$ is the base shear value for the case of optimum damping; $W_{seismic}$ is the seismic weight of the building; and γ_1, γ_2 are individual weightages for each of the two objectives.

The Table 4.2 shows the data of buildings for seismic analysis. Using shear modeling of the building frame, three key responses, i.e., displacement, interstory drifts, and story shears, are compared for each building before and after installation of the damping devices. The bare frame means without any devices, and responses of bare frame, with uniform damping and GA optimal damping, are compared in Figures 4.14 and 4.15. Figure 4.14 for a 5-storu building and Figure 4.15 for a 10-story building illustrate the reduction of vulnerability of the building frame without any devices to acceptable limits when supplemental damping is provided. Figures 4.16 and 4.17 compare the optimal allocation of damping, i.e., c_d, with that of uniform provision for 5- and 10-story buildings, respectively.

The GA optimizer allocates the damping devices, which are different from uniform damping provision, to effectively reduce the excessive interstory drifts, which is the most critical serviceability factor for buildings. The GA optimizer achieves about a reasonable 4% and 6% savings in damping provision compared with uniform provision in 5- and 10-story buildings, respectively. Besides, it can be noted from Figures 4.14 and 4.15 that the performance of GA allocated damping devices has been more effective in reducing the story shears, interstory drifts, and floor displacements than that shown by the uniform damping provision. In fact, this is quite obvious as the GA optimizer inherently tries to satisfy all the constraints to their optimum, which is impossible by plain uniform provision of devices. The convergence criteria was set as a stall limit of a maximum 1% variation in the average fitness and that of best

PSEUDOCODE FOR PASSIVE CONTROL PROBLEM: EXAMPLE 5

begin procedure

1. Initialize genetic algorithm parameters [Pop_Size=20,Cross_rate=0.9,mut_prob=0.1]
2. Initial population created randomly between the set limits of *cd=damping value*, the DV.
3. Initial evaluation of objective/fitness function [*Eq. 4.41*]
4. *while* stopping criterion is not fulfilled [*Max. iterations = 500*]
   i. Apply fitness proportional *selection for reproduction*
   ii. Apply *crossover* with crossover rate
   iii. Apply *mutation* with mutation probability
   iv. Check individuals for c/s area limits violation
   v. Evaluate Constraint function violation by Structural Seismic Analysis of each individual (i.e. building) [*Eq. 4.39 and Eq. 4.40*]
   vi. Evaluate objective/fitness function [*Eq. 4.41*]
   vii. Complete the pool of new parents using last and new population based on fitness ranking
   viii. The population is updated

end while

5. PLOT the convergence trend i.e. f_{obj} v/s No. of iterations
6. Report best f_{obj} and corresponding best DVs

end procedure

TABLE 4.2
Building Analysis Data

Building Frame	Lumped Floor Weight (kN)	Story Lateral Stiffness (kN/m)
	m1 = 848	k1 = 83445
	m2 = 848	k2 = 83445
5-story	m3 = 848	k3 = 83445
	m4 = 848	k4 = 7510
	m5 = 499	k5 = 6676
	m1 = 1060 m6 = 1060	k1 = 100134 k6 = 90121
	m2 = 1060 m7 = 1060	k2 = 100134 k7 = 80107
10-story	m3 = 1060 m8 = 1060	k3 = 100134 k8 = 80107
	m4 = 1060 m9 = 1060	k4 = 90121 k9 = 80107
	m5 = 1060 m10 = 530	k5 = 90121 k10 = 80107

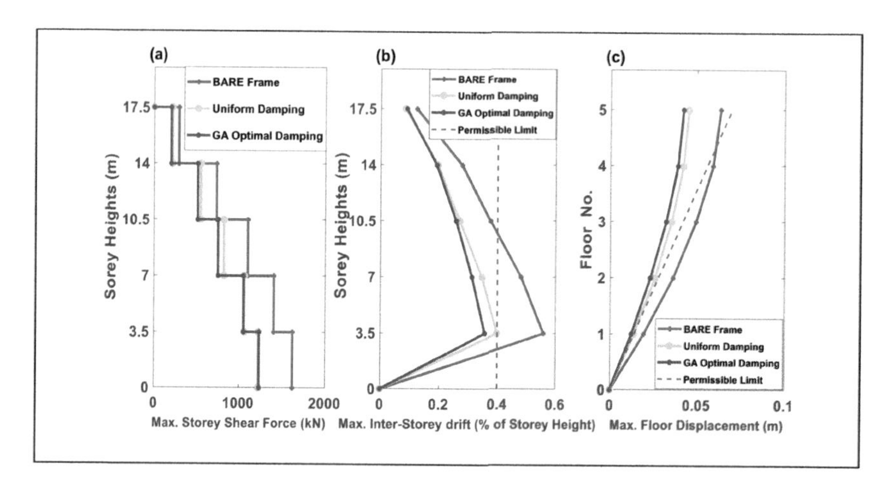

FIGURE 4.14 Comparison of GA optimal response for a 5-story building. (a) Story shear. (b) Maximum interstory drifts. (c) Maximum floor displacement.

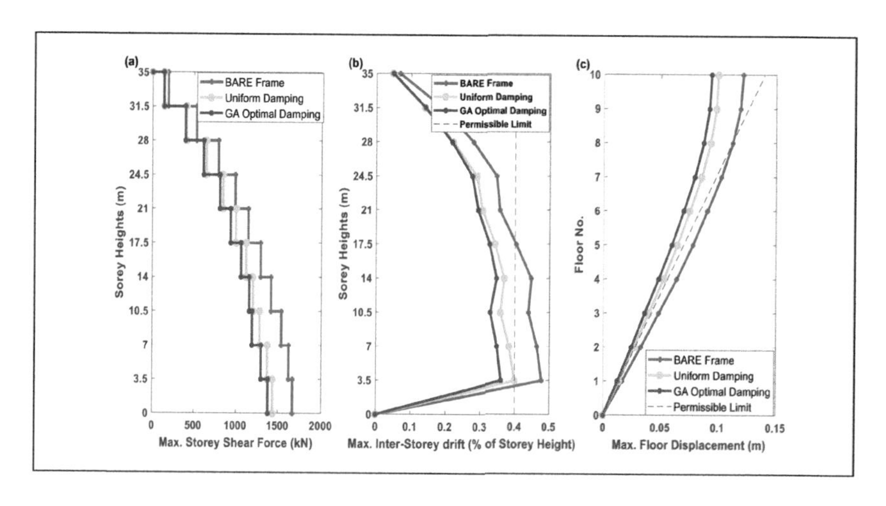

FIGURE 4.15 Comparison of GA optimal response for a 10-story building. (a) Story shear. (b) Maximum interstory drifts. (c) Maximum floor displacement.

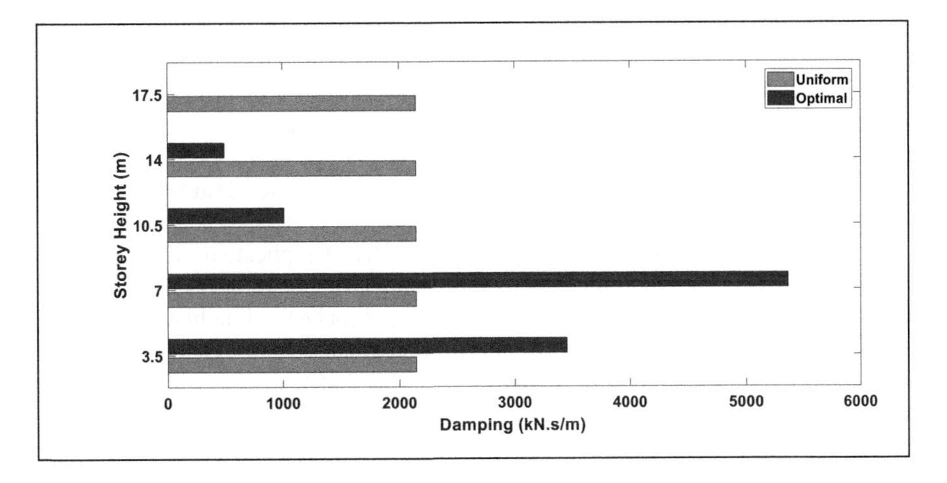

FIGURE 4.16 Comparison of GA optimal damping with uniform damping provision for a 5-story building.

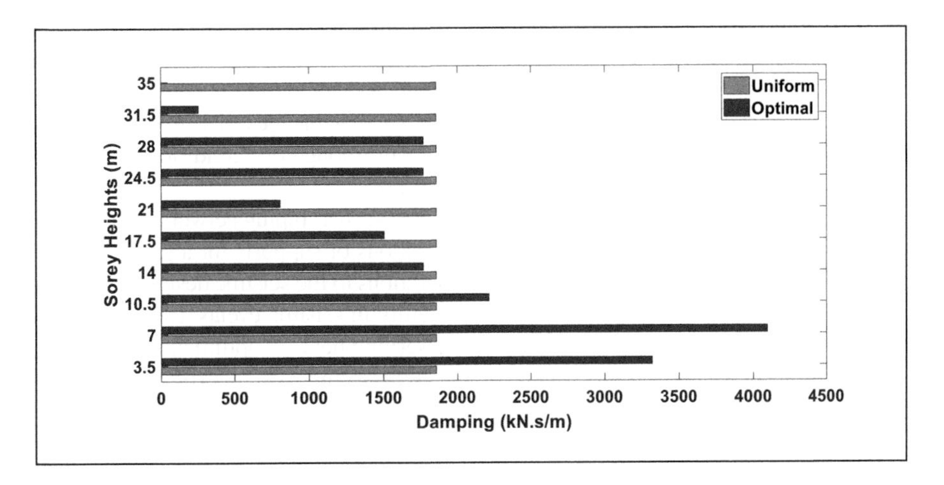

FIGURE 4.17 Comparison of GA optimal damping with the uniform damping provision for a 10-story building.

fitness of the population in the last reasonable number of cycles or maximum iteration limits of 500 cycles, whichever is reached first.

It has been found that the first criterion was met in both the buildings, and convergence occurred in 127 and 232 cycles for 5- and 10-story buildings, respectively, for the population size of 20. A large population size consumes more time for analyses, which is greater for buildings with increasing heights.

4.6 CHALLENGES AND FUTURE SCOPE

As the variety of problems for which GA has been successfully applied expands each year, the contents of this chapter in no way can be taken as to construe limitations in the application of GA. While appreciable work has been done in structural optimization using GA, it remains open to the imagination of researchers, designers, and system engineers as to what scope and extent the power of GA can be extended as an optimizer engine. If the efforts in structural optimization using GA can be classified in one of these categories (deterministic structural optimization [DSO] and reliability-based structural optimization [RBSO]), there remains many useful domains and areas in which GA can be employed. Moreover, a relatively new philosophy of performance-based structural design and optimization (PBSDO) is occupying a central place in structural design in the last two decades that can be assumed to borrow ideas from both DSO and RBSO. In the context of these three important categories of optimization, several challenging, fertile areas can be found that deserve the attention of the optimization community.

Pre-engineered buildings that have the typical character of replication of loading scenarios and topologies are ideal candidates for optimization using GA. The meager efforts taken toward consideration of 3D elastic time history analysis [137] and inelastic time history analysis [138] suggest there is more useful contribution to come. Similarly, it has been well known that a greater economy can be achieved when nonlinear analysis is utilized instead of linear elastic analysis as the design basis [139].

Large-scale structural systems optimization using GA in any context include DSO or RBSO, which require a meticulous consideration as the effort required to evaluate the fitness of individuals becomes significant. Methodologies that involve meta-modeling can form an efficient tool to alleviate such problems. The proposed methodology for RBSO using evolution strategies and Monte Carlo simulation [140] can be implemented similarly in GA. Design of the feedback-control systems for large-scale civil engineering structures is another prolific area of active research. The life cycle costs minimization is another very demanding area for optimization using GA The life cycle costs of large-scale structures such as bridges that involve maintenance and other costs can well be examined by GA for trade-offs among cost, condition, and safety over its life span. A few attempts in this area include [141–143].

Industrial facilities consist of relatively complex systems such as piping networks in nuclear plants and petroleum industries. Deciding the number and locations of supports for a piping network is a critical task. It is important to provide adequate considerations to the seismic design of the structural system for satisfactory performance of an industrial facility in seismic events. Research to this end has been shown to be very useful [144–146] and is a deserving field for optimization using GA.

Smart structures is an emerging concept of the decade and GA incorporating fuzzy logic [147–150] has been attracting researchers in this new field.

Setting up the right parameters for the GA (e.g., crossover probability, population size, mutation probability, crossover type, penalty parameters values such as multiplier and exponent) has been difficult in real-world engineering structures. Extensive research on these issues that can produce unified guidelines will help the future researcher in this direction. Similarly, formulation of guidelines on the problem-specific or customized chromosomal representations [151–153] to alleviate problems of variable string length, very long chromosomes, and other mechanics-specific problems can become very useful to future research. Parallel computing and visual simulation, for solving multi-objective optimization problems using GA, with more than two objective functions and several hundred decision variables is a future challenge.

GA blended with other metaheuristic algorithms such as artificial neural network, simulated annealing, or GA with hybrid strategies have been implemented effectively by several works, and the aim of the new research should be to use similar techniques in the context of auto decision or artificial intelligence. To this end, the readers will appreciate and find the works [154, 155] very useful.

Clearly, there is a need to design and develop a futuristic GA optimization mechanism that can handle the issues discussed above to adequately address the fundamental problems such as complexity, time duration and manufacturing conveniences in real-world optimization. The time has come now for artificial intelligence to take over as the next level of optimization in which all decision making will be done by bots similar to Alexa or Watson.

4.7 SUMMARY AND REMARKS

This chapter has provided a brief overview of the application of GA in structural optimization. In the opening section the structure of GA was explained, and how the principle of survival of the fittest is appropriately modeled through various operators in GA is discussed. The variation operators, their purpose, importance, and proper design to accommodate all design representations in binary and real coded GA was illustrated. Historical developments in structural optimizations using GA and its application in various areas such as steel structures, RCC, and pre-stressed structures, optimal passive and active control, and so forth were presented in the review. Constraint handling is a critical issue to attain near-best global optimal solutions, and different approaches followed successfully in this area have been discussed. It is important to mention that the approaches in the domain of multi-criteria optimization using GA deserves separate space under a wider and more dedicated scope. The three appendices included in this chapter provide a compact utility for working with GA. The working of GA has been demonstrated by several examples in the end to conclude that it offers a powerful metaheuristic search technique with demonstrated performance in almost all design optimizations. When combined with other soft computing algorithms, the promising features of GA possess the potential to become a member of the artificial intelligence system of the future to put structural optimization at the next level of sophistication.

APPENDIX 4.1 ILLUSTRATION OF BINARY IMPLEMENTATION OF GA: CYLINDRICAL CAN OPTIMIZATION PROBLEM

TABLE A1.1
Initialization of Population

x1	Xreal	Xint									x2	Xreal	Xint								
Candid 1	120	102	0	1	1	0	0	1	1	0	Candid 1	155	13	0	0	0	0	1	1	0	1
Candid 2	150	255	1	1	1	1	1	1	1	1	Candid 2	250	255	1	1	1	1	1	1	1	1
Candid 3	100	0	0	0	0	0	0	0	0	0	Candid 3	200	128	1	0	0	0	0	0	0	0
Candid 4	135	179	1	0	1	1	0	0	1	1	Candid 4	180	77	0	1	0	0	1	1	0	1
Candid 5	140	204	1	1	0	0	1	1	0	0	Candid 5	220	179	1	0	1	1	0	0	1	1
Candid 6	110	51	0	0	1	1	0	0	1	1	Candid 6	160	26	0	0	0	1	1	0	1	0

TABLE A1.2
Fitness Evaluation of Population and Selection

Candidate No.	Objective	%Fitness	Probability	Cumulative Probability	cp1	cp2	Random	Selection Frequency
1	148911.5	22.80757	0.228076	0.228076	0	0.228076	0.5334	1
2	259181.4	0.528933	0.005289	0.233365	0.228076	0.233365	0.5795	0
3	125663.7	27.5045	0.275045	0.50841	0.233365	0.50841	0.2896	2
4	190851.8	14.33407	0.143341	0.651751	0.50841	0.651751	0.3019	2
5	219911.5	8.462922	0.084629	0.73638	0.651751	0.73638	0.7747	0
6	131318.6	26.362	0.26362	1	0.73638	1	0.014	1

Note: First two random numbers fall in cumulative probability range cp1 and cp2 of Candidate No. 4. Hence it selection frequency is 2.

TABLE A1.3
Crossover and Mutation

		x1	x1									x2	x2								
Cross. Site	**Cycle = 1**	**Decimal**	**Integer**	**1**	**2**	**3**	**4**	**5**	**6**	**7**	**8**	**Decimal**	**Integer**	**9**	**10**	**11**	**12**	**13**	**14**	**15**	**16**
6	**Parert- 1**	120	102	0	1	1	0	0	1	1	0	155.098	13	0	0	0	0	1	1	0	1
	Parert- 2	100	0	0	0	0	0	0	0	0	0	200.196	128	1	0	0	0	0	0	0	0
9	**Parert- 3**	135.09	179	1	0	1	1	0	0	1	1	180.196	77	0	1	0	0	1	1	0	1
	Parert- 4	110	51	0	0	1	1	0	0	1	1	160.196	26	0	0	0	1	1	0	1	0
11	**Parert- 5**	100	0	0	0	0	0	0	0	0	0	200.196	128	1	0	0	0	0	0	0	0
	Parert- 6	135.09	179	1	0	1	1	0	0	1	1	180.196	77	0	1	0	0	1	1	0	1
Mutation																					
X	**Offspr- 1**	**119.60**	**100**	0	1	1	0	0	1	0	0	**200.196**	**128**	1	0	0	0	0	0	0	0
X	**Offspr- 2**	**100.39**	**2**	0	0	0	0	0	0	1	0	**155.098**	**13**	0	0	0	0	1	1	0	1
X	**Offspr- 3**	**135.09**	**179**	1	0	1	1	0	0	1	1	**160.196**	**26**	0	0	0	1	1	0	1	0
X	**Offspr- 4**	**110**	**51**	0	0	1	1	0	0	1	1	**180.196**	**77**	0	1	0	0	1	1	0	1
X	**Offspr- 5**	**100**	**0**	0	0	0	0	0	0	0	0	**205.294**	**141**	1	0	0	0	1	1	0	1
Yes	**Offspr- 6**	**135.09**	**179**	1	0	1	1	0	0	1	1	**175.098**	**64**	0	1	0	0	0	0	0	0
Mutated	**Site: 11**	**135.09**	**179**	1	0	1	1	0	0	1	1	**187.647**	**96**	0	1	1	0	0	0	0	0

Note: Shaded numbers indicate crossover location (site) in the bit string representing gene.

TABLE A1.4
Offspring Created in STEP-1 and Their Objective Function Values

x1	Xreal	Xinteger	x2	Xreal	Xinteger	Fobjective
Candid 1	119.6078	100	candid 1	200.1961	128	165113
Candid 2	100.3922	2	candid 2	155.098	13	**112242.2**
Candid 3	135.098	179	candid 3	160.1961	26	182668.3
Candid 4	110	51	candid 4	180.1961	77	138297.8
Candid 5	100	0	candid 5	205.2941	141	127326.9
Candid 6	135.098	179	candid 6	187.6471	96	194319.2

Note: The improved objective function shown highlighted.

APPENDIX 4.2 SELECTION METHODS AND THEIR CHARACTERISTICS

TABLE A2.1

	Type of Selection Method	Working Mechanics	Characteristics
1	***Proportional Selection: Canonical Selection*** Selection probability of any individual is defined as ratio of its fitness to average fitness of population. The individuals with high selection probability are chosen and put in the parent's pool by arranging the selection probability in descending order until the parent's pool is completed.	f_i = Fitness of *i*th individual Selection probability (SPi): $$SPi = \frac{f_i}{\left(\sum_i^{Np} f\right)/N_p}$$ *SPmax*............... *SPmin*	• This method has the same characteristics as that of roulette wheel selection.
2	***Proportional Selection: Roulette Wheel Selection*** Selection probability of an individual is defined as fitness of each individual normalized with respect to the sum of fitness of all individuals in the population. The cumulative probability is then calculated for each individual and the 360 degrees of the roulette wheel is divided in sectors with areas proportional to the cumulative probability and allocated to individuals. So, the higher the fitness is, the more area an individual occupies on the wheel claiming more range of cumulative probability. Pointer of a roulette wheel is represented by a random number between 0 and 1 because cumulative probability of all individuals also sums up to 1. The random number is generated as many times as there are individuals in the population. Finding in which range of cumulative probability the random number generated each time falls decides the winner of that attempt and is selected as a parent. The process is continued until the total number of parents selected is equal to the size of the population.	f_i = Fitness of *i*th individual Selection probability (SPi): $$SPi = \frac{f_i}{\sum_i^{Np} f_i}$$ Cumulative Probability (CPi) $$CP_i = CP_{i-1} + SP_i$$ 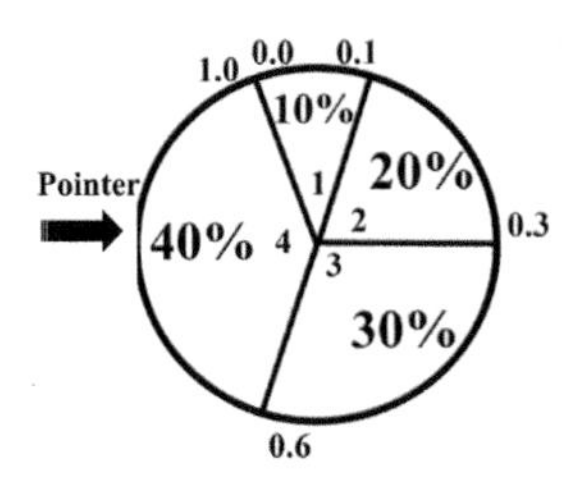 The roulette wheel is spun Np times to make the parent's pool full.	• The method favors high selection pressure due to the biased nature of the process that causes search stagnation, i.e., likely to end up with premature convergence. • It inherently follows uniform probability distribution of selection; therefore, variance of selection is high, which means a better individual may not get selected sometimes, or worst individuals may be selected. • So performance of this method is better when fitness is more or less uniformly distributed.

TABLE A2.1 (*Continued*)

	Type of Selection Method	Working Mechanics	Characteristics
	Illustration: Assume a population of four individuals with an individual fitness value of 1, 2, 3, and 4, respectively. The sum of fitness is 10 and the fitness probability of each individual can be calculated as 0.1, 0.2, 0.3, and 0.4 with a cumulative probability range for each individual of 0.0–0.1, 0.1–0.3, 0.3–0.6, and 0.6–1.0. A roulette wheel representing this entire range is shown below. Let us consider that four random numbers are generated, for the population size being four, which are 0.67, 0.17, 0.49, and 0.97. On comparing these with cumulative probability ranges we find that individuals selected as prospective parents are 4, 2, 3, and 4. The fourth individual gets selected twice due to its greater share on the wheel, i.e., higher individual fitness value.		
3	***Proportional Selection: Stochastic Universal Sampling*** In this process the offspring population is generated at once by drawing only one random number. All individuals of population are assumed to be represented on a single fitness strip such that each one occupies a segment of the strip proportional to its fitness. The fitness strip is then provided with Np number of equidistant pointers, on the scale of fitness, to select the individuals, as against one pointer of roulette wheel. The random number generated is used to set the first pointer at random offset from any one end of the fitness strip. All those individuals are picked up that have the pointer falling on their fitness segments. The number of copies of any individual are always equal to the number of pointers falling in their fitness segment. The selection probability of any individual has been thus defined by such universal stochastic sampling. Illustration: Assume a population of 10 individuals with individual fitness values of 0.8, 0.9, 1.6, 1.4, 0.2, 1.9, 0.6, 1.0, 0.4, and 1.2, respectively. All 10 individuals are shown represented on a fitness strip proportional to their fitness values as below. To select 10 parents, a train of 10 equidistant pointers has been constructed and the first pointer is placed at a random offset provided by a random number. One can see the individuals with fitness 0.2 and 0.8 are not selected, whereas the individuals with fitness 1.6 and 1.9 are copied twice each.	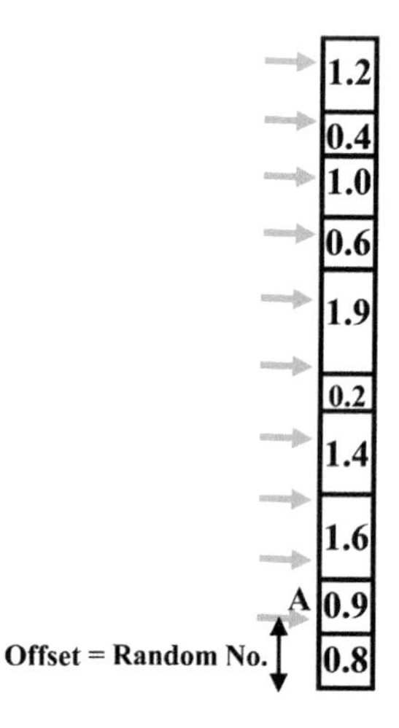 **A-The starting point of pointers-train at random offset from end of fitness strip.**	• This method is more efficient than the roulette wheel as it offers low selection variance of the population sampling. • The selection pressure and hence the rate of convergence is also moderate.

(*Continued*)

TABLE A2.1 (*Continued*)

	Type of Selection Method	Working Mechanics	Characteristics
4	***Rank-Based Selection:*** Unlike the fitness proportional-based selection followed by the usual method, rank-based selection follows an ordinal-based selection on the roulette wheel. It does not use fitness values directly for selection; instead, the individuals are ranked based on their fitness and ranks are used for selection. This eliminates, to a certain extent, the biased favor of the roulette wheel in the usual procedure and all individuals are given fair chance of competition. This allows more population diversity and reduced selection pressure. Worst fitness is given the lowest or first rank and best fitness gets the highest rank. Illustration: Consider a population of four individuals with individual fitness values of 0.7, 0.2, 0.06, and 0.04, respectively. On following the usual roulette wheel procedure the share of these individuals as per their fitness on the wheel are 70%, 20%, 6%, and 4%, respectively. The rank of these individuals according to rank-based procedure are 4, 3, 2, and 1, respectively. According to rank-based procedure, the same individuals will now occupy a 40%, 30%, 20%, and 10% share on the wheel. The difference in selection probability can be noted clearly in the two approaches. The rank-based procedure thus evens out the selection pressure and allows every individual to compete fairly as opposed to the usual fitness-based approach, which tends to favor highly fit individuals ignoring fair chance to every individual.	*Get Rank* Rank (ri) = 1 ....for worst fitness Rank (ri) = Np …for best fitness *Construct Roulette Wheel* % Share of individual on the roulette wheel defines Selection Probability (SPi): $SPi = \frac{r_i}{\sum_i^{Np} r_i}$ The roulette wheel is spun Np times to make the parent's pool full.	• Ranking automatically introduces a uniform scaling across the population. • Provides low selection pressure and high population diversity with low rate of convergence. It has higher exploration capabilities than roulette wheel and tournament selection.
5	***Tournament Selection:*** Tournament selection has been a straightforward application from sports to rank the population in an interesting and efficient way. This method sets the tournament between individuals of population and the fittest individual is declared the winner, which is then admitted to the parent's pool. In a single tournament a lot of z individuals, called tournament size, are considered and the best individual is selected as a prospective parent. The number of tournaments played thus equals the population size Np. It is clear, that large values of z tend to increase selective pressure and hence the typical value adopted is z = 2.	*Tournament Size (z):* A competition between z individuals to select the fittest individual. Steps: 1. Take a lot size of z individuals at random. 2. Choose the individual with the highest fitness value as the winner, i.e., prospective parent. 3. Repeat first two steps until the parent's pool contains Np individuals to stop.	• Provides selection pressure based on tournament size and moderate population diversity with the same rate of convergence. Works efficiently when fitness is highly scattered. • Computationally less intensive and therefore faster than both roulette wheel and rank-based selection scheme.

APPENDIX 4.3 REAL CODED VARIATION OPERATORS: CROSSOVER AND MUTATION

Crossover Operators

1. BLEND Crossover (BLX) ([C1]) {Interval schemata}
   X_child1 = (1- γ). X_parent1 + γ. X_parent2
   X_child2 = (1- γ). X_parent2 + γ. X_parent1

   OR

   X_child1 = X_parent1 + γ (X_parent2 - X_parent1)
   X_child2 = X_parent2 + γ (X_parent1 - X_parent2)

$$\gamma = (1+2\alpha).U_{0-1} - \alpha$$

with α = 0.5 as recommended value and U is Uniformly distributed random no. between 0-1.

   There exist other similar operators ([C2], [C3])
2. SIMULATED BINARY Crossover (SBX) ([C4])
   X_child1= 0.5{ (1+α).X_parent1 + (1-α).X_parent2}
   X_child2= 0.5{ (1-α).X_parent1 + (1+α).X_parent2}

$$\alpha = (2r)^{\frac{1}{\eta+1}} \quad \ldots\ldots\ldots r \le 0.5$$

$$\alpha = \left(\frac{1}{2(1-r}\right)^{\frac{1}{\eta+1}} \quad \ldots.r > 0.5$$

r is uniformly distributed random no. between 0-1. The η is distribution index. Large value of η gives higher probability to create "near-parent" offspring, while its small value implies distant-parent offspring.

Mutation Operators

1. Polynomial Mutation, ([C4]).
   X_child_new = X_child + δ.(X_upper - X_lower)

$$\alpha = (2r)^{\frac{1}{\lambda+1}} - 1 \quad \ldots..r < 0.5$$

$$\alpha = 1 - [2(1-r)]^{\frac{1}{\lambda+1}} \quad \ldots..r \ge 0.5$$

r is uniformly distributed random no. between 0-1. The λ is distribution index. Large value of λ gives higher probability to create "near-old" offspring, while its small value implies near-old offspring.

2. Normal distributed Mutation
   The zero mean Gaussian-Probability distribution is considered for creation of new offspring around the offspring to be mutated.
   X_child_new = X_child + N(0,σ)
   σ is standard deviation either pre-set for the problem or adaptively changed with generation advancement.
3. Non-uniform mutation
   In the initial phases of search the mutation follows uniform distribution while in later stage towards convergence it follows a Dirac's-delta function ([C5])
   X_child_new = X_child + τ.(X_upper - X_lower).$(1-[r]^{(1-t/tmax)^b})$
   The τ is Boolean value i.e. –1 or +1 with probability of 0.5.
4. Random mutation
   The solutions can be created randomly on entire search space ([C6])
   X_child_new = X_child + r.(X_upper - X_lower)
   The solutions can be created in the vicinity of parent solutions with uniform probability distribution.
   X_child_new = X_child + (r-0.5).Δ
   The Δ is user defined maximum perturbation value allowed in the decision variable.

Appendix 4.3 References

A3.1. Eshelman, L. J. and J. D. Schaffer. 1993. Real-coded genetic algorithms and internal schemata. Foundations of genetic algorithm. 2: 187–202.
A3.2. Michalewicz, Z. and C. J. Janikow. 1991. Handling constraints in genetic algorithms. In: Proceedings of the Fourth Conference on Genetic Algorithms, pp. 151–157.
A3.3. Voigt, H. M., H. Muhlenbein, and D. CvetKovic. 1995. Fuzzy recombination for the breeder genetic algorithm. In: Proceedings of the Fourth Conference on Genetic Algorithms, pp. 104–111.
A3.4. Deb, K. and S. Agrawal. 1995. Simulated binary crossover for continuous search space. Complex systems. 9(2): 115–148.
A3.5. Michalewicz, Z. 1992. Genetic Algorithms + Data Structures = Evolution Programs. Springer-Verlag, Berlin.
A3.6. Deb, K. and M. Goyal. 1996. A combined genetic adaptive search (GeneAS) for engineering design. Computer science and informatics. 26(4): 30–35.

REFERENCES

1. Holland, J. H. 1975. Adaption in Natural and Artificial Systems. MIT Press, Ann Arbor, MI.
2. Goldberg, D. E. 1989. Genetic Algorithms M Search, Optimization and Machine Learning. Addison-Wesley, Reading, MA.
3. Eshelman, L. J., and J. D. Schaffer. 1993. Real coded genetic algorithm and interval schemata. Foundations of genetic algorithms. 2: 187–202.
4. Deb, K. and S. Agrawal. 1995. Simulated binary crossover for continuous search space. Complex systems. 9(2): 115–148.
5. Michalewicz, Z. 1992. Genetic Algorithms + Data Structures = Evolution Programs. Springer-Verlag, Berlin.
6. Deb, K. and M. Goyal. 1996. A combined genetic adaptive search (GeneAS) for engineering design. Computer science and informatics. 26(4): 30–35.
7. Whitely, D. 1994. Genetic algorithm: a tutorial. Z. Michalewicz (Ed.). Statistics and computing. Special issue on evolutionary computation. 4: 65–85.
8. DeJong, K. A. 1985. Genetic algorithms: a 10 year perspective. In: Proceedings of the First International Conference on Genetic Algorithms, pp. 169–177. Lawrence Erlbaum Associates. Hillsdale, NJ.

9. J.E. Baker. 1987. Reducing bias and inefficiency in the selection algorithm. Proceedings of the Second International Conference on Genetic Algorithms, pp. 14–21. Lawrence Erlbaum Associates. Hillsdale, NJ.
10. Back, T., and F. Hoffmeister. 1991. Extended selection mechanisms in genetic algorithms. In: Proceedings of the Fourth International Conference, pp. 92–99. Morgan Kaufmann Publishers. Los Altos, CA.
11. Goldberg, D. E., K. Deb, and B. Korb. 1991. Do not worry, be messy. In: Proceedings of the Fourth International Conference on Genetic Algorithms, pp. 24–30. Morgan Kaufmann Publishers, Los Altos, CA.
12. Michalewicz, Z. 1995. A survey of constraint handling techniques in evolutionary computation methods. In: Proceedings of the 4th Annual Conference on Evolutionary Programming, pp. 135–155. MIT Press, Cambridge, MA.
13. Richardson, J. T., M. R. Palmer, G. Liepins et al. 1989. Some guidelines for genetic algorithms with penalty functions. In: Schaffer, J. D., editor. Proceedings of the Third International Conference on Genetic Algorithms, George Mason University, pp. 191–197. Morgan Kaufmann Publishers, Los Altos, CA.
14. Joines, J. A., and C. R. Houck. 1994. On the use of non-stationary penalty functions to solve nonlinear constrained optimization problems with GAs. In: Proceedings of the IEEE ICEC, pp. 579–584.
15. Michalewicz, Z., D. Dasgupta, R. G., Le Riche et al. 1996. Evolutionary algorithms for constrained engineering problems. Computers and industrial engineering journal. 4: 851–870.
16. Adeli, H., and N. T. Cheng 1994. Augmented Lagrangian genetic algorithm for structural optimization. Journal of aerospace engineering. 7(1): 104–118.
17. Adeli, H., and N. T. Cheng 1994. Concurrent genetic algorithms for optimization of large structures. Journal of aerospace engineering. 7(3): 276–296.
18. Hasançebi O., and F. Erbatur. 2000. Constraint handling in genetic algorithm integrated structural optimization. Acta mechanica. 139: 15–31.
19. Coello Coello, C. A. 2000. Use of a self-adaptive penalty approach for engineering optimization problems. Computers in industry. 41: 113–127.
20. Barbosa, H. J. C., and A. C. C. Lemonge. 2003. A new adaptive penalty scheme for genetic algorithms. Information sciences. 156: 215–251.
21. Lemonge, A. C. C., and H. J. C. Barbosa. 2004 An adaptive penalty scheme for genetic algorithms in structural optimization. International journal for numerical methods in engineering. 59: 703–36. http://dx.doi.org/10.1002/nme.899
22. Deb, K. 2000. An efficient constraint handling method for genetic algorithms. Computer methods in applied mechanics and engineering. 186(2–4): 311–338.
23. Runarsson, T. P., and X. Yao. 2000. Stochastic ranking for constrained evolutionary optimization. IEEE transactions of evolutionary computation. 4: 284–294.
24. Runarsson, T. P., and X. Yao. 2002. Continuous selection and self-adaptive evolution strategies. In: Proceedings of the 2002 Congress on Evolutionary Computation CEC'02. 1: 279–284.
25. Ho, P. Y., and K. Shimizu. 2007. Evolutionary constrained optimization using an addition of ranking method and a percentage-based tolerance value adjustment scheme. Information sciences. 177:2985–3004.
26. Rodrigues, M. C., B. S. L. P. de Lima, and S. Guimarães. 2016. Balanced ranking method for constrained optimization problems using evolutionary algorithms. Information Sciences. 327:71–90. http://dx.doi.org/10.1016/j.ins.2015.08.012.
27. de Paula Garcia, R., B. S. L. P. de Lima, and A. C. de Castro Lemonge et al. 2017. A rank-based constraint handling technique for engineering design optimization problems solved by genetic algorithms. Computers & structures. 187: 77–87.
28. Coello Coello, C. A. 2011. Constraint-handling in nature-inspired numerical optimization: past, present and future. Swarm and evolutionary computation. 1: 173–194.
29. Goldberg, D. E., and M. P. Samtani. 1986. Engineering optimization via genetic algorithm. In: Ninth Conference on Electronic Computation, pp. 471–82. New York. ASCE.
30. Jenkins, W. M. 1991. Towards structural optimization via the genetic algorithm. Computers & structures. 40(5): 1321–1327.
31. Hajela, P., and C. J. Shih. 1990. Multiobjective optimum design in mixed integer and discrete design variable problems. AIAA. 28(4): 670–675.
32. Rajeev, S., and C. S. Khrisnamoorthy. 1992. Discrete optimization of structures using genetic algorithms. Journal of structural engineering. 118(5): 1233–1250.

33. Lin, C. Y., and P. Hajela. 1993. EVOLVE: a genetic search based optimization code via multiple strategies. In: Hernandez, S., and C. A. Brebbia, editors. Computer Aided Optimum Design of Structures III: Optimization of Structural Systems and Applications, pp. 639–654. Amsterdam. Elsevier.
34. Schoenauer, M., and S. Xanthakis. 1993. Constrained GA optimization. In: Forrest, S., editor. Proceedings of the Fifth International Conference on Genetic Algorithms, pp. 573–580. Morgan Kaufmann Publishers. San Mateo, CA.
35. Hajela, P., and C. Y. Lin. 1992. Genetic search strategies in multi-criterion optimal design. Structural optimization. 4: 99–107.
36. Adeli, H., and N. T. Cheng. 1993. Integrated genetic algorithm for optimization of space structures. Journal of aerospace engineering. 6(4): 315–328.
37. Adeli, H., and N. T. Cheng. 1994. Augmented Lagrangian genetic algorithm for structural optimization. Journal of aerospace engineering. 7(1): 104–118.
38. Adeli, H., and S. Kumar. 1995. Distributed genetic algorithm for structural optimization. Journal of aerospace engineering. 8(3): 156–163.
39. Adeli, H., and S. Kumar. 1995. Concurrent structural optimization on massively parallel supercomputer. Journal of structural engineering. 121(11): 1588–1597.
40. Rajan, S. D. 1995. Sizing, shape, and topology design optimization of trusses using genetic algorithm. Journal of structural engineering. 121(10): 1480–1487.
41. Yeh, I. C. 1999. Hybrid genetic algorithms for optimization of truss structures. Microcomputers in civil engineering. 14(3): 199–206.
42. Tong, W. H., and G. R. Liu. 2001. An optimization procedure for truss structures with discrete design variables and dynamic constraints. Computers & structures. 79: 155–162.
43. Coello Coello, C. A., and A. D. Christiansen. 2000. Multiobjective optimization of trusses using genetic algorithms. Computers & structures. 75(6): 647–660. https://doi.org/10.1016/S0045-7949(99)00110-8
44. Koumousis, V. K., and P. G. Georgiou. 1994. Genetic algorithms in discrete optimization of steel truss roofs. Journal of computing in civil engineering. 8(3): 309–325. https://doi.org/10.1061/(ASCE)0887-3801(1994)8:3(309)
45. Erbatur, F., O. Hasançebi, İ. Tütüncü, and H. Kılıç. 2000. Optimal design of planar and space structures with genetic algorithms. Computers & structures. 75(2): 209–224.
46. Lagaros, N. D., M. Papadrakakis, and G. Kokossalakis. 2002. Structural optimization using evolutionary algorithms. Computers & structures. 80: 571–589.
47. Kaveh, A., and H. Rahami. 2006. Nonlinear analysis and optimal design of structures via force method and genetic algorithm. Computers & structures. 84(12): 770–778
48. Banichuk, N. V., M. Serra, and A. Sinitsyn. 2006. Shape optimization of quasi-brittle axisymmetric shells by genetic algorithm. Computers & structures. 84(29–30): 1925–1933.
49. Kelesoglu, O. 2007. Fuzzy multi-objective optimization of truss-structures using genetic algorithm. Advances in engineering software. 38(10): 717–721. https://doi.org/10.1016/j.advengsoft.2007.03.003
50. Šešok, D., and R. Belevičius. 2008. Global optimization of trusses with a modified genetic algorithm. Journal of civil engineering & management. 14(3): 147–154. https://doi.org/10.3846/1392-3730.2008.14.10
51. Koumousis, V. K., and S. J. Arsenis. 1994. Genetic algorithms in a multi-criterion optimal detailing of reinforced concrete members. Advances in Structural Optimization. pp. 233–240. Civil-Comp Ltd. Edinburgh Scotland.
52. Sarma, K. C., and H. Adeli. 1998. Cost optimization of concrete structures. Journal of structural engineering. 124(5):570–578.
53. Govindaraj, V. and J. V. Ramasamy. 2005. Optimum detailed design of reinforced concrete continuous beams using genetic algorithms. Computers & structures. 84(1–2): 34–48.
54. Camp, C. V., S. Pezeshk, and H. Hansson. 2003. Flexural design of reinforced concrete frames using a genetic algorithm. Journal of structural engineering. 129(1): 105–115. https://doi.org/10.1061/(ASCE)0733-9445(2003)129:1(105)
55. Lee, C. and J. Ahn. 2003. Flexural design of reinforced concrete frames by genetic algorithm. Journal of structural engineering. 129(6): 762–774. https://doi.org/10.1061/(ASCE)0733-9445(2003)129:6(762)

56. Govindaraj, V., and J. V. Ramasamy. 2007. Optimum detailed design of reinforced concrete frames using genetic algorithms. Engineering optimization. 39(4): 471–494. https://doi.org/10.1080/03052150601180767
57. Sahab, M. G., A. F. Ashour, and V. V. Toropov. 2005. A hybrid genetic algorithm for reinforced concrete flat slab buildings. Computers & structures. 83(8–9): 551–559.
58. Sahab, M. G., A. F. Ashour, and V. V. Toropov. 2005. Cost optimisation of reinforced concrete flat slab buildings. Engineering structures. 27: 313–322.
59. El Semelawy, M., A. O. Nassef, and A. A. El Damatty. 2012. Design of prestressed concrete flat slab using modern heuristic optimization techniques. Expert systems with applications. 29(5): 5758–5766.
60. Sgambi, L., K. Gkoumas, and F. Bontempi. 2014. Genetic algorithm optimization of precast hollow core slabs. Computers and concrete. 13: 389–409.
61. Aydın, Z. and Y. Ayvaz. 2013. Overall cost optimization of pre-stressed concrete bridge using genetic algorithm. KSCE Journal of civil engineering. 17(4):769–76.
62. Camp, C. V., S. Pezeshk, and H. Hansson. 2003 Flexural design of reinforced concrete frames using a genetic algorithm. Journal of structural engineering. 129(1): 105–15.
63. Jarmai, K., J. A. Snyman, J. Farkas, and G. Gondos. 2003. Optimal design of a welded I-section frame using four conceptually different optimization algorithms. Structural and multidisciplinary optimization. 25: 54–61.
64. Woon, S. Y., O. M. Querin, and G. P. Steven. 2001. Structural application of a shape optimization method based on a genetic algorithm. Structural and multidisciplinary optimization. 22(1): 57–64.
65. Alberdi, R., P. Murren, and K. Khandelwal. 2015. Connection topology optimization of steel moment frames using metaheuristic algorithms. Engineering structures. 100: 276–292.
66. Prendes-Gero, M., M. Álvarez-Fernández, and F. López-Gayarre. 2016. Cost optimization of structures using a genetic algorithm with eugenic evolutionary theory. Structural and multidisciplinary optimization. 54: 199–213. https://doi.org/10.1007/s00158-015-1249-5
67. Chau, K. W., and F. Albermani. 2002. Genetic algorithms for design of liquid retaining structure. In: International Conference on Industrial, Engineering and Other Applications of Applied Intelligent Systems, pp. 119–128. Springer, Berlin, Heidelberg.
68. Fu, K. C., Y. Zhai, and S. Zhou. 2005. Optimum design of welded steel plate girder bridges using a genetic algorithm with elitism. Journal of bridge engineering. 10(3): 291–301.
69. Kociecki, M., and H. Adeli. 2014. Two-phase genetic algorithm for topology optimization of free-form steel space-frame roof structures with complex curvatures. Engineering applications of artificial intelligence. 32: 218–227.
70. Kociecki, M., and H. Adeli. 2015. Shape optimization of free-form steel space-frame roof structures with complex geometries using evolutionary computing. Engineering applications of artificial intelligence. 38: 168–82.
71. Pezeshk, S., and C. V. Camp. 2002. State of the art on the use of genetic algorithms in design of steel structures. In: Burns, S, editor. Recent Advances in Optimal Structural Design, pp. 55–80. American Society of Civil Engineers, Reston, VA.
72. Sarma, K. C., and H. Adeli. 2000. Cost optimization of steel structures. Engineering optimization. 32(6): 777–802.
73. Kameshki, E. S., and M. P. Saka. 2003. Genetic algorithm based optimum design of nonlinear planar steel frames with various semi-rigid connections. Journal of constructional steel research. 59(1): 109–134.
74. Oskouei, A. V., S. S. Fard, and O. Aksogan. 2012. Using genetic algorithm for the optimization of seismic behavior of steel planar frames with semi-rigid connections. Structural and multidisciplinary optimization. 45: 287–302. https://doi.org/10.1007/s00158-011-0697-9
75. Aldwaik, M., and H. Adeli. 2014. Advances in optimization of highrise building structures. Structural multidisciplinary optimization. 50(6): 899–919. https://doi.org/10.1007/s00158-014-1148-1
76. Dehghani, S., A. R. Vosoughi, and M. R. Banan. 2019. The effects of rehabilitation objectives on near optimal trade-off relation between minimum weight and maximum drift of 2D steel X-braced frames considering soil-structure interaction using a cluster-based NSGA II. Structural multidisciplinary optimization. 59: 1703–1722. https://doi.org/10.1007/s00158-018-2153-6
77. Tu, X., Z. He, and G. Huang. 2020. Performance-based multi-objective collaborative optimization of steel frames with fuse-oriented buckling-restrained braces. Structural multidisciplinary optimization. 61: 365–379. https://doi.org/10.1007/s00158-019-02366-9

78. Baradaran, M., and M. Madhkhan. 2019. Determination of optimal configuration for mega bracing systems in steel frames using genetic algorithm. KSCE journal of civil engineering. 23: 3616–3627. https://doi.org/10.1007/s12205-019-2369-z
79. Talaslioglu, T. 2019. Optimal design of steel skeletal structures using the enhanced genetic algorithm methodology. Frontiers of structural and civil engineering. 13: 863–889. https://doi.org/10.1007/s11709-019-0523-9
80. Talaslioglu, T. 2019. Optimal dome design considering member-related design constraints. Frontiers of structural and civil engineering. 13: 1150–1170. https://doi.org/10.1007/s11709-019-0543-5
81. Camacho, V. T., N. Horta, M. Lopes et al. 2020. Optimizing earthquake design of reinforced concrete bridge infrastructures based on evolutionary computation techniques. Structural multidisciplinary optimization. 61: 1087–1105. https://doi.org/10.1007/s00158-019-02407-3
82. Singh, M. P., and L. M. Moreschi. 2001. Optimal seismic response control with dampers. Earthquake and structural dynamics. 30(4): 553–572. https://doi.org/10.1002/eqe.23
83. Singh, M. P., and L. M. Moreschi. 2002. Optimal placement of dampers for passive response control. Earthquake and structural dynamics. 30(4): 955–976. https://doi.org/10.1002/eqe.132
84. Moreschi, L. M., and M. P. Singh. 2003. Design of yielding metallic and friction dampers for optimal seismic performance. Earthquake and structural dynamics. 32(4): 1291–311. https://doi.org/10.1002/eqe.275
85. Lagaros, N. D., M. Papadrakakis, and G. Kokossalakis. 2002. Structural optimization using evolutionary algorithms. Computers & structures. 80: 571–589.
86. Liu, D. K., Y. L. Yang, and Q. S. Li. 2003. Optimum positioning of actuators in tall buildings using genetic algorithm. Computers & structures. 81: 2823–2827.
87. Wongprasert, N., and M. D. Symans. 2004. Application of a genetic algorithm for optimal damper distribution within the nonlinear seismic benchmark building. Journal of engineering mechanics. 30(4): 401–406.
88. Yoshida, O. and S. J. Dyke. Response control of full-scale irregular buildings using magnetorheological dampers. Journal of structural engineering. 131(5): 734–742.
89. Hejazi, F., I. Toloue, M. S. Jaafar, and J. Noorzaei. 2013. Optimization of earthquake energy dissipation system by genetic algorithm. Computer-aided civil and infrastructure engineering. 28(10): 796–810.
90. Lavan, O., G. F. Dargush, and A. M. Reinhorn. 2008. Multi-objective evolutionary optimization of passive Energy dissipation systems under seismic loading. In: The 14th World Conference on Earthquake Engineering, pp. 12–17, Beijing, China.
91. Lavan, O., and G. F. Dargush. 2009. Multiobjective evolutionary seismic design with passive energy dissipation systems. Journal of earthquake engineering. 13: 758–790.
92. Fallah, N., and S. Honarparast. 2013. NSGA-II based multi-objective optimization in design of Pall friction dampers. Journal of constructional steel research. 89: 75–85.
93. Deb, K., A. Pratap, S. Agarwal, and T. A. M. T. Meyarivan. 2002. A fast and elitist multiobjective genetic algorithm: NSGA-II. IEEE transactions on evolutionary computation. 6(2): 182–197.
94. Pourzeynali, S., S. Salimi, and K. H. Eimani. 2013. Robust multi-objective optimization design of TMD control device to reduce tall building responses against earthquake excitations using genetic algorithms. Scientia iranica. 20(2): 207–221.
95. Kim, Y. J., and J. Ghaboussi. 2001. Direct use of design criteria in genetic algorithm-based controller optimization. Earthquake engineering structural and dynamics. 30(9): 1261–1278.
96. Poh'Sie, G. H., C. Chisari, G. Rinaldin, C. Amadio, and M. Fragiacomo. 2016. Optimal design of tuned mass dampers for a multi-storey cross laminated timber building against seismic loads. Earthquake engineering structural and dynamics. 45(12): 1977–95.
97. Greco, R., G. C. Marano, and A. Fiore. 2016. Performance–cost optimization of tuned mass damper under low-moderate seismic actions. The structural design tall special buildings. 25(18): 1103–1122.
98. Desu, N. B., S. K. Deb, and A. Dutta. 2006. Coupled tuned mass dampers for control of coupled vibrations in asymmetric buildings. Structural control health monitoring. 13(5): 897–916. https://doi.org/10.1002/stc.64
99. Kim, H., and J. Kang. 2018. MOGA-based structural design method for diagrid structural control system subjected to wind and earthquake loads. International journal of steel structures. 18: 1598–1606. https://doi.org/10.1007/s13296-018-0055-5

100. Liu, X., D. W. Begg, and R. J. Fishwick. 1998. Genetic approach to optimal topology/controller design of adaptive structures. Numerical methods in engineering. 41(5): 815–830.
101. Liu, D. K., Y. L. Yang, and Q. S. Li. 2003. Optimum positioning of actuators in tall buildings using genetic algorithm. Computers & structures. 81: 2823–2827.
102. Kim, Y. J., and J. Ghaboussi. 2001. Direct use of design criteria in genetic algorithm-based controller optimization. Earthquake engineering and structural dynamics. 30: 1261–1278.
103. A. Alimoradi. 2001. Performance study of a GA-based active/hybrid control system under near source strong ground motion: a structural engineering odyssey. In: Proceedings of the 2001 Structures Congress & Exposition, Chang, P. C., editor. ASCE, Washington, D.C.
104. Arfiadi, Y., and M. N. S. Hadi. 2001. Optimal (direct) static output feedback controller using real-coded genetic algorithms. Computers & structures. 79: 1625–1634.
105. Arfiadi, Y., and M. N. S. Hadi. 2000. Passive and active control of three-dimensional buildings. Earthquake engineering and structural dynamics. 29: 377–396.
106. Wang A. P., and C. D. Lee. 2002. Fuzzy sliding model control for a building based on genetic algorithms. Earthquake engineering and structural dynamics. 31: 881–895.
107. Ahlawat A. S., and J. V. Ramaswamy. 2002. Multi-objective optimal design of FLC driven hybrid mass damper for seismically excited structures. Earthquake engineering and structural dynamics. 31: 1459–1479.
108. Ahlawat, A. S., and A. Ramaswamy. 2002. Multiobjective optimal FLC driven hybrid mass damper system for torsionally coupled, seismically excited structures. Earthquake engineering and structural dynamics. 31: 2121–2139.
109. Kim, H. S., and P. N. Roschke. 2006. Design of fuzzy logic controller for smart base isolation system using genetic algorithm. Engineering structures. 28: 84–96.
110. Kim, H. S., and P. N. Roschke. 2006. Fuzzy control of base-isolation system using mult-objective genetic algorithm. Computer- aided civil and infrastructure engineering. 21: 436–449.
111. Li, Q. S., P. Liu, N. Zhang, C. M. Tam, and L. F. Yang. 2001. Multi-level design model and genetic algorithm for structural control system optimization. Earthquake engineering and structural dynamics. 30: 927–942.
112. Ahlawat, A. S., and J. V. Ramaswamy. 2004. Multiobjective optimal fuzzy logic controller driven active and hybrid control systems for seismically excited structures. Journal of engineering mechanics. 130(4): 416–423.
113. Yang, Y., Z. Jin, and C. K. Soh. 2006. Integrated optimization of control system for smart cylindrical shells using modified GA. Journal of aerospace engineering. 19(2): 68–79.
114. Tan, P., S. J. Dyke, A. Richardson, and M. Abdullah. 2005. Integrated device placement and control design in civil structures using genetic algorithms. Journal of structural engineering. 131(10): 1489–1496.
115. Abdullah. M. M., A. Richardson, and J. Hanif. 2001. Placement of sensors/actuators on civil structures using genetic algorithms. Earthquake engineering and structural dynamics. 30: 1167–1184.
116. Fonseca, C. M., and P. J. Fleming. 1993. Genetic algorithms for Multiobjective optimization: formulation, discussion, and generalization. In: Forrest, S., (ed). Proceedings of the Fifth International Conference on Genetic Algorithms, Morgan Kaufmann Inc., USA, p. 416–423.
117. Horn, J., and N. Nafpliotis. 1993. Multiobjective optimization using the niched Pareto genetic algorithm (Technical Report No. IlliGAl Report 93005), University of Illinois at Urbana-Champaign Urbana, Illinois.
118. K. Deb. 1999. Evolutionary algorithms for multi-criterion optimization in engineering design. In: Miettinen, K., M. M. Makela, P. Neittaanmaki, and J. Periaux, editors. Evolutionary Algorithms in Engineering and Computer Science: Recent Advances in Genetic Algorithms, Evolution Strategies, Evolutionary Programming, Genetic Programming, and Industrial Applications. John Wiley & Sons, Chichester, New York.
119. K. Deb. 1999. Multi-objective genetic algorithms: Problem difficulties and construction of test problems. Evolutionary computation journal. 7(3): 205–230.
120. K. Deb. 2001. Multi-Objective Optimization Using Evolutionary Algorithms. Wiley, Chichester, New York.
121. Deb, K., and T. Goel. 2001. A hybrid multi-objective evolutionary approach to engineering shape design. In: Zitzler, E., K. Deb, L. Thiele, C. A. Coello Coello, and D. W. Corne, editors. Proceedings of the First International Conference on Evolutionary Multi-Criterion Optimization (EMO 2001), Zurich, Switzerland, Springer-Verlag, Heidelberg, Germany, p. 385–399.

122. Deb, K., S. Agrawal, A. Pratap, and T. Meyarivan. 2000. A fast elitist non-dominated sorting genetic algorithm for multi-objective optimization: NSGA-II. In: Schoenauer, M., K. Deb, G. Rudolph, X. Yao, E. Lutton, J. J. Merelo, and H. P. Schwefel, editors. Proceedings of the Sixth International Conference on Parallel Problem Solving from Nature (PPSN-VI), Paris, France, Lecture Notes in Computer Science, vol. 1917. Springer, Berlin, Heidelberg, pp. 849–58.
123. Coello Coello, C. A., and A. D. Christiansen. 1998. Two new GA-based methods for multiobjective optimization. Civil engineering and environmental systems. 15(3): 207–243.
124. Coello Coello, C. A., and A. D. Christiansen. 2000. Multiobjective optimization of trusses using genetic algorithms. Computers & structures. 75(6): 647–60.
125. Coello Coello, C. A., D. A., Van Veldhuizen, and G. B. Lamont. 2002. Evolutionary Algorithms for Solving Multi-Objective Problems. Kluwer Academic, New York.
126. Xu, C., S. Lin, and Y. Yang. 2015. Optimal design of viscoelastic damping structures using layer wise finite element analysis and multi-objective genetic algorithm. Computers & structures. 157: 1–8.
127. Greco, R., and G. C. Marano. 2016. Multi-objective optimization of a dissipative connection for seismic protection of wall-frame structures. Soil dynamics and earthquake engineering. 87: 151–63.
128. Wang, N. F., and K. Tai. 2010.Target matching problems and an adaptive constraint strategy for multi-objective design optimization using genetic algorithms. Computers & structures. 88: 1064–1076.
129. Coello Coello, C. A. 2000. An updated survey of GA-based multiobjective optimization techniques. ACM computer surveys. 32(2): 109–143
130. Kicinger, R., T. Arciszewski, and D. Jong. 2005. Evolutionary computation and structural design: A survey of the state-of-the-art. Computers & structures. 83: 1943–1978.
131. Chugh, T., K. Sindhya J. Hakanen et al. 2019. A survey on handling computationally expensive multi-objective optimization problems with evolutionary algorithms. Soft computing. 23: 3137–66. https://doi.org/10.1007/s00500-017-2965-0
132. Lee, K. S., and Z. W. Geem. 2004. A new structural optimization method based on the harmony search algorithm. Computers & structures. 82: 781–798
133. Perez, R. E., and K. Behdinan. 2007. Particle swarm approach for structural design optimization. Computers & structures. 85: 1579–1588.
134. Li, L. J., Z. B. Huang, F. Liu, and Q. H., Wu. 2007. A heuristic particle swarm optimizer for optimization of pin connected structures. Computers & structures. 85: 340–349.
135. Lamberti, L., and C. Pappalettere. 2009. An improved harmony-search algorithm for truss structure optimization. In: Topping B. H. V., editor. The Twelfth International Conference on Civil Structural and Environmental Engineering Computing. Civil-Comp Press, Stirlingshire, Scotland.
136. Degertekin, S. O., and M. S. Hayalioglu. 2013. Sizing optimization of truss structures using teaching-learning-based optimization. Computers & structures. 119: 177–188.
137. Kocer, F. Y., and J. S. Arora. 2002. Optimal design of latticed towers subjected to earthquake loading. Journal of structural engineering. 128(2): 197–204.
138. Foley, C. M., and D. Schinler. 2003. Automated design of steel frames using advanced analysis and object-oriented evolutionary computation. Journal of structural engineering. 129(5): 648–660.
139. Hayalioglu, M.S. 2000. Optimum design of geometrically non-linear elastic-plastic steel frames via genetic algorithm. Computers & structures. 77: 527–538.
140. Tsompanakis, Y., and M. Papadrakakis. 2004. Large-scale reliability-based structural optimization. Structural and multidisciplinary optimization. 26 (4): 429–440.
141. Liu, M., and D. M. Frangopol. 2005. Multiobjective maintenance planning optimization for deteriorating bridges considering condition, safety, and life-cycle cost. Journal of structural engineering. 131(5): 833–842.
142. Liu, M., and D. M. Frangopol. 2004. Optimal bridge maintenance planning based on probabilistic performance prediction. Engineering structures. 26: 991–1002.
143. Neves, L. A. C., D. M. Frangopol, and P. J. S. Cruz. 2005. Probabilistic lifetime-oriented multi-objective optimization of bridge maintenance: single maintenance type. Journal of structural engineering. 132(6): 991–1005.
144. Gupta, A., P. Kripakaran, G. Mahinthakumar, and J. W. Baugh. 2005. Genetic algorithm-based decision support for optimizing seismic response of piping systems. Journal of structural engineering. 131(3): 389–398.

145. Avrithi, K., and H. Hyung Min Kim. 2017. Optimization of piping supports and supporting structure. Journal of pressure vessel technology. 139 (4): 044503. https://doi.org/10.1115/1.4036144.
146. Kai, S., T. Watakabe, N. Kaneko, K. Tochiki, K. Tsukimori, and A. Otani. 2018. Study on piping seismic response under multiple excitation. Journal of pressure vessel technology. 140(3): 031801. https://doi.org/10.1115/1.4039453.
147. Ahlawat, A. S., and J. V. Ramaswamy. 2002. Multi-objective optimal design of FLC driven hybrid mass damper for seismically excited structures. Earthquake engineering and structural dynamics. 31: 1459–1479.
148. Kim, H.-S., and P. N. Roschke. 2006. Design of fuzzy logic controller for smart base isolation system using genetic algorithm. Engineering structures. 28: 84–96.
149. Faruque Ali, S. K., and A. Ramaswamy. 2009. Optimal fuzzy logic control for MDOF structural systems using evolutionary algorithms. Engineering applications of artificial intelligence. 22: 407–419.
150. Yazdi, H. M. 2016. Implementing designer's preferences using fuzzy logic and Genetic Algorithm in structural optimization. International journal of steel structures. 16: 987–995. https://doi.org/10.1007/s13296-014-0144-z.
151. Miyamoto, A., K. Kawamura, and H. Nakamura. 2000. Bridge management system and maintenance optimization for existing bridges. Computer-aided civil and infrastructure engineering. 15: 45–55.
152. Azid, I. A., A. S. K. Kwan, and K. N. Seetharamu. 2002. An evolutionary approach for layout optimization of a three-dimensional truss. Structural and multidisciplinary optimization. 24(4): 333–337.
153. Ohmori, K. H., and N. Kito. 2002. Truss topology optimization by modified genetic algorithm. Structural and multidisciplinary optimization. 23(6): 467–473.
154. Hadi, S., and R. Burgueño. 2018. Emerging artificial intelligence methods in structural engineering. Engineering structures. 171: 170–189. https://doi.org/10.1016/j.engstruct.2018.05.084.
155. Falcone, R., C. Lima, and E. Martinelli 2020. Soft computing techniques in structural and earthquake engineering: a literature review. Engineering structures. 207: 110269. https://doi.org/10.1016/j.engstruct.2020.110269.

Section II

Physics and Chemistry-Based Methods

5 Gravitational Search Algorithm
Theory, Literature Review, and Applications

Amin Hashemi, Mohammad Bagher Dowlatshahi, and Hossein Nezamabadi-pour

CONTENTS

5.1 INTRODUCTION

An optimization process is finding the best solution out of all the possible solutions. Optimization is widely used in real-world issues, for example, engineers aim to design a product with the best performance, traders seek to maximize profits from their transactions, investors try to minimize the investment risk, and so on (Nocedal and Wright, 2006). Today, real-world optimization issues face a problem called high-dimensional data (Hashemi, Dowlatshahi, and Nezamabadi-pour, 2020b). Optimization problems in high-dimensional spaces are very complex due to the lack of sufficient mathematical information to solve the problem (Hashemi, Dowlatshahi, and Nezamabadi-pour, 2020, 2020a). Most traditional heuristics cannot reach the optimal solution in high-dimensional spaces, thus in solving such problems more attention is paid to metaheuristic algorithms. Metaheuristic approaches evaluate potential solutions by performing an iterative heuristic procedure to achieve better and unique solutions (Rashedi, Rashedi, and Nezamabadi-pour, 2018). These algorithms offer no guarantee for finding optimal values or even a range for the solution that has been found. However, due to the capabilities that these algorithms have shown in practice, today they are used to solve various optimization problems. Unlike heuristic algorithms, metaheuristics can find good-quality solutions in a reasonable amount of time for medium-sized hard optimization problems (Talbi, 2009).

There are two important factors in all metaheuristic algorithms called exploration and exploitation. Exploration provides the diversity aspect in searching objective space, which gives the algorithm the ability to search all parts of the objective space and prevents trapping into local optima. On the other hand, exploitation is the ability to search precisely in a local area. Thus, the metaheuristic algorithms should balance these two factors to achieve a promising result and perform in a meaningful time (Lozano and García-Martínez, 2010; Rashedi, Rashedi, and Nezamabadi-pour, 2018).

In general, metaheuristic algorithms can be divided into two groups: population based versus single solution. Single-solution metaheuristic algorithms manipulate only one solution during the search process, whereas population-based ones manipulate a population of solutions during the search process. These two categories of metaheuristic algorithms have complementary features: single-solution metaheuristic algorithms usually emphasize utilization, whereas population-based metaheuristic algorithms focus on exploration (Talbi, 2009). The most popular population-based metaheuristic algorithms are evolutionary algorithms (Eiben and Smith, 2012) and swarm intelligence algorithms (Li and Clerc, 2019).

Evolutionary algorithms are a type of bio-inspired metaheuristics that try to solve optimization problems by natural genetic principles like mutation, crossover, and natural selection (Nezamabadi-pour and Barani, 2016). On the other hand, swarm intelligence algorithms are a group of random population-based metaheuristic algorithms inspired by collective behaviors in nature. The most important components in these algorithms are elements called particles, which are usually simple and uncomplicated factors (Talbi, 2009). Particle swarm optimization (PSO) (Eberhart and Kennedy, 1999) and ant colony optimization (ACO) (Dorigo and Socha, 2007; Paniri, Dowlatshahi, and Nezamabadi-pour, 2020; Dowlatshahi and Derhami, 2019), which simulate the collective behaviors of a flock of birds and a colony of ants, respectively, and gravitational search algorithm (GSA) (Rashedi, Nezamabadi-pour, and Saryazdi, 2009), which is based on the simulation of Newton's law of gravity and laws of motion, are the most popular swarm intelligence algorithms that have been used in various applications.

GSA is a population-based and iterative random-based metaheuristic algorithm that has been introduced to solve continuous optimization problems. The main idea of this algorithm is to simulate Newton's law of gravity and the laws of motion on a population of objects in a continuous space (Rashedi, Nezamabadi-pour, and Saryazdi, 2009). The search space of the GSA is the universe that all objects sense the force of gravity that the mass of objects and their distance from each other affect this force. Many improvements have been made to the base version of the algorithm, including versions for binary, discrete, continuous, single-objective, and multi-objective problems. The difference between these versions is the design variables and the objective functions they use. This algorithm has been used to solve many complex optimization problems in the engineering field (Rashedi, Rashedi, and Nezamabadi-pour, 2018).

In this chapter, we intend to review the basic GSA and its improved versions on various issues, how to design and select parameters, and its applications in engineering sciences.

The rest of this chapter is arranged as follows. In Section 5.2 we will introduce the basic GSA and Section 5.3 discusses the selection of parameters. Section 5.4 examines the improved versions of the GSA and Section 5.5 examines the convergence properties. In Section 5.6, we review the applications of the algorithm in various engineering fields. We conclude this chapter with Section 5.7.

5.2 THE BASICS OF GSA

The GSA is a population-based random algorithm that is inspired by Newton's law of gravity to solve optimization problems. This algorithm, first introduced by Rashedi, Nezamabadi-pour, and Saryazdi (2009), uses a set of search agents called masses and simulates Newton's law of gravitation and the laws of motion on this set of agents to provide a convenient way to search for sub-optimal solutions in the search space of complex optimization problems (Rashedi, Rashedi, and Nezamabadi-pour, 2018; Nezamabadi-pour and Barani, 2016).

By the gravitation law of Newton, each particle in the universe absorbs the other particles. In the GSA, researcher agents are a set of objects with a specific amount of mass. GSA simulates Newtonian gravity based on the interactions of these objects with each other. The mass of agents is determined by the impacts of gravity force and the movements of all agents. In GSA, good solutions are the heavier masses. This will lead the algorithm to the ability of exploitation.

There are four features considered by authors for the GSA:

1. *Position:* Objects possess positions that refer to candidate solutions for optimization problems.
2. *Active gravitational mass:* The criterion for measuring the power of a mass to produce a gravitational field.
3. *Passive gravitational mass:* The criterion for measuring the reaction of a mass in a gravitational field.
4. *Inertial mass:* The criterion for measuring the resistance of a mass to motion or stillness.

The system space in the GSA includes a multi-dimensional coordinate in the problem definition space. Each point in this space is a solution to the problem; therefore this space is usually called the solution space or decision space. In the GSA, search agents in the solution space are a set of factors. In the design of this algorithm, it is assumed that only the laws of gravity and motion prevail. The general form of these laws in GSA with minor variations is almost the same as the laws of nature, and they are considered as defined below.

- *The law of gravity:* Each agent in the artificial system absorbs all the other agents. The amount of this force for an agent is proportional to the product of the gravitational masses of that agent, and the second agent, and the inverse of the distance between them.
- *The law of motion:* The current velocity of each agent is equal to the sum of the coefficient of the previous agent velocity and its change in velocity. The change in velocity or acceleration of any agent is also equal to the force exerted on that agent, divided by its gravitational mass.

Let us assume an optimization problem includes an objective function *objf* and n bounded decision variables in which xl^d is the lower bound and xu^d is the upper bound of variable d. These boundaries form a search space with n dimensions.

$$xl^d \le x^d \le xu^d,\ d = 1,2,\ldots,n \tag{5.1}$$

GSA considers N objects for searching the search space randomly to achieve the sub-optimum of the objective function. The position of the ith object in the search space is defined as follows:

$$X_i = \left(x_i^1,\ldots,x_i^d,\ldots,x_1^n\right)\ i = 1,2,\ldots,N \tag{5.2}$$

where the position of the ith object in dimension d is x_i^d. According to Eq. 5.3, the inertia mass (M_{ii}), active mass (M_{ai}), and passive mass (M_{pi}) of agent i is computed, where $objf_i(t)$ refers to the objective value of agent i at the time t. For better objective function value, the value of mass will be greater.

$$M_{ii}(t),\ M_{ai}(t), M_{pi}(t) \propto obj\ f_i(t) \tag{5.3}$$

The acceleration of an agent is computed according to the total forces that other objects applied to that agent. Based on GSA, the applying force by object j to object i at time t is calculated based on Eq. 5.4.

$$F_{ij}^d(t) = G(t)\frac{M_{pi}(t)\times M_{aj}(t)}{R_{ij}(t)+\varepsilon}\left(x_j^d(t) - x_i^d(t)\right) \tag{5.4}$$

where $M_{pi}(t)$ is the passive gravitational mass of object i, $M_{aj}(t)$ refers to the active gravitational mass of object j, and $R_{ij}(t)$ presents the Euclidean distance between these objects. Therefore, if the Euclidean distance between objects i and j is equal to zero, then to prevent division by zero, parameter ε is used. This parameter is a very small value to avoid this error. Finally, $G(t)$ is a constant at times t to control the accuracy of the searching procedure. The value of this parameter is first initialized to G_0 and then the value will change in each iteration of the algorithm.

The overall applied force to the ith object from other agents in the dimension d is completed as follows:

$$F_i^d(t) = \sum_{j=1, j\ne i}^{N} rand_j F_{ij}^d(t) \tag{5.5}$$

where $rand_j$ is a value in the interval $[0,1]$ that will be chosen randomly. The acceleration of the agent i, in dimension d, at time t, is calculated as follows:

$$a_i^d(t) = \frac{F_i^d(t)}{M_{ii}(t)} \tag{5.6}$$

Based on the proposal of the authors, the inertial and gravitational masses are considered equally.
Then Eq. 5.6 can be rewritten as the following equation:

$$a_i^d(t) = \sum_{j=1, j \neq i}^{N} rand_j G(t) \frac{M_j(t)}{R_{ij}(t) + \varepsilon} \left(x_j^d(t) - x_i^d(t)\right) \tag{5.7}$$

The inertial and gravitational masses are uploaded by the map of fitness using the following equations:

$$m_i(t) = \frac{fit_i(t) - worst(t)}{best(t) - worst(t)} \tag{5.8}$$

$$M_i(t) = M_{pi}(t) = M_{ai}(t) = M_{ii}(t) = \frac{m_i(t)}{\sum_{j=1}^{N} m_j(t)} \tag{5.9}$$

where $best(t)$ and $worst(t)$ are the best and the worst objective values at time t and among all agents, and $fit_i(t)$ is also the fitness value for object i in time t.

The velocity of the ith object in the d dimension in the next iteration is calculated by Eq. 5.10. This value depends on its acceleration and current velocity.

$$v_i^d(t+1) = rand_i \times v_i^d(t) + a_i^d(t) \tag{5.10}$$

So, based on Eq. 5.10, we can say that the position of ith object in the d dimension can be calculated using the following equation:

$$x_i^d(t+1) = x_i^d(t) + v_i^d(t+1) \tag{5.11}$$

To control the balance between exploration and exploitation, the number of agents can be reduced with the lapse of time in Eq. 5.5. In the GSA, the authors chose several objects with bigger masses than the others to apply force to the objects. This leads to avoiding the fast convergence of the algorithm. Hence, Eq. 5.5 can be rewritten as follows:

$$F_i^d(t) = \sum_{j \in kbest, j \neq i} rand_j F_{ij}^d(t) \tag{5.12}$$

where the first K objects with greater masses and the best fitness value are placed in set $kbest$. The size of set $kbest$ depends on time and at first, starts with an initial value K_0 and decreases with time (Rashedi, Rashedi, and Nezamabadi-pour, 2018; Rashedi, Nezamabadi-pour, and Saryazdi, 2009; Nezamabadi-pour and Barani, 2016). The flowchart of the GSA algorithm is presented in Figure 5.1.

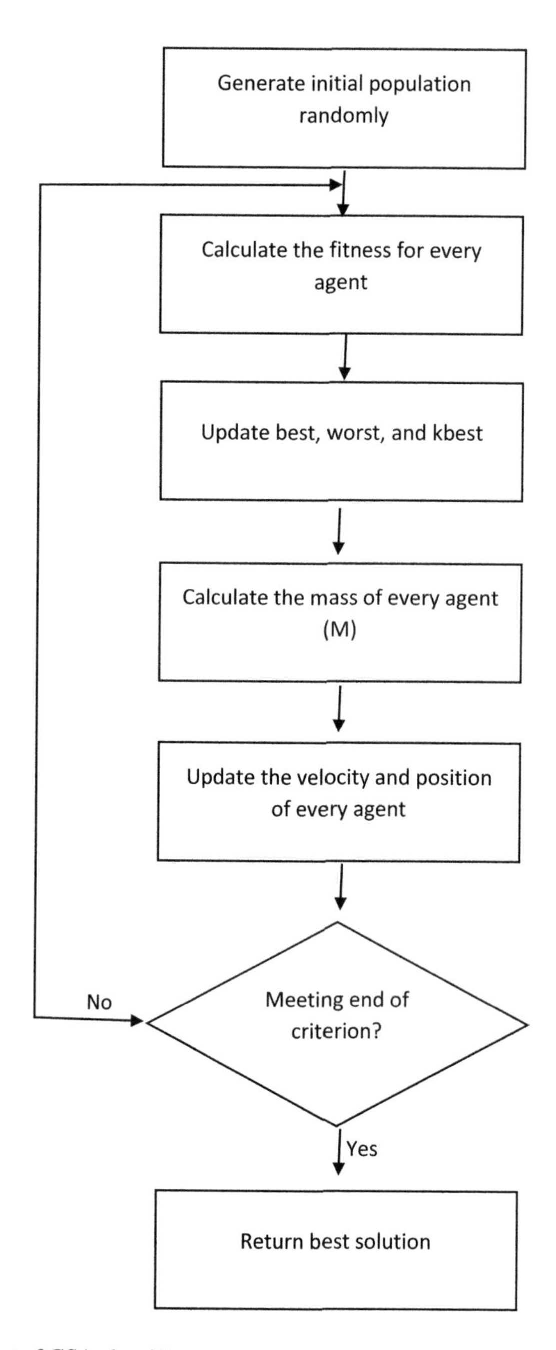

FIGURE 5.1 Flowchart of GSA algorithm.

The steps of the GSA are as follows:

1. The definition of the problem
2. Randomly produce an initial population of objects
3. Calculate the fitness for every agent
4. **While** the stopping criterion is not satisfied **do**

5. Update best, worst, *kbest*
6. Compute the mass of every agent using Eq. 5.8 and Eq. 5.9
7. Compute the velocity and acceleration of objects using Eq. 5.5, Eq. 5.6, and Eq. 5.10
8. Update the position of every agent using Eq. 5.11
9. Calculate the fitness for every agent
10. **Endwhile**

Output: Best solution found

5.3 PARAMETER SELECTION

In the GSA, there are parameters whose values must be specified before the implementation of the algorithm. These parameters include the size of *kbest* set (K), gravitational coefficient (G), population size (N), and mass values (M). The values of these parameters have a great impact on the performance of the algorithm, so in this section, we intend to analyze these parameters and their impact on the performance of the algorithm. Three approaches include deterministic, adaptive, and self-adaptive that are available as parameter controlling strategies (Eiben et al., 2007; Fister et al., 2015). A deterministic approach is a way in which parameters are changing by a distinct function. In most versions of the GSA, the deterministic approach is used and K and G parameters are decreased during the time.

K and G parameters in the GSA are the two main components that control the balance of exploitation and exploration abilities. In metaheuristic algorithms, the exploitation ability is used at the beginning of each iteration to avoid trapping in the local optimum. In GSA, similarly, high values are initially assigned to K_0 and G_0 parameters. Parameter K is the number of top agents that have the bigger mass and the value of fitness function. Given that the greater the number of agents in the solution space, increasing the exploration ability, we conclude that a high value should be assigned as the initial value of this parameter.

Parameter G is the gravitation coefficient that takes the initial value of G_0 and decreases over time. This parameter increases the exploitation ability of the algorithm by increasing the mobility of agents. As a result, by assigning high initial values to K and G parameters, the algorithm will be able to detect good areas in the solution space in the initial iterations. This causes these values to decrease in subsequent iterations over time, and the values assigned to parameters K and G become less and less (Dowlatshahi, Nezamabadi-pour, and Mashinchi, 2014; Rashedi, Rashedi, and Nezamabadi-pour, 2018).

As we discussed above, the values of K and G parameters decrease during iterations. In real GSA (Rashedi, Nezamabadi-pour, and Saryazdi, 2009), an exponential function that is presented in Eq. 5.13 and in Gu and Pan (2013), Eq. 5.14 is used as a linear function for this purpose.

$$G(t) = G_0 e^{-\frac{\alpha}{T}} \tag{5.13}$$

$$G(t) = G_0 \left(1 - \frac{1}{T}\right) \tag{5.14}$$

where α is the damping factor that is used to control the balance between exploration and exploitation capabilities. If α is a small value, it will approach G toward zero more slowly, thus increasing the exploration ability. If it is a large value, it will approach G toward zero more quickly, thus increasing the exploitation ability. Therefore, selecting an appropriate value is essential to balance the exploitation and exploration capabilities of the algorithm.

The self-adaptive approach is a way in which parameters are involved in problem space. Lei et al. (2020) proposed a new learning GSA called ALGSA, which uses a self-adaptive approach for parameter control. In ALGSA, a threshold and a probability are used to control the alteration of gravitational constants for each object in each iteration. Zhao et al. (2018) proposed a new algorithm that used a self-adaptive parameter control mechanism. They used Eqs. 5.15 and Eq. 5.16 to assign values to α and G_0.

$$\alpha(i) = randc_i(\mu\alpha, c) \tag{5.15}$$

$$G_0(i) = randc_i(\mu G_0, c) \tag{5.16}$$

where c is the learning rate ($c = 0.1$ in that research).

In each iteration, $\alpha(i)$ and $G_0(i)$ are saved as S_α, S_{G_0}, and $\mu\alpha$ and μG_0 are updated as follows:

$$\mu\alpha = (1-c)\times\mu\alpha + c\times mean(S_\alpha) \tag{5.17}$$

$$\mu G_0 = (1-c)\times\mu G_0 + c\times mean(\mu G_0) \tag{5.18}$$

Similarly, in some other research on GSA (Ji et al., 2017; Song et al., 2019) the self-adaptive mechanism is used.

The adaptive approach is a control parameter mechanism that updates the parameters based on the receiving feedback from the search. In some research, the fuzzy logic controller is used for parameter tuning (Askari and Zahiri, 2012; Saeidi-Khabisi and Rashedi, 2012). There are also other adaptive mechanisms to control parameters in the GSA (Precup et al., 2013; 2014). For example, in (Sun et al., 2018), an algorithm is presented that used an adaptive approach to update α parameter.

Another important parameter in GSA is N, which is usually tuned experimentally. The value of N is constant during the execution of the algorithm and different values are assigned to it depending on the applications in which the GSA is executed (Rashedi, Rashedi, and Nezamabadi-pour, 2018).

In GSA, the value of masses depends on their fitness values. In the original GSA (Rashedi, Nezamabadi-pour, and Saryazdi, 2009), Eqs. 5.8 and 5.9 as we discussed in Section 5.2, are used to obtain mass values.

There is also another strategy to calculate the value of the masses. Boltzmann and sigma scaling functions (Mood, Rashedi, and Javidi, 2015) that are presented by Eqs. 5.19 and 5.20 are other suggested formulations of mass calculation is GSA.

$$M_i(t) = \frac{exp\left(\frac{nobjf_i(t)}{temp}\right)}{< exp\left(\frac{nobjf_i(t)}{temp}\right) >_t} \tag{5.19}$$

$$M_i(t) = \begin{cases} 1 + \frac{nobjf_i(t) - < nobjf_i(t) >_t}{2\sigma(t)} & if\ \sigma(t) \neq 0 \\ 1 & if\ \sigma(t) = 0 \end{cases} \tag{5.20}$$

where $nobjf_i(t)$ is the normalized objective function, $\sigma(t)$ indicates the standard deviation of objective values at time t, $<.>_t$ shows the average values of t, and $temp$ indicates the temperature.

There is also another approach (Khajooei and Rashedi, 2016) that assigns negative mass values for objectives with low values.

5.4 EVOLUTION OF GSA

In this section, we intend to discuss the various operators and other variants of GSA for solving different optimization problems.

5.4.1 Operators of GSA

To achieve an efficient local and global search in metaheuristic algorithms, we should establish a balance between the exploitation and exploration abilities of the algorithm. This leads to solving the optimization problem efficiently. In metaheuristic algorithms, the operators are used to establish this balance. Some GSA operators, such as escape, Kepler, black hole, and disruption, are inspired by physical theories related to astronomy, gravity, and relativity. Besides, there are other GSA operators that we will discuss in this section. These operators include crossover, mutation, chaotic, and discrete local search (DLS).

5.4.1.1 Disruption

In physics, disruption occurs when a gravitational force exerted by a heavy mass on a solid object causes it to tear apart. Sarafrazi et al. (2011) proposed a disruption operator to improve the performance of GSA inspired by this phenomenon.

To simulate this operator in GSA, they considered the best agent as the star and other agents can be disrupted under the gravitational force of this star.

In this simulation, if the ratio of the distance between agent i and its nearest neighbor $\left(\frac{R_{i,j}}{R_{i,best}}\right)$ is smaller than a predefined threshold (C), then the position of agent i (X) will be altered by the disruption operator according to Eq. 5.21 and the mass value (D) updates by Eq. 5.22.

$$X_i(t) = X_i(t).D \tag{5.21}$$

$$D = \begin{cases} 1 + \rho.U(-0.5, 0.5) & \text{if } R_{i,best} < 1 \\ R_{i,j}.U(-0.5, 0.5) & \text{otherwise} \end{cases} \tag{5.22}$$

where $U(-0.5,0.5)$ is a uniformly distributed pseudorandom number in the interval $[-0.5,0.5]$ and ρ is a small number. If $R_{i,best} < 1$, the algorithm exploits around the best agent; otherwise, it explores the search space.

Liu et al. (2012) applied an improvement to the disruption operator that disrupts the agents so that their fitness values are improved. These methods decreased the computational complexity of the original version.

5.4.1.2 Mutation

The mutation is a phenomenon in which the objects with a close distance can affect the direction of the movement of each other. Nobahari et al. (2012) proposed two mutation operators in the original GSA to simulate the mutation phenomenon. These operators decrease the probability of trapping in local optima and improve diversity.

The first operator is called sign mutation and it changes the sign of velocity of each agent as follows:

$$\acute{v}_i^d(t+1) = sign_i^d v_i^d(t+1),\ d = 1,\dots,n,\ i = 1,\dots,s \tag{5.23}$$

$$sign_i^d = \begin{cases} -1 & rand < P_s \\ 1 & otherwise \end{cases} \tag{5.24}$$

$$x_i^d(t+1) = x_i^d(t) + \acute{v}_i^d(t+1) \tag{5.25}$$

where P_s is the probability of sign mutation, the mutated velocity by the sign mutation is presented by $\acute{v}_i^d(t+1)$, and *rand* is a uniform random number in the interval $[0,1]$.

The second operator is called recording mutation because by using a predefined probability, the velocity of some agents is altered randomly. In general, the mutation operators update the position of agents as follows:

$$v_i(t+1) = \omega(t) \times v_i(t) + a_0(t) \tag{5.26}$$

$$v_i'(t+1) = sign(v_i(t+1)) \tag{5.27}$$

$$v_i''(t+1) = reordering\_mutation(v_i'(t+1)) \tag{5.28}$$

$$x_i(t+1) = x_i(t) + v_i''(t+1) \tag{5.29}$$

where $\omega(t) = \omega_0 - (\omega_0 - \omega_1)\frac{t}{T}$ is a coefficient for improving the exploitation and exploration abilities, which varies by time. Tabatabaei (2014) proposed an improvement of the original GSA by four mutation operators.

5.4.1.3 Crossover

The nearest neighbor crossover operator (NCO) is proposed by Shang (2013) to improve the local search capacity of GSA. In this research, a scale factor, w, is established to balance NCO and GSA. We consider parameter r as a random number in the interval $[0,1]$. If $r < w$, then GSA searches the solution space. Otherwise, the NCO is applied to set a new position to the agents by the following equation:

$$X_i(t+1) = rand_i \times X_i(t) + U(-1,1) \times (rand_i \times X_i(t) - X_i(t)),\ i = 1,\dots,s \tag{5.30}$$

where $rand_i$ and $U(-1,1)$ are random numbers in the interval $[0,1]$ and $[-1,1]$, respectively. A scale for factor is also proposed as follows:

$$w = w_{max} - (w_{max} - w_{min}) \times \frac{t}{T} \tag{5.31}$$

where the maximum and minimum of the scale factor are presented by w_{max}, and w_{min}, respectively; T is the number of iterations; and t is the current iteration number.

Other researchers like Liu et al. (2019) also improved the search speed or the quality of the GSA by combining the genetic algorithm (GA) and crossover operator.

5.4.1.4 Chaotic

The chaotic operator is presented by Han and Chang (2012) to improve GSA. This operator decreases the probability to trap in local optima and premature convergence. It is applied in the phase of updating the position of agents. Let c_i be an n-dimensional random vector in the interval $[0,1]$. The new chaotic vectors are generated at each iteration as follows:

$$c_i^d = 4c_{i-1}^d\left(1-c_{i-1}^d\right)\ i=2,\ \ldots\ ,N \tag{5.32}$$

The velocity of each agent is also updated by the following equation:

$$v_i^d = \left[rand_i \times v_i^d(t) + \xi\left(c_i^d - 0.5\right)\right] + a_i^d(t) \quad i=2,\ldots,N \tag{5.33}$$

where c_i is a chaotic vector and ξ is a factor to control chaos scope. Also, the position of agents is updated as follows:

$$x_i^d(t+1) = x_i^d(t) + v_i^d(t+1) \quad i=2,\ldots,s \tag{5.34}$$

5.4.1.5 Escape

Another operator for GSA that is inspired by the concept of escape velocity in physics is the escape operator. This operator was proposed by Rashedi et al. (2013) for an image segmentation algorithm based on GSA, called SGISA. The escape velocity is an undirected form of speed, which means the minimum speed for an object to escape from gravity.

In SGISA, each agent belongs to a cluster and by a probability it can escape from that cluster and be absorbed by the nearest cluster. The escape velocity for agent i to escape from cluster k and with mass M_k is calculated as follows:

$$Vc_i^k = \sqrt{\frac{2GM_k}{r_i}} \tag{5.35}$$

where r_i refers to the Euclidean distance between agent i and cluster k or we can say agent k. The escape probability of agent i is calculated by the following equation:

$$pc_i^k = \begin{cases} 1 & Vc_i^k \le Vc_{min} \\ \dfrac{Vc_{max} - Vc_i^k}{Vc_{max} - Vc_{min}} & Vc_{min} < Vc_i^k < Vc_{max} \\ 0 & Vc_i^k \ge Vc_{max} \end{cases} \tag{5.36}$$

where Vc_{max} is the escape velocity of an agent close to the cluster center less than a given value d_{min} that is calculated according to Eq. 5.37, and Vc_{min} is the escape velocity of an agent far from the cluster center more than a given value d_{min} that is calculated according to Eq. 5.38.

$$Vc_{max} = \sqrt{\frac{2G(t)}{d_{min}}} \tag{5.37}$$

$$Vc_{min} = \sqrt{\frac{2G(t)}{d_{max}}} \tag{5.38}$$

5.4.1.6 Black Hole

The black hole operator is proposed by Doraghinejad and Nezamabadi-pour (2014) for the GSA inspired by the black hole phenomenon in astronomy. For the simulation of this operator, the best agent is assumed as the black hole that absorbs other agents. In this approach, agents are grouped into heavy and light agents based on the mass. Based on Eq. 5.39, the position of each agent will be updated.

$$x_i^d(t+1) = \begin{cases} x_i^d(t) + rand.\left(x_{BH}^d(t) - x_i^d(t)\right) & \textit{if agent i is heavy and } r < R_s \\ rand.\left(x_i^d(t).\dfrac{r'}{R_s'}\right) & \textit{if agent i is light and } r' < R_s' \end{cases} \tag{5.39}$$

where r is the distance of every heavy agent from the black hole and r' is the distance of every light agent from it. The term *rand* is a random number in the interval $[0,1]$ and x_{BH}^d indicates the position of the black hole in dimension d. R_s and R_s' also present the Schwarzschild radiuses of heavy and light agents, respectively, that are calculated as follows:

$$R_s = GM\frac{v^2}{t} \tag{5.40}$$

$$R_s' = M.log(t) \tag{5.41}$$

where G is the gravity constant and M is the mass of the black hole, v is the velocity of the agent, and t indicates the iteration number.

5.4.1.7 Kepler

The Kepler operator is presented by Sarafrazi, et al.(2015); it is inspired by Kepler's first law to improve the performance of GSA. Kepler's first law states that the path taken by a planet is elliptical, parabolic, hyperbolic, or other conical sections in which the center body (for example, the sun) is at one of its focuses. We can say that the planets have different distances from the sun. In the simulation of Kepler's law for optimization problems, the best solution is considered as the sun and the other solution as planets. Orbiting the planets around the sun causes us to explore and exploit the solution space.

Based on Kepler's law, the position of planets is updated by Eq. 5.42. Also, to avoid the best solution from trapping into the local optimum, its position is altered by Eq. 5.43.

$$X_{i,new}(t+1) = X_{best}(t) + R_{i,best} \times U(-2,2) \; i = 2, \dots, s \tag{5.42}$$

$$X_{best,new}(t+1) = X_{best}(t) \times U(-2,2) \tag{5.43}$$

where $U(-2,2)$ is a random number in the interval [-2,2]. If the number generated by $U(-2,2)$ is close to 1, the operator will perform exploitation, otherwise, exploration will be performed.

The next position of the planets is calculated by Eq. 5.44 based on the elitism strategy.

$$X_i(t+1) = \begin{cases} X_{i,new}(t) & \textit{if } fit\left(X_{i,new}\right) > fit\left(X_i\right) \\ X_i(t) & \textit{otherwise} \end{cases} \tag{5.44}$$

5.4.1.8 Discrete Local Search (DLS) Operator

The DLS operator is presented by Doraghinejad et al. (2014) to solve the channel assignment problem in mesh networks by GSA. The DLS operator is suggested to enhance the exploiting ability around the best solution. This operator first selects a few nodes of the best solutions and then alters randomly a few channels. It is noteworthy that the DLS operator can be applied to problems that have integer-valued variables.

5.4.2 Variants of GSA

The optimization problems differ from each other based on the form of values of their variables. The problem variables can be in the form of real-value (continuous), discrete, binary, and mixed. In the same way, the GSA's variants are proposed by these categorizations. Also, multi-objective and multimodal are two forms of optimization problems that vary by their objective functions. Hence, we can consider another categorization for the GSA that includes multi-objective, multimodal, and constraint. In addition to these categorizations, other modified versions of GSA are also presented like quantum GSA. In this section, we intend to discuss the modified versions of GSA.

5.4.2.1 Real GSA

The original GSA was proposed by Rashedi et al. (2009) to deal with continuous optimization problems in which each variable is assigned with real value. In some works (Mallick et al., 2013; Bounar, Labdai, and Boulkroune, 2019; Eappen and T, 2020; Priyadarshi et al., 2020; Sahib et al., 2020; Song, Xiao, and Xu, 2020) the PSO velocity term is combined with GSA to propose an algorithm to decrease the memory usage by original GSA. For example, Mallick et al. (2013) used Eq. 5.45 to modify the velocity of GSA based on the combination of PSO and GSA.

$$V_i^d(t+1) = \omega(t+1) \times rand \times V_i^d(t) + c_1 \times rand \times a_i^d(t) + c_2 \times rand \times \left(x_{best}^d(t) - x_i^d(t)\right) \quad (5.45)$$

where ω is the decreasing function, c_1 and c_2 are constant values, and $x_{best}(t)$ presents the best solution in time t.

Yin et al. (2018) used a crossover operator to propose a modified version of GSA. This algorithm, called CROGSA, applied crossover to update the position of solutions and enhance the exploitation ability. Opposition-based GSA is another form of GSA (Shaw, Mukherjee, and Ghoshal, 2012, 2013; Özyön et al., 2015). An opposition-based GSA was proposed by Shaw et al. (2012), in which the candidate solutions and their opposition are used to find a good approximation.

Other improved versions of GSA were also proposed based on negative mass (Khajooei and Rashedi, 2016) and according to both attractive and repulsive forces (Zandevakili, Rashedi, and Mahani, 2019).

5.4.2.2 Binary GSA

Many real-world problems correspond with variables in binary format, in which every variable can be assigned by 0 or 1 value. Therefore, these kinds of problems need to be solved by binary versions of the optimization algorithms. BGSA (Rashedi, Nezamabadi-pour, and Saryazdi, 2010) is the binary version of the GSA. In BGSA, the force and velocity are updated like continuous versions, and the distance (R) is calculated by a hamming distance.

The velocity and the position of each agent in BGSA are updated by Eqs. 5.10 and 5.46, respectively.

$$S(V_i^d(t)) = \left|\tanh\left(V_i^d(t)\right)\right| \quad (5.46)$$

After the calculation of $S(V_i^d)$, the agent's movements are based on Eq. 5.47, where the complement reverses the bits.

$$if\ rand() < S(V_i^d(t+1))\ then\ x_i^d(t+1) = Complement\left(x_i^d(t)\right) else\ x_i^d(t+1) = x_i^d(t) \quad (5.47)$$

To control the convergence rate, V_i^d is limited by $V_{max}\left(\left|V_i^d\right| < V_{max}\right)$. V_{max} is set to 6 in BGSA literature (Rashedi, Nezamabadi-pour, and Saryazdi, 2010) and is also improved by Rashedi and Nezamabadi-pour (2014).

BGSA is applied to many optimization problems including unit commitment (Ji et al., 2014; Yuan et al., 2014; Barani et al., 2017), feature selection (Barani, Mirhosseini, and Nezamabadi-pour, 2017; Nagpal et al., 2017; Bardamova et al., 2018; Sarhani, El Afia, and Faizi, 2018; Guha et al., 2020), scheduling for multi-processor computing systems (Thakur, Biswas, and Kuila, 2020), and speech processing (Sheikhan, 2014).

5.4.2.3 Discrete GSA

DGSA is the discrete version of GSA that was proposed by Shamsudin et al. (2012) to tackle the optimization problems, in which the variables are discrete. DGSA modeled the discrete optimization problems into vector of integer values. In this algorithm, the acceleration and velocity are calculated the same as the original GSA, but the position of each agent is updated by Eq. 5.48. This equation is dependent on the direction of the velocity.

$$X_i(t+1) = \begin{cases} random\left(x_i^d(t).x_i^d(t)+1.x_i^d(t)+2.......x_i^d(t)+v_i^d(t+1)\right)\ if\ v_i^d(t+1) < 0 \\ random\left(x_i^d(t).x_i^d(t)-1.x_i^d(t)-2.......x_i^d(t)-v_i^d(t+1)\right)\ if\ v_i^d(t+1) \geq 0 \end{cases} \quad (5.48)$$

A version of DGSA is also presented to solve the knapsack problem (Sajedi and Razavi, 2017). In other literature, by using an undirected graph for modeling dependent movement, Dowlatshahi et al. (2014) proposed a new DGSA to solve the traveling salesman problem (TSP). The proposed DGSA uses a path re-linking (PR) strategy instead of the classic way in which the agents of GSA usually move from their current position to the position of other agents.

Gao et al. (2014) proposed a triple-valued optimization problem based on GSA to solve the graph planarization problem.

5.4.2.4 Mixed GSA

In some problems, the variables are in both binary and continuous forms. Sarafrazi and Nezamabadi-pour (2013) proposed a mixed version of GSA to deal with this kind of problem. In this algorithm, each dimension based on the type of variables has its movement equation. This method is designed for problems with continuous and binary variables (Rashedi, Nezamabadi-pour, and Saryazdi, 2013).

GSA-GA (Garg, 2019) is a version of GSA that is proposed for problems with mixed variables. In this algorithm, the solution is first tuned up with the GSA and then the solutions are updated based on GAs. This algorithm also can be practical in medical applications (Shirazi and Rashedi, 2016).

5.4.2.5 Quantum GSA

Quantum-based GSA (QGSA) (Soleimanpour-Moghadam and Nezamabadi-pour 2012; Soleimanpour-Moghadam, Nezamabadi-pour, and Farsangi 2014; Thakur, Biswas, and Kuila 2020) is another form of GSA that is constructed based on quantum complexity in which a new generation of computers are used according to quantum theory. In QGSA, each agent acts like quantum waves. The formulation for position updating in this algorithm is as follows:

$$\begin{cases} x_i^d(t+1) = c_i^d + g\left|c_i^d - x_i^d(t)\right| \times ln\left(\frac{1}{rand}\right) & S \geq 0.5 \\ x_i^d(t+1) = c_i^d - g\left|c_i^d - x_i^d(t)\right| \times ln\left(\frac{1}{rand}\right) & otherwise \end{cases} \tag{5.49}$$

where *rand* and S are two random values and c_i^d is the dimension d of a position randomly selected from *kbest* set.

QBGSA (Ibrahim, Mohamed, and Shareef, 2012) is another quantum-based algorithm, which is the combination of BGSA and quantum computing to improve the exploration ability.

BQIGSA (Nezamabadi-pour, 2015) is another improvement of BGSA inspired by quantum computation. In this algorithm, $q_i(t)$ is a binary vector with L bits that represent each object. This vector is defined as Eq. 5.50, where $\left|\alpha_i^d\right|^2 + \left|\beta_i^d\right|^2 = 1$. In this algorithm, at first randomly the population is initialized by producing random numbers in the range $[-1,1]$ for α_i^d. Then, based on Eq. 5.51, the binary bits are generated for each bit. According to these binary values, each object $q_i(t)$ is transformed into $X_i(t)$, which presents a solution to the problem. The angular velocity (w) acceleration is calculated by Eqs. 5.52 and 5.53, respectively. The mass values are calculated based on the objective functions. The bit movement between 0 and 1 is called angular velocity. $\Delta\theta = w \times \Delta t$ shows the value of movement in a circular system. If we consider $\Delta t = 1$, the movement value is calculated by Eq. 5.54, and $q_i(t)$ for each agent is updated by Eq. 5.55. This procedure will be repeated for some iterations.

$$q_i(t) = \left[q_i^1(t), q_i^2(t), \ldots, q_i^L(t)\right] = \left[\left|\begin{matrix}\alpha_i^1(t) \\ \beta_i^1(t)\end{matrix}\right| \cdots \left|\begin{matrix}\alpha_i^2(t) \\ \beta_i^2(t)\end{matrix}\right| \cdots \left|\begin{matrix}\alpha_i^L(t) \\ \beta_i^L(t)\end{matrix}\right|\right] \tag{5.50}$$

$$if\ rand < \left(\alpha_i^d(t)\right)^2\ then\ x_i^d = 0\ else\ x_i^d = 1 \tag{5.51}$$

$$w_i^d(t+1) = rand_i \times w_i^d(t) + a_i^d(t) \tag{5.52}$$

$$a_i^d(t) = \sum_{j \in kbest} G(t)\frac{M_j(t)}{R_{i,j} + \varepsilon}\left(x_i^d(t) - x_i^d(t)\right) \tag{5.53}$$

$$\Delta\theta_1^d(t) = \begin{cases} w_i^d(t+1) & if\ a_i^d(t+1)\beta_i^d(t+1) \geq 0 \\ -w_i^d(t+1) & if\ a_i^d(t+1)\beta_i^d(t+1) < 0 \end{cases} \tag{5.54}$$

$$\begin{bmatrix} a_i^d(t+1) \\ \beta_i^d(t+1) \end{bmatrix} = \begin{bmatrix} cos\left(\Delta\theta_1^d(t)\right) & -sin\left(\Delta\theta_1^d(t)\right) \\ sin\left(\Delta\theta_1^d(t)\right) & cos\left(\Delta\theta_1^d(t)\right) \end{bmatrix} \begin{bmatrix} a_i^d(t) \\ \beta_i^d(t) \end{bmatrix} \tag{5.55}$$

QBGSA is used to solve various optimization problems such as unit commitment (Barani et al., 2017), quality monitor placement (Ibrahim, Mohamed, and Shareef, 2012), failure prediction (Lou et al., 2018), and scheduling problems (Singh and Raza, 2017, 2018).

5.4.2.6 Constraint GSA

A constraint optimization problem is defined as follows:

$$\begin{aligned} &optimize\ objf\left(x^1,x^2,\dots,x^m\right) \quad subject\ to\ g_i\left(x^1,x^2,\dots,x^m\right)\le 0,\\ &i=1,2,\dots,N_g\ and\ h_j\left(x^1,x^2,\dots,x^m\right)\le 0, j=1,2,\dots,N_g \end{aligned} \tag{5.56}$$

If we try to perform an optimization procedure for an objective function under N_h number of quality constraint functions (h), and N_g the number of inequality constraint functions (g), then we are dealing with a constraint optimization. In this kind of problem with constraints, usually, the penalty functions are used. These functions are added to the objective function in a minimization problem to evaluate solution X. Eq. 5.57 presents an example of a penalty function (Poole, Allen, and Rendall, 2014).

$$evaluation(X)=objf(X)+\sum_{i=1}^{N_g} max\left\{g_i(t),0\right\}+\sum_{i=1}^{N_h}\left|h_j(X)\right| \tag{5.57}$$

The solution that does not violate the constraints is called a feasible solution. A version of GSA is proposed to deal with constraint problems (Yadav and Deep, 2013) by parameter-exempt constraint dealing. A dynamic constrained optimization problem is also solved by this method. This algorithm first created a reference set and then updated it during iterations. This set represents the best-founded feasible solution, then the point S' will be calculated based on Eq. 5.58 and replaced. This value is the combination of R and S, where R belongs to the reference set with the least distance from S.

$$generate\ S'=r\times R+(1-r)\times S \qquad until\ S'\ is\ feasible \tag{5.58}$$

where r is a random number.

Also, other approaches to deal with constraint problems are used based on GSA (Amoozegar and Nezamabadi-pour, 2012; Poole, Allen, and Rendall, 2014; Mahdad and Srairi, 2015; Chen et al., 2017; Garg, 2019; Geyun et al., 2016).

5.4.2.7 Multimodal GSA

Some real-world optimization problems correspond with objective functions with multimodal behavior that need to be optimized by finding several global optimal solutions. Therefore, there is a version of GSA to deal with this kind of optimization problem.

The Niching methods are used in this kind of problem because the optimization algorithms tend to converge into a single optimal solution. The Niching methods partition the population into some groups and in each group, the focus is on finding a single feasible solution. By using this strategy, the location of optimal solutions is determined in multimodal solutions.

The Niching methods are combined with GSA (called NGSA; Yazdani, Nezamabadi-pour, and Kamyab, 2014) to solve multimodal optimization problems. The calculation of the velocity and the acceleration is the same as the real GSA. But in NGSA, each agent just applies force to its K-nearest neighbors. Hence the active gravitational force is obtained as follows:

$$M_{ai}=\left\{M_{ai1},\dots,M_{aiN}\right\} \quad where\ M_{i,j}=\begin{cases}\dfrac{fit_j(t)-worst_i(t)}{best_i(t)-worst_i(t)} & if\ agent\ j\in KNN\\ 0 & otherwise\end{cases} \tag{5.59}$$

where KNN_i is the K-nearest neighbors of agent i.

Nearest neighbor GSA (NNGSA) (Haghbayan, Nezamabadi-pour, and Kamyab, 2017) is another version of GSA that used a hill valley algorithm to find the niches. In this method, first, the population is divided into several partitions. Then in each iteration, the hill valley algorithm is applied to find the niches and the GSA is applied to each partition. Also, a niching co-swarm GSA (Yadav and Kim, 2015) is used for solving multimodal optimization problems using GSA and differential evaluation.

Other literature also used GSA to solve multimodal optimization problems (Dowlatshahi, Derhami, and Nezamabadi-pour, 2018; Golzari et al., 2018; Bala and Yadav, 2020). Dowlatshahi, Derhami, and Nezamabadi-pour (2018) proposed a locally informed particle swarm with gravitational velocity updating rule (GLIPS). In GLIPS, each particle successively adjusts its position toward the best positions of its local neighbors using laws of gravity and motion. The local neighbors with a higher quality get a greater gravitational mass; therefore they are allowed to apply the higher gravity force to other particles to attract them. In this case, the particles near good solutions try to attract the other particles, which are exploring the search space. Bala and Yadav (2020) introduced a new comprehensive learning gravitational search algorithm (CLGSA). This algorithm improved the searching process of GSA. This is done by an intensive comprehensive learning methodology that leads to choosing good elements and reduces the chance of trapping at a local optimum. Golzari et al. (2018) proposed a new version of GSA using the *k*-means algorithm to solve multimodal optimization problems. This clustering strategy is applied to GSA to expand the search space.

Dowlatshahi, Derhami, and Nezamabadi-pour (2019) proposed the nearest-better neighborhoods in swarm intelligence algorithms and then used the GSA to solve multimodal optimization problems. For this purpose, they define two neighborhoods, called topological nearest-better (TNB) and distance-based nearest-better (DNB), and then these two structures are used separately in the GSA. Two different versions of the GSA for multi-modal optimization problems are provided.

5.4.2.8 Multi-Objective GSA

In many real-world problems, we have to optimize several objectives simultaneously to find an optimal solution. Such problems are called multi-objective optimization, and several versions of GSA have been proposed to solve these kinds of problems. To do this, we should find non-dominated solutions called Pareto optimal solutions.

Hassanzadeh and Rouhani (2010) proposed a multi-objective GSA (MOGSA) using a uniform mutation operator and an elitist policy to save Pareto optimal solutions. NGSA is an algorithm to find non-dominated solutions that have been proposed by Nobahari et al. (2012). MLGSA is a multi-leader GSA to deal with multi-objective optimization problems. A clustering-based archive MOGSA (CA-MOGSA) is presented by Abbasian et al. (2015) that stores Pareto optimal solutions in an external archive. At first, a clustering algorithm is applied to the archive, and then a random cluster is assigned to each agent to apply the gravitational force and to absorb that specific agent. Solutions with less crowding distance and extra solutions are eliminated from the archive. MOGSA algorithms have been applied in many optimizations of different engineering problems (Bhowmik and Chakraborty, 2018; Feng et al., 2019; Zhang et al., 2019; Eid, 2020; Mosa, 2020).

5.4.2.9 Gravitational Search Programming (GSP)

Automated programming is the process by which computer programs are automatically found or generated by searching techniques. This programming method has become very popular due to its use in various applications as well as the widespread use of genetic programming (GP). Automated programming searches the space of all possible computer programs to select the best one. Therefore, this can be considered as a search problem that involves a population of solutions, and population-based metaheuristic algorithms can perform this search for automated programming. The first technique for automated programming is GP that uses a GA to perform the searching process on program space. After GP other automated programming techniques have been proposed based on population-based metaheuristic algorithms such as ACO, bee colony optimization, and PSO.

Gravitational search programming (GSP) (Mahanipour and Nezamabadi-pour, 2019b) is an automatic programming technique with the GSA. GSP is so much simpler and faster than GP because the operators like selection, replacement, and crossover are eliminated in this algorithm. In GSP there is a tree structure with a smaller depth compared to the tree in GSP.

5.4.3 Hybridization of GSA

One of the most common techniques used to solve problems is the hybridization of the two algorithms. The reason for this hybridization is to use the advantages of both algorithms and eliminate their disadvantages so that the new method improves the exploration and exploitation abilities and in general has a better performance than the two basic algorithms (Rashedi, Rashedi, and Nezamabadi-pour, 2018).

One reason for using hybridization is the low precision of the algorithm, which is usually significantly improved by a local search. The hybridization methods are categorized into two main groups: collaborative and integrative. In integrative methods, one algorithm acts as part of another algorithm and is embedded in it. A collaborative hybrid is a way in which two algorithms exchange information and run in parallel or sequentially (Lozano and García-Martínez, 2010).

Many combinations of GSAs with other algorithms have been proposed so far, including the hybridization with PSO (Sarhani, El Afia, and Faizi, 2018; Eappen and T, 2020; Mosa, 2020), *K*-means (Priyadarshini et al., 2020), simulated annealing (Chen, Li, and Tang, 2011), artificial bee colony (Guo, 2012; Zhang, 2018), GA (Garg, 2014; 2019; Khadanga and Satapathy, 2015), neural network (Mehdizadeh et al., 2020), pattern search (Chiranjeevi and Jena, 2017; Khadanga and Satapathy, 2018), and so on.

5.5 GSA CONVERGENCE PROPERTIES

The initial population of objects in GSA is randomly chosen and then moved to the search space. One of the important issues is that we determine whether the algorithm can converge to a solution in the search space.

Ghorbani and Nezambadi-pour (2012) analyze the convergence properties of GSA. To prove the convergence of GSA, the position of each object should be traced based on their interaction with the other objects. Random selection and time varying are the other important parameters.

To prove the convergence of a sequence $\{A(t)\}$, $t = 0,1,\ldots,$ if $\lim_{t\to\infty} A(t) = A$, where A is a constant number (called convergent value), the convergence is reached. Thus, for GSA, if we could show that $\lim_{t\to\infty} x_i^d(t) = C$, which means $\lim_{t\to\infty}\left(x_i^d(t+1) - x_i^d(t)\right) = 0$, then the GSA is converged.

To prove this convergence, there are two theorems as follows:

Theorem 1. Assume $E(t)$ and $B(t)$ as two real value functions and for $N > 0, |B(t)| < N$ for all t, and $\lim_{t\to\infty} E(t) = 0$, then we can claim:

$$\lim_{t\to\infty} E(t).B(t) = 0 \tag{5.60}$$

Proof. If we consider $\varepsilon > 0$, then $T \geq 0$, so that for each $t > T$, $|E(t)| < \frac{\varepsilon}{N}$. Thus, for each $t > T$:

$$E(t).B(t) < N \times \frac{\varepsilon}{N} = \varepsilon \tag{5.61}$$

Theorem 2. If $\lim_{t\to\infty} G(t)=0$, then:

$$\lim_{t\to\infty}\left(x_i^d(t+1)-x_i^d(t)\right)=0 \tag{5.62}$$

Proof. Let $R_{ij}(t)$ be the Euclidean distance between agent i and agent j.

$$R_{ij}=\left\|X_i(t),X_j(t)\right\|_2 \tag{5.63}$$

Hence, we can claim:

$$R_{ij}=\left\|X_i(t),X_j(t)\right\|_2 \ge \left|x_j^d(t)-x_i^d(t)\right| \Rightarrow R_{ij}+\varepsilon > \left|x_j^d(t)-x_i^d(t)\right| \Rightarrow \left|\frac{x_j^d(t)-x_i^d(t)}{R_{ij}+\varepsilon}\right|<1 \tag{5.64}$$

According to Eqs. 5.8 and 5.9, it can be concluded that $M_j(t)\ge 0$ and $\sum_{j=1}^{N} M_j>0$, where k can be a constant value bigger than 1.

Based on Eq. 5.64, it can be discovered:

$$\left|\sum_{j=1,j\ne i}^{S}\left(r_j(t).M_j(t).\frac{x_j^d(t)-x_i^d(t)}{R_{ij}+\varepsilon}\right)\right| \le \left|\sum_{j=1,j\ne i}^{S}\left(M_j(t).\frac{x_j^d(t)-x_i^d(t)}{R_{ij}+\varepsilon}\right)\right| < \left|\sum_{j=1,j\ne i}^{S} M_j(t)\right|$$

$$<\left|\sum_{j=1}^{S} M_j(t)\right|<K \tag{5.65}$$

If we limit the set of agents that apply force to the *kbest*, we can also write:

$$\left|\sum_{j=kbest,j\ne i}^{S}\left(r_j(t).M_j(t).\frac{x_j^d(t)-x_i^d(t)}{R_{ij}+\varepsilon}\right)\right| \le \left|\sum_{j=kbest,j\ne i}^{S}\left(M_j(t).\frac{x_j^d(t)-x_i^d(t)}{R_{ij}+\varepsilon}\right)\right|$$

$$<\left|\sum_{j=kbest,j\ne i}^{S} M_j(t)\right|<\left|\sum_{j=1}^{S} M_j(t)\right|<K \tag{5.65}$$

$$a_i^d(t)=G(t).\sum_{j=1,j\ne i}^{S}\left(r_j(t).M_j(t).\frac{x_j^d(t)-x_i^d(t)}{R_{ij}+\varepsilon}\right) \tag{5.66}$$

According to Eq. 5.65 and Theorem 1, we can conclude that $\lim_{t\to\infty} a_i^d(t)=0$.

Based on Eq. 5.26, it can be observed that:

$$a_i^d(t)=v_i^d(t+1)-r_i(t)v_i^d(t) \tag{5.67}$$

So, we can say:

$$\lim_{t\to\infty} a_i^d(t)=0 \overset{1.67}{\Rightarrow} \lim_{t\to\infty} v_i^d(t+1)-r_i(t)v_i^d(t)=0 \Rightarrow \lim_{t\to\infty} v_i^d(t)-r_i(t)v_i^d(t)=0 \Rightarrow$$
$$\lim_{t\to\infty}\left(1-r_i(t)\right)v_i^d(t)=0 \tag{5.68}$$

We have noted before that $r_i(t)$ is a random number in the interval $[0,1]$, and it is time-independent. Hence $(1-r_i(t))$ cannot converge to zero by lapse of time. So, let $r_i'(t)=(1-r_i(t))$ and according to ti Eq. 5.68, we can claim:

$$\lim_{t\to\infty} r_i'(t).v_i^d(t)=0 \tag{5.69}$$

where both $r_i(t)$ and $r_i'(t)$ are random numbers in the interval $[0,1]$. Therefore, we can use $r_i(t)$ instead of $r_i'(t)$. So, we have:

$$\lim_{t\to\infty} r_i(t).v_i^d(t)=0 \tag{5.70}$$

On the other hand:

$$\begin{aligned}&v_i^d(t)=v_i^d(t).(1-r_i(t)+r_i(t))=(1-r_i(t)).v_i^d(t)+r_i(t).v_i^d(t)\Rightarrow\\&\lim_{t\to\infty} v_i^d(t)=\lim_{t\to\infty}(1-r_i(t)).v_i^d(t)+\lim_{t\to\infty} r_i(t).v_i^d(t)\overset{1.69,1.70}{\Rightarrow}\lim_{t\to\infty} v_i^d(t)=0\end{aligned} \tag{5.71}$$

Based on Eq. 5.11, we can say:

$$x_i^d(t+1)-x_i^d(t)=v_i^d(t)\overset{1.71}{\Rightarrow}\lim_{t\to\infty}\left(x_i^d(t+1)-x_i^d(t)\right)=\lim_{t\to\infty} v_i^d(t)=0 \tag{5.72}$$

Hence, we can conclude that $x_i^d(t)$ is a convergent sequence and the proof of GSA convergence is now accomplished. This means that each object will converge to a stable point.

Also, other literature investigated the convergence of GSA by discrete time-invariant linear system theory (Jiang, Wang, and Ji, 2014) and Lyapundov stability theorem (Farivar and Shoorehdeli, 2016). Recently, another convergence analysis was performed on GSA based on linear and quadratic function (Yadav, Anita, and Kim, 2019).

5.6 APPLICATION OF GSA

The GSA has been applied in many engineering applications. In such problems, the key is to correctly define the variables and objective functions as well as the appropriate formulation of the problem. In this section, we have categorized the usage of GSA in various applications in different categories and individually review the literature in that area.

5.6.1 Power Engineering

The process of optimizing the operations of a power system is called optimal power flow (OPF), which is a nonlinear process. GSA is used to set and place the devices in the system to minimize the cost and power loss in a system.

Shilaja and Arunprasath (2019) proposed a hybrid algorithm called MSA-GSA by combining GSA and moth swarm algorithm (MSA) that tried to supply the power with the help of wind energy source. The wind source is used as an alternative to renewable energy sources.

In other literature (Radosavljević et al., 2014) the OPF problem is solved in a distributed network by the GSA. The control variables in this problem are voltage magnitudes, active and reactive power outputs, root node voltage, and shunt compensators. Manuel and Shivkumar (2019) proposed a GSA-based power flow control algorithm. The GSA is used to optimize the parameters of the power controller according to the active and reactive powers in the system.

There are also other kinds of literature (Bhowmik and Chakraborty, 2014; 2015; 2018; Bhowmik, Chakraborty, and Babu, 2014) that used the multi-objective version of GSA for solving the OPF optimization problem. This multi-objective strategy has been used due to the existence of conflicting objectives in the problem such as the maximization of voltage stability and bus voltage deviation versus the minimization of total emission, active power loss, and fuel cost.

In power systems, the study and analysis of the system, which is done to determine the important parameters of the system in normal or emergency conditions, is called economic load dispatch. In load dispatch, a schedule is provided to fulfill load demanding, so the fuel cost is minimized. Therefore, GSA is also helpful to solve this optimization efficiently. Chen et al. (2017) used the combination of GSA and PSO to deal with economic load dispatch. In other literature, the GSA has shown promising results for this optimization problem (Saleh et al., 2018; Hardiansyah, 2020).

Power system stabilizer (PSS) refers to the damping control in a real multi-machine power system. GSA has been used to stabilize the power in this system. Lal and Balachander (2020) proposed a GSA-based algorithm to deal with this issue.

5.6.2 Pattern Recognition

The procedure of finding patterns in data is called pattern recognition. This process is applied to various kinds of data such as image, voice, medical, and so on. Many machine learning algorithms are used to recognize these patterns.

Clustering is a common pattern recognition technique that places similar objects in the same cluster. GSA is widely used for data clustering (Diwakar, Kumar, and Gupta, 2018; Sun et al., 2019). GGSA (Dowlatshahi and Nezamabadi-pour, 2014) is a clustering algorithm based on grouping GSA. GGSA differs from the standard GSA in two important aspects. First, a special encoding scheme called grouping encoding is used to make the relevant structures of clustering problems become parts of solutions. Second, given the encoding, special GSA updating equations suitable for the solutions with grouping encoding are used. A modified version of GSA is also used for data clustering based on a bird flock strategy to add diversity to the algorithm.

GSA is also used in feature subset selection (Sarhani, El Afia, and Faizi, 2018; Taradeh et al., 2019). CPBGSA (Guha et al., 2020) is a clustering-based population feature selection algorithm that used BGSA for feature evaluation. In other literature (Nagpal et al., 2017) a feature subset selection method is proposed for biomedical data. Mahanipour and Nezamabadi-pour (2019a) proposed a multiple feature construction (MFC) algorithm based on GSP to achieve a population of good features.

Metaheuristic methods are usually helpful in image processing. These methods can select salient features and tune the parameters of the algorithm. Liu and Chen (2020) and Cataloluk and Çelebi (2018) proposed GSA-based algorithms for image segmentation. A feature selection algorithm for hyperspectral images is also presented by Paliwal et al. (2017) using the combination of GSA, PSO, and support vector machine (SVM).

Classification is also a famous technique in pattern recognition that assigns each object in the data to a class (Mosavi, Khishe, and Ghamgosar, 2016; Dowlatshahi and Rezaeian, 2016). Xue et al. (2019) proposed a classification algorithm using the fusion of GSA and PSO. An image classifier is also offered by Maria Jenifer et al. (2018) based on the GSA.

5.6.3 Communication Engineering

GSA is also applied to many communication engineering problems such as Wireless sensor network (WSN) routing and clustering (Rafsanjani, Dowlatshahi, and Nezamabadi-pour, 2015;

Gupta, Pandey, and Nandi, 2018; Javidi and Ebrahimi, 2019), WSN localization (Rafsanjani and Dowlatshahi, 2012), and channel assignment (Balusu, Pabboju, and Narsimha, 2020).

Lalwani et al. (2017) proposed an algorithm for routing in two-tiered WSNs. In this algorithm, GSA-based routing is used to optimize the distance and residual energy. QQIGA (Mirhosseini, Barani, and Nezamabadi-pour, 2017) used the quantum version of GSA for designing an optimal adaptive WSN. In another article (Kumar and Shanmugam, 2018) a cloned node detection process is used for WSNs by GSA-based simulated annealing.

5.6.4 Control and Electrical Engineering

GSA has been used to design circuits in electrical engineering (Roberts et al., 2019). Optimizing the size of the CMOS analog amplifier using GSA is performed in several articles (Mallick et al. 2017; Asaithambi, Rajappa, and Ravi 2019; Majeed and Rao 2019).

GSA has been used to design control systems as well. GSA-based fuzzy control tuning is a common procedure in control systems (Chao et al., 2016; Bounar, Labdai, and Boulkroune, 2019). In other literature (Safi et al., 2018), a proportional integral derivative (PID) design is proposed based on GSA for controlling load frequency in a two-area multi-source power system.

5.6.5 Civil Engineering

GSA is applied to civil engineering problems for optimizing the design parameters. Khatibinia and Yazdani (2018) proposed a modified version of GSA to avoid slow convergence and weak local exploitation as two weaknesses of GSA. They used two crossover and mutation parameters to balance the exploitation and exploration capabilities of GSA. This method is proposed for the size optimization of truss structures. A hybridization of GSA and PSO (Mirzai, Zahrai, and Bozorgi, 2017) is also proposed for uniformity and drift reduction. Khajehzadeh et al. (2012) presented a modified version of GSA for slope stability analysis. In this method, a constraint is defined for maximum velocity to control global exploration and improve the convergence rate. Momeni et al. (2020) proposed an efficient optimal neural network based on the GSA in predicting the deformation of geogrid-reinforced soil structures.

5.6.6 Software Engineering

Amoozegar and Nezamabadi-pour (2012) proposed a software performance optimization based on a constrained GSA to explore automatically the software design space and determine the best configuration for performance evaluation. Later, they employed MOGSA (Amoozegar and Nezamabadi-Pour, 2015) to determine the best configuration of a software model to minimize the cost, response time, and the strength of the bottleneck. This algorithm is utilized based on the performance of a layered queening network (LQN) and to analyze the bottleneck of the layers.

5.6.7 Mechanical Engineering

Rather and Bala (2020) investigated chaotic gravitational search algorithm (CGSA) performance in solving mechanical engineering designs such as pressure vessel design (PVD), compression spring design (CSD), and welded beam design (WBD). Bostanian (2017) proposed a fuzzy-GSA-based control method to control the vehicle distance and speed to the proceeding vehicle in the same lane. The control of velocity and distance is achieved by fuzzy logic and parameters are tuned by GSA

5.6.8 Biology

GSA is also used in biology to solve the DNA sequence analysis problem (González-Álvarez et al., 2013; Chang and Tian, 2016). A hybrid algorithm based on GSA and *K*-means is proposed (Priyadarshini et al., 2020) for the prediction of diabetes mellitus. In this algorithm, instead of considering random samples as cluster centers, the GSA finds the best centroids based on two classes. These two classes are diabetic and non-diabetic people.

5.7 CONCLUSION

The GSA is one of the most powerful metaheuristic algorithms currently used to solve various optimization problems in different scientific applications. Due to the variety of problems in different applications, several versions of GSA have been proposed to solve binary, discrete, continuous, constraint, multimodal, and multi-objective optimization problems. Researchers have thus far tried to improve the performance of the algorithm and cover its weaknesses by combining it with other algorithms and providing operators for this algorithm such as the black hole, Kepler, disruption, and escape.

GSA is an algorithm with unique properties and operators that no other metaheuristic algorithms can replace in its absence. GSA and its variants are new types of metaheuristic algorithms that are not similar to other methods. This algorithm has been used in various engineering applications including power engineering, electronics, pattern recognition, control engineering, biology, mechanical engineering, civil engineering, and other engineering applications.

Given that different real-world issues require different exploration and exploitation procedures, many versions of GSA have been offered to improve GSA by implying the setting of parameters in a particular application, the effects of the parameters on the convergence rate, and the prevention of local optimum.

If we want to mention the limitations of the basic GSA, it suffers from long computational time and slow searching speed in the last iterations. Other problems also existed such as using complex operators and the difficulty in finding suitable values for the gravitational constant parameter (G). Despite these limitations, GSA has been very effective when solving optimization problems in the various engineering areas. GSA has shown better performance in terms of avoiding the local minima and converging speed compared with metaheuristic methods such as ACO and PSO. Also, it could produce a better quality solution compared with the GA and PSO in terms of stable convergence characteristics and shorter computational time.

Finally, one of the challenges facing the GSA is the study of its convergence characteristics. Also, this algorithm can be used to solve optimization problems in other engineering applications such as complex industrial and economic problems as well as many-objective problems. Large-scale and dynamic problems are also among the challenges facing GSA. Although many operators have been designed for the GSA so far, there is still great potential for designing new operators based on concepts such as planet motions, anti-gravity, star clustering, and other gravity-related concepts. These high potentials can be considered for researchers who intend to improve the algorithm and overcome limitations or use it to solve various problems.

REFERENCES

Abbasian, Mohammad Amir, Hossein Nezamabadi-pour, and Maryam Amoozegar. 2015. "A Clustering Based Archive Multi Objective Gravitational Search Algorithm." *Fundamenta Informaticae* 138 (4): 387–409. https://doi.org/10.3233/FI-2015-1218.

Amoozegar, M., and H. Nezamabadi-pour. 2015. "A Multi-Objective Approach to Model-Driven Performance Bottlenecks Mitigation." *Scientia Iranica* 22 (3): 1018–1030.

Amoozegar, Maryam, and Hossein Nezamabadi-pour. 2012. "Software Performance Optimization Based on Constrained GSA." In *AISP 2012 – 16th CSI International Symposium on Artificial Intelligence and Signal Processing*, 134–139. https://doi.org/10.1109/AISP.2012.6313732.

Asaithambi, Sasikumar, Muthaiah Rajappa, and Logesh Ravi. 2019. "Optimization and Control of CMOS Analog Integrated Circuits for Cyber-Physical Systems Using Hybrid Grey Wolf Optimization Algorithm." *Journal of Intelligent and Fuzzy Systems* 36 (5): 4235–4245. https://doi.org/10.3233/JIFS-169981.

Asrul, Ahmad, Ibrahim Azah Mohamed, and Hussain Shareef. 2014. "Optimal Power Quality Monitor Placement in Power Systems Using an Adaptive Quantum-Inspired Binary Gravitational Search Algorithm." *International Journal of Electrical Power and Energy Systems* 57: 404–413. https://doi.org/10.1016/j.ijepes.2013.12.019.

Bala, Indu, and Anupam Yadav. 2020. "Comprehensive Learning Gravitational Search Algorithm for Global Optimization of Multimodal Functions." *Neural Computing and Applications* 32 (11): 7347–7382. https://doi.org/10.1007/s00521-019-04250-5.

Balusu, Nandini, Suresh Pabboju, and G. Narsimha. 2020. "Swarm Optimization Based Gravitational Search Approach for Channel Assignment in MCMR Wireless MESH Network." *International Journal of Computer Networks and Communications* 12 (3): 41–54. https://doi.org/10.5121/ijcnc.2020.12303.

Barani, Fatemeh, Mina Mirhosseini, and Hossein Nezamabadi-pour. 2017. "Application of Binary Quantum-Inspired Gravitational Search Algorithm in Feature Subset Selection." *Applied Intelligence* 47 (2): 304–318. https://doi.org/10.1007/s10489-017-0894-3.

Barani, Fatemeh, Mina Mirhosseini, Hossein Nezamabadi-pour, and Malihe M. Farsangi. 2017. "Unit Commitment by an Improved Binary Quantum GSA." *Applied Soft Computing Journal* 60: 180–189. https://doi.org/10.1016/j.asoc.2017.06.051.

Bardamova, Marina, Anton Konev, Ilya Hodashinsky, and Alexander Shelupanov. 2018. "A Fuzzy Classifier with Feature Selection Based on the Gravitational Search Algorithm." *Symmetry* 10 (11): 609. https://doi.org/10.3390/sym10110609.

Bhowmik, Arup Ratan, and A. K. Chakraborty. 2014. "Solution of Optimal Power Flow Using Nondominated Sorting Multi Objective Gravitational Search Algorithm." *International Journal of Electrical Power and Energy Systems* 62: 323–334. https://doi.org/10.1016/j.ijepes.2014.04.053.

Bhowmik, Arup Ratan, and Ajoy Kumar Chakraborty. 2015. "Solution of Optimal Power Flow Using Non Dominated Sorting Multi Objective Opposition Based Gravitational Search Algorithm." *International Journal of Electrical Power and Energy Systems* 64: 1237–1250. https://doi.org/10.1016/j.ijepes.2014.09.015.

Bhowmik, Arup Ratan, and Ajoy Kumar Chakraborty. 2018. "Non Dominated Sorting Based Multi Objective GSA for Solving Optimal Power Flow Problems." In *IEEE International Conference on Power, Control, Signals and Instrumentation Engineering, ICPCSI 2017*, 2195–2200. https://doi.org/10.1109/ICPCSI.2017.8392108.

Bhowmik, Arup Ratan, Ajoy K. Chakraborty, and K. Narendra Babu. 2014. "Multi Objective Optimal Power Flow Using NSMOGSA." In *2014 International Conference on Circuits, Power and Computing Technologies, ICCPCT 2014*, 84–88. https://doi.org/10.1109/ICCPCT.2014.7054776.

Bostanian, M. 2017. "Fuzzy-GSA Based Control Approach for Developing Adaptive Cruise Control." *ADMT Journal* 10 (4): 7–19.

Bounar, N., S. Labdai, and A. Boulkroune. 2019. "PSO–GSA Based Fuzzy Sliding Mode Controller for DFIG-Based Wind Turbine." *ISA Transactions* 85: 177–188. https://doi.org/10.1016/j.isatra.2018.10.020.

Çataloluk, Hatice, and Fatih Vehbi Çelebi. 2018. "A Novel Hybrid Model for Two-Phase Image Segmentation: GSA Based Chan–Vese Algorithm." *Engineering Applications of Artificial Intelligence* 73: 22–30. https://doi.org/10.1016/j.engappai.2018.04.027.

Chang, Billy Heung Wing, and Weidong Tian. 2016. "GSA-Lightning: Ultra-Fast Permutation-Based Gene Set Analysis." *Bioinformatics* 32 (19): 3029–3031. https://doi.org/10.1093/bioinformatics/btw349.

Chao, Chun Tang, Ming Tang Liu, Juing Shian Chiou, Yi Jung Huang, and Chi Jo Wang. 2016. "A GSA-Based Adaptive Fuzzy PID-Controller for an Active Suspension System." *Engineering Computations (Swansea, Wales)* 33 (6): 1659–1667. https://doi.org/10.1108/EC-08-2015-0240.

Chen, Gonggui, Lilan Liu, Zhizhong Zhang, and Shanwai Huang. 2017. "Optimal Reactive Power Dispatch by Improved GSA-Based Algorithm with the Novel Strategies to Handle Constraints." *Applied Soft Computing Journal* 50: 58–70. https://doi.org/10.1016/j.asoc.2016.11.008.

Chen, H., S. Li, and Z. Tang. 2011. "Hybrid GSA with Random-Key Encoding Scheme Combined with Simulated Annealing." *International Journal of Computer Science and Network Security* 11 (6): 208–217.

Chiranjeevi, Karri, and Umaranjan Jena. 2017. "Hybrid Gravitational Search and Pattern Search–Based Image Thresholding by Optimising Shannon and Fuzzy Entropy for Image Compression." *International Journal of Image and Data Fusion* 8 (3): 236–269. https://doi.org/10.1080/19479832.2017.1338760.

Diwakar, Maya, Jeetendra Kumar, and Indresh Kumar Gupta. 2018. "A Hybrid GA-GSA Noval Algorithm for Data Clustering." In *Proceedings of the 4th IEEE International Conference on Recent Advances in Information Technology, RAIT 2018*, 1–6. https://doi.org/10.1109/RAIT.2018.8389064.

Doraghinejad, Mohammad, and Hossein Nezamabadi-pour. 2014. "Black Hole: A New Operator for Gravitational Search Algorithm." *International Journal of Computational Intelligence Systems* 7 (5): 809–826. https://doi.org/10.1080/18756891.2014.966990.

Doraghinejad, Mohammad, Hossein Nezamabadi-pour, and Ali Mahani. 2014. "Channel Assignment in Multi-Radio Wireless Mesh Networks Using an Improved Gravitational Search Algorithm." *Journal of Network and Computer Applications* 38 (1): 163–171. https://doi.org/10.1016/j.jnca.2013.04.007.

Dorigo, Marco, and Krzysztof Socha. 2007. "Ant Colony Optimization." In *Handbook of Approximation Algorithms and Metaheuristics*. New York: Chapman and Hall/CRC. https://doi.org/10.1201/9781420010749.

Dowlatshahi, M. B., V. Derhami, and H. Nezamabadi-pour. 2018. Gravitational Locally Informed Particle Swarm Algorithm for Solving Multimodal Optimization Problems." *Tabriz Journal of Electrical Engineering* 48 (85): 1131–1140. https://tjee.tabrizu.ac.ir/article_8209.html.

Dowlatshahi, Mohammad Bagher, and Hossein Nezamabadi-pour. 2014. "GGSA: A Grouping Gravitational Search Algorithm for Data Clustering." *Engineering Applications of Artificial Intelligence* 36: 114–121. https://doi.org/10.1016/j.engappai.2014.07.016.

Dowlatshahi, Mohammad Bagher, Hossein Nezamabadi-pour, and Mashaallah Mashinchi. 2014. "A Discrete Gravitational Search Algorithm for Solving Combinatorial Optimization Problems." *Information Sciences* 258: 94–107. https://doi.org/10.1016/j.ins.2013.09.034.

Dowlatshahi, Mohammad Bagher, and Mehdi Rezaeian. 2016. "Training Spiking Neurons with Gravitational Search Algorithm for Data Classification." In *Proceedings of the 1st Conference on Swarm Intelligence and Evolutionary Computation, CSIEC 2016*, 53–58. https://doi.org/10.1109/CSIEC.2016.7482125.

Dowlatshahi, Mohammad Bagher, and Vali Derhami. 2019. "Winner Determination in Combinatorial Auctions Using Hybrid Ant Colony Optimization and Multi-Neighborhood Local Search." *Journal of AI and Data Mining* 5 (2): 169–181. https://doi.org/https://dx.doi.org/10.22044/jadm.2017.880.

Dowlatshahi, Mohammad Bagher, Vali Derhami, and Hossein Nezamabadi-pour. 2019. "Gravitational Search Algorithm with Nearest-Better Neighborhood for Multimodal Optimization Problems." *Journal of Soft Computing and Information Technology (JSCIT)* 8 (3): 10–19.

Eappen, Geoffrey, and Shankar T. 2020. "Hybrid PSO-GSA for Energy Efficient Spectrum Sensing in Cognitive Radio Network." *Physical Communication* 40: 101091. https://doi.org/10.1016/j.phycom.2020.101091.

Eberhart, Russell, and James Kennedy. 1999. "A New Optimizer Using Particle Swarm Theory." *Sixth International Symposium on Micro Machine and Human Science,* 39–43. https://doi.org/10.1.1.470.3577.

Eiben, A. E., Zbigniew Michalewicz, Marc Schoenauer, and James E. Smith. 2007. "Parameter Control in Evolutionary Algorithms." *Studies in Computational Intelligence* 54: 19–46. https://doi.org/10.1007/978-3-540-69432-8_2.

Eiben, Ágoston E., and James E. Smith. 2012. "Evolutionary Algorithms." *Studies in Computational Intelligence* 379: 9–27. https://doi.org/10.1007/978-3-642-23247-3_2.

Eid, Ahmad. 2020. "Allocation of distributed generations in radial distribution systems using adaptive PSO and modified GSA multi-objective optimizations." *Alexandria Engineering Journal* 59 (6): 4771–4786. https://doi.org/10.1016/j.aej.2020.08.042.

Farivar, Faezeh, and Mahdi Aliyari Shoorehdeli. 2016. "Stability Analysis of Particle Dynamics in Gravitational Search Optimization Algorithm." *Information Sciences* 337–338: 25–43. https://doi.org/10.1016/j.ins.2015.12.017.

Feng, Zhong Kai, Shuai Liu, Wen Jing Niu, Zhi Qiang Jiang, Bin Luo, and Shu Min Miao. 2019. "Multi-Objective Operation of Cascade Hydropower Reservoirs Using TOPSIS and Gravitational Search Algorithm with Opposition Learning and Mutation." *Water (Switzerland)* 11 (10): 2040. https://doi.org/10.3390/w11102040.

Fister, Iztok, Damjan Strnad, Xin She Yang, and Iztok Fister. 2015. "Adaptation and Hybridization in Nature-Inspired Algorithms." *Adaptation, Learning, and Optimization* 18: 3–50. https://doi.org/10.1007/978-3-319-14400-9_1.

Gao, Shangce, Yuki Todo, Tao Gong, Gang Yang, and Zheng Tang. 2014. "Graph Planarization Problem Optimization Based on Triple-Valued Gravitational Search Algorithm." *IEEJ Transactions on Electrical and Electronic Engineering* 9 (1): 39–48. https://doi.org/10.1002/tee.21934.

Garg, Harish. 2014. "A Hybrid GA-GSA Algorithm for Optimizing the Performance of an Industrial System by Utilizing Uncertain Data." In *Handbook of Research on Artificial Intelligence Techniques and Algorithms*, 620–654. IGI Global. https://doi.org/10.4018/978-1-4666-7258-1.ch020.

Garg, Harish. 2019. "A Hybrid GSA-GA Algorithm for Constrained Optimization Problems." *Information Sciences* 478: 499–523. https://doi.org/10.1016/j.ins.2018.11.041.

Ghorbani, Farzaneh, and Hossein Nezamabadi-pour. 2012. "On the Convergence Analysis of Gravitational Search Algorithm." *Journal of Advances in Computer Research* 3 (2): 45–51. www.jacr.iausari.ac.ir.

Golzari, Shahram, Mohammad Nourmohammadi Zardehsavar, Amin Mousavi, Mahmoud Reza Saybani, Abdullah Khalili, and Shahaboddin Shamshirband. 2018. "KGSA: A Gravitational Search Algorithm for Multimodal Optimization Based on K-Means Niching Technique and a Novel Elitism Strategy." *Open Mathematics* 16 (1): 1582–1606. https://doi.org/10.1515/math-2018-0132.

González-Álvarez, David L., Miguel A. Vega-Rodríguez, Juan A. Gómez-Pulido, and Juan M. Sánchez-Pérez. 2013. "Comparing Multiobjective Swarm Intelligence Metaheuristics for DNA Motif Discovery." *Engineering Applications of Artificial Intelligence* 26 (1): 314–326. https://doi.org/10.1016/j.engappai.2012.06.014.

Gu, Binjie, and Feng Pan. 2013. "Modified Gravitational Search Algorithm with Particle Memory Ability and Its Application." *International Journal of Innovative Computing, Information and Control* 9 (11): 4531–4544.

Guha, Ritam, Manosij Ghosh, Akash Chakrabarti, Ram Sarkar, and Seyedali Mirjalili. 2020. "Introducing Clustering Based Population in Binary Gravitational Search Algorithm for Feature Selection." *Applied Soft Computing Journal* 93: 106341. https://doi.org/10.1016/j.asoc.2020.106341.

Guo, Zhifeng. 2012. "A Hybrid Optimization Algorithm Based on Artificial Bee Colony and Gravitational Search Algorithm." *International Journal of Digital Content Technology and Its Applications* 6 (17): 620–626. https://doi.org/10.4156/jdcta.vol6.issue17.68.

Gupta, Rohan Kumar, Ashish Pandey, and Arnab Nandi. 2018. "Lifetime Enhancement of WSN Using Evolutionary Clustering and Routing Algorithms." In *2018 IEEE International Students' Conference on Electrical, Electronics and Computer Science, SCEECS 2018*. https://doi.org/10.1109/SCEECS.2018.8546977.

Haghbayan, Pourya, Hossein Nezamabadi-pour, and Shima Kamyab. 2017. "A Niche GSA Method with Nearest Neighbor Scheme for Multimodal Optimization." *Swarm and Evolutionary Computation* 35: 78–92. https://doi.org/10.1016/j.swevo.2017.03.002.

Han, Xiaohong, and Xiaoming Chang. 2012. "A Chaotic Digital Secure Communication Based on a Modified Gravitational Search Algorithm Filter." *Information Sciences* 208: 14–27. https://doi.org/10.1016/j.ins.2012.04.039.

Hardiansyah, Hardiansyah. 2020. "Hybrid Psogsa Technique for Solving Dynamic Economic Emission Dispatch Problem." *Engineering Review* 40 (3): 96–104. https://doi.org/10.30765/er.40.3.10.

Hashemi, Amin, Mohammad Bagher Dowlatshahi, and Hossein Nezamabadi-pour. 2020. "A Bipartite Matching-Based Feature Selection for Multi-Label Learning." *International Journal of Machine Learning and Cybernetics* 12: 459–475. https://doi.org/10.1007/s13042-020-01180-w.

Hashemi, Amin, Mohammad Bagher Dowlatshahi, and Hossein Nezamabadi-pour. 2020a. "MFS-MCDM: Multi-Label Feature Selection Using Multi-Criteria Decision Making." *Knowledge-Based Systems* 206: 106365. https://doi.org/10.1016/j.knosys.2020.106365.

Hashemi, Amin, Mohammad Bagher Dowlatshahi, and Hossein Nezamabadi-pour. 2020b. "MGFS: A Multi-Label Graph-Based Feature Selection Algorithm via PageRank Centrality." *Expert Systems with Applications* 142 (March): 113024. https://doi.org/10.1016/j.eswa.2019.113024.

Hassanzadeh, Hamid Reza, and Modjtaba Rouhani. 2010. "A Multi-Objective Gravitational Search Algorithm." In *Proceedings, 2nd International Conference on Computational Intelligence, Communication Systems and Networks, CICSyN 2010*, 7–12. https://doi.org/10.1109/CICSyN.2010.32.

Ibrahim, Ahmad Asrul, Azah Mohamed, and Hussain Shareef. 2012. "A Novel Quantum-Inspired Binary Gravitational Search Algorithm in Obtaining Optimal Power Quality Monitor Placement." *Journal of Applied Sciences* 12 (9): 822–830. https://doi.org/10.3923/jas.2012.822.830.

Javidi, Mohammad Masoud, and Sepehr Ebrahimi. 2019. "A Novel Version of GSA and Its Application in the K-of-N Lifetime Problem in Two-Tiered WSNs." *AUT Journal of Modeling and Simulation*. https://doi.org/https://dx.doi.org/10.22060/miscj.2019.15273.5129.

Ji, Bin, Xiaohui Yuan, Xianshan Li, Yuehua Huang, and Wenwu Li. 2014. "Application of Quantum-Inspired Binary Gravitational Search Algorithm for Thermal Unit Commitment with Wind Power Integration." *Energy Conversion and Management* 87: 589–598. https://doi.org/10.1016/j.enconman.2014.07.060.

Ji, Junkai, Shangce Gao, Shuaiqun Wang, Yajiao Tang, Hang Yu, and Yuki Todo. 2017. "Self-Adaptive Gravitational Search Algorithm with a Modified Chaotic Local Search." *IEEE Access* 5: 17881–1795. https://doi.org/10.1109/ACCESS.2017.2748957.

Jiang, Shanhe, Yan Wang, and Zhicheng Ji. 2014. "Convergence Analysis and Performance of an Improved Gravitational Search Algorithm." *Applied Soft Computing Journal* 24: 363–384. https://doi.org/10.1016/j.asoc.2014.07.016.

Khadanga, Rajendra Ku, and Jitendriya Ku Satapathy. 2015. "A New Hybrid GA-GSA Algorithm for Tuning Damping Controller Parameters for a Unified Power Flow Controller." *International Journal of Electrical Power and Energy Systems* 73: 1060–1069. https://doi.org/10.1016/j.ijepes.2015.07.016.

Khadanga, Rajendra Kumar, and Jitendriya Kumar Satapathy. 2018. "A Hybrid Gravitational Search and Pattern Search Algorithm for Tuning Damping Controller Parameters for a Unified Power Flow Controller—A Comparative Approach." *International Journal of Numerical Modelling: Electronic Networks, Devices and Fields* 31 (3): e2312. https://doi.org/10.1002/jnm.2312.

Khajehzadeh, Mohammad, Mohd Raihan Taha, Ahmed El-Shafie, and Mahdiyeh Eslami. 2012. "A Modified Gravitational Search Algorithm for Slope Stability Analysis." *Engineering Applications of Artificial Intelligence* 25 (8): 1589–1597. https://doi.org/10.1016/j.engappai.2012.01.011.

Khajooei, Fatemeh, and Esmat Rashedi. 2016. "A New Version of Gravitational Search Algorithm with Negative Mass." In *Proceedings of the 1st Conference on Swarm Intelligence and Evolutionary Computation, CSIEC 2016*, 1–5. https://doi.org/10.1109/CSIEC.2016.7482123.

Khatibinia, Mohsen, and Hessam Yazdani. 2018. "Accelerated Multi-Gravitational Search Algorithm for Size Optimization of Truss Structures." *Swarm and Evolutionary Computation* 38: 109–119. https://doi.org/10.1016/j.swevo.2017.07.001.

Kumar, D. Rajesh, and A. Shanmugam. 2018. "A Hyper Heuristic Localization Based Cloned Node Detection Technique Using GSA Based Simulated Annealing in Sensor Networks." In *Cognitive Computing for Big Data Systems Over IoT*, 307–335. Cham, Switzerland: Springer. https://doi.org/10.1007/978-3-319-70688-7_13.

Lal, Channu, and K. Balachander. 2020. "Design and Analysis of PSS and SVC Controller Using Crossover GSA Approach." *Test Engineering and Management* 82: 6220–6227.

Lalwani, Praveen, Haider Banka, and Chiranjeev Kumar. 2017. "GSA-CHSR: Gravitational Search Algorithm for Cluster Head Selection and Routing in Wireless Sensor Networks." In *Applications of Soft Computing for the Web*, 225–252. Singapore: Springer Singapore. https://doi.org/10.1007/978-981-10-7098-3_13.

Lei, Zhenyu, Shangce Gao, Shubham Gupta, Jiujun Cheng, and Gang Yang. 2020. "An Aggregative Learning Gravitational Search Algorithm with Self-Adaptive Gravitational Constants." *Expert Systems with Applications* 152: 113396. https://doi.org/10.1016/j.eswa.2020.113396.

Li, Xiaodong, and Maurice Clerc. 2019. "Swarm Intelligence." In *Handbook of Metaheuristics*, vol. 272, pp. 353–84. Cham, Switzerland: Springer. https://doi.org/10.1007/978-3-319-91086-4_11.

Liu, Hao, Guiyan Ding, and Huafei Sun. 2012. "An Improved Opposition-Based Disruption Operator in Gravitational Search Algorithm." In *Proceedings of the 2012 5th International Symposium on Computational Intelligence and Design, ISCID 2012*, 2: 123–126. https://doi.org/10.1109/ISCID.2012.183.

Liu, Qing, Jin Chang Hu, Yan Wang, and Yao Hua Wu. 2019. "An Improved Gravitational Search Algorithm for the Vehicle Routing Problem." In *Proceedings of the 8th International Conference on Logistics and Systems Engineering 2018*, 421–430.

Liu, Zilong, and Guobin Chen. 2020. "Remote Sensing Image Landmark Segmentation Algorithm Based on Improved GSA and PCNN Combination." *International Journal of Electrical Engineering Education.* https://doi.org/10.1177/0020720920936826.

Lou, Jungang, Yunliang Jiang, Qing Shen, and Ruiqin Wang. 2018. "Failure Prediction by Relevance Vector Regression with Improved Quantum-Inspired Gravitational Search." *Journal of Network and Computer Applications* 103: 171–177. https://doi.org/10.1016/j.jnca.2017.11.013.

Lozano, M., and C. García-Martínez. 2010. "Hybrid Metaheuristics with Evolutionary Algorithms Specializing in Intensification and Diversification: Overview and Progress Report." *Computers and Operations Research* 37 (3): 481–497. https://doi.org/10.1016/j.cor.2009.02.010.

Mahanipour, Afsaneh, and Hossein Nezamabadi-pour. 2019a. "A Multiple Feature Construction Method Based on Gravitational Search Algorithm." *Expert Systems with Applications* 127: 199–209. https://doi.org/10.1016/j.eswa.2019.03.015.

Mahanipour, Afsaneh, and Hossein Nezamabadi-pour. 2019b. "GSP: An Automatic Programming Technique with Gravitational Search Algorithm." *Applied Intelligence* 49 (4): 1502–1516. https://doi.org/10.1007/s10489-018-1327-7.

Mahdad, Belkacem, and Kamel Srairi. 2015. "Interactive Gravitational Search Algorithm and Pattern Search Algorithms for Practical Dynamic Economic Dispatch." *International Transactions on Electrical Energy Systems* 25 (10): 2289–2309. https://doi.org/10.1002/etep.1961.

Majeed, M. A. Mushahhid, and Patri Sreehari Rao. 2019. "Optimal Design of CMOS Amplifier Circuits Using Whale Optimization Algorithm." *Communications in Computer and Information Science*, 839: 590–605. https://doi.org/10.1007/978-981-13-2372-0_53.

Mallick, S., R. Kar, D. Mandal, and S. P. Ghoshal. 2017. "Optimal Sizing of CMOS Analog Circuits Using Gravitational Search Algorithm with Particle Swarm Optimization." *International Journal of Machine Learning and Cybernetics* 8 (1): 309–331. https://doi.org/10.1007/s13042-014-0324-3.

Mallick, Sourav, S. P. Ghoshal, P. Acharjee, and S. S. Thakur. 2013. "Optimal Static State Estimation Using Improved Particle Swarm Optimization and Gravitational Search Algorithm." *International Journal of Electrical Power and Energy Systems* 52 (1): 254–265. https://doi.org/10.1016/j.ijepes.2013.03.035.

Manuel, Raju, and Poorani Shivkumar. 2019. "Power Flow Control Model of Energy Storage Connected to Smart Grid in Unbalanced Conditions: A GSA-Technique-Based Assessment." *International Journal of Ambient Energy* 40 (4): 417–426. https://doi.org/10.1080/01430750.2017.1405280.

Maria Jenifer, L., T. Sathiya, and B Sathiyabhama. 2018. "GSA Based Classification of Lung Nodules in CT Images." *SSRN Electronic Journal.* https://doi.org/10.2139/ssrn.3134259.

Mehdizadeh, Saeid, Babak Mohammadi, Quoc Bao Pham, Dao Nguyen Khoi, and Nguyen Thi Thuy Linh. 2020. "Implementing Novel Hybrid Models to Improve Indirect Measurement of the Daily Soil Temperature: Elman Neural Network Coupled with Gravitational Search Algorithm and Ant Colony Optimization." *Measurement: Journal of the International Measurement Confederation* 165: 108127. https://doi.org/10.1016/j.measurement.2020.108127.

Mirhosseini, Mina, Fatemeh Barani, and Hossein Nezamabadi-pour. 2017. "QQIGSA: A Quadrivalent Quantum-Inspired GSA and Its Application in Optimal Adaptive Design of Wireless Sensor Networks." *Journal of Network and Computer Applications* 78: 231–241. https://doi.org/10.1016/j.jnca.2016.11.001.

Mirzai, Nadia M., Seyed Mehdi Zahrai, and Fatemeh Bozorgi. 2017. "Proposing Optimum Parameters of TMDs Using GSA and PSO Algorithms for Drift Reduction and Uniformity." *Structural Engineering and Mechanics* 63 (2): 147–160. https://doi.org/10.12989/sem.2017.63.2.147.

Momeni, Ehsan, Akbar Yarivand, Mohamad Bagher Dowlatshahi, and Danial Jahed Armaghani. 2020. "An Efficient Optimal Neural Network Based on Gravitational Search Algorithm in Predicting the Deformation of Geogrid-Reinforced Soil Structures." *Transportation Geotechnics* 26: 100446. https://doi.org/10.1016/j.trgeo.2020.100446.

Mood, Sepehr Ebrahimi, Esmat Rashedi, and Mohammad Masoud Javidi. 2015. "New Functions for Mass Calculation in Gravitational Search Algorithm." *Journal of Computing and Security* 2 (3): 233–246. http://www.jcomsec.org.

Mosa, Mohamed Atef. 2020. "A Novel Hybrid Particle Swarm Optimization and Gravitational Search Algorithm for Multi-Objective Optimization of Text Mining." *Applied Soft Computing Journal* 90: 106189. https://doi.org/10.1016/j.asoc.2020.106189.

Mosavi, Mohammad Reza, Mohammad Khishe, and Abolfazl Ghamgosar. 2016. "Classification Of Sonar Data Set Using Neural Network Trained By Gray Wolf Optimization." *Neural Network World* 26 (4): 393–415. https://doi.org/10.14311/nnw.2016.26.023.

Nagpal, Sushama, Sanchit Arora, Sangeeta Dey, and S. Shreya. 2017. "Feature Selection Using Gravitational Search Algorithm for Biomedical Data." *Procedia Computer Science* 115: 258–265. https://doi.org/10.1016/j.procs.2017.09.133.

Nezamabadi-pour, Hossein. 2015. "A Quantum-Inspired Gravitational Search Algorithm for Binary Encoded Optimization Problems." *Engineering Applications of Artificial Intelligence* 40: 62–75. https://doi.org/10.1016/j.engappai.2015.01.002.

Nezamabadi-pour, Hossein, and Fatemeh Barani. 2016. "Gravitational Search Algorithm: Concepts, Variants, and Operators." In *Handbook of Research on Modern Optimization Algorithms and Applications in Engineering and Economics*, 700–750. IGI Global. https://doi.org/10.4018/978-1-4666-9644-0.ch027.

Nobahari, Hadi, Mahdi Nikusokhan, and Patrick Siarry. 2012. "A Multi-Objective Gravitational Search Algorithm Based on Non-Dominated Sorting." *International Journal of Swarm Intelligence Research* 3 (3): 32–49. https://doi.org/10.4018/jsir.2012070103.

Nocedal, J., and S. J. Wright. 2006. *Numerical Optimization*. New York: Springer.

Özyön, Serdar, Celal Yaşar, Burhanettin Durmuş, and Hasan Temurtaş. 2015. "Opposition-Based Gravitational Search Algorithm Applied to Economic Power Dispatch Problems Consisting of Thermal Units with Emission Constraints." *Turkish Journal of Electrical Engineering and Computer Sciences* 23: 2278–2288. https://doi.org/10.3906/elk-1305-258.

Paliwal, K. K., Surjeet Singh, and Purnima Gaba. 2017. "Feature Selection Approach of Hyperspectral Image Using GSA-FODPSO-SVM." In *Proceedings of the IEEE International Conference on Computing, Communication and Automation, ICCCA 2017*, 1070–1075. https://doi.org/10.1109/CCAA.2017.8229954.

Paniri, Mohsen, Mohammad Bagher Dowlatshahi, and Hossein Nezamabadi-pour. 2020. "MLACO: A Multi-Label Feature Selection Algorithm Based on Ant Colony Optimization." *Knowledge-Based Systems* 192: 105285.

Poole, Daniel J., Christian B. Allen, and Thomas C. S. Rendall. 2014. "Analysis of Constraint Handling Methods for the Gravitational Search Algorithm." In Proceedings of the 2014 IEEE Congress on Evolutionary Computation, CEC 2014, 2005–2012. https://doi.org/10.1109/CEC.2014.6900271.

Precup, Radu Emil, Radu Codruţ David, Emil M. Petriu, Stefan Preitl, and Mircea Bogdan Rădac. 2013. "Experiments in Fuzzy Controller Tuning Based on an Adaptive Gravitational Search Algorithm." *Proceedings of the Romanian Academy Series A - Mathematics Physics Technical Sciences Information Science* 14 (4): 360–367.

Precup, Radu Emil, Radu Codrut David, Emil M. Petriu, Mircea Bogdan Radac, and Stefan Preitl. 2014. "Adaptive GSA-Based Optimal Tuning of PI Controlled Servo Systems with Reduced Process Parametric Sensitivity, Robust Stability and Controller Robustness." *IEEE Transactions on Cybernetics* 44 (11): 1997–2009. https://doi.org/10.1109/TCYB.2014.2307257.

Priyadarshi, Neeraj, Mahajan Sagar Bhaskar, Sanjeevikumar Padmanaban, Frede Blaabjerg, and Farooque Azam. 2020. "New CUK–SEPIC Converter Based Photovoltaic Power System with Hybrid GSA–PSO Algorithm Employing MPPT for Water Pumping Applications." *IET Power Electronics* 13 (13): 2824–2830. https://doi.org/10.1049/iet-pel.2019.1154.

Priyadarshini, Rojalina, Rabindra Kumar Barik, Nilamadhab Dash, Brojo Kishore Mishra, and Rachita Misra. 2020. "A Hybrid GSA-K-Mean Classifier Algorithm to Predict Diabetes Mellitus." *Cognitive Analytics*, 589–603. IGI Global. https://doi.org/10.4018/978-1-7998-2460-2.ch030.

Radosavljević, Jordan, Miroljub Jevtić, Nebojša Arsić, and Dardan Klimenta. 2014. "Optimal Power Flow for Distribution Networks Using Gravitational Search Algorithm." *Electrical Engineering* 96 (4): 335–345. https://doi.org/10.1007/s00202-014-0302-5.

Rafsanjani, Marjan Kuchaki, and Mohammad Bagher Dowlatshahi. 2012. "Using Gravitational Search Algorithm for Finding Near-Optimal Base Station Location in Two-Tiered WSNs." *International Journal of Machine Learning and Computing*, 377–380. https://doi.org/10.7763/ijmlc.2012.v2.148.

Rafsanjani, Marjan Kuchaki, Mohammad Bagher Dowlatshahi, and Hossein Nezamabadi-pour. 2015. "Gravitational Search Algorithm to Solve the K-of-N Lifetime Problem in Two-Tiered WSNs." *Iranian Journal of Mathematical Sciences and Informatics* 10 (1): 81–93. https://doi.org/10.7508/ijmsi.2015.01.006.

Rashedi, Esmat, and Hossein Nezamabadi-pour. 2014. "Feature Subset Selection Using Improved Binary Gravitational Search Algorithm." *Journal of Intelligent and Fuzzy Systems* 26 (3): 1211–1221. https://doi.org/10.3233/IFS-130807.

Rashedi, Esmat, Hossein Nezamabadi-pour, and Saeid Saryazdi. 2009. "GSA: A Gravitational Search Algorithm." *Information Sciences* 179 (13): 2232–2248. https://doi.org/10.1016/j.ins.2009.03.004.

Rashedi, Esmat, Hossein Nezamabadi-pour, and Saeid Saryazdi. 2010. "BGSA: Binary Gravitational Search Algorithm." *Natural Computing.* 9: 727–745. https://doi.org/10.1007/s11047-009-9175-3.

Rashedi, Esmat, Hossein Nezamabadi-pour, and Saeid Saryazdi. 2013. "A Simultaneous Feature Adaptation and Feature Selection Method for Content-Based Image Retrieval Systems." *Knowledge-Based Systems* 39: 85–94. https://doi.org/10.1016/j.knosys.2012.10.011.

Rashedi, Esmat, Elaheh Rashedi, and Hossein Nezamabadi-pour. 2018. "A Comprehensive Survey on Gravitational Search Algorithm." *Swarm and Evolutionary Computation* 41: 141–158. https://doi.org/10.1016/j.swevo.2018.02.018.

Rather, Sajad Ahmad, and P. Shanthi Bala. 2020. "Swarm-Based Chaotic Gravitational Search Algorithm for Solving Mechanical Engineering Design Problems." *World Journal of Engineering* 17 (1): 97–114. https://doi.org/10.1108/WJE-09-2019-0254.

Roberts, Jarred M., Gary S. Varner, Patrick Allison, Brendan Fox, Eric Oberla, Ben Rotter, and Stefan Spack. 2019. "LAB4D: A Low Power, Multi-GSa/s, Transient Digitizer with Sampling Timebase Trimming Capabilities." *Nuclear Instruments and Methods in Physics Research Section A: Accelerators, Spectrometers, Detectors and Associated Equipment* 925 (May): 92–100. https://doi.org/10.1016/j.nima.2019.01.091.

Saeidi-Khabisi, Fatemeh Sadat, and Esmat Rashedi. 2012. "Fuzzy Gravitational Search Algorithm." In *2012 2nd International EConference on Computer and Knowledge Engineering, ICCKE 2012*, 156–160. https://doi.org/10.1109/ICCKE.2012.6395370.

Safi, Shah Jahan, Suleyman Sungur Tezcan, Ibrahim Eke, and Zakirhussain Farhad. 2018. "Gravitational Search Algorithm (GSA) Based PID Controller Design for Two Area Multi-Source Power System Load Frequency Control (LFC)." *Gazi University Journal of Science* 31 (1): 139–153.

Sahib, Naji Mutar, Abtehaj Hussein, Suha Falih, and Hafeth I. Naji. 2020. "Construction Projects Problems Optimization Using PSO and GSA." *Applied Mechanics and Materials* 897: 147–151. https://doi.org/10.4028/www.scientific.net/amm.897.147.

Sajedi, Hedieh, and Seyedeh Fatemeh Razavi. 2017. "DGSA: Discrete Gravitational Search Algorithm for Solving Knapsack Problem." *Operational Research* 17 (2): 563–591. https://doi.org/10.1007/s12351-016-0240-2.

Saleh, Muhammad, Fitri Imansyah, Usman A. Gani, and H. Hardiansyah. 2018. "Hybrid PSO-GSA Technique for Environmental/Economic Dispatch Problem." *International Journal of Emerging Research in Management and Technology* 6 (9): 112. https://doi.org/10.23956/ijermt.v6i9.94.

Sarafrazi, S., H. Nezamabadi-pour, and S. Saryazdi. 2011. "Disruption: A New Operator in Gravitational Search Algorithm." *Scientia Iranica* 18 (3 D): 539–548. https://doi.org/10.1016/j.scient.2011.04.003.

Sarafrazi, Soroor, and Hossein Nezamabadi-pour. 2013. "Facing the Classification of Binary Problems with a GSA-SVM Hybrid System." *Mathematical and Computer Modelling* 57 (1–2): 270–278. https://doi.org/10.1016/j.mcm.2011.06.048.

Sarafrazi, Soroor, Hossein Nezamabadi-pour, and Saeid R. Seydnejad. 2015. "A Novel Hybrid Algorithm of GSA with Kepler Algorithm for Numerical Optimization." *Journal of King Saud University - Computer and Information Sciences* 27 (3): 288–296. https://doi.org/10.1016/j.jksuci.2014.10.003.

Sarhani, Malek, Abdellatif El Afia, and Rdouan Faizi. 2018. "Facing the Feature Selection Problem with a Binary PSO-GSA Approach." *Operations Research/Computer Science Interfaces Series* 62: 447–462. https://doi.org/10.1007/978-3-319-58253-5_26.

Shamsudin, Hasrul Che, Addie Irawan, Zuwairie Ibrahim, Amar Faiz Zainal Abidin, Sophan Wahyudi, Muhammad Arif Abdul Rahim, and Kamal Khalil. 2012. "A Fast Discrete Gravitational Search Algorithm." In *Proceedings of International Conference on Computational Intelligence, Modelling and Simulation*, 24–28. https://doi.org/10.1109/CIMSim.2012.28.

Shang, Zhongping. 2013. "Neighborhood Crossover Operator: A New Operator in Gravitational Search Algorithm." *International Journal of Computer Science* 10 (5): 116–126.

Shaw, Binod, V. Mukherjee, and S. P. Ghoshal. 2012. "A Novel Opposition-Based Gravitational Search Algorithm for Combined Economic and Emission Dispatch Problems of Power Systems." *International Journal of Electrical Power and Energy Systems* 35 (1): 21–33. https://doi.org/10.1016/j.ijepes.2011.08.012.

Shaw, Binod, V. Mukherjee, and Sakti Prasad Ghoshal. 2013. "Solution of Optimal Reactive Power Dispatch by an Opposition-Based Gravitational Search Algorithm." In *Swarm, Evolutionary, and Memetic Computing. SEMCCO 2013. Lecture Notes in Computer Science*, vol 8297. Cham, Switzerland: Springer. https://doi.org/10.1007/978-3-319-03753-0_50.

Sheikhan, Mansour. 2014. "Generation of Suprasegmental Information for Speech Using a Recurrent Neural Network and Binary Gravitational Search Algorithm for Feature Selection." *Applied Intelligence* 40 (4): 772–790. https://doi.org/10.1007/s10489-013-0505-x.

Shilaja, C., and T. Arunprasath. 2019. "Optimal Power Flow Using Moth Swarm Algorithm with Gravitational Search Algorithm Considering Wind Power." *Future Generation Computer Systems* 98: 708–715. https://doi.org/10.1016/j.future.2018.12.046.

Shirazi, Fatemeh, and Esmat Rashedi. 2016. "Detection of Cancer Tumors in Mammography Images Using Support Vector Machine and Mixed Gravitational Search Algorithm." In *Proceedings in the 1st Conference on Swarm Intelligence and Evolutionary Computation, CSIEC 2016*, 98–101. https://doi.org/10.1109/CSIEC.2016.7482133.

Singh, Krishan Veer, and Zahid Raza. 2017. "A Quantum-Inspired Binary Gravitational Search Algorithm–Based Job-Scheduling Model for Mobile Computational Grid." *Concurrency Computation* 29 (12): e4103. https://doi.org/10.1002/cpe.4103.

Singh, Krishan Veer, and Zahid Raza. 2018. "Resource Scheduling Approach Using a Quantum Variant of Gravitational Search Algorithm in Computational Mobile Grid." In *Proceedings of the 8th International Conference Confluence 2018 on Cloud Computing, Data Science and Engineering, Confluence 2018*, 636–640. https://doi.org/10.1109/CONFLUENCE.2018.8444137.

Soleimanpour-Moghadam, Mohadeseh, and Hossein Nezamabadi-pour. 2012. "An Improved Quantum Behaved Gravitational Search Algorithm." In *ICEE 2012 – 20th Iranian Conference on Electrical Engineering*, 711–714. https://doi.org/10.1109/IranianCEE.2012.6292446.

Soleimanpour-Moghadam, Mohadeseh, Hossein Nezamabadi-pour, and Malihe M. Farsangi. 2014. "A Quantum Inspired Gravitational Search Algorithm for Numerical Function Optimization." *Information Sciences* 267: 83–100. https://doi.org/10.1016/j.ins.2013.09.006.

Song, Baoye, Yihui Xiao, and Lin Xu. 2020. "Design of Fuzzy PI Controller for Brushless DC Motor Based on PSO–GSA Algorithm." *Systems Science and Control Engineering* 8 (1): 67–77. https://doi.org/10.1080/21642583.2020.1723144.

Song, Zhenyu, Cheng Tang, Xingqian Chen, Shuangyu Song, and Junkai Ji. 2019. "A Self-Adaptive Mechanism Embedded Gravitational Search Algorithm." In *Proceedings of the 2019 12th International Symposium on Computational Intelligence and Design, ISCID 2019*, 1:108–112. https://doi.org/10.1109/ISCID.2019.00031.

Sun, Genyun, Ping Ma, Jinchang Ren, Aizhu Zhang, and Xiuping Jia. 2018. "A Stability Constrained Adaptive Alpha for Gravitational Search Algorithm." *Knowledge-Based Systems* 139: 200–213. https://doi.org/10.1016/j.knosys.2017.10.018.

Sun, Genyun, Aizhu Zhang, Xiuping Jia, Xiaodong Li, Shengyue Ji, and Zhenjie Wang. 2016. "DMMOGSA: Diversity-Enhanced and Memory-Based Multi-Objective Gravitational Search Algorithm." *Information Sciences* 363: 52–71. https://doi.org/10.1016/j.ins.2016.05.007.

Sun, Liping, Tao, Xiaoyao Zheng, Shuting Bao, and Yonglong Luo. 2019. "Combining Density Peaks Clustering and Gravitational Search Method to Enhance Data Clustering." *Engineering Applications of Artificial Intelligence* 85: 865–873. https://doi.org/10.1016/j.engappai.2019.08.012.

Tabatabaei, Sajad. 2014. "A New Gravitational Search Optimization Algorithm to Solve Single and Multiobjective Optimization Problems." *Journal of Intelligent and Fuzzy Systems* 26 (2): 993–1006. https://doi.org/10.3233/IFS-130791.

Talbi, El Ghazali. 2009. *Metaheuristics: From Design to Implementation. Metaheuristics: From Design to Implementation*. Hoboken, NJ: Wiley. https://doi.org/10.1002/9780470496916.

Taradeh, Mohammad, Majdi Mafarja, Ali Asghar Heidari, Hossam Faris, Ibrahim Aljarah, Seyedali Mirjalili, and Hamido Fujita. 2019. "An Evolutionary Gravitational Search-Based Feature Selection." *Information Sciences* 497: 219–239. https://doi.org/10.1016/j.ins.2019.05.038.

Thakur, Abhijeet Singh, Tarun Biswas, and Pratyay Kuila. 2020. "Binary Quantum-Inspired Gravitational Search Algorithm-Based Multi-Criteria Scheduling for Multi-Processor Computing Systems." *Journal of Supercomputing* 77: 796–817. https://doi.org/10.1007/s11227-020-03292-0.

Xue, Hongxin, Yanping Bai, Hongping Hu, Ting Xu, and Haijian Liang. 2019. "A Novel Hybrid Model Based on TVIW-PSO-GSA Algorithm and Support Vector Machine for Classification Problems." *IEEE Access* 7: 27789–27801. https://doi.org/10.1109/ACCESS.2019.2897644.

Yadav, Anupam, and Joong Hoon Kim. 2015. "A Niching Co-Swarm Gravitational Search Algorithm for Multi-Modal Optimization." *Advances in Intelligent Systems and Computing* 335: 599–607. https://doi.org/10.1007/978-81-322-2217-0_48.

Yadav, Anupam, Anita, and Joong Hoon Kim. 2019. "Convergence of Gravitational Search Algorithm on Linear and Quadratic Functions." In *Decision Science in Action*, 31–39. Singapore: Springer. https://doi.org/10.1007/978-981-13-0860-4_3.

Yadav, Anupam, and Kusum Deep. 2013. "Constrained Optimization Using Gravitational Search Algorithm." *National Academy Science Letters* 36 (5): 527–534. https://doi.org/10.1007/s40009-013-0165-8.

Yazdani, Sajjad, Hossein Nezamabadi-pour, and Shima Kamyab. 2014. "A Gravitational Search Algorithm for Multimodal Optimization." *Swarm and Evolutionary Computation* 14: 1–14. https://doi.org/10.1016/j.swevo.2013.08.001.

Yin, Baoyong, Zhaolu Guo, Zhengping Liang, and Xuezhi Yue. 2018. "Improved Gravitational Search Algorithm with Crossover." *Computers and Electrical Engineering* 66: 505–516. https://doi.org/10.1016/j.compeleceng.2017.06.001.

Yuan, Xiaohui, Bin Ji, Shuangquan Zhang, Hao Tian, and Yanhong Hou. 2014. "A New Approach for Unit Commitment Problem via Binary Gravitational Search Algorithm." *Applied Soft Computing Journal* 22: 249–260. https://doi.org/10.1016/j.asoc.2014.05.029.

Zandevakili, Hamed, Esmat Rashedi, and Ali Mahani. 2019. "Gravitational Search Algorithm with Both Attractive and Repulsive Forces." *Soft Computing* 23 (3): 783–825. https://doi.org/10.1007/s00500-017-2785-2.

Zhang, Aizhu, Sihan Liu, Genyun Sun, Hui Huang, Ping Ma, Jun Rong, Hongzhang Ma, Chengyan Lin, and Zhenjie Wang. 2019. "Clustering of Remote Sensing Imagery Using a Social Recognition-Based Multi-Objective Gravitational Search Algorithm." *Cognitive Computation* 11 (6): 789–798. https://doi.org/10.1007/s12559-018-9582-9.

Zhang, Lingling. 2018. "A Gravitational Artificial Bee Colony Optimization Algorithm and Application." In *Proceedings of the 8th International Conference on Instrumentation and Measurement, Computer, Communication and Control, IMCCC 2018*, 1839–1842. https://doi.org/10.1109/IMCCC.2018.00379.

Zhao, Fuqing, Feilong Xue, Yi Zhang, Weimin Ma, Chuck Zhang, and Houbin Song. 2018. "A Hybrid Algorithm Based on Self-Adaptive Gravitational Search Algorithm and Differential Evolution." *Expert Systems with Applications* 113: 515–530. https://doi.org/10.1016/j.eswa.2018.07.008.

6 Stochastic Diffusion Search

Andrew Owen Martin

CONTENTS

ALGORITHM 6.1 STOCHASTIC DIFFUSION SEARCH AS INTRODUCED BY BISHOP [1]

```
INITIALISE (mappings);
REPEAT
    TEST (mappings);
    DIFFUSE (mappings);
UNTIL TERMINATION;
```

6.1 INTRODUCTION TO SDS

First described in 1989 by J.M. Bishop, stochastic diffusion search (SDS) was introduced as an evolution of the study of Hinton mapping [2]. Where Hinton mapping identifies objects by testing for all possible combinations of stimuli, SDS stochastically tests certain hypotheses against partial combinations of stimuli, and subsequently performs a diffusion process where more promising hypotheses are more likely to be tested against further features. At the time, and at least as late as 1998 [3], the term "stochastic search networks" was used, but SDS will be used henceforth.

SDS has a number of important features including robustness in the presence of noise, a simplicity that lends itself well to mathematical modeling, and stable global state emerging from the stochastic behavior of individuals. Its robustness makes SDS particularly suited to real-world tasks, which may be noisy, and may change over time. The relative ease of modeling SDS mathematically has led to the ability to predict a number of aspects of its behavior. The stability of SDS means it may also be an important model for methods of guiding intelligent behavior without a central executive control.

The operation of SDS is most intuitively explained by analogy, a number of which have been used, most commonly *The Restaurant Game* [4], and more recently *The Mining Game* [5]. These are described in the following sections, followed by the pseudocode of a single iteration of SDS.

6.1.1 The Restaurant Game

The Restaurant Game, reproduced here from 2003 [4], describes how a group of people may employ SDS to perform a search for the best restaurant in an unfamiliar town.

> A group of delegates attends a long conference in an unfamiliar town. Each night they have to find somewhere to dine. There is a large choice of restaurants, each of which offers a large variety of meals. The problem the group faces is to find the best restaurant, that is the restaurant where the maximum number of delegates would enjoy dining. Even a parallel exhaustive search through the restaurant and meal combinations would take too long to accomplish. To solve the problem delegates decide to employ a Stochastic Diffusion Search.
>
> Each delegate acts as an agent maintaining a hypothesis identifying the best restaurant in town. Each night each delegate tests his hypothesis by dining there and randomly selecting one of the meals on offer. The next morning at breakfast every delegate who did not enjoy his meal the previous night, asks one randomly selected colleague to share his dinner impressions. If the experience was good, he also adopts this restaurant as his choice. Otherwise he simply selects another restaurant at random from those listed in "Yellow Pages."
>
> Using this strategy it is found that very rapidly [a] significant number of delegates congregate around the best restaurant in town.

The process of the restaurant game has a number of notable features. Within minimal centralized control the group of delegates acts together to solve a problem that could not be quickly solved by an individual. The delegates will efficiently move to the next best restaurant if the current one has a significant drop in standards or closes down entirely. The restaurants, the menus, or the individual meals need to be directly comparable, all that is required is for each agent to decide for himself whether the experience was good. Delegates will find themselves enjoying many evenings in a relatively high-quality restaurant long before all of the meals in all of the restaurants in town could have been evaluated.

This analogy has been criticized on the grounds that delegates are likely to have differing dining preferences; hence, it is possible for a delegate to locate a restaurant in which they enjoy all of the meals on offer, but which is unsatisfying to all other delegates. In the case, where only one delegate, or a small proportion of the group, remains permanently at such a restaurant, the rest of the group will proceed largely as usual and so the majority will still converge on the best restaurant. Taken to the extreme, however, all agents may find themselves dining alone, even when there exists a single superior restaurant that would satisfy the largest portion of the delegates. This superior restaurant would never be located as all delegates are satisfied with the meals at their current restaurant and, hence, never select a new restaurant. This critique led to the development of The Mining Game, which depends on the less subjective notion of locating gold rather than dining preferences.

6.1.2 The Mining Game

Originally defined in 2010 [6], the Mining Game uses an analogy of searching a landscape for the best mining prospect. A slightly improved version from 2013 [7] is quoted below.

> A group of friends (miners) learn that there is gold to be found on the hills of a mountain range but have no information regarding its distribution. On their maps the mountain range is divided into a set of discrete hills and each hill contains a discrete set of seams to mine. Over time, on any day the probability of finding gold at a seam is proportional to its net wealth.
>
> To maximize their collective wealth, the miners need to identify the hill with the richest seams of gold so that the maximum number of miners can dig there (this information is not available a-priori). In order to solve this problem, the miners decide to employ a simple Stochastic Diffusion Search.
>
> - At the start of the mining process each miner is randomly allocated a hill to mine (his hill hypothesis, *h*).
> - Every day each miner is allocated a randomly selected seam on his hill to mine.
> - At the end of each day, the probability that a miner is happy[1] is proportional to the amount of gold he has found.
> - At the end of the day the miners congregate and over the evening each miner who is unhappy selects another miner at random to talk to. If the chosen miner is happy, he happily tells his colleague the identity of the hill he is mining (that is, he communicates his hill hypothesis, *h*, which thus both now maintain). Conversely, if the chosen miner is unhappy he says nothing and the original miner is once more reduced to selecting a new hypothesis — identifying the hill he is to mine the next day — at random.
>
> In the context of SDS, agents take the role of miners; active agents being "happy miners," inactive agents being "unhappy miners" and the agent's hypothesis being the miner's "hill-hypothesis." It can be shown that this process is isomorphic to SDS, and thus that the miners will naturally self-organize and rapidly congregate over hill(s) on the mountain range with a high concentration of gold.

The happiness of the miners can be measured probabilistically, or represented with an absolute Boolean value, as long as each miner is either happy or unhappy by the end of each day [8]. Furthermore, if the gold is modeled as a finite resource, reducing over time, then the search is sufficiently adaptive that miners change where they congregate as the location with the most gold changes.

6.1.3 Pseudocode

In this example, agents have access to a function for randomly selecting a hypothesis for the set of all possible hypotheses, and for randomly selecting a microtest from a given set. Each microtest evaluates a specific part of a hypothesis and returns a Boolean value indicating whether or not the test passed.

ALGORITHM 6.2 A SINGLE ITERATION OF STANDARD STOCHASTIC DIFFUSION SEARCH

1: **for each** agent **in** swarm **do** ▷ Diffusion phase
2: **if** agent is inactive **then**
3: polled ← randomly selected agent in swarm
4: **if** polled is active **then**

```
5:         agent assumes the hypothesis of polled
6:      else
7:         agent randomly selects a new hypothesis
8: for each agent in swarm do        ▷Test phase
9:   microtest ← randomly selected microtest
10:  agent performs the microtest against their hypothesis    ▷Partial evaluation
11:  if the microtest passes then
12:      agent becomes active
13:  else
14:      agent becomes inactive
```

To recap definitions, an *agent* is an individual member of a swarm and a *swarm* is a collection of agents. A *search space* is the set of all possible hypotheses and a *hypothesis* is an element of the search space, or an index referring to an element of the search space. A *microtest* is a function that partially evaluates a hypothesis. The *score* of a hypothesis is equal to the probability of an agent with that hypothesis being active after the test phase.

SDS lends itself well to mathematical analysis. In the next section a simple but powerful model of the convergence behavior of SDS is described. The features of SDS, which have previously been modeled, and the insights into its behavior using this model are then described. Following this a formalism for describing variants of SDS is developed, and some of the advantages and disadvantages of all the main variants that have been published are identified.

6.2 MATHEMATICAL ANALYSIS OF SDS

To appreciate the mathematical analysis of SDS, which has been published, it is required that first the model is described.

This analysis will always assume the search space contains exactly one hypothesis with a score larger than any other hypothesis (called the optimal hypothesis) and many hypotheses with a significantly lower score, the collective effect of which can be described by a single value [9, p. 118] known collectively as homogeneous background noise. Furthermore the search space size, the number of hypotheses in a search space, is represented by the symbol S. The swarm size, the number of agents in a swarm, is represented by the symbol A. The proportion of the swarm that is active at a given time is known as the global activity.

6.2.1 Discrete Dynamical Systems Model

Introduced in [10] and explored in [11], the discrete dynamical system model (DDSM) of SDS has been used to clearly express results previously reached via other methods as well as new models of the behavior of SDS. Similar to previous models, an assumption is made of homogeneous background noise. The DDSM introduces the following notation: α is the score of the optimal hypothesis, β is the mean probability of an agent at any non-optimal hypothesis becoming active, and $\bar{c}_i$ is the mean number of active agents maintaining the optimal hypothesis as a proportion of the total population.

From these definitions of α and β, it holds that $0 \leq \beta < \alpha \leq 1$, and the following expressions can be immediately derived, as represented in Figure 6.1.

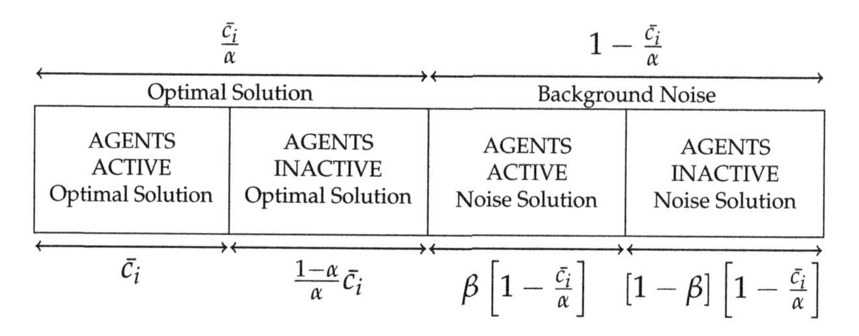

FIGURE 6.1 Myatt's discrete dynamical systems model of SDS.

Notes:

$\frac{\bar{c}_i}{\alpha}$ = proportion of active and inactive agents at the optimal hypothesis
$\frac{1-\alpha}{\alpha}\bar{c}_i$ = proportion of inactive agents at the optimal hypothesis
$1-\frac{\bar{c}_i}{\alpha}$ = proportion of active and inactive agents at any noise hypothesis
$\beta\left(1-\frac{\bar{c}_i}{\alpha}\right)$ = proportion of active agents at any noise hypothesis
$(1-\beta)\left(1-\frac{\bar{c}_i}{\alpha}\right)$ = proportion of inactive agents at any noise hypothesis

The properties of SDS that have been modeled can be split into three categories: the convergence criteria, which describe the set of parameters over which SDS can be expected to converge; the convergence time, which measures the mean number of iterations before a cluster forms at the optimal hypothesis; and the steady state, which describes the number of agents at the optimal hypothesis after convergence.

6.2.2 Minimum Convergence Criteria

Using a Markov chain model, it was shown that for a stable cluster to form it must hold that

$$2\alpha(1-p_\alpha)<1 \tag{6.1}$$

where p_α is the probability of selecting the optimal hypothesis at random [9, p. 72]. As the search space is often large, p_α can be assumed to be small, in which case one obtains

$$\alpha \geq \frac{1}{2} \tag{6.2}$$

A similar result was shown using a probabilistic model [12], where the minimum value of optimal hypothesis score at which a stable cluster will form in the presence of homogeneous background noise was shown to be

$$\alpha_{\min}=\frac{1}{2-\beta} \tag{6.3}$$

This result was confirmed using the DDSM. The mean number of inactive agents adopting the optimal hypothesis during the diffusion phase is described by the function $g(\alpha, \beta, \bar{c}_i)\bar{c}_i$, where g yields the number of inactive agents for a given iteration. From Figure 6.1, g can be written as

$$g(\alpha, \beta, \bar{c}_i)=\frac{1-\alpha}{\alpha}\bar{c}_i+(1-\beta)\left(1-\frac{\bar{c}_i}{\alpha}\right) \tag{6.4}$$

Therefore, the function *f*, which defines the one-step evolution function, is

$$\begin{aligned} \bar{c}_{i+1} &= f(\alpha, \beta, \bar{c}_i) \\ &= \alpha(\bar{c}_i + g(\bar{c}_i, \alpha, \beta)\bar{c}_i) \\ &= \bar{c}_i(\beta\bar{c}_i + 2\alpha - \alpha\bar{c}_i - \alpha\beta) \end{aligned} \tag{6.5}$$

For SDS to converge, the repeated iteration of *f* must stabilize at a non-zero value. It therefore must hold that $\frac{df}{dc_i} > 1$ when $\bar{c}_i = 0$. Differentiating *f* by $\bar{c}_i$ and solving for α when $\frac{df}{dc_i} > 1$ yields $\alpha_{\min}$.

$$\alpha_{\min} = \frac{1}{2-\beta} \tag{6.6}$$

It has been shown experimentally that an SDS with a value of α such that $\alpha = \alpha_{\min} + 0.01$ would reliably converge, and if $\alpha = \alpha_{\min} - 0.01$ there would be no stable clusters [10]. Hence, the experimental value of $\alpha_{\min}$ was shown to be within ±0.01 of the theoretical, validating the theory.

An advantage of the homogeneous background noise model is that an estimate of β can be achieved by taking the mean value of activity over repetitions of the test phase. This value can be substituted into Eq. 6.6 to provide a lower bound for convergence.

6.2.2.1 Robustness

By taking all combinations of values of α and β, the minimum convergence criteria can be used to identify each combination as falling into one of three categories: (1) invalid search spaces, where $\beta \geq \alpha$; (2) search spaces that successfully converge as $\alpha > \alpha_{\min}$; and (3) search spaces that do not converge [11]. The proportion of valid search spaces for which standard SDS (SSDS) will successfully converge is named the robustness (ζ) and has been calculated to be approximately 0.614 [11, p. 54].

$$\zeta = 2(1 - \ln 2) \approx 0.614 \tag{6.7}$$

6.2.3 Convergence Time

The convergence of SDS can be described in three phases. In the first phase few agents are active and many new hypotheses are generated at random each iteration; the number of iterations spent in this phase is described by the time to first hit (TH) (Section 6.2.3.1). In the second phase a cluster forms once an agent has assumed an appropriate hypothesis. This phase is relatively short as it is dominated by the positive feedback effect of the diffusion (Section 6.2.3.2). In the third phase, after a stable cluster has formed, many agents are active and few new hypotheses are selected. This phase can be practically indefinite, because of the very high probability of a cluster persisting (Section 6.2.3.3). A Markov chain model was used to show that the convergence time increases dramatically when there was a small relative discrimination between α and β [13].

6.2.3.1 Time to First Hit (TH)

The TH describes the number of iterations after which the probability that the optimal hypothesis has been located reaches a given probability p. In cases where $\beta = 0$ the TH has been shown to be [3, 1]

$$\text{TH} = \frac{\ln(1-p)}{A \ln\left(\frac{S-1}{S}\right)} \tag{6.8}$$

In cases where $\beta > 0$, the TH will be larger as some agents will be maintaining sub-optimal hypotheses; hence, it will not generate new random hypotheses in the diffusion phase. The TH has therefore been shown to be calculated as [14]

$$\text{TH} = S \frac{\ln \frac{1}{p}}{A(1-\beta)^2} \tag{6.9}$$

which is linear with respect to S [1, 14].

6.2.3.2 Positive Feedback Effect of Cluster Formation

Once an agent has assumed the optimal hypothesis a process of positive feedback occurs. This occurs because the most important factor in the probability of an agent assuming the optimal hypothesis is the number of agents currently maintaining that hypothesis. As the number of agents maintaining the optimal hypothesis increases, the rate at which new agents assume that hypothesis also increases, as long as there are sufficient inactive agents remaining. This very quickly reaches a steady state in which the number of agents that were maintaining the optimal hypothesis becoming inactive equals the number of agents assuming the optimal hypothesis in the diffusion phase [15]. This phase lasts for a relatively small number of iterations compared with the TH, which is a linear with respect to search space size and the steady state, which will be seen to be practically indefinite.

6.2.3.3 Stability of Convergence

The stability of convergence may describe the mean number of iterations for which a cluster can be expected to persist once it has formed as well as the property of SDS where clusters are able to form at the optimal hypothesis in the presence of other hypotheses that have similar scores.

6.2.3.3.1 Strong Convergence Criterion

It has been proved mathematically that in cases where $\beta = 0, \alpha = 1$ and at least one agent is maintaining the optimal hypothesis, that the probability of all agents maintaining that hypothesis approaches 1 [9, p.56; 11, p.63; 16]. These conditions taken as together define the strong convergence criterion.

In cases where $\alpha < 1$, SDS will not strongly converge and may weakly converge [13]. Weakly converging means that SDS is only ever statistically stable, there is always a non-zero probability that the swarm will transition into any state from any other state. This means any apparently stable cluster may collapse at any iteration, although it is likely to exist long enough to be detected. An advantage of this behavior is that an apparently stable cluster will collapse once a superior hypothesis has been located; hence, SDS can be applied successfully to problem spaces that change over time.

The stability of the converged state of SDS was derived using a Markov chain model [17]. Using some selected values $A = 1000, p_\alpha = 0.001, \alpha = 0.8, \beta = 0$ the mean number of iterations before a stable cluster will collapse (m_0) was calculated to be

$$m_0 \approx \frac{1}{(0.25)^{1000}} \propto 10^{602}$$

thus this state is visited very rarely and as such the search is very stable, even when $A = 10$ $m_0 \propto 10^6$.

As the stability of the search depends exponentially on the number of agents in the swarm, the ratio of logarithms of mean return times to the "all-inactive" state for a pair of instances of SDS is equal to the ratio of the numbers of agents involved in the search. Hence, the number of agents in a swarm may efficiently control the stability of solutions in practical applications [17].

It has been stated that the rate of convergence is geometric "for some constant" [13]. The DDSM has been used to show the constant to be

$$\alpha(2-\beta) \tag{6.10}$$

as such, it can be observed that an increase in α will reduce convergence time more than decreasing β by the same amount [11].

6.2.4 Sensitivity of Convergence

A distractor is a single hypothesis with a score (δ) such that $\beta < \delta < \alpha$. Using an Ehrenfest urn model of SDS, it has been shown that even in the presence of a strong distractor there is a significant increase in the order of magnitude of the probability of an agent maintaining the optimal hypothesis against that of an agent maintaining the distractor [9, p. 101]. In a three-dimensional plot of the optimal hypothesis score α and the distractor hypothesis score δ against the probability of an agent maintaining the optimal hypothesis three distinct surfaces of equal probability were visible. One surface represented the case when δ was larger than α and so can be ignored by definition ($\alpha > \delta$). The second surface occurred at $\delta < 0.5$ and $\alpha < 0.5$, which represented that an agent was equally likely to maintain either hypothesis when both hypotheses have a score that is too small to form a cluster. The third surface occurred when, $\alpha > \max(\delta, 0.5)$ which represented when optimal hypothesis had a score that could form a cluster.

An important feature of these surfaces was the steep transition between them, and this represented that SDS was very sensitive to differences between α and δ, so an agent would most likely be found at the optimal hypothesis even when δ was within 1% of α. Another important feature of these surfaces was their flatness, which described that the probability of convergence was not affected by the presence of the distractor, only the ability of the optimal hypothesis to form a cluster. This means that SDS is particularly insensitive to signal-to-noise ratio, in cases where the signal (the optimal hypothesis), is strong enough.

6.2.5 Steady-State Resource Allocation

In the case where $\beta = 0$ and $\alpha < 1$, there is a model of the cluster size mean and standard deviation [15]. The mean number of active agents ($E[n]$) is

$$E[n] = A\pi_1 \tag{6.11}$$

and the standard deviation is

$$\sigma = \sqrt{A\pi_1\pi_2} \tag{6.12}$$

where

$$\pi_1 = \frac{\pi_3 - 1 + \sqrt{[\pi_3 - 1]^2 + 2\pi_3(1-\beta)p_\alpha}}{\pi_3} \tag{6.13}$$

$$\pi_2 = \frac{1 - \sqrt{[\pi_3 - 1]^2 + 2\pi_3(1-\beta)p_\alpha}}{\pi_3} \tag{6.14}$$

$$\pi_3 = 2\alpha(1 - p_\alpha) \tag{6.15}$$

It has also been shown that the attractors are never strange attractors for valid values of α and β. This proves that the convergence behavior is not chaotic; hence, the level of activity in SDS will always tend toward a constant level [11].

For cases where $\alpha > \alpha_{\min}$ the mean stationary state will be [11]

$$\gamma = \frac{\alpha(2-\beta)-1}{\alpha-\beta} \tag{6.16}$$

6.2.5.1 Estimating Optimal Hypothesis Score from Steady State

The value for α can be estimated by dividing the number of active agents at the largest cluster by the sum of active and inactive agents at the largest cluster [11].

$$\alpha = \frac{c_{\text{active}}}{c_{\text{active}} + c_{\text{inactive}}} \tag{6.17}$$

This is notable as this expression does not depend on the level of homogeneous background noise.

It is also observed that the same estimate can be achieved using the global activity, and not just the activity at the largest cluster. Given the stationary state for the optimal cluster (Equation 6.16) and Figure 6.1, the stationary state for all active agents must be equal to

$$\begin{aligned} \gamma' &= \gamma + \beta\left[1 - \frac{\gamma}{\alpha}\right] \\ &= 2 - \frac{1}{\alpha} \end{aligned} \tag{6.18}$$

thus α can be estimated by

$$\alpha = \frac{1}{2-\hat{\gamma}} \tag{6.19}$$

6.3 A FORMAL SPECIFICATION FOR SDS

Given the initial description of SDS (Algorithm 6.1) each line may be considered as indicating potential for variation. Some of these dimensions of variation are relevant only to the specific implementation and do not pertain to the variant themselves. All remaining dimensions are described and their major characteristics identified. As SDS can be modified in a practically unlimited number of ways, a strict definition of what is and is not a variant cannot be given, instead a specification that succinctly describes the majority of the published variants is presented. Where any variant has been mathematically analyzed, it is presented here.

6.3.1 Areas of Potential Variation in SDS

In the original description (Algorithm 6.1) SDS is defined in five lines. Using these five lines and following the framework defined in [18], the main areas of variation of SDS can be identified. The first of these areas is *initialization*, which is concerned with variations in the initial hypotheses and activities of agents in the swarm. Second, the text REPEAT indicates variations in the *iteration* of the test phase and diffusion phase. Third, there are variations in the *test phase*, where hypotheses are partially evaluated. Fourth, there are variations in the *diffusion phase*, where new hypotheses are selected and potentially distributed among the swarm. Fifth, there are variations in the *termination*, or *halting* conditions, in which the aim is to halt the procedure once SDS has converged to a solution. Finally, one more area of variation has been identified, which is the method of *extraction*, where a solution is indicated by the formation of clusters [18, p. 117]. A formalism is developed to systematically distinguish and compare the most significant variations of SDS. A variant of SDS will be defined as a combination of its modes of iteration, diffusion, test, and halting.

6.3.2 Formalism

An instance of SDS can be described as a combination of its mode of iteration (I) and halting (H). Before each iteration the halting function is evaluated and the process continues to perform iterations until the halting function evaluates as *true*. I defines the mode of iteration, and each evaluation of the I function will count as one iteration.

ALGORITHM 6.3 STOCHASTIC DIFFUSION SEARCH AS FORMALIZED

```
1: function SDS(I, H)
2:    function SDS′ ()
3:        while not H() do
4:            I()
5:    return SDS′
```

6.3.2.1 Mode of Iteration *I*

The mode of iteration in the original description of SDS is known as a synchronous operation. A single iteration is defined as having passed once all agents perform the diffusion phase followed by all agents performing the test phase. This mode of iteration, named $I^{\text{SYNCHRONOUS}}$, can be seen in Algorithm 6.4. The definition of the mode of iteration therefore requires the parameters D to define the mode of diffusion, T to define the mode of testing, and a reference to the swarm. The returned function has no arguments and each call effects a single iteration on the swarm. Algorithm 6.4 shows synchronous iteration, which defines the mode of iteration of SSDS, for the reasons explained in Section 6.3.4.1, the diffusion phase is performed before the test phase.

ALGORITHM 6.4 $I^{\text{SYNCHRONOUS}}$: SYNCHRONOUS ITERATION

```
1: function I^SYNCHRONOUS (D, T, swarm)
2: function I′()
3: for each agent in swarm do
4: D (agent) ▷Diffusion phase
5: for each agent in swarm do
6: T (agent) ▷Test phase
7: return I′
```

6.3.2.2 Mode of Diffusion *D*

The mode of diffusion is the mechanism by which agents share hypotheses or select new hypotheses, each search space therefore requires a unique method by which hypotheses are selected. The definition of the mode of diffusion therefore requires the parameter *DH*, which defines a function for selecting new hypotheses, and a reference to the swarm. The returned function takes an agent as an argument and each call effects the mode of diffusion on that agent. Algorithm 6.5 shows passive diffusion, which relies on the inactive agents requesting hypotheses from active agents and is the mode of diffusion of SSDS.

ALGORITHM 6.5 $D^{PASSIVE}$: PASSIVE DIFFUSION

1: **function** $D^{PASSIVE}$ (*DH*, swarm)
2:     **function** *D′*(agent)
3:         **if** agent is inactive **then**
4:             polled ← random agent in swarm
5:             **if** polled is active **then**
6:                 hypothesis of agent ← hypothesis of polled
7:             **else**
8:                 hypothesis of agent ←*DH*()
9:     **return** *D′*

6.3.2.3 Mode of Hypothesis Selection, *DH*

The hypothesis selecting function is the method by which agents select new hypotheses. The definition of *DH*, therefore, requires the parameter *hypotheses*, which represents the set of all the possible hypotheses. When implementing this function the set of all possible hypotheses may be fully enumerated, or selected as required by an underlying function. Where full enumeration is likely to require less computational resources, it will likely require more memory and conversely selecting hypotheses from a function will likely require more computational resources but less memory. The returned function takes no arguments and each call returns a randomly selected hypothesis.

As originally introduced [1], the standard method for selecting hypotheses is to uniformly randomly distribute new hypotheses over the search space. This does not require any specific a priori knowledge of the task, and it will evenly distribute the resources through the search space. This technique does not modify its behavior as a result of previous activity, so it has the advantage of never being trapped indefinitely in locally optimal areas of the search space, but it has the disadvantage of not responding to potentially helpful information. Algorithm 6.6 shows uniformly random hypothesis selection.

ALGORITHM 6.6 $DH^{UNIFORM}$: UNIFORMLY RANDOM HYPOTHESIS SELECTION

1: **function** $DH^{UNIFORM}$ (hypotheses)
2:     **function** *DH′*()
3:         hypothesis ← element selected uniformly at random from hypotheses
4:         **return** hypothesis
5:     **return** *DH′*

6.3.2.4 Mode of Testing, *T*

The mode of testing for each agent involves selecting a microtest, using that microtest to perform a partial evaluation of their hypothesis and updating their activity correspondingly. The definition of the mode of testing therefore requires the parameter *TM,* which defines a function for selecting a microtest. The returned function takes an agent as an argument and each call effects the mode of testing on that agent. Algorithm 6.7 shows Boolean testing, which defines the mode of testing of SSDS. It is important to note that while the mechanism of Boolean testing is simple, it requires that a method of evaluating a hypothesis is defined as a set of Boolean functions. This is a significant requirement as the solutions to many search tasks cannot be evaluated with a set of Boolean functions, or the functions themselves may include the use of predetermined thresholds.

ALGORITHM 6.7 $T^{BOOLEAN}$: BOOLEAN TESTING

1: **function** $T^{BOOLEAN}(TM)$
2: **function** T'(agent)
3: microtest ← TM ()
4: partial evaluation ← microtest(hypothesis of agent) ▷partial evaluation $\in \mathbb{B}$
5: **if** partial evaluation = *true* **then**
6: agent becomes active
7: **else**
8: agent becomes inactive
9: **return** T'

6.3.2.5 Microtest Selection, *TM*

The microtest selecting function is the method by which agents select how they will partially evaluate their hypothesis. The definition of *TM* therefore requires the parameter *microtests*, which represents the set of all possible microtests. As with the *hypotheses* parameter to *DH*, the set of microtests may be fully enumerated or selected as required; however, the number of microtests is often small compared with other features of a search space, so full enumeration is more common. The returned function takes no arguments and each call returns a randomly selected microtest. Algorithm 6.8 shows uniformly random microtest selection, which defines the microtest selection of SSDS.

ALGORITHM 6.8 $TM^{UNIFORM}$: UNIFORM MICROTEST SELECTION

1: **function** $TM^{UNIFORM}$ (microtests)
2: **function** TM'()
3: microtest ← element selected uniformly at random from microtests
4: **return** microtest
5: **return** TM'

6.3.2.6 Mode of Halting, *H*

The halting function is the method by which an instance of SDS determines whether or not to continue to call *I* and hence perform further iterations. A halting function can therefore be defined as a function that takes no parameters and returns a Boolean value.

$$H = f : () \rightarrow \mathbb{B} \tag{6.20}$$

There are many different halting functions; a very simple example is to halt after a predetermined number of iterations. This method, called fixed iterations halting, requires the parameter *maximum iterations* to define the maximum number of iterations that will elapse. The returned function takes no arguments and will return *false* until it has been called a number of times greater than *maximum iterations*. Other halting functions perform more complicated operations over the current and previous states of the swarm, thus requiring a more complicated set of parameters. Algorithm 6.9 shows fixed iterations halting.

ALGORITHM 6.9 H^{FIXED}: FIXED ITERATION COUNT HALTING

```
1: function H^FIXED (maximum iterations)
2:     iteration count ← 0
3:     function H′()
4:         iteration count ← iteration count + 1
5:         halt ← iteration count > maximum iterations        ▷halt ∈ 𝔹
6:         return halt
7:     return H′
```

6.3.3 Standard SDS (SSDS)

The SSDS can be therefore denoted as in Eq. 6.21.

$$\text{SDS}^{\text{STANDARD}} = \text{SDS}\left(I^{\text{SYNCHRONOUS}}\left(D^{\text{PASSIVE}}(DH^{\text{UNIFORM}}), T^{\text{BOOLEAN}}(TM^{\text{UNIFORM}})\right), H^{\text{FIXED}}\right) \quad (6.21)$$

To simplify the description of further variants, all the features of SSDS will be denoted as standard forms as this allows further variants to be defined in terms of which features differ from SSDS and which features remain unchanged. These standard features are I^{STANDARD}, D^{STANDARD}, DH^{STANDARD}, T^{STANDARD}, and TM^{STANDARD} and can be seen in Eqs. 6.22–6.26.

$$I^{\text{STANDARD}} = I^{\text{SYNCHRONOUS}}(D^{\text{STANDARD}}, T^{\text{STANDARD}}) \quad (6.22)$$

$$D^{\text{STANDARD}} = D^{\text{PASSIVE}}(DH^{\text{STANDARD}}) \quad (6.23)$$

$$DH^{\text{STANDARD}} = DH^{\text{UNIFORM}} \quad (6.24)$$

$$T^{\text{STANDARD}} = T^{\text{BOOLEAN}}(TM^{\text{STANDARD}}) \quad (6.25)$$

$$TM^{\text{STANDARD}} = TM^{\text{UNIFORM}} \quad (6.26)$$

6.3.4 Omissions of the Formalism

In the formalism defined in Section 6.3.2, certain aspects of a variant of SDS are not defined. The initialization behavior of a variant is omitted as the effect of initialization is similar to the effect of the diffusion phase (as described in Section 6.3.4.1); hence, any duplication in description is avoided. The set of microtests that agents use in the test phase is also ignored as this is specific to each application of SDS. To consider each instance of SDS with a unique set of microtests to be a distinct variant would prohibit any meaningful grouping. When describing a specific instance of SDS it may be necessary to describe the set of microtests, but this is not necessary when describing a variant. The number of agents in a swarm is not considered to define a variant, and it is considered to be a parameter that must be manually selected. Outside of extreme values, which are so small that stable cluster formation is impossible, or so large that processing the swarm is impractical, swarm size does not affect any qualitative differences in performance.

6.3.4.1 Initialization

The purpose of initialization is to set the hypotheses of all agents in preparation for the first test phase, which is the same purpose as the diffusion phase. Therefore the initialization phases can usually by entirely omitted if the diffusion phase will be performed before the first test phase. Reversing the order of the diffusion phase and test phase makes no significant difference in the dynamical behavior of SDS [19]. As noted by Bishop [1], in cases where knowledge of the search space is available a priori this can be used to distribute the hypotheses accordingly, such as weighting the initial distribution toward one end of a one-dimensional search space if solutions have most commonly been located there. Although this is a valid observation, it can also be considered to be an effect of the diffusion phase. Any knowledge about the probable distribution of solutions in a search space may be incorporated into the hypothesis selection function, which will therefore be utilized during any diffusion phase instead of during only the first iteration.

Consider, for example, an instance of string-search SDS in which the target string has a larger probability appearing toward the end of the search space toward the beginning. One could initialize hypotheses of agents with a probability distribution weighted toward the end of the search space, but this would only be beneficial if one of the agents was initialized with the hypothesis that pointed to the solution. A better solution would be to similarly weight the probability distribution for selection of new hypotheses and potentially benefit from the knowledge at every diffusion phase. Even in cases where the knowledge only applies during the first iteration an effect identical to that of the initialization phase may be achieved by a method of new hypothesis selection that behaves differently during the first iteration.

Furthermore, by removing the initialization phase the process of SDS can be seen more clearly as a procession of alternating diffusion phases and test phases. This interpretation makes it easier to see how a swarm resulting from any variant of SDS may be subsequently modified by any other variant of SDS. In other words, each iteration of any variant of SDS can be understood as the initialization phase of another variant of SDS.

Using the final state of a swarm in one instance of SDS as the initial state of the swarm in another instance of SDS has been investigated by Grech-Cini and McKee [20]. In the case studied, facial features were being tracked in the individual frames of a video. The location of the target in one image is likely to be close to its location in the previous image; hence, the distribution of the swarm from the previous instance can be used as the initial distribution in a subsequent instance. This technique was found to be successful in the task of locating the ear-nose region in a video of a person reading a passage of a text.

6.3.4.2 Extraction

The method of extraction is also omitted as this mechanism remains largely the same over all variants. Many variants of SDS converge to a single large cluster, some converge to a number of clusters with sizes proportional to the score of the hypothesis, and some distribute agents most densely in the region of the search space near the highest scoring hypotheses. All of these can be interpreted as indicating solution quality with cluster size; the exact way this information will be used is specific to the nature of the task.

6.4 VARIANTS OF SDS

The following sections each describe a category of variants of SDS. First, implementation variants are those that modify something about the operation of SDS without having a significant impact on its performance as a search algorithm to enable SDS to be implemented in situations otherwise unsuitable. The remaining variants can be seen as biasing the behavior of SDS toward exploration, evaluating further hypotheses, or exploitation, converging to currently maintained hypotheses, but there are many dimensions in which this bias can be effected.

Convergence variants affect behavior of agents to modify the set of hypotheses from a given search space at which clusters may form. Exploration variants affect the interactions between agents to modify the diffusion of hypotheses through the swarm, therefore, modifying the probability of the optimal hypothesis being located. Heuristic variants use external knowledge of a search task to modify the behavior of agents to be more suitable for that task, the range of possible optimizations is very large as each search task may have its own peculiarities to be optimized against, and for all heuristic variants the performance is only improved as long as the assumption underlying the heuristic is accurate.

Each variant is described in a number of aspects; the variation in individual behavior, the resulting variation in the population behavior, and the change in performance of the algorithm. Each change in performance of SDS has both advantages and disadvantages, in accordance with the no free lunch (NFL) theorems. Introduced in [21] and applied to search algorithms in [22],[2] the NFL theorems state that over the set of all possible search spaces, there can be no algorithm which performs better on average than random search. The implication of the NFL theorems in the context of comparing SDS variants is that over the set of all possible search spaces there are as many search spaces for which SSDS outperforms a given variant than there are search spaces for which the same variant outperforms SSDS. This claim can also be applied to any two variants of SDS and also to any variant of SDS and any other search algorithm. Given the NFL theorems, the emphasis of any variant should not be evaluated solely on the advantage it introduces, but rather the trade-off it represents, and how large the effect of the trade-off should be to maximize the net gain in performance after the negative impact of the disadvantage is realized against the positive effect of the advantage. The optimal choice of variant and the associated parameters should therefore be considered in the context of the search space of the given task and the requirements of the solution.

6.4.1 IMPLEMENTATION VARIANTS

Some variants of SDS have been shown to exhibit only a small performance decrease when compared with SSDS, but enable SDS to be implemented in significantly different ways. For example, requirements for SDS to be implemented in hardware have been identified as (1) the internal state of agents is unavailable to other agents, (2) asynchronous mode of iteration, (3) limited connectivity between agents, and (4) a one-way communication protocol [19]. Other variants described here, which affect the implementation of SDS but do not significantly affect performance, are active diffusion, and deterministic SDS.

6.4.1.1 Asynchronous Iteration

The behavior of SDS has been studied in three distinct modes of iteration: synchronous, where all agents perform the mode of testing and then all agents perform the mode of diffusion; asynchronous [19], where each agent individually performs the mode of diffusion followed by the mode of testing; and parallel [19], where all agents perform mode of diffusion and mode of testing in their own time. Although these variants are significant, they do not significantly change the performance of most variants except for the timescale [18].

In asynchronous iteration, each agent individually performs the diffusion behavior followed by the test behavior, shown in Algorithm 6.10. It is one of the requirements for implementing SDS in hardware, and yet it does not affect the resource allocation behavior of SDS [19].

ALGORITHM 6.10 $I^{\text{ASYNCHRONOUS}}$: ASYNCHRONOUS ITERATION

1: **function** $I^{\text{ASYNCHRONOUS}}$ (D, T, swarm)
2: **function** $I'()$
3: **for each** agent **in** swarm **do**
4: D(agent)
5: T(agent)
6: **return** I'

In parallel iteration each agent has a probability distribution over time to update its state, shown in Algorithm 6.11.

ALGORITHM 6.11 I^{PARALLEL}: PARALLEL ITERATION

1: **function** I^{PARALLEL} (D, T, swarm)
2: **function** $I'()$ ⊳any agent may perform either action at any time
3: **for each** agent **in** swarm **do**
4: after some time perform D(agent)
5: after some time perform T (agent)
6: **return** I'

One of the aims of formulating asynchronous variants of SDS is to cast the algorithm in such a way that it may be more naturally implemented in electronic hardware, while retaining the essential characteristics of the algorithm. Though formulated with hardware implementation in mind, there have been software implementations of each of the individual requirements, which retained the characteristic behavior of SDS. Asynchronous iteration allocates resources in the same way as synchronous iteration, as does parallel iteration because it is a special case of asynchronous iteration [19]. Similarly, the strong convergence criterion holds for asynchronous iteration [19] and for parallel iteration. When $\alpha = 1$ all agents will converge on the optimal hypothesis with probability 1 [19, p. 24]. In which case the number of iterations before all agents maintain the optimal hypothesis can be computed as

$$TC = \sum_{a=0}^{A-1} \frac{A^2}{(A-a)(a+(A-a)p_\alpha)} \tag{6.27}$$

6.4.1.2 Private Internal State

One of the features required for a hardware implementation of SDS was that agents may not directly access the internal state of other agents [19]. A mode of diffusion in which agents do not need to access the activity value of other agents (though still need to access the hypothesis of other agents) is described in Algorithm 6.12. Note that the action of the polling agent is not affected by the activity of the polled agent. The polling agent always assumes the hypothesis of the polled agent, and the polled agent then does or does not generate a new hypothesis depending on its own activity.

ALGORITHM 6.12 $D^{PASSIVE}$: PRIVATE INTERNAL STATE DIFFUSION

```
1: function D^PASSIVE (DH, swarm)
2:   function D′(agent)
3:     if agent is inactive then
4:       hypothesis of agent ←DH()
5:       polled ← random agent in swarm
6:       hypothesis of agent ← hypothesis of polled
7:       if polled is inactive then
8:         hypothesis of polled ←DH()
9:   return D′
```

It has been shown that the probability of an inactive agent assuming the optimal hypothesis during diffusion is identical for passive diffusion and private internal state diffusion, though the convergence time of an SDS using private internal state diffusion is marginally longer [19].

6.4.1.3 Limited Connectivity

The notion of limited connectivity, in which agents may only communicate with a subset of the swarm, was introduced with the original description of SDS [1]. Where in SSDS each agent may poll any other agent in the swarm, variants have been developed in which an agent may only poll a subset of the swarm.

If an agent may poll a second agent then the first agent can be thought of as being connected to the second agent. It is not required that all connections are reciprocal, so one agent being connected to another does not necessitate that both agents are connected to each other. The set of all agents to which one agent is connected is named that agent's neighborhood. The neighborhoods of all agents can therefore be represented as the adjacency matrix of an unweighted directed graph. The number of possible connection configurations of a swarm of size A is therefore 2^{A^2}. Of all the possible connection configurations, a few types have been studied for their effects on the performance of SDS.

Reciprocal connectivity in an orthogonal grid, where each agent was bi-directionally connected to their four neighbors, has been shown to be sufficient for the effective performance of SDS [1]. While the reduced number of connections necessarily slows the diffusion of hypotheses through the swarm, the reduced complexity of the swarm significantly increases the feasibility of implementing SDS in hardware.

The performance of swarms with limited connectivity has been investigated experimentally [19]. A variant of SDS using $I^{ASYNCHRONOUS}$ was implemented with swarms of 16, 32, and 64 agents called "lattices," and the number of connections between each agent was varied over the experiments. A fully connected lattice is equivalent to SSDS; hence it can be described by existing models. In comparing the convergence time for a fully connected lattice with that of the convergence time of simulations of partially connected lattices, it was shown that convergence is possible and with only a relatively small increase in convergence time even for lattices with few connections. For example, performance was "acceptable" with 4–6 connections in a lattice of 16–64 agents [19].

Further experimental work showed that performance is only slightly degraded if connectivity is significantly reduced, as long as the lattice exhibited random or small-world connectivity [23]. A lattice exhibits random connectivity when each agent is connected to a small number of other agents selected uniformly randomly. A lattice exhibits small-world connectivity when the majority of each agent's connections are to nearby agents, with a small number of connections to randomly chosen agents. Relatively long-distance connections are more costly in terms of materials and engineering complexity, so small-world networks have the advantage of requiring mostly short connections, but the small number of random and potentially long-distance connections ensures that the most nodes can be reached from all other nodes in a small number of steps. Small-world networks are therefore the preferred network topology for hardware implementation. An implementation of SDS on a chip was designed, which used limited connectivity, resulting in 64 agents arranged in four layers of 4×4 [24]; no performance results have been published. It was noted that the good performance of SDS under limited connectivity has further implications because biological neural structures have been observed as exhibiting analogous small-world connectivity; hence, relaxing the requirements of full connectivity in SDS leads to a more biologically plausible architecture.

The most important aspect regarding limited connectivity is that the performance of SDS is negligibly affected unless the neighborhoods of all agents are extremely limited. This therefore enables SDS to be implemented in various architectures in which the full connectivity of SSDS would be impractical.

6.4.1.4 Active Diffusion

Where passive diffusion defines that inactive agents poll random agents with the possibility of receiving a new hypothesis from an active agents, active diffusion relies on the active agents polling for inactive agents to which they pass their current hypothesis (D^{ACTIVE}, Algorithm 6.13). The resulting behavior is very similar to passive diffusion, though slightly less robust due to being more sensitive to the effect of noise [11, p. 164]. The cause of the slightly reduced robustness is that in certain cases clusters increase in size slower than under passive polling. As each active agent may poll at most one other inactive agent, a cluster cannot increase in size faster than doubling each iteration, where in passive diffusion there is a possibility that all inactive agents will assume the same hypothesis in a single iteration.

There is a minor issue in implementing active diffusion; it is not clearly defined what should happen when an active agent attempts to impose its hypothesis on an agent that has already received a hypothesis from another active agent. Either the inactive agent should assume the new hypothesis, and hence slightly reduce the number of diffused hypotheses, or a mechanism needs to be implemented to avoid this feature, which would increase algorithmic complexity. There is also a question of when inactive agents should select a new hypothesis. One option (T^{ACTIVE}, Algorithm 6.13) is for all agents to select a new hypotheses whenever they become inactive. There is a risk that under this scheme agents may select a new hypothesis that is then discarded should an active agent immediately transmit its hypothesis to the agent. If new hypothesis selection is an expensive action then this extra work may be significant. Alternatively, once all active agents have performed their polling action, any inactive agents that were not polled could select new hypotheses. This method requires that the set of polled inactive agents is recorded somehow, at the cost of some algorithmic complexity. In practice, the effect is often negligible [11, p.148].

ALGORITHM 6.13 D^{ACTIVE}, ACTIVE DIFFUSION AND T^{ACTIVE}, TESTING FOR ACTIVE DIFFUSION

1: **function** D^{ACTIVE} (swarm) ▷No *DH* required as agents select a new hypothesis immediately on becoming inactive
2: **function** *D′*(agent)
3: **if** agent is active **then**
4: polled ← random agent in swarm
5: **if** polled is inactive **then**
6: hypothesis of agent ← hypothesis of polled
7: **return** *D′*
8: **function** T^{ACTIVE} (*T*, *DH*)
9: **function** *T′*(agent)
10: microtest ← *TM*()
11: partial evaluation ← microtest(hypothesis of agent) ▷partial evaluation ∈ $\mathbb{B}$
12: **if** partial evaluation = ***true*** **then**
13: agent becomes active
14: **else**
15: agent becomes inactive
16: hypothesis of agent ← *DH*()
17: **return** *T′*

Active diffusion is shown to evolve in a "very similar" [11] way to SSDS, though significantly more difficult to model mathematically, due to complications of the case of an inactive agent selected by more than one active agent. The one-step evolution function for active diffusion us shown in Eq. 6.28 and minimum convergence criteria is shown in Eq. 6.29 [11].

$$f_{\text{active}}\left(\alpha,\beta,\bar{c}_i\right)=\alpha\left[\bar{c}_i+\frac{\bar{c}_i\left(1-\beta\right)-\bar{c}_i^{\,2}\left(1-\frac{\beta}{\alpha}\right)}{\bar{c}_i+\beta\left(1-\frac{\bar{c}_i}{\alpha}\right)}\left(1-e^{-\left(\bar{c}_i+\beta\left(1-\frac{\bar{c}_i}{\alpha}\right)\right)}\right)\right] \tag{6.28}$$

$$\alpha_{\min}=\begin{cases}\dfrac{\beta}{1-(1-\beta)e^{-\beta'}} & \text{if } \beta>0\\[2ex] 0.5, & \text{if } \beta=0\end{cases} \tag{6.29}$$

The trade-off of active diffusion is that an arguably more natural behavior is employed, but at the expense of slightly increased complexity in implementing and analyzing the algorithm, and slightly reduced robustness. As the resulting practical behavior is similar to passive diffusion, most variants of SDS employ passive diffusion unless active diffusion is a requirement for a specific implementation.

The combination of active diffusion and passive diffusion is known as dual diffusion. Each inactive agent attempts to assume a hypothesis from an active agent and each active agent attempts to impose their hypothesis on an inactive agent. As dual diffusion performs the combined behavior of two different diffusion variants, it can be considered to be a form of multi-diffusion, which is described in Section 6.4.2.2.

6.4.1.5 Deterministic SDS

SDS relies on a random number generator in a number of steps, notably the selection of random agents in the diffusion phase and the selection of random microtests in the test phase, in a computational context this is assumed to be a pseudorandom number generator as is included in the standard libraries of modern programming languages. This is in some sense already not a random system as most pseudorandom number generators can be seeded with a constant value and will always produce the same sequence of values. The random number generator can be omitted entirely and any step that normally requires a random action can be replaced with a deterministic and regular process. It has been shown, using a simple iterative process, that SDS still performs a characteristic search, with exploration of the search space followed by rapid convergence [18]. This deterministic variant was observed to potentially offer improved performance if the deterministic sampling processes were quicker to compute than the usual pseudorandom process, and that the deterministic sampling process is guaranteed to provide samples evenly. As the performance gains of speeding up the random sampling process are likely to be small, and modern pseudorandom number generators already provide fairly uniform sampling, the behavior of a deterministic SDS is likely to be indistinguishable from SDS. The advantage of not requiring a pseudorandom number generator is that it reduces circuit complexity in the case of hardware implementation of SDS.

This group of variants shows that there are diverse ways in which SDS can be implemented without losing the characteristic behavior of a non-greedy search utilizing partial evaluation and exhibiting a rapid convergence to a solution once identified by means of a positive feedback loop. Each variant has its own trade-off, but each are likely to have a negligible impact on performance in practice outside of extreme cases.

6.4.2 Convergence Variants

There are a number of variants that significantly affect the convergence of SDS. The aim of these variants is therefore to enable convergence for search spaces in which SSDS would not converge, or to reduce the amount of the mean probability of an agent at any non-optimal hypothesis becoming active for search spaces in which SSDS would converge so that convergence comes sooner. As anticipated by the NFL theorems, each variant represents a trade-off. Each variant introduces some algorithmic complexity to implement the additional behavior, and each variant introduces the requirement to select at least one extra parameter. For each application of a convergence variant any extra parameters introduced must be carefully selected or else the performance may be worse than SSDS applied to the same problem.

6.4.2.1 Multi-Testing

When investigating the effect of noise on the convergence time, it was observed that the increase in time that resulted from the increase in noise could be reduced exponentially, as agents each performed more than one microtest during the test phase [1]. This mechanism was hence named multi-testing. Multi-testing introduces two new parameters: "amount," which is the number of tests that each agent should perform in each test phase, and "combinator," which is the method for combining the multiple results into a single activity value for the agent. There are many possible combinators, including the two that have been investigated either return *true* if all partial evaluations are *true* (called AND-multi-testing), or return *true* if any partial evaluations are *true* (called OR-multi-testing). An amount of 1 or a combinator that simply selects one of the partial evaluation results at random is equivalent to SSDS [11].

It has been shown that the effect of multi-testing is that the effective score of the optimal hypothesis is increased (in the case of OR-multi-testing) or decreased (in the case of AND-multi-testing), and the relative score of all other hypotheses is preserved [25]. The term "effective score" is used here as the fundamental quality of any hypothesis remains unchanged, but the probability of an agent remaining active at any hypothesis after the test phase changes.

The advantage of this mechanism is that, in being able to modify the effective score of all hypotheses, values may be selected in which otherwise convergence would be impossible due to $\alpha < \alpha_{\text{min}}$ or to reduce the amount of homogeneous background noise (β). For example, OR-multi-testing increases the robustness factor of SDS [11, p. 88]. The disadvantage of this mechanism is the increased computational complexity of the test phase, as the number of microtests performed increases as a multiple of the number of agents. Multi-testing is therefore most useful in cases where the optimal hypothesis scores too poorly to form a stable cluster, or where too many sub-optimal hypotheses score highly enough to form a stable cluster. Multi-testing is described in Algorithm 6.14.

ALGORITHM 6.14 $T^{\text{MULTI-TESTING}}$: MULTI-TESTING AND EXAMPLE COMBINATORS

1: **function** $T^{\text{MULTI-TESTING}}$ (*TM*, amount, combinator)
2: **function** *T*′(agent)
3: evaluations ← empty list
4: **for** once per amount **do**
5: partial evaluation ← microtest(hypothesis of agent)
6: append partial evaluation to evaluations
7: multi-evaluation ← combinator(evaluations) ▷multi-evaluation $\in \mathbb{B}$
8: **if** multi-evaluation = *true* **then**
9: agent becomes active
10: **else**
11: agent becomes inactive
12: **return** *T*′
13: **function** AND-COMBINATOR(evaluations)
14: **if** all partial evaluations in evaluations are *true* **then**
15: **return** *true*
16: **else**
17: **return** *false*
18: **function** OR-COMBINATOR(evaluations)
19: **if** any partial evaluations in evaluations are *true* **then**
20: **return** *true*
21: **else**
22: **return** *false*

6.4.2.2 Multi-Diffusion

Multi-diffusion is similar to multi-testing, except the mode of diffusion includes polling of multiple agents. The effect of multi-diffusion has not been studied directly, but the studies of three variants of polling, each of which enabled differing amounts of diffusion in each iteration, do provide some information. Active diffusion, which enabled the least diffusion, was shown to have the lowest robustness (ζ) [11, p. 161], with passive diffusion having a higher robustness and dual

TABLE 6.1
Summarizing the Robustness Factors of Each Recruitment Polling Method Over All Homogeneous Background Noise Search Spaces [11]

Polling method	ζ
Active polling	0.543
Passive polling	0.614
Dual polling	0.769

polling, which combined passive and active polling, to enable the most diffusion in a single iteration; hence it had the highest robustness (see Table 6.1). In an experiment utilizing 1000 agents, a search space of size 10,000, $\beta = 0.1$ and a variant denoted as

$$SDS(I^{SYNCHRONOUS}(D^{MULTI-DIFFUSION}(DH^{STANDARD}, 2, or-combinator), T^{STANDARD}), H^{FIXED}(1000))) \quad (6.30)$$

The variant converged to the optimal hypothesis 77 times out of 100 when $\alpha = 0.4$ and 100 times out of 100 when $\alpha = 0.5$, whereas the same conditions utilizing SSDS converged to the optimal hypothesis 0 times out of 100. Multi-diffusion is described in Algorithm 6.15.

ALGORITHM 6.15 $D^{MULTI-DIFFUSION}$: MULTI-DIFFUSION AND EXAMPLE COMBINATORS

1: **function** $D^{MULTI-DIFFUSION}$ (*DH*, amount, combinator)
2: **function** *D′*(agent)
3: **if** agent is inactive **then**
4: polled ← list of random agents in swarm of length 'amount'
5: result ← combinator(polled)
6: **if** combined = *false* **then**
7: hypothesis of agent ←*DH*()
8: **else**
9: hypothesis of agent ← combined
10: **return** *D′*
11: **function** AND-COMBINATOR(polled)
12: **if** all agents in polled are active **then**
13: **return** random agent in polled
14: **else**
15: **return** *false*
16: **function** OR-COMBINATOR(polled)
17: **if** any agents in polled are active **then**
18: active polled ← list of active agents in polled
19: **return** random agent in active polled
20: **else**
21: **return** *false*

The advantage of multi-diffusion is that convergence may be possible in search spaces in which SSDS will not converge due to the optimal hypothesis having an insufficiently high score. The disadvantage is that in cases where convergence was already likely, the steady-state cluster sizes will be larger under multi-diffusion than under SSDS. The effect is that there will be fewer agents generating new hypotheses once a cluster has formed. In cases where the first cluster forms at a sub-optimal hypothesis this reduces the probability of an agent locating the optimal hypothesis. Multi-diffusion is therefore a method that biases the action of SDS toward being greedy, but with the benefit of clusters being more stable.

6.4.2.3 The Hermit

Hermit diffusion is the name given to a variant in which agents will occasionally refuse to communicate their hypothesis (D^{HERMIT}, Algorithm 6.16) [12]. An equivalent mechanism is to include a number of extra agents in the swarm that are permanently inactive and perform no action other than to be available for polling during the diffusion phase. All clusters will therefore have fewer inactive agents assuming their hypothesis each iteration, but the rate at which agents are becoming inactive and leaving the cluster will remain unaffected; hence, all hypotheses will appear to have a lower score than in SSDS.

ALGORITHM 6.16 D^{HERMIT}: HERMIT DIFFUSION

```
1: function D^HERMIT (DH, hermitage, swarm)
2:   function D'(agent)
3:     if agent is inactive then
4:       polled ← random agent in swarm
5:       p ← true with probability P(hermitage)      ▷p ∈ 𝔹
6:       if polled is active and not p then
7:       hypothesis of agent ← hypothesis of polled
8:     else
9:       hypothesis of agent ←DH()
10:  return D'
```

The effect of the hermit is catastrophic in situations where the effective score of the optimal hypothesis is reduced below α_{min}, therefore, making convergence impossible. The effect of the hermit is advantageous in situations where the search space contains many sub-optimal hypotheses with a high enough score to form a stable cluster. If the effective score of the sub-optimal hypotheses is reduced below minimum convergence criteria but the score of the optimal hypothesis remains greater than minimum convergence criteria, then convergence time will be improved as there will be more inactive agents generating new hypotheses.

The opposite method, in which some agents that are permanently active are added to the swam is not considered to be worth investigating. First, any active agent needs to maintain a particular hypothesis, although how this hypothesis is selected is unclear, and second, the effect is likely to be similar to that of the secret optimist.

6.4.2.4 The Secret Optimist

One variant introduces a probabilistic element into activity updating. Any agent whose microtest failed will only become inactive a certain proportion of the time and hence will perform further evaluations of their current hypothesis; this behavior is known as the secret optimist [12].

The secret optimist behavior has the effect of increasing the effective test scores of all hypotheses. This is useful in the case where the score of the optimal solution is naturally low to form a stable cluster, but has the negative effect of increasing the activity at sub-optimal solutions, reducing the number of inactive agents exploring the search space.

This will be most useful in tasks where there is known to be a single optimal solution that will score significantly higher than all other solutions, but has a sufficiently high score to form a stable cluster on its own.

This has the opposite effect on the hermit (Section 6.4.2.3) in that it raises the effective scores of all hypotheses.

ALGORITHM 6.17 T^{OPTIMIST} — SECRET OPTIMIST TESTING

```
1: function T^OPTIMIST (TM, optimism)
2:   function T′(agent)
3:     microtest ← TM()
4:     optimistic ← true with probability P(optimism)
5:     partial evaluation ← microtest(hypothesis of agent)    ▷partial evaluation ∈ B
6:     if partial evaluation = true or optimistic = true then
7:       agent becomes active
8:     else
9:       agent becomes inactive
10:  return T′
```

The effect on the minimum convergence criteria (α_{min}) of the T^{OPTIMIST} is given in Eq. 6.31 [12], where i is the average number of iterations an agent will retain a failed hypothesis.

$$\alpha_{\min} = \frac{1+i\beta}{2+i-\beta} \tag{6.31}$$

$$\alpha_{\text{secret optimist}} = \frac{i\alpha}{i+1} \tag{6.32}$$

Although this is a larger value than with SSDS, the effective score of the optimal hypothesis will also increase. The advantage of this variant is that α_{min} will always be lower than the equivalent case using SSDS as long as $i > 0$. When $i = 0$ the behavior is identical to SSDS. The disadvantage of this variant is that the effective score of all hypotheses increases, so there will be an increase in the homogeneous background noise.

6.4.3 Exploration Variants

These variants modify the exploration behavior of SDS, which does not change which clusters may form, but in modifying the resource allocation during the search it will affect the average number of iterations taken to locate the optimal hypothesis. The trade-off is that for there to be more agents selecting new hypotheses there must be fewer agents evaluating existing hypotheses, leading to less stable clusters and potentially collapsing altogether.

6.4.3.1 Context-Free Diffusion and Context-Sensitive Diffusion

The SSDS has been criticized for forming a single large cluster, leaving few agents to test hypotheses in the remainder of the search space and possibly missing superior solutions [9].

To decrease the number of active agents and increase the number of agents generating new hypotheses, two variants were defined in which active agents may become inactive during the diffusion phase. The variants are named context-free diffusion and context-sensitive diffusion.

Context-free diffusion ($D^{\text{CONTEXT-FREE}}$, Algorithm 6.18) introduces a mechanism in which active agents poll an agent at random from the swarm. In the case that the polled agent is active, the polling agent will become inactive and will generate a new hypothesis using *DH*. Context-sensitive diffusion ($D^{\text{CONTEXT-SENSITIVE}}$, Algorithm 6.19) introduces a similar mechanism but adds an extra clause. Active agents poll an agent at random from the swarm, but the polling agent only becomes inactive and generates a new hypothesis using *DH* in the case where the polled agent is active and both agents share the same hypothesis.

ALGORITHM 6.18 $D^{\text{CONTEXT-FREE}}$: CONTEXT-FREE DIFFUSION

```
1: function D^CONTEXT-FREE(DH, swarm)
2:   function D'(agent)
3:     polled ← random agent in swarm
4:     if agent is inactive or polled is active then
5:       if agent is inactive and polled is active then
6:         hypothesis of agent ← hypothesis of polled
7:       else
8:         agent becomes inactive
9:         hypothesis of agent ←DH()
10:  return D'
```

The one step-evolution functions, minimum convergence criteria (α_{min}), robustness (ζ), and steady state (γ), for context-free SDS and context-sensitive SDS have been derived from developments of the DDSM [11] and can be seen in Table 6.2.

ALGORITHM 6.19 $D^{\text{CONTEXT-SENSITIVE}}$: CONTEXT-SENSITIVE DIFFUSION

```
1: function D^CONTEXT-SENSITIVE (DH, swarm)
2:   function D'(agent)
3:     polled ← random agent in swarm
4:     if agent is inactive and polled is active then
5:       hypothesis of agent ← hypothesis of polled
6:     else
7:       a ← agent is inactive
8:       b ← polled is active and hypothesis of agent = hypothesis of polled
9:       if a or b then
10:        agent becomes inactive
11:        hypothesis of agent ←DH()
12:  return D'
```

TABLE 6.2
Comparison of the One-Step Evolution Function ($\bar{c}_{i+1}$), Minimum Convergence Criteria (α_{min}), Robustness (ζ), and Steady State (γ) of Context-Free SDS and Context-Sensitive SDS

	Context-Free SDS	Context-Sensitive SDS
$\bar{c}_{i+1}$	$\bar{c}_i\left[2\alpha(1-\beta)\right]-\bar{c}_i^{\,2}\left[2(\alpha-\beta)\right]$	$\bar{c}_i\left[\alpha(2-\beta)\right]-\bar{c}_i^{\,2}(2\alpha-\beta)$
α_{min}	$\frac{1}{2(1-\beta)}$	$\frac{1}{2-\beta}$
ζ	$1-\ln 2 \approx 0.307$	$2(1-\ln 2) \approx 0.614$
γ	$\frac{2\alpha(1-\beta)-1}{2(\alpha-\beta)}$	$\frac{\alpha(2-\beta)-1}{\alpha-\beta}$

These mechanisms effectively cap the number of agents that can be globally active, in the case of context-free SDS, or active in the same cluster, in the case of context-sensitive SDS. This is useful in cases where the first hypothesis at which a stable cluster forms is unlikely to be the optimal solution.

Some publications [26–29] refer to a *relate phase* in which the behaviors specific to context-free SDS and context-sensitive SDS take place; rather than defining a new phase, these will be considered modifications of the diffusion phase to avoid unnecessary complexity.

Three scenarios have been proposed in which context-free SDS might be a better choice than SSDS [9]:

1. There are multiple hypotheses that share the same score as the optimal hypothesis and the aim is to identify them all.
2. The aim is to find the optimal hypothesis and the remaining hypotheses with the highest score.
3. The search space and/or microtests are changing over time, so the best solution at one time may not be so some time later.

Where SSDS will equally allocate agents to two hypotheses of equal score over many iterations, with one cluster being orders of magnitude larger than the other for some time, and then a switch to the other one being larger context-free SDS, it is able to simultaneously maintain multiple swarms. In the case where there are two hypotheses with equal score in SSDS, one cluster will be orders of magnitude larger than the other for some time, after which a switch will occur and the other cluster will be the larger. Context-free SDS also has the ability to maintain multiple swarms at hypotheses of different scores, but the cluster size is dramatically smaller for hypotheses of only a slight difference in score, due to the very nonlinear gain amplification.

In summary, context-free SDS distributes its resources more evenly than SSDS, but it maintains a comparable stability of clusters supporting solutions.

This makes the modeling assumption that agents at the noise assumption will never become inactive in the diffusion phase, but this is an imperfect assumption as the diffusion phase will result in noise agents duplicating identical hypotheses.

6.4.3.2 Hypothesis Transmission Noise

Uniformly random hypothesis selection has been described as a strength and a weakness of SDS It is a strength as SDS does not require a usable gradient in the objective function of the search task, but it is a weakness as SDS is unable to utilize the information should one exist [18].

In this variant, when an agent assumes a hypotheses from another agent, the hypothesis is copied with some random perturbation, known as hypothesis transmission noise. This has the effect that a form of hill-climbing behavior emerges in the locality of the transmitted hypothesis [18, 30]. This has the advantage of exploiting self-similarity of a search space, which is when similarly located hypotheses have similar scores. This can be understood intuitively by imagining a cluster near an optimal solution in a continuous search space. The time for which agents at the cluster will remain active is a function of the score of that hypothesis. Agents that diffuse to the side further away from the optimal will remain active for a shorter time, and agents on the side toward the optimal will remain active for longer. In this way the cluster will appear to flow toward the optimal until it is centered on it. This mechanism makes it very rare for agents to have identical hypotheses, so a cluster must be considered to be a collection of agents with similar hypotheses rather than identical hypotheses. The result may be "extracted" by calculating the average hypothesis of all active agents. A mode of diffusion utilizing hypothesis transmission noise can be seen in Algorithm 6.20; it requires an extra parameter "noise," which is a function that takes a hypothesis as a parameter and returns a perturbed hypothesis.

ALGORITHM 6.20 D^{NOISE}: NOISY DIFFUSION AND GAUSSIAN NOISE

1: **function** $D^{NOISE}(DH$, noise, swarm)
2: **function** D'(agent)
3: **if** agent is inactive **then**
4: polled ← random agent in swarm
5: **if** polled is active **then**
6: h ← hypothesis of polled
7: n ← noise(h)
8: hypothesis of agent ← n
9: **else**
10: hypothesis of agent ←$DH()$
11: **return** D'
12: **function** NOISE(hypothesis, sigma)
13: d ← Gaussian distribution with standard deviation = sigma
14: x ← random value from d
15: **return** hypothesis + x

There are many functions that could be used to perturb a hypothesis, and an example using a Gaussian distribution is given in Algorithm 6.20. The amount of noise introduced, by using a larger value for sigma in the example using Gaussian distribution, has an effect on the speed with which a cluster will appear to move through a search space. The trade-off in this context is that a faster moving cluster may miss smaller details of the search space and effectively evaluate multiple hypotheses at once, with potentially misleading results. A small amount of noise will lead to slower convergence and increases the probability that the cluster will not move out of any local optima in the search space. Of course a zero or negligible amount of noise is equivalent to SSDS, where clusters do not appear to move at all. A mechanism has been suggested for estimating a value for the transmission error online from the standard deviation of the test scores of an "elite sample" of the best scoring agents [31].

The feature where, in the case of a large amount of noise, a cluster may effectively evaluate multiple solutions may be an advantage or disadvantage depending on the context. For some tasks it may be useful information that a certain area of the search space contains many high-scoring hypotheses, whereas in other tasks it may be essential that a single hypothesis is evaluated in isolation. For example, consider the following variant of SDS. In a search space with a wide "spike" of high-scoring hypotheses around one area, and a much thinner and slightly taller "spike" in another area, this variant of SDS would consistently converge to the hypothesis representing the peak of the wider spike if the value for noise was above a certain value. It would also consistently converge to the hypothesis representing the peak of the thinner spike if the value for noise was lower than a certain value.

6.4.3.3 Comparative Partial Evaluations

In cases where the partial evaluations for a search task are not easily described as a set of Boolean functions, the functions may instead return scalar values. When using SSDS each microtest can be treated as if it were a Boolean function by providing a threshold for each microtest such that results over a certain value are considered to be *true* and results under the value are considered to be *false*. Although this will work for many cases, it is required that a threshold is chosen. At whichever level the threshold is selected hypotheses that score just under the threshold for many of the microtests, or significantly over the threshold in a small number of microtests, will appear to have a poor score. Similarly is the case where hypotheses with the opposite properties may appear to be high-scoring hypotheses even though the overall quality of the hypothesis may be lower than a hypothesis with a lower effective score.

An alternative approach, which does not require a threshold, is for agents to convert the scalar values of the microtests to Boolean values by comparing them with the scalar value obtained by an agent at a different hypothesis.

ALGORITHM 6.21 $T^{\text{COMPARATIVE}}$: COMPARATIVE TESTING

```
1: function T^COMPARATIVE (TM)
2:   function T′ (agent)
3:     polled ← randomly selected agent in swarm
4:     microtest ←TM()
5:     partial evaluation a ← microtest(hypothesis of agent)
6:     partial evaluation p ← microtest(hypothesis of polled)
7:     if partial evaluation a > partial evaluation p then
8:       agent becomes active
9:     else
10:      agent becomes inactive
11:  return T′
```

The advantage of this approach is that a threshold does not need to be chosen; the threshold is effectively selected by the action of the evolving swarm. Each iteration effectively removes half of the swarm, which evaluates lower than average. The disadvantage is that as there is no minimum threshold for activity, the highest scoring hypotheses located so far will always maintain a cluster, convergence therefore cannot be detected from the global activity alone. There is a further disadvantage: it is possible for the superiority of one hypothesis to be intransitive. This means that there may be a set of three or more hypotheses such that each hypothesis consistently scores higher than one other hypothesis. This can be seen in the case where three hypotheses A, B, and C each score [2,2,6,6,7,7], [1,1,5,5,9,9], and [3,3,4,4,8,8], respectively. If a partial evaluation is to select at random a single value from the list pertaining to that hypothesis, then *A* will most commonly evaluate higher than *B*, *B* will most commonly evaluate higher than *C*, and *C* will most commonly evaluate higher than *A*. In such cases the global activity of an SDS remains high, but the hypothesis of the largest cluster remains unstable.

6.4.4 Heuristic Variants

In variants that employ heuristics of a specific task, as is the nature of heuristics, the performance of the search will be improved as long as the heuristic is veridical.

6.4.4.1 Fixed Heuristics

If some information is known about the distribution of the score of the hypothesis in a search space then new hypothesis selection may be modified to select hypotheses more frequently from areas with a higher probability of containing hypotheses with a higher score. If the information is accurate, then the higher probability of selecting hypotheses with a higher score will effectively increase the homogeneous background noise as more randomly selected hypotheses will have a higher score, but the probability of selecting the optimal hypothesis will also be higher. Similarly, if one microtest is known to be significantly more valuable in determining the quality of a solution represented by a hypothesis the mode of microtest selection can employ a probability distribution to bias selection toward the more informative microtests. A related technique is to select a hypothesis that is furthest away from any previously selected hypothesis [30]. This method is almost the opposite of selecting a hypothesis from an area of the search space that has been proven to be relatively dense with high-scoring hypotheses, and this method is more likely to perform a search that is well distributed through the search space.

In cases where the search space is too large to practically enumerate all hypotheses, a mode of hypothesis selection may be employed that uses an existing heuristic. Consider, for example, the traveling salesman problem (TSP); the number of possible routes is of the order $n!$ where n is the number of cities, selecting one route from the set of all possible routes may be impractical. Therefore a heuristic may be employed, such as a stochastic greedy method, or a heuristic that guarantees to only select paths with certain features. Similarly, in the task of estimating a hyperplane that best describes a set of points a hypothesis can be stochastically constructed by fitting a hyperplane to a small randomly selected subset of the points. Heuristics such as these have the potential to dramatically reduce the size of a search space, and to speed up the convergence time of SDS. They achieve this, however, by leaving significant sections of the search space unexplored. It is therefore most suitable to apply a fixed heuristic in cases where a uniformly random search of the search space would be impractical, or when strong heuristics exist for the given task.

SDS is not limited to using existing heuristics, however; a method of adapting a heuristic as the search evolves is readily available.

6.4.4.2 Adaptive Heuristics

In all the variants examined so far, when an agent selects a new hypothesis it does so entirely independently of the rest of the swarm. As there are likely to be many agents maintaining a range of hypotheses, an agent may use the distribution of hypotheses over the set of active agents as information on how to bias the selection of their next hypothesis. A hypothesis may also be constructed from the combination of parts of the hypotheses of various agents. This is analogous to the technique employed by ant colony optimization (ACO), where a path is constructed probabilistically by combining sections of the route that were previously evaluated. An agent is therefore no longer required to generate the optimal hypothesis entirely at random. A swarm can be imagined to be evaluating hypotheses not just for their standalone value, but for the potential that they consist of strong parts suitable for recombination into other hypotheses. Care should be taken not to lean too heavily on the hypotheses of active agents, as it is effectively a form of multi-diffusion, which increases the "greediness" of the search. Consider the extreme case in which a new hypothesis is "constructed" by simply taking the hypothesis of an active agent this is exactly equivalent to passive diffusion.

As with fixed heuristics the performance will be improved as long as the underlying assumption is valid, which in this case is that a strong hypothesis can be combined from the features of other strong hypotheses. This kind of polling-based heuristic is therefore most suitable in cases where a uniformly random search of the search space is impractical, and where no strong heuristics are applicable for the task.

6.4.5 Halting Variants

There are many techniques to halt SDS, and most can be combined. Some use detection of cluster formation as a means to determine when to finish.

6.4.5.1 Time-Based Halting

6.4.5.1.1 Fixed Iteration Halting and Fixed Wall-Clock Time Halting

A very simple method of halting, fixed iteration count halting (H^{FIXED}, Algorithm 6.9), has already been described. This method simply keeps track of the number of iterations elapsed and halts once a given threshold is reached. A suitable value for the maximum number of iterations must be defined beforehand, which requires striking a balance between more thorough evaluations of a search space, at the expense of longer execution times, and timely results, at the expense of a higher change of optimal hypothesis. Halting after a certain amount of real time is also a valid method, but this method is rarely explored in academic studies to avoid any results being dependent on specifics of the host hardware, the software which is implementing the SDS, and the task-specific microtests. Real-time threshold halting has analogous advantages and disadvantages to fixed iterations halting.

ALGORITHM 6.22 H^{TIME}: FIXED WALL-CLOCK TIME HALTING

```
1: function H^TIME(maximum time)
2:   elapsed time ← 0
3:   function H′()
4:     update elapsed time
5:     halt ← elapsed time > maximum time          ▷halt ∈ B
6:     return halt
7:   return H′
```

Fixed iteration count halting and fixed wall-clock time halting are the only halting methods guaranteed to halt over all search spaces; whether this is an advantage or disadvantage is dependent on the search task to which it is being applied. This is an advantage in cases where halting before convergence, or before convergence to a hypotheses with a sufficient score, is unacceptable. In such cases it may be preferable to loop indefinitely than to proceed with an unsuitable solution. This is a disadvantage in cases where any decision is better than no decision. In such cases even the poorest of hypotheses represents a better case than failing to proceed at all.

ALGORITHM 6.23 $H^{\text{INDEFINITE}}$: INDEFINITE HALTING

```
1: function H^INDEFINITE()
2:   function H'()
3:     return false
4:   return H'
```

6.4.5.1.2 Indefinite Halting

An instance of SDS does not need to halt to be useful; at any time in its execution a separate process may read the state of the swarm and calculate the clusters. This has been described as the preferred method of halting for dynamically changing problems, where SDS is used to maintain a solution to an objective function that is changing over time [18, p.115]. The obvious disadvantage is that this method potentially requires an unlimited amount of resources. If a cluster has formed at the optimal hypothesis, and the search space is not evolving over time, then there is no value to be gained in further searching, and there is a small but ever-present risk that the cluster will spontaneously collapse. Furthermore a sampling strategy will need to be defined, which will introduce a number of extra configuration parameters and thresholds. The advantage of this method is that a series of solutions could be reported from an ever-changing search space.

6.4.5.2 Cluster Size-Based Halting

6.4.5.2.1 Unique Hypothesis Count

Unique hypothesis count is a method that halts when the number of unique hypotheses maintained by the entire swarm is lower than a certain threshold. This method was used in an application of SDS to realistically place objects in a visual scene [32]. As convergence is characterized by the formation of a single large cluster, this necessitates that many agents will be maintaining the same hypothesis and therefore only the relatively few remaining agents may maintain distinct hypotheses. The advantage of this method is that the chosen value for the threshold parameter controls the minimal quality of a hypothesis that will induce a halt. The disadvantage of this method is that the threshold must be chosen with respect to the background noise, as search spaces with high homogeneous background noise will quickly converge to states with relatively few hypotheses, and search spaces with very low homogeneous background noise will have many distinct hypotheses even after the formation of a large cluster.

ALGORITHM 6.24 H^{UNIQUE}: UNIQUE HYPOTHESIS COUNT HALTING

```
1: function H^UNIQUE (unique count) ▷unique count ∈ N
2:   function H'()
3:     hypotheses ← hypotheses of agents in swarm
4:     unique ← unique values in hypotheses
5:     halt ← unique < unique count          ▷halt ∈ B
6:     return halt
7:   return H'
```

6.4.5.2.2 *Largest Cluster Size and Global Activity Threshold Halting*

There are two closely related halting methods that halt when a certain value reaches a given threshold. The values are (1) the largest cluster size and (2) the global activity. These methods are similar as global activity is a proxy for the largest cluster size. This can be understood intuitively as once a stable cluster has formed, the level of global activity is equal to the size of the largest cluster plus the activity of all background noise hypotheses. As the background noise is naturally stable, the level of global activity will correlate with the size of the largest cluster. The advantage of these halting methods is that, like unique hypothesis count halting, the chosen value for the threshold parameter controls the minimal quality of a hypothesis, which will induce a halt. These halting methods, like unique hypothesis count halting, also require that the threshold is chosen with respect to the amount of noise in a search space as higher values of global activity will be encountered in search spaces with higher homogeneous background noise, even if the score of the optimal hypothesis is the same. The disadvantage of largest cluster size halting in comparison with global activity halting is that the calculation of the largest cluster size is somewhat more computationally complex than the calculation of the global activity as size of all clusters must be calculated, rather than simply counting the number of active agents.

ALGORITHM 6.25 $H^{LARGEST}$: LARGEST CLUSTER SIZE HALTING

1: **function** $H^{LARGEST}$ (swarm, threshold) ▷threshold ∈ [0,1]
2: **function** $H'()$
3: cluster ← largest cluster
4: count ← number of active agents in cluster ▷count ∈ N
5: largest ← $\frac{\text{count}}{\text{swarmsize}}$ ▷largest ∈ R
6: halt ← largest >= threshold ▷halt ∈ B
7: **return** halt
8: **return** H'

ALGORITHM 6.26 $H^{ACTIVITY}$: GLOBAL ACTIVITY THRESHOLD HALTING

1: **function** $H^{ACTIVITY}$ (swarm, threshold) ▷threshold ∈ [0,1]
2: **function** $H'()$
3: active count ← number of active agents in swarm ▷active count ∈ N
4: activity ← $\frac{\text{active count}}{\text{swarm size}}$ ▷activity ∈ R
5: halt ← activity >= threshold ▷halt ∈ B
6: **return** halt
7: **return** H'

A disadvantage of all threshold-based halting methods is that halting as soon as the threshold is reached introduces the possibility that clusters that have begun forming at superior hypotheses are missed. A way to mitigate this possibility is to detect stable convergence, which requires that clusters form and subsequently stabilize, as this will ensure that the clusters detected were stable, and that superior clusters were not in the process of forming.

6.4.5.3 Convergence-Based Halting

Convergence is detected by a stability in the swarm; theoretically this refers to a stability in the sizes of the largest cluster. In some variants the global activity is a suitable proxy [11] and a more convenient value to calculate. In some variants, such as $D^{\text{CONTEXT-SENSITIVE}}$, multiple significant stable clusters may form, and in others such as $T^{\text{COMPARATIVE}}$ global activity is always stable. In these cases convergence must be detected using other methods.

6.4.5.3.1 *Strong and Weak Halting Criteria*

One convergence-based halting method is described in the original definition of SDS [1]. This method, known as the strong halting criterion [13, p. 8], halts when the size of the largest cluster is determined to be stable over a number of iterations. Stability is defined by two parameters, a and b, which define the size of largest cluster c, which will induce a halt and a tolerance of error in which the largest cluster size may fluctuate and still be considered stable. The strong halting criteria are defined as

$$\underset{a,b>0}{\exists}(2b < A \wedge b + a \leq A \wedge a - b \geq 0) \underset{n_0}{\exists} \underset{n>n_0}{\forall} (|c - a| < b) \tag{6.33}$$

Given the values, a, the cluster size, which is sufficiently large to indicate convergence, and b, an amount the cluster size may vary and still be considered stable, then the algorithm has reached equilibrium under the strong halting criteria if there exists an instant n_0 after which the largest cluster size c is sufficiently stable, such that $a - b \leq c_n \leq a + b$, otherwise written as $|c_n - a| < b$. In practice this requires a further variable, t, which defines a number of iterations that $|c_n - a| < b$ must hold for the algorithm to halt. This ensures that even when the search converges on a solution that there will be more searching for a hypothesis with a potentially higher score before the search halts. The weak halting criteria, identical in all respects except that c_n, refers to the global activity rather than the size of the largest cluster, exhibits identical performance in detecting convergence.

ALGORITHM 6.27 H^{STRONG}: STRONG HALTING

1: **function** H^{STRONG} (a, b, T)
2:   t←0
3:   **function** $H'()$
4:     $c \leftarrow$ largest cluster size
5:     **if** $|c - a| < b$ **then**
6:       $t \leftarrow t + 1$
7:     **else**
8:       $t \leftarrow 0$
9:     halt $\leftarrow t > T$ ▷halt ∈ B
10:     **return** halt
11:   **return** H'

ALGORITHM 6.28 H^{WEAK}: WEAK HALTING

1: **function** H^{WEAK} (a, b, T)
2:   t←0
3:   **function** $H'()$

```
4:        c ← global activity
5:        if |c − a| < b then
6:            t ← t + 1
7:        else
8:            t ← 0
9:        halt ← t > T              ▷halt ∈ B
10:           return halt
11:   return H′
```

The advantage of these methods is that the algorithm will not halt while a cluster is in the process of forming. As with cluster size-based halting methods, the requirement for the largest cluster size to be consistently within the region defined by $|c_n - a| < b$ ensures that the process will only halt when a hypothesis with a sufficiently high score has been located. A disadvantage of this method is that the region of stability defined by a and b has to be selected carefully, if $b < 1 - a$ holds then there will be a value for c, which is outside of the region, but also representing a relatively high-scoring hypothesis. For example, if $a = 0.8$ and $b = 0.1$ then to converge it must hold that $0.7 < c < 0.9$. The process would therefore never halt if c remained stable around 0.999, even though this represents convergence to a hypothesis with a high score.

6.4.5.3.2 *Running-Mean Stability*

An alternative method to detect convergence without selecting a global activity threshold is to record the global activity for a given number of iterations and halt when the standard deviation of the recorded values is within a given threshold. This method requires the selection of a number of parameters: the maximum memory size, which determines the number of iterations to record at any one time, after which each new value recorded causes the oldest recorded value to be forgotten; the minimum stability, which is the value that standard deviation of the recorded values must be lower than to be considered stable; and minimum stable iterations, which is the number of consecutive iterations that must be considered stable before halting. The advantage of this method is that convergence will be detected for a hypothesis of any score sufficient to form a stable cluster. There is a disadvantage that this method will halt too soon if no solution is located within the first few iterations, as the largest cluster size being stable at very low values will induce a halt.

ALGORITHM 6.29 H^{STABLE}: STABLE GLOBAL ACTIVITY HALTING

```
1: function H^STABLE (max memory size, min stability, min stable iterations)
2:    memory ← empty queue of maximum length 'max memory size'
3:    stable iterations ← 0
4:    halt ← false
5:    function H′()
6:        activity ← activeagentsinswarm / swarmsize
7:        push activity into memory
8:        stability ← standard deviation of all values in memory
9:        if stability <= min stability then
10:           stable iterations ← stable iterations + 1
11:           if stable iterations >= min stable iterations then
```

```
12:            halt ← true
13:        else
14:            stable iterations ← 0
15:        return halt
16:    return H′
```

This method therefore benefits from being combined with a cluster size-based halting method to avoid the case in which it halts before a stable cluster has formed.

6.4.5.4 Combined Halting Methods

As the output of each halting method is a Boolean value, they can be combined using Boolean logical operators to form new halting methods. Any such combined halting method will exhibit the advantages and disadvantages of the component halting methods but in working together may produce more consistent performance than any individual halting method.

For example, the combined halting method

$$H^{\text{AND}}\left(H^{\text{FIXED}}(1000), H^{\text{OR}}\left(H^{\text{FIXED}}(10000), H^{\text{ACTIVITY}}(0.5)\right)\right)$$

guarantees at least 1000 iterations, after which will only halt once the global activity reaches 0.5 or a maximum of 10,000 iterations elapse. Such a combined halting method will avoid the disadvantage of global threshold halting, where the algorithm may halt after the early formation of a cluster before significant exploration of the search space is performed, as long as a superior solution is found within 10,000 iterations. Similarly, the disadvantage of global activity threshold halting potentially never halting is also mitigated by the fixed iterations halting method. These mitigations come at the cost of some increased algorithmic complexity and the requirement to select values for the three thresholds (minimum iterations, global activity threshold, and maximum iterations) with all the considerations already described.

A practical combination of the halting methods described in this section is defined in Algorithm 6.30. This halting function is guaranteed to halt after a defined upper limit of iterations, and it will halt earlier if a stable cluster forms over a given size and will ignore any activity from clusters that form before a defined lower limit. As with the simpler combined halting method described above, there is increased algorithmic complexity, and the requirement of five arguments to be manually determined.

ALGORITHM 6.30 EXAMPLE OF COMBINED HALTING

```
1: function H^combined(minimum iterations, maximum iterations, memory, stable iterations,
   minimum activity)
2:   return H^OR(
3:     H^AND(
4:       H^FIXED(minimum iterations),
5:       H^STABLE(memory, stable iterations),
6:       H^ACTIVITY(minimum activity),
7:     ),
8:     H^FIXED(maximum iterations)
9:   )
```

6.4.6 Summary of Variants of SDS

The behavior introduced by each variant brings with it advantages and disadvantages. Each variant therefore has superior performance than SSDS in certain circumstances and inferior performance than SDS in other circumstances. This trade-off is to be expected and is in accordance with the NFL theorems. For example, the effect of multi-testing is to increase the effective score of all hypotheses, increasing the ability of the optimal hypothesis to maintain a larger cluster but also increasing the homogeneous background noise. Multi-testing therefore improves performance in cases where a stable cluster would otherwise not form, and impairs performance in cases where the increase in noise increases the time for the optimal hypothesis to be located. Just as with SSDS, over all possible search spaces any multi-testing variant will outperform random search in as many search spaces as random search outperforms the multi-testing variant. If any task can be considered to be a search over a subset of all the possible search spaces, then to select a variant of SDS without any analysis of such a subset is to make the assumption that by chance alone the subset of search spaces is one for which on average the variant outperforms SSDS, or indeed, random search [33]. This observation raises the question of how one may identify which subset of search spaces are represented in a given task and which variant is most likely to outperform SSDS. A number of features of search problems have been identified in which SSDS exhibits greatest performance [18], and this can be interpreted as an identification of the search spaces in which SSDS outperforms random search. For all other variants, some features of search spaces that lead to improved or impaired performance have been identified.

The NFL theorems in context of halting variants are distinct from other variants as they do not change the search or convergence behavior of the algorithm. Time-based halting functions merely change the number of iterations that will elapse before halting. The trade-off is therefore how long the algorithm must run before a user is confident that the optimal hypothesis has been located, and while the probability of locating the optimal hypothesis grows over time for any non-greedy variant, optimality can never be guaranteed. A related trade-off occurs in the case of halting methods that examine the state of the swarm to detect convergence; every method of detecting convergence introduces the potential for the method to be deceived and hence to erroneously halt. As with time-based halting the user will have to decide whether the risk of an erroneous halt is sufficiently low for each application of SDS.

SDS is most suitable for search tasks where there is little, or no, information available about the distribution of solutions in the search space, and the evaluation of each element of the search space can be decomposed into the evaluation of a number of separate functions. SDS is less suitable for tasks where it is absolutely essential that a single, globally optimal solution is found, in which case an exhaustive search may be preferable, or when features of the search space can be exploited to calculate the location of strong solutions rather than relying on random hypothesis generation.

Like ACO, SDS is a stochastic procedure, where the locally guided action of individuals gradually leads to the globally optimal behavior of the swarm. Unlike ACO, the state of an instance of SSDS can be described with a list of the hypotheses of the active agents, and a count of the inactive agents, whereas ACO requires that the modifications to the search space are represented somehow, which adds some memory requirements.

Furthermore some of the characteristics of swarm intelligence are demonstrated here, as variants that modify individual behavior in simple ways enable useful and varied behavior to emerge at the population level. Regarding the hermit, we see how a mechanism that is equivalent to a swarm, including a number of completely ineffective agents, can still result in an effective search, and in specific circumstances may even improve performance. An investigation into the precise conditions in which the hermit improves performance will be of some benefit. Similarly with the secret optimist, the simple case of agents remaining active after a failed microtest could be interpreted as discarding valuable information, but the effect is that search spaces in which SSDS would not converge due to the optimal hypothesis having an insufficient score can be expected to converge with an appropriate probability of "optimism."

Regarding multi-diffusion, we have only observed that a variant in which agents poll two agents increases the robustness of the variant, at the expense of reduced exploration of the search space by inactive agents. No investigation has been performed into an exact formulation of the effect on robustness. As conceived the multi-diffusion behavior is related to the logical *OR* function, as diffusion will occur if *any* polled agents are active. This alternative strategy would be related to the logical *AND* function, which would require that *all* of the polled agents were active. This alternative would reduce the amount of diffusion, and hence presumably reduce the robustness of the variant, and increase the amount of exploration of the search space.

6.5 NATURAL ANALOGUES OF SDS

A number of investigations into natural swarm intelligence have described processes that are analogous to various aspects of SDS. Make sure this list is up to date. These processes include Robinson ant nest-site selection.

6.5.1 ANT NEST-SITE SELECTION

In 2013, J.M. Bishop and the author attended a lecture given by E.J.H. Robinson called "Distributed Decisions: New Insight from Radio Tagged Ants" [34, 35] on the nest-site selection behavior of ants of the genus *Temnothorax*. In this lecture Robinson showed the observed behavior of individual ants engaged in the activity of evaluating a candidate site for a new nest.

Individual ants were observed to occasionally perform a search of the area outside of their nest that resembled a random walk. If, during this process, the ant encountered a location that may be a suitable site for a new nest then it would enter the candidate nest site and perform an evaluation. When evaluating a candidate nest site an ant will explore inside the nest and eventually reach an exit; ants did not necessarily visit all of the candidate nest site, and it did not follow any fixed path of exploration. If the candidate site was evaluated to be of sufficient quality then the ant would return to its nest, recruit another ant from the swarm, and lead it to the candidate nest site using a technique known as tandem running. While tandem running, the leader ant will move a short distance in the direction of the destination and then wait until the following ant has reached the leading ant. This process is repeated until both ants arrive at the destination. Tandem running not only brings two ants to a destination but also provides an opportunity for the following ant to learn the route from the original nest. When the ants reach the destination both perform the evaluating action and, if the candidate nest site is determined to be of sufficient quality, further recruitment ensues. The result is that a strong candidate has an increasing number of ants performing evaluations as the ants, which were recruited into the evaluation process, begin to perform tandem runs of their own. Robinson observed that if the site was of sufficient quality such that at any time approximately one-third of the whole nest had been recruited into the evaluation and recruitment process, then those individuals would undergo a behavioral phase change. In the new behavior, rather than recruiting further ants into the procedure, they would simply lift unrecruited ants (or eggs) in their mandibles and deposit them at the candidate site. Ants transported in this way would not have made the journey to the new nest themselves and no pheromone trail is formed that could be followed, so they would not know the route between the two sites. This means transported ants cannot perform the recruiting behavior. Transported ants simply adopt the new location as their new home, and they play no further part in the process.

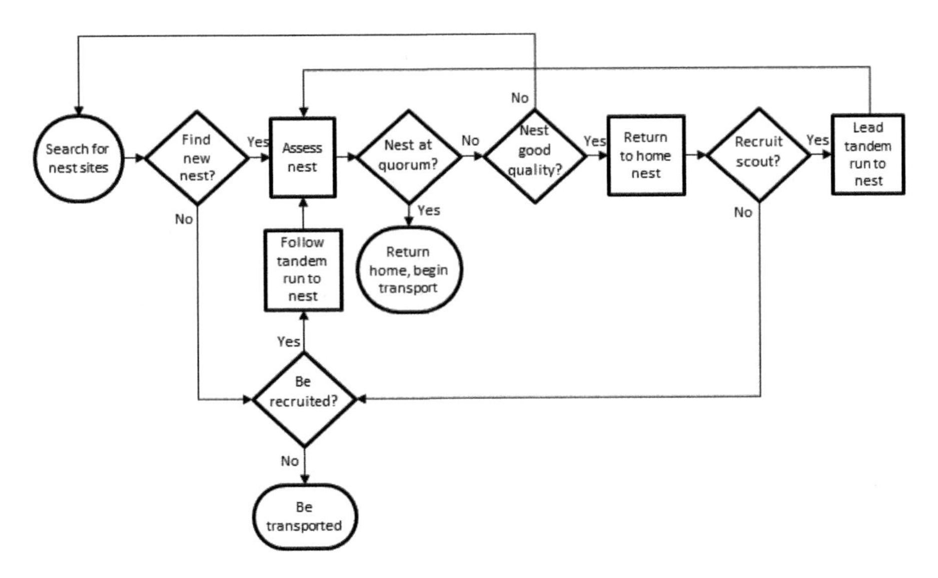

FIGURE 6.2 A flowchart of the emigration behavior as observed by Robinson. Begin at "Search for nest sites."

Once all ants and eggs had either been recruited into the evaluation procedure or transported to the new site, the nest has been fully migrated to its new site. Similar observations have been made in experiments in which migrations were induced in colonies of the ant *Leptothorax albipennis* [36]. This procedure is described in a flowchart in Figure 6.2.

A similar experiment showed that repeated partial evaluation effectively implemented a measure of the average quality of a resource, which fluctuates over time [37]. In the experiment, a colony was presented with two candidate nest sites, one of a quality sufficient to initiate a migration and another that was initially significantly superior, but which was occasionally modified to be of significantly inferior. The observed behavior was that there was a higher probability that the colony would migrate to whichever nest was superior on average; in cases where the fluctuating nest was superior for half the time and inferior for half the time there was an average probability of approximately 0.5 that the colony would migrate to either nest.

6.5.2 Bee Nest-Site Selection

Similar experiments were performed involving the migration behavior of colonies of bees [38]. Primarily, the behavior described had many similarities between the behavior observed in ants. The same process was followed, where scouts would search an area, and on locating a candidate nest, recruit others to also evaluate it, leading to the formation of a large swarm at the best located candidate. The most distinct difference is the method by which one bee would locate a candidate nest site for another bee. The tandem running of ants was replaced by the waggle dance, and the transporting behavior of ants is replaced by a mass migration as a swarm.

The waggle dance is a behavior in which a bee communicates information on various attributes of a feature outside of the nest to other nearby bees. A feature may be a source of nectar, or other features of interest, but in this context it is a candidate nest site. The bee repeatedly walks a path describing a short line in the nest while "waggling" its abdomen; on reaching the end of the line the bee alternately circles back to the beginning turning left or right. The orientation of the line relative to gravity indicates the direction of the feature, the length of the line is indicative of the distance of the feature from the nest, and features such as the speed of waggling, and the number of times the bee walks along the line are indicative of the bee's evaluation of the quality of the feature.

6.5.3 Comparisons between SDS and Nest-Site Selection in Ants and Bees

The similarities between the recruiting behaviors of SDS and recruiting behaviors in ants have been previously recorded [4, 18], though this was in the context of ants foraging for food, rather than selecting a nest site. Nevertheless a process that employed tandem running was observed. This similarity was used as evidence that suggested ants do not require the cognitive capacity to directly compare two nest sites as, in SDS, the selection of one hypothesis over another occurs without any direct comparison of hypotheses.

The processes described above have many close analogies with SDS. An ant or bee that is not currently evaluating a nest site and is exploring its search space with a path resembling a random walk is similar to an inactive agent selecting a hypothesis at random. An ant or bee performing a partial evaluation of a candidate nest site by exploring the nest using a path resembling a random walk is similar to an agent partially evaluating a hypothesis by applying a randomly selected microtest. An ant or bee subsequently performing one of two actions after a partial evaluation of a nest site is similar to an agent becoming either active or inactive after performing a partial evaluation. An ant or bee that was satisfied with the partial evaluation of a candidate nest site and recruiting a random ant or bee to the evaluation process is similar to the hypothesis of an active agent being copied by an inactive agent. An ant or bee that was not satisfied with the partial evaluation of a candidate nest site and abandoning the evaluation of the candidate nest site is similar to an inactive agent selecting a new hypothesis. A significant proportion of the ant or bee colony simultaneously evaluating the same candidate nest site is similar to a large number of agents maintaining the same hypothesis.

The experiments that showed the ability of ants to effectively calculate an average score for a candidate nest site that fluctuated in value is analogous to the behavior of SDS resulting from partial evaluations. Similarly in ant nest-site selection, the effective score of a candidate nest site is a result of the average result of many partial evaluations. This reflects the behavior of SDS under certain circumstances, depending on the frequency at which the score of hypothesis fluctuated between two values. If the number of iterations between a fluctuation was large in compared with the number of iterations required for a cluster to stabilize, for example, once every thousand iterations, then the largest cluster would similarly fluctuate between the superior hypothesis and the stable hypothesis. If the number of iterations between a fluctuation was small compared with the number of iterations required for a cluster to stabilize, for example, the hypothesis alternated in quality every time an agent performed a microtest, then the fluctuating hypothesis would be evaluated as if it had a score between the superior and inferior values. This latter case of frequent fluctuations is what was observed in ants [37], as the time for two fluctuations was always 10 minutes, and the time for a population to achieve quorum at a nest site was of the order of 100 minutes. This is further evidence that SDS models aspects of the nest-site selection behavior of ants, and that this successful evaluation can be conducted by a swarm of agents with no global knowledge.

There are also a few differences between SDS and the process of nest-site selection in ants and bees. Candidate nest sites are not discovered with a uniform probability distribution; as ants and bees explore their search space with a path resembling a random walk there is a greater probability of locating candidate nest sites that are closer to their current nest site than candidate nest sites that are further away from the their current nest site. Ants or bees that abandon the evaluation of one candidate nest site are not required to immediately begin evaluation on another candidate nest site, whereas an agent that becomes inactive immediately selects a new hypothesis, either from an active agent, or uniformly at random. Ants or bees that return to the nest recruit other ants or bees they encounter, which is unlike the recruitment action of agents in two ways. First, the selection of other ants or bees is arbitrary rather than uniformly at random. Second, "active" ants, i.e., ants that have recently performed a successful partial evaluation, recruit other ants, whereas in SDS inactive agents assume the hypothesis of other agents. This can be seen as more analogous to active diffusion (D^{ACTIVE}) than passive diffusion (D^{PASSIVE}) as it is individuals that have discovered a potential solution that perform the activity of recruitment; ants and bees that have located a good candidate nest site recruit other ants and bees. The equivalent behavior for bees is mode analogous to passive diffusion, as a small portion of the colony observing a waggle dance is similar to a number of inactive agents polling the same active agent.

There is no close analogy in SDS with the behavioral change when ants or bees moved from a recruiting behavior to a migration behavior. The aspect of SDS closest to the behavioral change is the mode of halting, especially when applying a convergence-based halting variant, though the significant difference is that in SDS the decision to halt is performed by a single function with access to the state of the entire swarm, whereas the decision of an ant or a bee to change behavior must be informed by their individual experience.

What remained unclear was the mechanism by which individual ants or bees decided to move from recruiting behavior to migration behavior. This ability, called quorum sensing, is essential for any swarm to act as a unit. Individuals may make decisions based on local information, or on contacts with other members of the swarm, but decisions that require mass action also require that a certain proportion of the swarm is in agreement with the decision. Investigations into this area are described in the next section.

6.5.4 Mechanism of Quorum Sensing in Ants

There have been a number of investigations into the mechanism that enables quorum sensing in ants. A simple hypothesis is that ants follow a fixed program of a certain number of tandem runs, followed by transporting behavior. However, there is evidence against this hypothesis as the tandem running phase has been observed to be completely absent in cases of migration over very short distances [36].

In experiments involving the migration behavior of colonies of the ant *L. albipennis* [36] the switch from tandem running to transporting was observed to be determined by the population size at the new nest. Reverse tandem running was also observed, where an ant would lead another from the new nest back to the old nest. It was suggested that this was to induce idle workers at the new site to join them in taking items from the old nest. If a potential weakness was identified, then ants from one colony may be transported to two different sites if two separate groups commence transporting at similar times. It was indicated that a requirement for a level of quorum, named a quotum threshold, had to be reached before initiating transport would mitigate this possibility. It was concluded that the entire process allowed a colony to indirectly compare the quality of multiple candidate nest sites by initially performing the relatively slow process of tandem running, and later switching to the significantly faster process of transporting at a time determined by the number of ants at the new site. Evidence was provided against the hypothesis that the switch from tandem running to transporting was determined by the distance between the old and new nest; a colony was observed to perform the same number of tandem runs for a site only 10 cm away from the old nest as for nests 60 cm away.

Further experiments investigated the mechanisms underlying ant nest-site selection [39]. An apparatus that enabled the experimenter to grant and restrict access to a new nest at will allowed the experimenter to separate ants arriving at the nest from any ants currently evaluating it. The experimenter could therefore determine the importance of direct contact with other ants in contrast to the pheromonal effect of highly populated nests with no direct contact. They found that ants only transitioned to their transporting behavior after having made direct contact with other ants in the process of evaluating the same nest. They also found that once an ant had begun transporting, it would not continue to evaluate the criteria for quorum. This was achieved by manually removing the ants from nests where transporting had begun and observing that ants would remain transporting when the new nest contained many fewer other ants that would be required to initiate transporting an individual ant. It was observed that the time an ant that has begun transporting spends at a candidate nest drops from 2 to 1 minute, also indicating that it has ceased performing a certain action. To test the hypothesis that quorum sensing was based on the absolute number of ants evaluating a nest, the time before ants began transporting was compared in two cases with different sized nests. An equal number of ants evaluating two nests of different sizes would lead to a lower rate of direct contact in the larger nest than in the smaller nest due to a lower population density. The ants were seen to reach quorum sooner at the smaller nest hence encounter rate, not the absolute number of ants at a candidate nest, was predictive of quorum sensing [39]. The experimental evidence collected supported two hypotheses for the mechanism of quorum sensing, and both were founded on direct contact with other ants at a candidate nest side. The first hypothesis, average encounter rate, hypothesized that an ant would sense quorum when the average rate at which they encountered other ants at a candidate nest was required to be over a certain threshold. The second hypothesis, time for initial contact, hypothesized that an ant would sense quorum when the average time before their first direct contact with another ant after arriving at a candidate nest was below a certain threshold. Neither hypothesis was determined to be superior to the other, though they are closely linked as any nest with a higher encounter rate would, on average, lead to shorter times before the first encounter.

6.5.4.1 Selection of Quorum Threshold in Ants

Further experiments observing ant migration manipulated the size of the migrating colony. The quotum threshold was observed to be proportional to the number of ants in the colony [40]. When a large colony of ants was artificially split in two, the individual ants were observed to adopt a proportionately reduced quotum threshold. It was hypothesized that this adaptation may be achieved by comparing interaction rates at the new and old nests [40]. Comparison with natural colonies showed that colony size was not the only factor for the selection of a quotum threshold; age could be relevant, for example, as well as level of experience of the scouts.

Similar experiments induced migration under various conditions. The emigration behavior of a nest was compared under benign conditions and harsh conditions. Ants were observed to sense quorum and hence select a new nest site more quickly, and with less discrimination, in harsh conditions than in benign ones. In this context benign conditions were still air and an enclosed nest with adequate humidity, and harsh conditions had high winds and an exposed nest with poor humidity. It was shown that a lower quotum threshold determined whether a decision would be taken quickly at the risk of selecting a sub-optimal solution, or accurately, at the cost of a more lengthy search process [41, 42]. This showed that the quotum threshold can modulate behavior depending on the current circumstances.

6.5.5 Mechanism of Quorum Sensing in Bees

In the previously described experiment involving bee migrations (Section 6.5.2) the mechanism of quorum sensing was described in two stages. First individual bees needed to sense that quorum had been reached and second a signal needed to reach the entire colony to induce migration as a swarm.

The experiment compared the time for sensing quorum when evaluating nest sites of equal quality, but at which different numbers of other bees were present. The method by which this was achieved was to compare the speed at which quorum was sensed for a nest of a certain quality and compare it with the speed in a situation where five nests of equal quality were located close together but at such a distance from the swarm that the accuracy of the waggle dance was insufficient to unambiguously indicate a single nest. In this sense a bee indicating a candidate nest site was effectively indicating five sites, from which one would be randomly selected by any bee attempting to locate the indicated nest. As the nests were all of equal quality the recruitment process continued as normal, but as the bees were spread randomly across five nests the population at each nest was on average one-fifth of its size. The swarm was observed to take significantly longer before swarming to the location of the new nest in the case with five nests than in the case with just one.

Two important observations were drawn from this experiment: the size of the population at the candidate nests is important, and the distinction between quorum and consensus. Where consensus requires that all hypotheses held by a population are in agreement, quorum requires that a single hypothesis is held by a certain threshold number of the population. This experiment showed that consensus was neither necessary nor sufficient for a colony to swarm to a candidate location. It was not sufficient, as in the case where there were multiple nests of identical quality all waggle dances would have indicated the same location, yet in this case the colony did not swarm to the new nest site until a sufficient population had gathered at one of the sites. It was not even necessary as the requirement for the population at a candidate nest site to reach a threshold is not contingent on there being no bees indicating other sites.

There are a number of mechanisms by which bees could detect the population level at a nest site. Visual perception is a possibility as bees may be able to spot other bees on the surface of a nest site and near the entrance as long as the light level is sufficient. Alternatively, bees may use direct contact as do ants because bees are seen frequently coming into contact with each other during the process of evaluating a nest site. There may also be olfactory cues, and while this has not been tested for directly it would be analogous to observations of ants utilizing pheromones.

Once individual bees had sensed quorum, they would begin to induce the migratory behavior in the colony with an activity called worker piping. Worker piping is described as "the prepare-for-takeoff signal" [38]. Worker bees will perform this vibrational signal while running over and through the bees inhabiting the outer edge of the nest known as the mantle. Piping has a stimulatory effect on a shivering behavior which leads to an increase in the temperature at the mantle. The mantle is usually cooler than the core so an increase in temperature at the mantle leads to a uniformity of temperature throughout the nest. Once a sufficient uniform temperature has been achieved the colony will swarm to a location previously encountered by the scout bees.

6.5.6 Mechanism of Quorum Sensing in Bacteria

Some bacteria use quorum sensing as a mechanism for regulating gene expression, and others use it to coordinate a bioluminescent behavior, ensuring that the appropriate behavior is performed by the population in unison [43]. The unison action is important for bacteria as their behavior, if individually expressed, is often too little to be of any value. In the example of bioluminescence a single bacterium will emit very little light, but an entire population fluorescing at once may emit enough light to enable a host organism to hunt more effectively, as is the case with the splitfin flashlightfish *Anomalops katoptron* [44], or to camouflage the host organism, as with the bobtail squid *Euprymna scolopes* [45]. The bacteria, *Vibrio fischeri*, which colonize the light organ of the bobtail squid, use two quorum sensing systems [46]. Both systems are characterized by the secretion of signal molecules by individuals. The frequency with which individuals encounter the signal molecules is an effective proxy for the density of the bacteria population. Under low population density the production of the luminescence signal molecule is suppressed by a negative modulator, minimizing any energy being expended on bioluminescence before it is beneficial to the population. Under a medium population density, achieved when the bacteria are attached to the mucus secreted by the light organ of a bobtail squid, a second signal molecule will induce colonization behavior, where the bacteria expend energy moving into the light organ itself in an attempt to locate and colonize "crypts." Once colonization behavior has been successful, the population density inside a crypt is much larger than outside the organ. The higher population density means that a second effect of the colonization signal molecule becomes significant; it suppresses the production of the negative modulator, effectively stimulating the production of the luminescence signal molecule. A sufficient density of the luminescence signal molecule therefore suggests that colonization has been successful and that producing light will now be a beneficial use of energy.

These two mechanisms show that by the simple method of detecting the presence of a molecule it is possible to effectively detect a related occurrence, which in this case is the presence of a certain population density. Furthermore, detecting subsequent molecules is indicative of very specific occurrences, which in this case is the individual's location within a crypt. It is significant that as the second system requires a population density that is not achieved outside of very specific circumstances, that even in such a single "bit" of information the presence of a molecule is greater than a given threshold. This can be taken as a rich message about the state of the world of a bacterium, namely the successful colonization of a light organ. This is an indication of how the simple "Boolean" signals between members of any swarm can provide valuable information to an individual as long as the signals are restricted to only being active in specific situations meaningful to the individual. In other words, the bacteria do not "need to know" that they have colonized the light organ of a flashlightfish, they simply detect that a situation has occurred in which a specific behavior may be expressed. It is also worth noting that one of the effects of quorum sensing is to stimulate a behavior, not by direct induction, but by suppressing the secretion of an inhibitor. Although it is true that in the domain of logic a negated false value is equivalent to a true value, in nature there may be reasons why one behavior is preferred over another. This may also simply be a result of the stochastic nature of evolution. It should be showing how the detection can lead to previously defined behavior being expressed only in the correct situations.

6.5.7 Comparisons between SDS and Quorum Sensing in Nature

There are some analogues with SDS and the mechanisms of quorum sensing observed in ants, bees, and bacteria. Direct contact between ants and bees has been identified as critical in the quorum sensing process; in the terms of SDS two ants encountering each other at their candidate nest site is analogous to one agent polling another agent with the same hypothesis. One mode of diffusion that depends on two agents maintaining the same hypothesis is context-sensitive diffusion ($D^{\text{CONTEXT-SENSITIVE}}$, Algorithm 6.19). The resulting behavior is that one agent becomes inactive and hence may change its hypothesis. This is superficially analogous to one ant in a candidate nest site encountering another ant and leading it back to the original nest, though it is no evidence that either ant changes its behavior as a result.

It has been determined that the distance between candidate nests and the original nest did not directly influence the mechanism of quorum sensing, as the same number of tandem runs was observed between candidate nest sites of different distances from the original nest. Initially this appears to be analogous to the behavior of SDS as the evaluation hypotheses is not affected by the position of the hypothesis in the search space. However, it should be noted that tandem running is a relatively slow process, so any number of tandem runs over a longer distance will take longer than the same number of tandem runs over a shorter distance. The closest analogy to this in SDS would be in the situation where parallel iteration is employed and certain hypotheses took much longer to partially evaluate than others. In this case clusters would still form at various hypotheses after the same number of evaluations, but one hypothesis may form a cluster sooner due to quicker evaluations.

The closest analogue between SDS and quorum sensing is its mode of halting as both are mechanisms in which the search procedure yields its result. Quorum sensing can be interpreted as mode of halting enacted by the individuals that make up a swarm. The mechanisms of natural quorum sensing described above are all guided by the level of population at a candidate nest site, or in the case of bacteria, in the local area. These mechanisms are related to some mode of halting methods described in Section 6.4.5. In the simplest sense, population at a candidate nest site may be modeled byglobal activity threshold halting (H^{ACTIVITY}, Algorithm 6.26), but this considers all activity pertaining to the same hypothesis. A better model is largest cluster size halting (H^{LARGEST}, Algorithm 6.25) as population at a single candidate nest site is modeled as the size of a cluster at a single hypothesis. This variant is not a good model of observed behavior on its own as the ants or bees were not observed to immediately change their behavior once a sufficient population had been detected. Another variant, which also models some of the behavior of natural quorum sensing, is stable global activity halting (H^{STABLE}, Algorithm 6.29) as this method requires that a hypothesis forms a stable cluster, and that the cluster remains stable for a given number of iterations. This halting method is therefore analogous to the behavior of an ant or a bee requiring that their candidate nest site has a sufficiently high population for a number of visits before they switch behavior. This variant is also not a good model of observed behavior on its own as an instance of zero, or very low, global activity is considered stable, but it represents a situation that would not induce migration in ants or bees. The closest model to natural quorum sensing of the variants described in Section 6.4.5 would therefore be a combination of both algorithms, so both requirements would be modeled. Those are a sufficiently high population that is stable over time. This variant may be denoted as $H^{\text{AND}}(H^{\text{LARGEST}}, H^{\text{STABLE}})$.

The most significant difference between mode of halting and quorum sensing in nature is that the quorum sensing is an action that results from the interaction of individuals, whereas each variant of mode of halting results from an analysis of the state of the entire swarm. Rather than being enabled by the individuals of the swarm, it is a separate process, outside of the swarm. As such, certain aspects of individual behavior are not represented in SDS. There is no time when an agent in SDS switches into a new behavior, or modifies its internal state to be more or less likely to switch into the behavior in the future. These aspects of quorum sensing will be added into a variant of SDS and explored in the next chapter.

6.6 CONCLUSION

This article introduced a formalism that defined SSDS as a set of interacting functions. This enables variants of SDS to be described in terms of behavior that is distinct from SSDS, and in doing so the description of variants of SDS are more concise and the aspect of variation is more clearly identified than if the entire variant needed to be described. This demonstrates the modularity of SDS and the implementation of the variants for use in experiments. The formalism therefore facilitates the expression of variants of SDS in literature and in algorithmic implementation.

A number of variants of SDS are described. For each variant the individual level rules that define the variant are described, as is the resulting population level behavior. The advantages and disadvantages of the behavior of the variant are described including the situation in which such behavior is useful. No variant should be considered objectively superior to any other. Each variant represents a trade-off of one beneficial behavior for another, so certain variants are simply more suited to certain situations than others.

The final section described a number of investigations into the group decision-making behaviors of natural swarms. Many of the individual-level and population-level behaviors had analogues in SDS. One aspect present in natural swarms but not in SDS was the apparent ability to sense when a certain proportion of the swarm had reached consensus. The modes of halting all relied on a function that analyzed the entire state of the swarm and did not emerge from the action of the individuals. SDS is therefore a suitable model of the decision-making behavior of the natural swarms described, and it provided a means to explore potential mechanisms for quorum sensing.

NOTES

1. Though "happy" is a term that is similarly subjective as one's dining preferences, in this case it is used in an objective sense. All miners share an identical process in which the amount of gold they locate on a single day defines a probability that the miner will declare themselves happy at the end of the day when they congregate to potentially share the identity of the hills they are mining.
2. The apparent anomaly of a 1995 work building on a 1996 work is that the former was based on a preprint of the latter.

REFERENCES

1. J. M. Bishop. "Anarchic techniques for pattern classification." PhD thesis. University of Reading, 1989.
2. J. M. Bishop. "Stochastic searching networks." In: *Proceedings of 1st IEE Conf. on Artificial Neural Networks*. 1989, pp. 329–331.
3. S. J. Nasuto, J. M. Bishop, and S. Lauria. "Time complexity analysis of the stochastic diffusion search." In: *Proceedings of the International ICSC/IFAC Symposium on Neural Computation (NC 1998)*, 1998, pp. 260–266.
4. K. De Meyer, J. M. Bishop, and S. J. Nasuto. "Stochastic diffusion: using recruitment for search." In: *Evolvability and Interaction: Evolutionary Substrates of Communication, Signalling, and Perception in the Dynamics of Social Complexity* (eds. P. McOwan, K. Dautenhahn & C.L. Nehaniv) Technical Report 393, 2003, pp. 60–65.
5. M. M. al-Rifaie, J. M. Bishop, and T. Blackwell. "An investigation into the merger of stochastic diffusion search and particle swarm optimization." In: *Proceedings of the 13th Annual Conference on Genetic and Evolutionary Computation*. ACM, 2011, pp. 37–44.
6. M. M. al-Rifaie and J. M. Bishop. "The mining game: a brief introduction to the stochastic diffusion search metaheuristic." *Q: The Magazine of AISB* 130, (2010), pp. 8–9.
7. M. M. al-Rifaie and J. M. Bishop. "Stochastic diffusion search review." *Paladyn, Journal of Behavioral Robotics* 4.3 (2013), pp. 155–173.
8. M. M. al-Rifaie. "Information sharing impact of stochastic diffusion search on population-based algorithms." PhD thesis. Goldsmiths, University of London, 2011.
9. S. J. Nasuto. "Resource allocation analysis of the stochastic diffusion search." PhD thesis. University of Reading, 1999.
10. D. R. Myatt, J. M. Bishop, and S. J. Nasuto. "Minimum stable convergence criteria for stochastic diffusion search." *Electronics Letters* 40.2 (2004), pp. 112–113.
11. D. R. Myatt. "Analysis of stochastic diffusion search and its application to robust estimation." PhD thesis. University of Reading, 2005.
12. H. Grech-Cini. "Locating facial features." PhD thesis. University of Reading, 1995.
13. S. J. Nasuto and J. M. Bishop. "Convergence analysis of stochastic diffusion search." *Parallel Algorithms and Applications* 14.2 (1999), pp. 89–107.
14. D. R. Myatt, S. J. Nasuto, and J. M. Bishop. "Exploration and exploitation in stochastic diffusion search." In: *Exploration vs Exploitation in Naturally Inspired Search. Symposium on Nature Inspired Systems*. AISB, 2006.

15. S. J. Nasuto and J. M. Bishop. "Steady state resource allocation analysis of the Stochastic Diffusion Search." *Biologically Inspired Cognitive Architectures* 12 (2015), pp. 65–76.
16. J. M. Bishop and P. Torr. "The stochastic search network." In: *Neural Networks for Vision, Speech and Natural Language*. Springer, the Netherlands, 1992, pp. 370–387.
17. S. J. Nasuto and J. M. Bishop. "Stabilizing swarm intelligence search via positive feedback resource allocation." In: *Nature Inspired Cooperative Strategies for Optimization (NICSO 2007)*. Springer, 2008, pp. 115–123.
18. K. De Meyer. "Foundations of stochastic diffusion search." PhD thesis. University of Reading, 2004.
19. K. De Meyer. Explorations in stochastic diffusion search: soft-and hardware implementations of biologically inspired spiking neuron stochastic diffusion networks. Technical Report KDM/JMB/2000, 2000.
20. H. Grech-Cini and G. T. McKee. "Locating the mouth region in images of human faces." In: *Optical Tools for Manufacturing and Advanced Automation. International Society for Optics and Photonics*, 1993, pp. 458–465.
21. D. H. Wolpert. "The lack of a priori distinctions between learning algorithms." *Neural Computation* 8.7 (1996), pp. 1341–1390.
22. D. H. Wolpert, W. G. Macready, et al. No free lunch theorems for search. Technical Report SFI-TR-95-02-010, Santa Fe Institute, 1995.
23. K. De Meyer, J. M. Bishop, and S. J. Nasuto. "Small-world effects in lattice Stochastic Diffusion Search." In: *Artificial Neural Networks-ICANN 2002*. Springer, Berlin, Heidelberg, 2002, pp. 147–152.
24. M. Warriner. Hardware stochastic diffusion search. Technical Report. Department of Cybernetics, University of Reading, 2000.
25. D. R. Myatt and J. M. Bishop. Data driven stochastic diffusion search for robust estimation. In: *Proceedings of the School Conference for Annual Research Projects (SCARP)*, University of Reading, Reading, UK, 2003.
26. D. Jones. "Constrained stochastic diffusion search." In: SCARP 2002. University of Reading, 2002.
27. Q. Nguyen and J. M. Bishop. *"Exploration of data driven SDS vs. coupled SDS."* Cognitive Computing MSc. MA thesis. Goldsmiths, University of London, 2008.
28. H. Williams. *"Stochastic diffusion search processes."* Cognitive Computing MSc. MA thesis. Goldsmiths, University of London, 2010.
29. H. Williams and J. M. Bishop. "Stochastic diffusion search: a comparison of swarm intelligence parameter estimation algorithms with RANSAC." *Algorithms* 7.2 (2014), pp. 206–228.
30. M. G. H. Omran et al. "Stochastic diffusion Search for continuous global optimization." In: *ICSI 2011: International Conference on Swarm Intelligence*. 2011.
31. M.G. H. Omran and A. Salman. "Probabilistic stochastic diffusion search." In: *Swarm Intelligence*. Springer, 2012, pp. 300–307. http://link.springer.com/chapter/10.1007/978-3-642-32650-9_31.
32. R. J. Cant and C. S. Langensiepen. "Methods for automated object placement in virtual scenes." In: *UKSIM'09. 11th International Conference on Computer Modelling and Simulation, 2009*. IEEE, 2009, pp. 431–436.
33. D. H. Wolpert. "What the no free lunch theorems really mean; how to improve search algorithms." Santa Fe Institute, 2012.
34. E. J. H. Robinson. "Distributed decisions: new insight from radio tagged ants." Whitehead Lecture at Goldsmiths, University of London, 2013. https://www.gold.ac.uk/calendar/?id=6158.
35. E. Robinson and W. Mandecki. "Distributed decisions: new insights from radio-tagged ants." In: *Ant Colonies: Behavior in Insects and Computer Applications*. Nova Science Publishers, Hauppauge, NY, 2011, pp. 109–128.
36. S. C. Pratt et al. "Quorum sensing, recruitment, and collective decision-making during colony emigration by the ant *Leptothorax albipennis*." *Behavioral Ecology and Sociobiology* 52.2 (2002), pp. 117–127.
37. N. R. Franks et al. "How ants use quorum sensing to estimate the average quality of a fluctuating resource." *Scientific Reports* 5 (2015), p. 11890.
38. T. D. Seeley and P. K. Visscher. "Quorum sensing during nest-site selection by honeybee swarms." *Behavioral Ecology and Sociobiology* 56.6 (2004), pp. 594–601.

39. S. C. Pratt. "Quorum sensing by encounter rates in the ant *Temnothorax albipennis*." *Behavioral Ecology* 16.2 (2005), pp. 488–496.
40. A. Dornhaus and N. R. Franks. "Colony size affects collective decision-making in the ant *Temnothorax albipennis*." *Insectes Sociaux* 53.4 (2006), pp. 420–427. https://doi.org/10.1007/s00040-006-0887-4.
41. N. R. Franks et al. "Speed versus accuracy in collective decision making." *Proceedings of the Royal Society of London B: Biological Sciences* 270.1532 (2003), pp. 2457–2463. DOI: 10.1098/rspb.
42. N. R. Franks et al. "Speed versus accuracy in decision-making ants: Expediting politics and policy implementation." *Philosophical Transactions of the Royal Society of London B: Biological Sciences* 364.1518 (2009), pp. 845–852.
43. M. B. Miller and B. L. Bassler. "Quorum sensing in bacteria." *Annual Reviews in Microbiology* 55.1 (2001), pp. 165–199.
44. J. Hellinger et al. "The flashlight fish Anomalops katoptron uses bioluminescent light to detect prey in the dark." *PLoS One* 12.2 (2017)
45. B. Jones and M. Nishiguchi. "Counterillumination in the Hawaiian bobtail squid, Euprymna scolopes Berry (Mollusca: Cephalopoda)." *Marine Biology* 144.6 (2004), pp. 1151–1155.
46. C. Lupp and E. G. Ruby. "*Vibrio fischeri* uses two quorum-sensing systems for the regulation of early and late colonization factors." *Journal of Bacteriology* 187.11 (2005), pp. 3620–3629.

Section III

Socio-inspired Methods

7 The League Championship Algorithm

Applications and Extensions

Ali Husseinzadeh Kashan, Alireza Balavand, Somayyeh Karimiyan, and Fariba Soleimani

CONTENTS

7.1 INTRODUCTION

Optimization is the art of finding the best solution among the set of solutions and the optimization techniques are tools that make the search for the best solution faster. Defining the best solution in mathematical problems has been a challenge for researchers, especially in complicated problems in the field of engineering. Therefore, finding the best solution in the least time with high accuracy has always been interesting to researchers in the field of optimization. Metaheuristic algorithms are one of the best tools that have been used in many papers for solving diverse optimization problems in an iterative manner. These algorithms start with one or a set of random initial solutions, and the solutions are updated in each iteration by some particular operators that often make a better solution. These operations have been divided into exploration and exploitation in terms of structure. The proper structure of most of them makes them successful in achieving reasonable results in an acceptable time. There are various operations in each metaheuristic algorithm for which the proper combination can increase performance. Balancing between exploration and exploitation has a great impact on each metaheuristic algorithm. The experiences show that in the initial iterations, the ability of exploration should be increased and in the final iterations, the power of exploitation should be increased. Exploration operations are a set of techniques that search for new solutions in the undiscovered solution space and operations of exploitation search the vicinity of solutions that have been already discovered. Based on the no free lunch (NFL) theorem [1], no metaheuristic algorithm can be the best option for solving all optimization problems. In other words, a metaheuristic algorithm in a given problem provides satisfactory results, and the same algorithm may have poor performance in other problems. NFL makes it possible to provide many algorithms in different categories with new approaches for solving various problems. The main characteristics of metaheuristics methods can be stated as follows. Unlike heuristics methods, the main purpose of these methods is to search the solution space effectively and efficiently instead of just finding the optimal or near-optimal solution; trans-heuristics methods are the policies and strategies that guide the search process; and metaheuristic methods are approximate and often uncertain. These methods may use mechanisms to prevent the search process from being trapped in local optima.

Unlike heuristic methods, metaheuristic algorithms do not depend on the type of problem; in other words, they can be used to solve a wide range of optimization problems. More advanced metaheuristics methods use the experiences and information gained during the search process in the form of memory to direct the search toward more promising areas of the solution space. In short, metaheuristic algorithms are advanced and general search methods that suggest steps and criteria that are very effective in escaping from the trap of local optimization. An important factor in these methods is the dynamic balance between exploration and exploitation strategies. Therefore, by creating a dynamic balance between these two strategies, the search is directed to areas of the solution space in which better solutions are generated. It does not waste more time in a part of the solution space that was already checked or contains the worse solution. One of the distinguishing features of metaheuristic algorithms is the number of solutions that are generated and examined in each iteration. Accordingly, the metaheuristic algorithms can be divided into four main categories. The first group consists of evolutionary algorithms that most often simulate the concept of evolution in nature. In this category of algorithms, crossover and mutation operators are considered as exploitation and exploration mechanisms, that are used to generate new solutions. The genetic algorithm [2], the differential evolution algorithm [3], and evolution strategy [4] are the classes of efficient evolution algorithms that are easy to conceptualize and implement. The second category consists of trajectory algorithms. These algorithms start with an initial solution and for each middle solution, a neighboring solution is defined that this neighboring solution determines either randomly or based on a particular rule. Tabu search algorithm is placed in this category [5]. The variable neighborhood search algorithm [6] and simulated annealing [7] are in the trajectory methods category. The third category consists of collective intelligence-based algorithms that have been called swarm intelligence [8] algorithms.

Direct and indirect relationships exist between different solutions in these algorithms that are inspired by the collective behavior of animals, insects, and other living organisms in nature. The particle swarm optimization algorithm [9] has been modeled based on the social behavior of the bird group. The ant colony optimization algorithm is one of the most prominent examples of swarm intelligence methods [10]. This algorithm is inspired based on the collective behavior of the ants that find the shortest path from nest to food sources. The artificial bee colony algorithm [11] simulates the search behavior of honey bee groups to find food. The shuffled frog-leaping algorithm [12] stemmed from the search for food by the frog group. The main idea of the firefly algorithm [13] is the optical connection between fireflies. A cuckoo's bird life and laying is the main idea in the cuckoo optimization algorithm [14]. The seeker optimization algorithm [15] simulates the human intelligent search. The gray wolf optimization algorithm [16] is a nature-inspired metaheuristic algorithm simulating the behavior of gray wolves and the hierarchy of leadership and their hunting methods. The symbiotic organisms search algorithm [17] was modeled on the coexistence of different organisms in their habitats and their interactions. The whale optimization algorithm [18] simulates the social behavior of humpback whales. This algorithm is inspired by the strategy of hunting using the bubble network. Crocodile hunting strategy (CHS) [19] simulates the crocodile's strategy in hunting fish.

The fourth category includes algorithms based on politics, music, sports, physics, and so on. The main idea of the imperialist competitive algorithm [20] is to simulate the imperialist political process. The F3EA algorithm [21] imitates the process of monitoring the enemy's positions to destroy them in the battle. From modeling and simulating a process that a composer tracks to harmonize a piece of music, the harmony search algorithm [22] was proposed. The gravitational search algorithm [23] is inspired by the gravitational search rules of the planets. The league championship algorithm (LCA) [24] simulates the process of competition considering the amount of target function as the strength of the teams that play against each other, as well as considering the solutions as the team's makeup system when playing against each other. The optics inspired optimization algorithm [25] uses Snell's law of refraction when the light enters a darker environment from a brighter one. The artificial chemical reaction optimization algorithm [26] is inspired by the process and molecular reactions. The optics-inspired optimization algorithm [27] uses the law of reflection to model optical phenomena of curved mirrors to show different image types depending on the distance between the object and the mirror. The teaching-learning-based optimization algorithm [28] was inspired by the learning and training process.

The aim of this chapter is to investigate all aspects of LCA applications from 2012 to the latest applications. Therefore, this chapter is organized as follows. In Section 7.2, the background of LCA will be investigated. In Section 7.3, the LCA's idea is presented. In Section 7.4, the application of LCA in various research is investigated. Section 7.5 reviews some modifications performed on LCA to either improve it or adapt it for solving problems with a special structure.

7.2 BACKGROUND OF LCA

LCA is a population-based algorithm for solving problems in continuous space that has been proposed by Husseinzadeh Kashan [29] in 2009. LCA, like other population-based algorithms, attempts to find the best solution iteratively. It uses operators to generate solutions in a variety of iterations in a promising area.

There are some specific definitions and notations in this algorithm as follows.

- *Sport league:* A sport league is an organization that provides competition in a specific sport during a time range. A league consists of a set of sports teams that play with each other to determine the loser and winner. The results of some competitions may end in a draw in a league. Therefore, at the end of a league, the sports teams are ranked from the strongest teams to the weakest teams. Like another metaheuristic, LCA generates an initial population that has been shown with L.
- *Formations:* In each sport how to place players on the sports fields is defined as a formation. There are some formations for each sport that are usually determined by coaches. For example, the most common formations in soccer are variations of 4-4-2, 4-3-3, 3-2-3-2, 5-3-2, and 4-5-1. The structure of each solution array simulates the formation in LCA.
- *Match analysis:* Analyzing the performance of a team in terms of weaknesses and strengths during or after a competition refers to match analysis. Analyzing strengths, weaknesses, opportunities, threats (SWOT) is the main objective of match analysis. Strengths and weaknesses are internal factors that are inside each team and must be identified by the coaches and other members of the technical group. Opportunities and threats refer to external factors that are not controllable, but identifying them can have a big effect on team performance.

There are several axioms by which LCA rationale is formed:

1. It is more likely that a team with better playing strength wins the game.
2. The outcome of a game is not predictable given the teams' playing strength perfectly.

3. The probability that team i beats team j is assumed equal to both teams' points of view.
4. When team i beats team j, any strength helped team i to win has a dual weakness that caused team j to lose. In other words, any weakness is a lack of particular strength.
5. Teams only focus on their upcoming match without regard for the other future matches. Formation settings are done based on the previous week's events.

There are four main strategies in the LCA algorithm, which are based on the following:

- *S/T matches* show the strengths in light of major threats from competitors. The team should use its strengths to avoid or defuse threats.
- *S/O matches* show strengths and opportunities. The team should attempt to use its strengths to exploit opportunities.
- *W/T matches* show the weaknesses against existing threats. The team must attempt to minimize its weaknesses and avoid threats. Such strategies are generally defensive.
- *W/O matches* illustrate the weaknesses coupled with major opportunities. The team should try to overcome its weaknesses by taking advantage of opportunities.

7.3 LCA MODELING

Most metaheuristics have been simulated based on creatures in nature. These creatures include mammals, insects, marine creatures, magnifying glass creatures, and a variety of viruses. Some kinds of metaheuristics are based on some other fields like sports, music, and politics. The most prominent algorithm in the field of sport is LCA. This algorithm was the first algorithm inspired by sport and was proposed in 2009 [30]. After proposing LCA, a new approach to sports algorithms was developed and presented [31, 32]. LCA mimics the sporting competitions in a sport league. The main idea of LCA is based on an artificial match analysis process in which a metaphorical SWOT analysis is used for generating new solutions. To investigate the LCA in detail, the pseudocode is presented according to Figure 7.1.

In the initial step, some parameters must be determined. These parameters include the number of iterations, number of seasons, and control parameters. L is the number of teams/individuals. The stopping criteria satisfy a certain number of seasons (S) in which a season includes $L-1$ week. Therefore, LCA continues for $S\times(L-1)$ iterations, where L is an even number. Each solution in LCA is considered a team formation. A solution in LCA is a vector $(1\times n)$ where n is the number of variables. The value of the objective function of each solution is the team strength. In the initial phase of LCA, a population of teams is generated and evaluated. After initialization, the league schedule is generated based on a single round-robin schedule [28]. The LCA will be stopped after S successive seasons.

Let $f(X=(x_1,x_2,\ldots,x_n))$ be an n variable numerical function that should be minimized over the decision space. A team formation (a potential solution) for team i at week t can be represented by $X_i^t=(x_{i1}^t,x_{i2}^t,\ldots,x_{in}^t)$, with $f(X_i^t)$ indicating the fitness/function value resulted from X_i^t. By $B_i^t=(b_{i1}^t,b_{i2}^t,\ldots,b_{in}^t)$ we denote the best previously experienced formation of team i until week t, yielding the best function value for team i. To determine B_i^t, we employ a greedy selection between X_i^t and B_i^{t-1} based on the value of $f(X_i^t)$ and $f(B_i^{t-1})$.

There is a procedure to identify the loser or winner team in the LCA algorithm according to the following equation:

$$p_i^t=\frac{f(X_j^t)-\hat{f}}{f(X_j^t)+f(X_i^t)-2\hat{f}}. \tag{7.1}$$

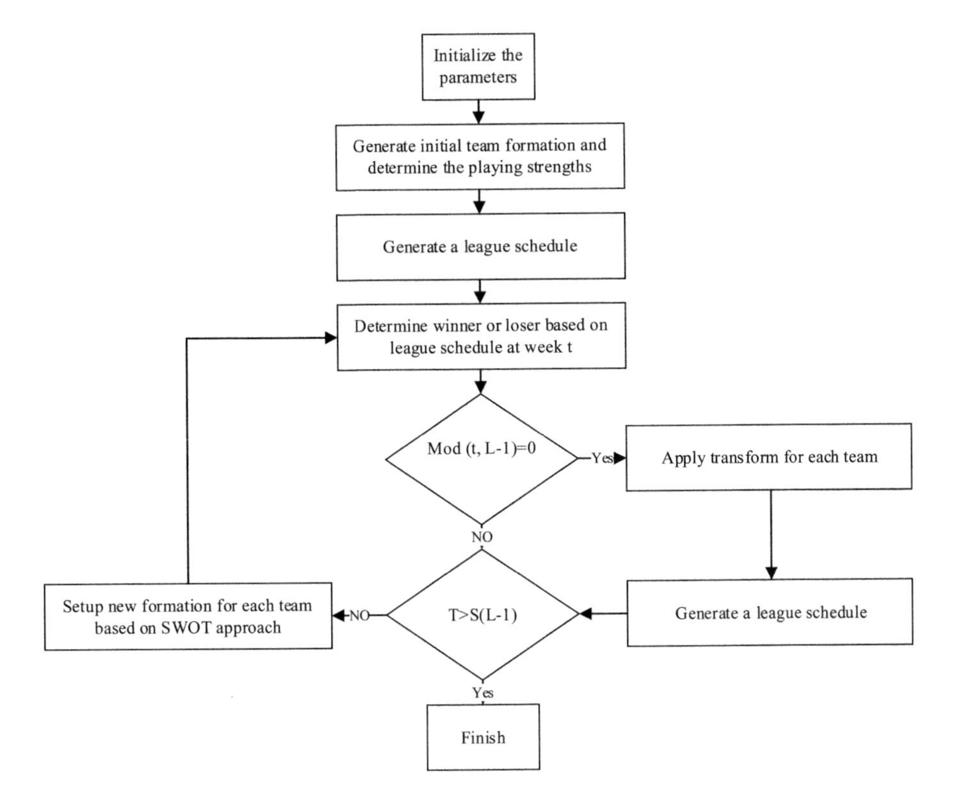

FIGURE 7.1 Flowchart of LCA.

According to Eq. 7.1, consider teams i and j playing at week t, with their formations X_i^t and X_j^t and playing strengths $f(X_i^t)$ and $f(X_j^t)$, respectively. Let p_i^t denote the chance of team i to beat team j at week t (p_j^t can be defined accordingly). Also let $\hat{f}$ be an ideal value (e.g., a lower bound on the optimal function value).

According to Eq. 7.1, a random number between zero and one is generated. If this number is less than or equal to p_i^t, then team i wins, otherwise, team j wins. The main part of LCA is related to match analysis that was denoted already. LCA uses a SWOT strategy to generate a new solution. According to Figure 7.2, SWOT shows that each team benefits from its internal evaluation and external evaluation of other teams. In this way, the match analysis for each team will be done before playing in two ways. First, the performance of the team is considered based on its previous game; second, the performance of the opponent team is investigated based on its previous game.

Based on the previous week's events, the possible actions for team i derived from the artificial match analysis can be summarized in the hypothetical SWOT matrix of Figure 7.2. Given the basic S/T, S/O, W/T, and W/O strategies explained earlier in Section 7.2, Figure 7.2 demonstrates the conditions under which each of these metaphorical strategies is applicable for team i. For example, if team i had won and team l had lost, then it is reasonable that team i focuses on the strengths that made it capable of winning. At the same time, it should focus on the weaknesses that brought the loss for team l. These weaknesses may open opportunities for team i. Therefore, adopting an S/O type strategy would be a proper action for team i. The matrix presented in Figure 7.2 is a metaphorical demonstration of the SWOT matrix that is typically used in planning.

	Adopt S/T *i had won* *l had won* *Focusing on ...*	Adopt S/O *i had won* *l had lost* *Focusing on ...*	Adopt W/T *i had lost* *l had won* *Focusing on ...*	Adopt W/O *i had lost* *l had lost* *Focusing on ...*
S	own strengths (or weaknesses of *j*)	own strengths (or weaknesses of *j*)	–	–
W	–	–	own weaknesses (or strengths of *j*)	own weaknesses (or strengths of *j*)
O	–	weaknesses of *l* (or strengths of *k*)	–	weaknesses of *l* (or strengths of *k*)
T	strengths of *l* (or weaknesses of *k*)	–	strengths of *l* (or weaknesses of *k*)	–

FIGURE 7.2 Hypothetical SWOT matrix derived from the artificial match analysis.

Let us introduce the following indices:

l: Index of the team that will play with team i at week $t+1$ based on the league schedule.
j: Index of the team that has played with team i at week t based on the league schedule.
k: Index of the team that has played with team l at week t based on the league schedule.

Four states occur according to in Eqs. 7.2–7.5:

If *i had won and l had won too, then the new formation is generated based on the adaptation of S/T strategy:*

$$\left(S\,/\,T\,equation\right):\ x_{id}^{t+1} = b_{id}^{t} + y_{id}^{t}(\psi_1 r_1(b_{id}^{t} - b_{kd}^{t}) + \psi_1 r_2(b_{id}^{t} - b_{jd}^{t})) \qquad \forall d = 1,\ldots,n \tag{7.2}$$

Else if *i had won and l had lost, then the new formation is generated based on the adaptation of S/O strategy:*

$$\left(S\,/\,O\ equation\right):\ x_{id}^{t+1} = b_{id}^{t} + y_{id}^{t}(\psi_2 r_1(b_{kd}^{t} - b_{id}^{t}) + \psi_1 r_2(b_{id}^{t} - b_{jd}^{t})) \qquad \forall d = 1,\ldots,n \tag{7.3}$$

Else if *i had lost and l had won, then the new formation is generated based on the adaptation of W/T strategy:*

$$\left(W\,/\,T\ equation\right):\ x_{id}^{t+1} = b_{id}^{t} + y_{id}^{t}(\psi_1 r_1(b_{id}^{t} - b_{kd}^{t}) + \psi_2 r_2(b_{jd}^{t} - b_{id}^{t})) \qquad \forall d = 1,\ldots,n \tag{7.4}$$

Else if *i had lost and l had lost too, then the new formation is generated based on the adaptation of W/O strategy:*

$$\left(W\,/\,O\ equation\right):\ x_{id}^{t+1} = b_{id}^{t} + y_{id}^{t}(\psi_2 r_1(b_{kd}^{t} - b_{id}^{t}) + \psi_2 r_2(b_{jd}^{t} - b_{id}^{t})) \qquad \forall d = 1,\ldots,n \tag{7.5}$$

End if

According to the Eqs. 7.2–7.5, to generate a new solution the strength and weakness of team i is investigated by considering the performance of team j. The opportunity and threat are investigated by considering the performance of team k. To take more information from the above equations, please refer to [29]. In the above equations, d is the variable or dimension index, r_1 and r_2 are uniform random numbers in [0,1], and ψ_1 and ψ_2 are coefficients used to scale the contribution of "retreat" or "approach" components, respectively. Note that the difference sign in parentheses results in acceleration toward the winner or retreat from the loser. y_{id}^t is a binary change variable that indicates whether the dth element in the current best formation will change or not. Only $y_{id}^t = 1$ allows making a change in the value of b_{id}^t. Let us define $Y_i^t = (y_{i1}^t, y_{i2}^t, ..., y_{in}^t)$ as the binary change array in which the number of ones is equal to q_i^t. It is not usual that coaches do changes in all or many aspects of their team. Normally a small number of changes are recommended. By analogy, it is sensible that the number of changes made in B_i^t (i.e., the value of q_i^t) are small. To simulate the number of changes, LCA uses a truncated geometric distribution [33]. Using a truncated geometric distribution, we can set the number of changes dynamically with more emphasis given to the smaller rate of changes. The following formula simulates the random number of changes made in B_i^t to get the new team formation X_i^{t+1}.

$$q_i^t = \left\lceil \frac{\ln(1-(1-(1-p_c)^{n-q_0+1})r)}{\ln(1-p_c)} \right\rceil + q_0 - 1 \quad : \quad q_i^t \in \{q_0, q_0+1, ..., n\} \tag{7.6}$$

where r is a random number in [0,1] and $p_c \in (0,1)$ is a control parameter and q_0 is the least number of changes realized during the artificial match analysis. We assume that the number of changes made in the best formation is at least one, (i.e., $q_0 = 1$). Typically, p_c is known as the probability of success in the truncated geometric distribution. The greater the value of p_c, the smaller number of changes are recommended. After simulating the number of changes by Eq. 7.6, q_i^t the number of elements is selected randomly from B_i^t and their value changes according to one of the Eqs. 7.2–7.5.

7.4 THE APPLICATIONS OF LCA

In this section, we peruse how passable the LCA algorithm was over the years within researches. Making an easy search in some of the major academic databases revealed that the amount of research evidence (both research articles and LCA articles) using the terms "League Championship Algorithm" and "Champions League Algorithm" is augmenting yearly. Below we reported the outcome of such investigations in Google Scholar, Scopus, IEEE Xplore, and in the Web of Science.

The results from Figure 7.3 and Table 7.1 show that the number of documents containing the terms "League Championship Algorithm" or "Champions League Algorithm" in the Google Scholar database is increasing from 2012 to 2020. Especially in 2017 and 2018, the number of articles related to the algorithm (LCA) in the Google Scholar database has increased significantly. This statistic shows that LCA is important in the scientific community and has been accepted by academicians, which means that the algorithm has an effective design to solve the problems. According to Table 7.1, most of the applications of LCA include scheduling and clustering. According to Figure 7.4, From 29 papers, 8 were studied on scheduling; 6 were studied on engineering problems; modified LCA was introduced in 4 papers; clustering fields were studied in 3 papers; stock market and transportation were studied in 2 papers; and prediction, review, and image processing each had an article.

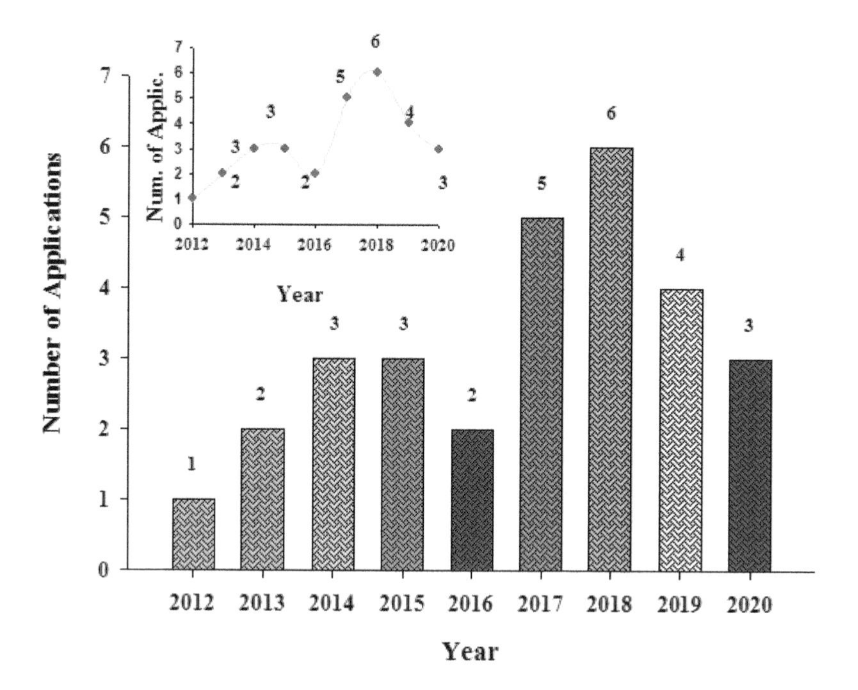

FIGURE 7.3 The number of researches that have applied LCA.

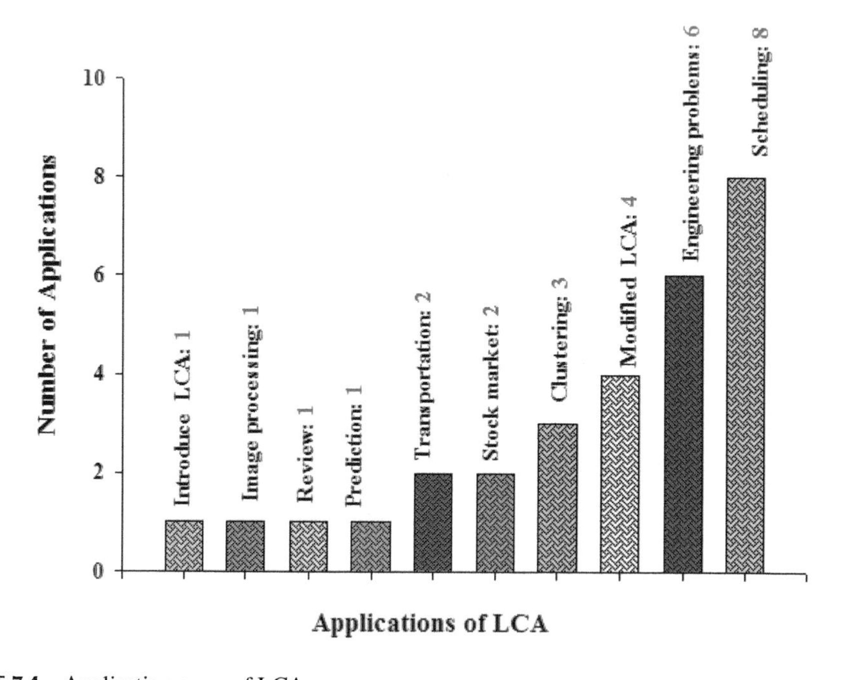

FIGURE 7.4 Application areas of LCA.

TABLE 7.1
Research Trend of "League Championship Algorithm" in Google Scholar Database from 2012 to 2020

Row	Author/Authors	Application	Year	Title	Journal	Reference
1	A. H. Kashan, S. Karimiyan, M. Karimiyan, and M. H. Kashan	Modified algorithm	2012	A modified league championship algorithm for numerical function optimization via artificial modeling of the "between two halves analysis"	The 6th International Conference on Soft Computing and Intelligent Systems, and the 13th International Symposium on Advanced Intelligence Systems	[34]
2	A. H. Kashan, B. Rezaee, and S. Karimiyan	Clustering	2013	An efficient approach for unsupervised fuzzy clustering based on grouping evolution strategies	Pattern Recognition	[35]
3	M. J. Stephen	Introduce LCA	2013	Simple league championship algorithm	International Journal of Computer Applications	[36]
4	H. Bouchekara, M. Abido, A. Chaib, and R. Mehasni	Engineering problems	2014	Optimal power flow using the league championship algorithm: a case study of the Algerian power system	Energy Conversion and Management	[37]
5	S. M. Sajadi, A. H. Kashan, and S. Khaledan	Scheduling	2014	A new approach for permutation flow-shop scheduling problem using league championship algorithm	Proceedings of CIE44 and IMSS	[38]
6	S. M. Abdulhamid and M. S. A. Latiff	Scheduling	2014	League championship algorithm based job scheduling scheme for infrastructure as a service cloud	arXiv preprint arXiv:1410.2208	[39]
7	S. M. Abdulhamid, M. S. A. Latiff, and I. Idris	Scheduling	2015	Tasks scheduling technique using league championship algorithm for makespan minimization in IAAS cloud	arXiv preprint arXiv:1510.03173	[40]
8	S. Yadav and S. J. Nanda	Clustering	2015	League championship algorithm for clustering	2015 IEEE Power, Communication and Information Technology Conference (PCITC)	[41]
9	M. Abdullahi	Scheduling	2015	Job scheduling technique for infrastructure as a service cloud using an enhanced league championship algorithm	Presented in the Second International Conference on Advanced Data and	[42]

(Continued)

TABLE 7.1 (*Continued*)
Research Trend of "League Championship Algorithm" in Google Scholar Database from 2012 to 2020

Row	Author/Authors	Application	Year	Title	Journal	Reference
10	H. Bingol and B. Alatas	Modified algorithm	2016	Chaotic league championship algorithms	Arabian Journal for Science and Engineering	[43]
11	F. Khoshalhan	Modified algorithm	2016	A new play-off approach in league championship algorithm for solving large-scale support vector machine problems	International Journal of Industrial Engineering & Production Research	[44]
12	M. S. Abd Latiff	Scheduling	2017	A checkpointed league championship algorithm-based cloud scheduling scheme with secure fault tolerance responsiveness	Applied Soft Computing	[45]
13	S. Jalili, A. H. Kashan, and Y. Hosseinzadeh	Engineering problems	2017	League championship algorithms for optimum design of pin-jointed structures	Journal of Computing in Civil Engineering	[46]
14	H. R. Bouchekara, M. Nahas, and H. M. Kaouach	Engineering problems	2017	Optimal design of electromagnetic devices using the league championship algorithm	Applied Computational Electromagnetics Society Journal	[47]
15	T. Wangchamhan, S. Chiewchanwattana, and K. Sunat	Clustering	2017	Efficient algorithms based on the *k*-means and chaotic league championship algorithm for numeric, categorical, and mixed-type data clustering	Expert Systems with Applications	[48]
16	M. Ebrahimi, S. Vakilipour, and M. E. Inanlou Shahverdi	Modified algorithm	2017	A novel computational procedure based on league championship algorithm for solving an inverse heat conduction problem	Journal of Computational Applied Mechanics	[49]
17	M. R. Alimoradi and A. H. Kashan	Stock market	2018	A league championship algorithm equipped with network structure and backward Q-learning for extracting stock trading rules	Applied Soft Computing	[50]

TABLE 7.1 (*Continued*)
Research Trend of "League Championship Algorithm" in Google Scholar Database from 2012 to 2020

Row	Author/Authors	Application	Year	Title	Journal	Reference
18	W. Xu, R. Wang, and J. Yang	Scheduling	2018	An improved league championship algorithm with free search and its application on production scheduling	Journal of Intelligent Manufacturing	[51]
19	E. Abbasi, M. Ghayour, M. Danesh, and M. H. Yoosefian	Transportation	2018	Optimal path tracking of a quadrotor in the presence of obstacle using the league championship algorithm	2018 6th RSI International Conference on Robotics and Mechatronics (IcRoM)	[52]
20	R. Soto, B. Crawford, R. Olivares, and J. R. Fernández	Engineering problems	2018	Solving the MCDP using a league championship algorithm	International Conference on Industrial, Engineering and Other Applications of Applied Intelligent Systems	[53]
21	A. Husseinzadeh Kashan, S. Jalili, and S. Karimiyan	Engineering problems	2018	Optimum structural design with discrete variables using league championship algorithm	Civil Engineering Infrastructures Journal	[54]
22	M. M. Karakoc	Prediction	2018	Prediction of electroencephalogram time series via artificial neuro-fuzzy inference system trained by league championship algorithm	Nature-Inspired Intelligent Techniques for Solving Biomedical Engineering Problems	[55]
23	A. H. Kashan, A. Abbasi-Pooya, and S. Karimiyan	Transportation	2019	A rig-based formulation and a league championship algorithm for helicopter routing in offshore transportation	Proceedings of the 2nd International Conference on Data Engineering and Communication Technology	[56]
24	N. Alizadeh and A. H. Kashan	Scheduling	2019	Enhanced grouping league championship and optics inspired optimization algorithms for scheduling a batch processing machine with job conflicts and non-identical job sizes	Applied Soft Computing	[57]
25	S. Subbaraj, R. Thiagarajan, and M. Rengaraj	Scheduling	2019	Multi-objective league championship algorithm for real-time task scheduling	Neural Computing and Applications	[58]

(*Continued*)

TABLE 7.1 (*Continued*)
Research Trend of "League Championship Algorithm" in Google Scholar Database from 2012 to 2020

Row	Author/Authors	Application	Year	Title	Journal	Reference
26	D. Maharana, S. Maheshka, and P. Kotecha	Engineering problems	2019	Multi-objective league championship algorithm and its applications to optimal control problems	Smart Innovations in Communication and Computational Sciences	[59]
27	S. K. Abdulateef, T.-A. N. Abdali, M. D. S. Alroomi, and M. A. A. Altaha	Image processing	2020	An optimise ELM by league championship algorithm based on food images	Indonesian Journal of Electrical Engineering and Computer Science	[60]
28	A. H. Kashan, M. Eyvazi, and A. Abbasi-Pooya	Stock market	2020	An effective league championship algorithm for the stochastic multi-period portfolio optimization problem	Scientia Iranica	[61]
29	H. Moayedi, D. Tien Bui, A. Dounis, and P. T. T. Ngo	Review	2020	A novel application of league championship optimization (LCA): hybridizing fuzzy logic for soil compression coefficient analysis	Applied Sciences	[62]

7.5 MODIFICATIONS ON LCA

Since the introduction of LCA, some modifications have been applied to LCA for practical applications. We investigate some of these modifications in the following section.

7.5.1 Modeling Tie Outcome

The LCA does not investigate the tie outcome as the result between two teams. Investigation of tie results can improve the efficiency of LCA [33]. Different interpretations can be extracted from a tie outcome during the artificial match analysis. As the result, some equations can be obtained under each interpretation for generating new solutions.

- Tie outcome is interpreted as the consequence of the SWOT.
- The outcome is neutral. There is nothing to capture from a game with a tie outcome.
- Tie outcome is randomly interpreted as win or loss.
- Tie outcome is interpreted as a win.
- Tie outcome is interpreted as a loss.

The details of the required equations have been explained in [28]. Experiments showed that the second interpretation has had satisfactory results compared with others based on [28] and [45]. For this type of interpretation, the updating equations are changed from Eqs. 7.2–7.5 to the following:

***If** i had won and l had won too, then*

$$x_{id}^{t+1} = b_{id}^{t} + y_{id}^{t}(\psi_1 r_1 (b_{id}^{t} - b_{kd}^{t}) + \psi_1 r_2 (b_{id}^{t} - b_{jd}^{t})) \quad \forall d = 1,\ldots,n \tag{7.7}$$

***Else if** i had won and l had lost, then*

$$x_{id}^{t+1} = b_{id}^t + y_{id}^t(\psi_2 r_1(b_{kd}^t - b_{id}^t) + \psi_1 r_2(b_{id}^t - b_{jd}^t)) \quad \forall d = 1,\ldots,n \tag{7.8}$$

***Else if** i had lost and l had won, then*

$$x_{id}^{t+1} = b_{id}^t + y_{id}^t(\psi_1 r_1(b_{id}^t - b_{kd}^t) + \psi_2 r_2(b_{jd}^t - b_{id}^t)) \quad \forall d = 1,\ldots,n \tag{7.9}$$

***Else if** i had lost and l had lost too, then*

$$x_{id}^{t+1} = b_{id}^t + y_{id}^t(\psi_2 r_1(b_{kd}^t - b_{id}^t) + \psi_2 r_2(b_{jd}^t - b_{id}^t)) \quad \forall d = 1,\ldots,n \tag{7.10}$$

***Else if** i had won and l had tied*

$$x_{id}^{t+1} = b_{id}^t + y_{id}^t(\psi_1 r_1(b_{id}^t - b_{jd}^t)) \quad \forall d = 1,\ldots,n \tag{7.11}$$

***Else if** i had tied and l had won*

$$x_{id}^{t+1} = b_{id}^t + y_{id}^t(\psi_1 r_1(b_{id}^t - b_{kd}^t)) \quad \forall d = 1,\ldots,n \tag{7.12}$$

***Else if** i had tied and l had tied too,*

$$x_{id}^{t+1} = b_{id}^t \quad \forall d = 1,\ldots,n \tag{7.13}$$

***Else if** i had lost and l had tied*

$$x_{id}^{t+1} = b_{id}^t + y_{id}^t(\psi_2 r_1(b_{jd}^t - b_{id}^t)) \quad \forall d = 1,\ldots,n \tag{7.14}$$

***Else if** i had tied and l had lost*

$$x_{id}^{t+1} = b_{id}^t + y_{id}^t(\psi_2 r_1(b_{kd}^t - b_{id}^t)) \quad \forall d = 1,\ldots,n \tag{7.15}$$

End if

Two random numbers are generated to simulate a tie outcome; if these numbers are less or equal to p_i^t (see Eq. 7.1), then team i wins/team j loses; in another state if these numbers are greater than p_i^t, then j wins/i loses. If one of these numbers is smaller than p_i^t and the other one is greater than p_i^t, then the result is considered a tie outcome.

7.5.2 Modeling Artificial "Between Two Halves Analysis"

The notion of post-match analysis is used just to generate a new solution in the original version of LCA. Generating new sets of updating equations as the simulation of "between two halves analysis" has been modeled in [33]. Here, it is presumed that coaches change their current formation just when they lose the first half; otherwise, they retain their formation style. LCA is modified by changing the winner/loser recognition part and a new formation (solution) generator mechanism is added. If the game is being accomplished between teams i and j, we can determine the first-half winner/loser, using the stochastic winner/loser recognition mechanism described in Eq. 7.1. If i loses the first half to j, it should devise an alternative formation (solution), vice versa. Here, there is only an internal source of evaluation.

Let $X_i^{t-second\ half}$ and $X_j^{t-second\ half}$ be the new formation for team i, during its game in week t. These formations may be generated as follows:

***If** i had won the first half from j*

$$x_{jd}^{t-second\ half} = b_{jd}^t + y_{id}^t(\psi_2 r(x_{id}^t - b_{jd}^t)) \quad \forall d = 1,\ldots,n \tag{7.16}$$

$$X_j^t \leftarrow X_j^{t-second\ half} \tag{7.17}$$

***Else if** j had won the first half from i*

$$x_{id}^{t-second\ half} = b_{id}^t + y_{id}^t(\psi_2 r(x_{jd}^t - b_{id}^t)) \quad \forall d = 1,\ldots,n \tag{7.18}$$

$$X_i^t \leftarrow X_i^{t-second\ half} \tag{7.19}$$

End if

In Figure 7.5 the modified winner/loser diagnosis part has been shown.

7.5.3 Modifying LCA for Grouping Problems

There are many combinatorial optimization problems in which the partition of a set of objects into disjoint groups is an intention. The bin packing problem, the graph coloring problem, the parallel machine scheduling problem, and the data clustering problem are all grouping problems. For these problems, continuous optimization algorithms like LCA cannot be used directly because such algorithms do their computations in the real-valued space [63]. However, the building blocks of the solution are the groups/sets of items in grouping problems. Here the arithmetic subtraction operator cannot be used for groups/sets and suitable operators should be devised for them.

A group dissimilarity measure $Dis(G,G')$ quantifies the degree of difference or dissimilarity between groups G and G'. According to [56], the equation of LCA can be adapted to solve grouping problems as follows:

***If** i had won and l had won too, then*

$$Dis\left(x_{id}^{t+1}, b_{id}^t\right) \approx \psi_2 r_{1id} Dis\left(b_{id}^t, b_{kd}^t\right) + \psi_1 r_{2id} Dis\left(b_{id}^t, b_{jd}^t\right) \quad \forall d = 1, 2, \ldots, D \tag{7.20}$$

- *Determine the winner/loser of the first half as described in section* III-B *using* X_i^t *and* X_j^t.
- *Use equation* (16) *or* (18) *to set up a new team formation for the loser of the first half. Evaluate the playing strength along with the resultant formation.*
- *Determine the final winner/loser of the game as described in section* III-B *using* $X_i^{t-\text{second half}}$ *and* X_j^t *or* X_i^t *and* $X_j^{t-\text{second half}}$.
- *Treat the resultant formation as the team formation at the end of week t (see* (17) *and* (19)*). If the new formation is the fittest one, hereafter, consider the new formation as the team's current best formation.*

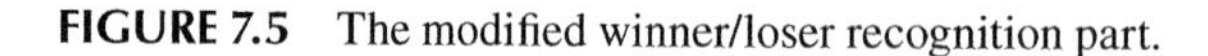

FIGURE 7.5 The modified winner/loser recognition part.

***Else if** i had won and l had lost, then*

$$Dis\left(x_{id}^{t+1}, b_{id}^{t}\right) \approx \psi_1 r_{1id} Sim\left(b_{id}^{t}, b_{kd}^{t}\right) + \psi_1 r_{2id} Dis\left(b_{id}^{t}, b_{jd}^{t}\right) \quad \forall d = 1, 2, \ldots, D \tag{7.21}$$

***Else if** i had lost and l had won, then*

$$Dis\left(x_{id}^{t+1}, b_{id}^{t}\right) \approx \psi_1 r_{1id} Dis\left(b_{id}^{t}, b_{kd}^{t}\right) + \psi_2 r_{2id} Sim\left(b_{id}^{t}, b_{jd}^{t}\right) \quad \forall d = 1, 2, \ldots, D \tag{7.22}$$

***Else if** i had lost and l had lost too, then*

$$Dis\left(x_{id}^{t+1}, b_{id}^{t}\right) \approx \psi_2 r_{1id} Sim\left(b_{id}^{t}, b_{kd}^{t}\right) + \psi_2 r_{2id} Sim\left(b_{id}^{t}, b_{jd}^{t}\right) \quad \forall d = 1, 2, \ldots, D \tag{7.23}$$

End if

where

$$Dis\left(G, G''\right) = 1 - \frac{\left|G \cap G''\right|}{\left|G \cup G''\right|} \tag{7.24}$$

$$Sim\left(G, G''\right) = \frac{\left|G \cap G''\right|}{\left|G \cup G''\right|} \tag{7.25}$$

In the above equations, all variables denote groups. For instance, x_{id}^{t} indicates group d of the solution relevant to team i at iteration t. The symbol $\approx$, stating semi-equality, implies that it may be impossible to obtain equal sides in the equations. D is the number of groups in solution B_i^t. To access more details on the implementation of the algorithm, please refer to [56].

7.6 CONCLUSION

In this chapter, we gave a brief description of the origins of the LCA algorithm. We outlined the synopsis of some of the existing applications and experimental results in LCA literature. We provided a survey on recent research direction as the prospects of future LCAs. The application of LCA was investigated in some major categories including scheduling, engineering problems, clustering, stock market, transportation, prediction, and image processing. According to the NFL theorem, no metaheuristic algorithm can be the best option for solving all optimization problems. This conclusion is also true for LCA. Like any other metaheuristic algorithm, LCA has its own set of parameters. Choosing a proper level for these parameters would have a direct effect on improving the performance. Self-adaptive strategies can be exploited to manage the global behavior of the algorithm during the search. LCA has shown a good ability for balancing exploration and exploitation. Therefore, hybridizing it with other algorithms would be worth investigating. For future research, the applications of LCA are worth investigating in clustering [64] or classification of medical images [65].

REFERENCES

1. Wolpert, D.H. and W.G. Macready, *No free lunch theorems for optimization.* IEEE transactions on evolutionary computation, 1997. **1**(1): pp. 67–82.
2. Holland, J.H., *Adaptation in natural and artificial systems. an introductory analysis with applications to biology, control and artificial intelligence*, Vol. **1**. Ann Arbor: University of Michigan Press, 1975.
3. Storn, R. and K. Price, *Differential evolution–a simple and efficient heuristic for global optimization over continuous spaces.* Journal of global optimization, 1997. **11**(4): pp. 341–359.

4. Hansen, N., S.D. Müller, and P. Koumoutsakos, *Reducing the time complexity of the derandomized evolution strategy with covariance matrix adaptation (CMA-ES).* Evolutionary computation, 2003. **11**(1): pp. 1–18.
5. Glover, F., *Future paths for integer programming and links to artificial intelligence.* Computers & operations research, 1986. **13**(5): pp. 533–549.
6. Mladenović, N. and P. Hansen, *Variable neighborhood search.* Computers & operations research, 1997. **24**(11): pp. 1097–1100.
7. Kirkpatrick, S., C.D. Gelatt, and M.P. Vecchi, *Optimization by simulated annealing.* Science, 1983. **220**(4598): pp. 671–680.
8. Beni, G. and J. Wang, *Swarm intelligence in cellular robotic systems,* in *Robots and biological systems: towards a new bionics?* 1993, Springer, Berlin, Heidelberg. pp. 703–712.
9. Eberhart, R. and J. Kennedy. A new optimizer using particle swarm theory. in Proceedings of the *s*ixth *i*nternational *s*ymposium on *m*icro *m*achine and *h*uman *s*cience, 1995. MHS'95. 1995. IEEE.
10. Socha, K. and M. Dorigo, *Ant colony optimization for continuous domains.* European journal of operational research, 2008. **185**(3): pp. 1155–1173.
11. Karaboga, D., *An idea based on honey bee swarm for numerical optimization.* 2005, Technical report-tr06, Erciyes University, Engineering Faculty, Computer Engineering Department.
12. Eusuff, M., K. Lansey, and F. Pasha, *Shuffled frog-leaping algorithm: a memetic meta-heuristic for discrete optimization.* Engineering optimization, 2006. **38**(2): pp. 129–154.
13. Yang, X.-S. Firefly algorithms for multimodal optimization. in International symposium on stochastic algorithms. 2009. Springer.
14. Yang, X.-S. and S. Deb. Cuckoo search via Lévy flights. in *World congress on n*ature & *b*iologically *i*nspired *c*omputing, 2009. 2009. IEEE.
15. Dai, C., et al., *Seeker optimization algorithm: a novel stochastic search algorithm for global numerical optimization.* Journal of systems engineering and electronics, 2010. **21**(2): pp. 300–311.
16. Mirjalili, S., S.M. Mirjalili, and A. Lewis, *Grey wolf optimizer.* Advances in engineering software, 2014. **69**: pp. 46–61.
17. Cheng, M.-Y. and D. Prayogo, *Symbiotic organisms search: a new metaheuristic optimization algorithm.* Computers & structures, 2014. **139**: pp. 98–112.
18. Mirjalili, S. and A. Lewis, *The whale optimization algorithm.* Advances in engineering software, 2016. **95**: pp. 51–67.
19. Balavand, A., *A new feature clustering method based on crocodiles hunting strategy optimization algorithm for classification of MRI images.* The visual computer, 2021. **56**: pp. 1–30.
20. Atashpaz-Gargari, E. and C. Lucas. Imperialist competitive algorithm: an algorithm for optimization inspired by imperialistic competition. in IEEE congress on evolutionary computation, 2007. CEC 2007. 2007. IEEE.
21. Kashan, A.H., R. Tavakkoli-Moghaddam, and M. Gen. *A warfare inspired optimization algorithm: the find-fix-finish-exploit-analyze (F3EA) metaheuristic algorithm.* in *Proceedings of the tenth international conference on management science and engineering management.* 2017. Springer.
22. Geem, Z.W., J.H. Kim, and G. Loganathan, *A new heuristic optimization algorithm: harmony search.* Simulation, 2001. **76**(2): pp. 60–68.
23. Rashedi, E., H. Nezamabadi-Pour, and S. Saryazdi, *GSA: a gravitational search algorithm.* Information sciences, 2009. **179**(13): pp. 2232–2248.
24. Kashan, A.H. *League championship algorithm: a new algorithm for numerical function optimization.* in *International conference of soft computing and pattern recognition, 2009. SOCPAR'09.* 2009. IEEE.
25. Kaveh, A. and M. Khayatazad, *A new meta-heuristic method: ray optimization.* Computers & structures, 2012. **112**: pp. 283–294.
26. Alatas, B., *ACROA: artificial chemical reaction optimization algorithm for global optimization.* Expert systems with applications, 2011. **38**(10): pp. 13170–13180.
27. Kashan, A.H., *A new metaheuristic for optimization: optics inspired optimization (OIO).* Computers & operations research, 2015. **55**: pp. 99–125.
28. Rao, R.V., V.J. Savsani, and D. Vakharia, *Teaching–learning-based optimization: a novel method for constrained mechanical design optimization problems.* Computer-aided design, 2011. **43**(3): pp. 303–315.

29. Kashan, A.H., *League championship algorithm (LCA): an algorithm for global optimization inspired by sport championships.* Applied soft computing, 2014. **16**: pp. 171–200.
30. Kashan, A.H. *League championship algorithm: a new algorithm for numerical function optimization.* in *2009 international conference of soft computing and pattern recognition.* 2009. IEEE.
31. Alatas, B., *Sports inspired computational intelligence algorithms for global optimization.* Artificial intelligence review, 2019. **52**(3): pp. 1579–1627.
32. Bouchekara, H., *Most valuable player algorithm: a novel optimization algorithm inspired from sport.* Operational research, 2020. **20**(1): pp. 139–195.
33. Kashan, A.H., B. Karimi, and F. Jolai, *Effective hybrid genetic algorithm for minimizing makespan on a single-batch-processing machine with non-identical job sizes.* International journal of production research, 2006. **44**(12): pp. 2337–2360.
34. Kashan, A.H., et al. *A modified league championship algorithm for numerical function optimization via artificial modeling of the "between two halves analysis".* in *the 6th international conference on soft computing and intelligent systems, and the 13th international symposium on advanced intelligence systems.* 2012. IEEE.
35. Kashan, A.H., B. Rezaee, and S. Karimiyan, *An efficient approach for unsupervised fuzzy clustering based on grouping evolution strategies.* Pattern recognition, 2013. **46**(5): pp. 1240–1254.
36. Stephen, M.J., *Simple league championship algorithm.* International journal of computer applications, 2013. **75**(6): pp. 28–32.
37. Bouchekara, H., et al., *Optimal power flow using the league championship algorithm: a case study of the Algerian power system.* Energy conversion and management, 2014. **87**: pp. 58–70.
38. Sajadi, S.M., A.H. Kashan, and S. Khaledan, *A new approach for permutation flow-shop scheduling problem using league championship algorithm.* Proceedings of CIE44 and IMSS, 2014. **14**: p. 2014.
39. Abdulhamid, S.M. and M.S.A. Latiff, *League championship algorithm based job scheduling scheme for infrastructure as a service cloud.* arXiv preprint arXiv:1410.2208, 2014.
40. Abdulhamid, S.M., M.S.A. Latiff, and I. Idris, *Tasks scheduling technique using league championship algorithm for makespan minimization in IAAS cloud.* arXiv preprint arXiv:1510.03173, 2015.
41. Yadav, S. and S.J. Nanda. *League championship algorithm for clustering.* in *2015 IEEE power, communication and information technology conference (PCITC).* 2015. IEEE.
42. Abdullahi, M., and A.L.M. Shafie, *Job scheduling technique for infrastructure as a service cloud using an enhanced league championship algorithm.* in *Second international conference on advanced data and information engineering (DaEng-2015)*, 2015.
43. Bingol, H. and B. Alatas, *Chaotic league championship algorithms.* Arabian journal for science and engineering, 2016. **41**(12): pp. 5123–5147.
44. Khoshalhan, F., *A new play-off approach in league championship algorithm for solving large-scale support vector machine problems.* International journal of industrial engineering & production research, 2016. **27**(1): pp. 61–68.
45. Abd Latiff, M.S., *A checkpointed league championship algorithm-based cloud scheduling scheme with secure fault tolerance responsiveness.* Applied soft computing, 2017. **61**: pp. 670–680.
46. Jalili, S., A.H. Kashan, and Y. Hosseinzadeh, *League championship algorithms for optimum design of pin-jointed structures.* Journal of computing in civil engineering, 2017. **31**(2): pp. 04016048.
47. Bouchekara, H.R., M. Nahas, and H.M. Kaouach, *Optimal design of electromagnetic devices using the league championship algorithm.* Applied computational electromagnetics society journal, 2017. **32**(6): pp. 487–489.
48. Wangchamhan, T., S. Chiewchanwattana, and K. Sunat, *Efficient algorithms based on the k-means and chaotic league championship algorithm for numeric, categorical, and mixed-type data clustering.* Expert systems with applications, 2017. **90**: pp. 146–167.
49. Ebrahimi, M., S. Vakilipour, and M.E. Inanlou Shahverdi, *A novel computational procedure based on league championship algorithm for solving an inverse heat conduction problem.* Journal of computational applied mechanics, 2017. **48**(2): pp. 285–296.
50. Alimoradi, M.R. and A.H. Kashan, *A league championship algorithm equipped with network structure and backward Q-learning for extracting stock trading rules.* Applied soft computing, 2018. **68**: pp. 478–493.

51. Xu, W., R. Wang, and J. Yang, *An improved league championship algorithm with free search and its application on production scheduling*. Journal of intelligent manufacturing, 2018. **29**(1): pp. 165–174.
52. Abbasi, E., et al. *Optimal path tracking of a quadrotor in the presence of obstacle using the League Championship Algorithm*. in *2018 6th RSI international conference on robotics and mechatronics (IcRoM)*. 2018. IEEE.
53. Soto, R., et al. *Solving the MCDP using a league championship algorithm*. in *international conference on industrial, engineering and other applications of applied intelligent systems*. 2018. Springer.
54. Husseinzadeh Kashan, A., S. Jalili, and S. Karimiyan, *Optimum structural design with discrete variables using league championship algorithm*. Civil engineering infrastructures journal, 2018. **51**(2): pp. 253–275.
55. Karakoc, M.M., *Prediction of electroencephalogram time series via artificial neuro-fuzzy inference system trained by league championship algorithm*, in Nature-inspired intelligent techniques for solving biomedical engineering problems. 2018, IGI Global, Turkey pp. 232–248.
56. Kashan, A.H., A. Abbasi-Pooya, and S. Karimiyan. *A rig-based formulation and a league championship algorithm for helicopter routing in offshore transportation*. in *Proceedings of the 2nd international conference on data engineering and communication technology*. 2019. Springer.
57. Alizadeh, N. and A.H. Kashan, *Enhanced grouping league championship and optics inspired optimization algorithms for scheduling a batch processing machine with job conflicts and non-identical job sizes*. Applied soft computing, 2019. **83**: pp. 105657.
58. Subbaraj, S., R. Thiagarajan, and M. Rengaraj, *Multi-objective league championship algorithm for real-time task scheduling*. Neural computing and applications, 2019: pp. 1–12.
59. Maharana, D., S. Maheshka, and P. Kotecha, *Multi-objective league championship algorithms and its applications to optimal control problems*, in *Smart innovations in communication and computational sciences*. 2019, Springer, Singapore. pp. 35–46.
60. Abdulateef, S.K., et al., *An optimise ELM by league championship algorithm based on food images*. Indonesian journal of electrical engineering and computer science, 2020. **20**(1): pp. 132–137.
61. Kashan, A.H., M. Eyvazi, and A. Abbasi-Pooya, *An effective league championship algorithm for the stochastic multi-period portfolio optimization problem*. Scientia iranica (E), 2020. **27**(2): pp. 829–845.
62. Moayedi, H., et al., *A novel application of league championship optimization (LCA): hybridizing fuzzy logic for soil compression coefficient analysis*. Applied sciences, 2020. **10**(1): pp. 67.
63. Kashan, A.H., B. Karimi, and A. Noktehdan, *A novel discrete particle swarm optimization algorithm for the manufacturing cell formation problem*. The international journal of advanced manufacturing technology, 2014. **73**(9–12): pp. 1543–1556.
64. Balavand, A., A.H. Kashan, and A. Saghaei, *Automatic clustering based on crow search algorithm-Kmeans (CSA-Kmeans) and data envelopment analysis (DEA)*. International journal of computational intelligence systems, 2018. **11**(1): pp. 1322–1337.
65. Balavand, A. and A.H. Kashan, *A package including pre-processing, feature extraction, feature reduction, and classification for MRI classification*, in *Optimization in machine learning and applications*. 2020, Springer, Singapore. pp. 51–68.

8 Cultural Algorithms for Optimization

Carlos Artemio Coello Coello and
Ma Guadalupe Castillo Tapia

CONTENTS

Evolutionary algorithms have been successfully applied in a wide variety of optimization problems [1–3]. However, when used as optimizers, evolutionary algorithms are "blind" techniques in the sense that they do not require specific information about the problem but only a way of estimating how good a solution is with respect to the others (the so-called *fitness function*, which is, in general, a variation of a normalized version of the objective function to be minimized or maximized).

The use of domain knowledge within an evolutionary algorithm with the aim of improving its performance has been a research topic for several years [4, 5]. However, the incorporation of domain-specific knowledge evidently removes generality to an evolutionary algorithm, since such knowledge is specific for a particular problem (or class of problems). Additionally, the incorporation of domain-specific knowledge also replaces some of the stochastic nature of the evolutionary algorithm by deterministic information. Such extra information will certainly increase the selection pressure and will normally speed up convergence, although the risk of having premature convergence will also increase.

Nevertheless, it is worth noticing that the incorporation of domain-specific knowledge into an evolutionary algorithm is one of the choices that have been suggested [6] to circumvent the limitations imposed by the famous "no free lunch" theorem, which roughly states that all heuristics are equally efficient when assessing their performance over all possible problems [7].

Cultural algorithms are evolutionary computation techniques that extract domain knowledge (which is normally stored) during the evolutionary process with the aim of improving performance (normally, by providing a biased behavior for the evolutionary operators). In this chapter, we provide a review of the use of cultural algorithms for both single- and multi-objective optimization.

The remainder of this chapter is organized as follows. Section 8.1 provides an introduction to cultural algorithms. Section 8.2 reviews the most relevant work done on the use of cultural algorithms for single-objective optimization. The most relevant work on the use of cultural algorithms in multi-objective optimization is described in Section 8.3. Some sample applications of cultural algorithms in real-world problems are provided in Section 8.4. Some possible paths for future research from the authors' perspective are briefly described in Section 8.5. Finally, our conclusions are provided in Section 8.6.

8.1 CULTURAL ALGORITHMS

Cultural algorithms were originally proposed by Robert Reynolds in the mid-1990s [8], as an approach that tries to add domain knowledge to an evolutionary algorithm during the search process, avoiding the need to add it a priori.

According to Reynolds [9] cultural algorithms were developed as a complement to the metaphor that inspired evolutionary algorithms (natural selection and genetic concepts). Thus, cultural algorithms are based on some sociological and anthropological theories that have tried to model the phenomenon called *cultural evolution*. Such theories propose that the evolution of societies where culture exists is slightly more complicated than only genetic evolution, and it can be seen as a process of inheritance at two levels: the *micro-evolutionary level*, which consist on the genetic material inherited by parents to their descendants, and the *macro-evolutionary level*, which is conformed by the knowledge acquired by the individuals through their experiences, that once encoded and stored, it is useful for guiding the behavior of new individuals in a population (not only descendants in a genetic line) [10, 11].

Culture, then, can be seen as a set of ideological phenomena shared by a population that influences the way in which an individual interprets its experiences and decides its behavior (i.e., how to act). Culture affects the success and survival of individuals and groups, leading to evolutionary processes that are every bit as real and important as those that shape genetic variation [12]. In these models, it is easy to appreciate the component of the system that is shared by the population: the knowledge, collected by the members of a society, but encoded in a way that is potentially accessible for all the population. Similarly, the individual components of the system are the experiences, and the way they can contribute to the shared knowledge, for the other individuals to learn them indirectly.

Reynolds adopted this phenomenon of double inheritance as inspiration to create *cultural algorithms* [8]. The aim is to increase the convergence or learning rates, and therefore, that the system responds better to a variety of problems [13].

The following are the components of cultural algorithms:

- *The population space:* This space maintains a set of individuals (potential solutions to the problem). Each individual possesses characteristics independently from other individuals, and these characteristics define its fitness in the environment (the problem to solve). Throughout generations, individuals can be replaced by their descendants, obtained by means of the application of operators that somehow affect the population.
- *The belief space:* In this space is where the knowledge, acquired by the individuals through the generations, will be stored. This knowledge must be accessible to any individual in the population, and it can be used to influence its behavior (modify its characteristics and then modify its fitness). It is worth noting that the belief space is normally designed for a specific problem or class of problems to be solved.

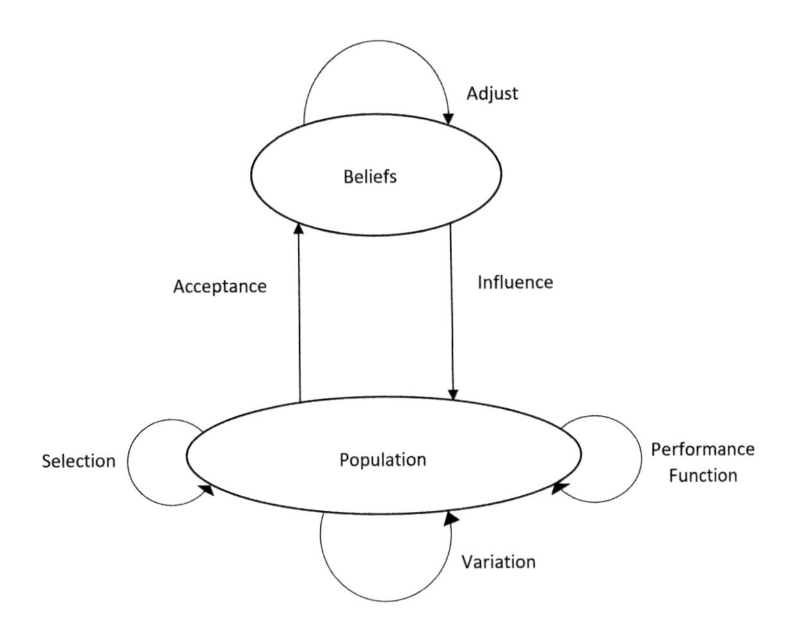

FIGURE 8.1 Spaces of a cultural algorithm.

- *A communication protocol:* This is necessary to link both spaces, defining rules about the type of information that the spaces will interchange (i.e., which information will pass from the population space to the belief space and vice versa).

At each generation, a cultural algorithm selects some individuals from the population to extract information from them that can be useful during the search.

Such information is used to update the belief space. The belief space will then influence the operators of the evolutionary algorithm to transform them in informed operators and enhance the search process. These interactions between the spaces of a cultural algorithm are depicted in Figure 8.1.

In his original proposal, Reynolds [8] adopted the population of a genetic algorithm (and its associated operators) as the population space and the version spaces [14] were used as the belief space. This cultural algorithm was called the *version space guided genetic algorithm* (VGA), and it was applied to solve some instances of the *Boole problem* (this problem consists of inferring the characteristic function for an unknown Boolean multiplexer) [15], with encouraging results. In this early application, a graph (the belief space) of the solutions in the population was built and classified based on the fitness of each particular instance. This was a very illustrative approach, because the graph's dynamics reflects the discovering of good and bad solutions. Based on those findings, Reynolds argued about the usefulness of cultural algorithms with an adaptation of the schema theorem, taking advantage of the genetic algorithm-based approach previously introduced. The *schema theorem* is an expression that bounds the propagation of the best solutions within the population of a genetic algorithm [16]. This modification indicates that a genetic algorithm, with the addition of a belief space, can improve its performance by increasing its convergence rates.

Cultural algorithms have had a limited use as optimizers in the specialized literature, but they have been adopted for both single- and multi-objective optimization as reviewed in the next two sections.

8.2 CULTURAL ALGORITHMS FOR SINGLE-OBJECTIVE OPTIMIZATION

The general (single-objective) global optimization problem is defined as follows (without loss of generality, we will assume minimization):

$$\begin{array}{ll} \text{Minimize} & f(\vec{x}) \\ \text{Subject to} & lb_j \leq x_j \leq ub_j, \quad j = 1,2,...,D \end{array} \tag{8.1}$$

where $\vec{lb}$, $\vec{ub} \in \mathbb{R}^D$, are the lower bound and the upper bound of the decision variables $\vec{x}$, respectively, and f: $\mathbb{R}^D \rightarrow \mathbb{R}$ is the objective function. The feasible solution space is defined as $\Omega = \left\{ \vec{x} \in \mathbb{R}^D \middle| lb_j \leq x_j \leq ub_j, \ \forall j \in \{1,2,3,...,D\} \right\}$.

Single-objective optimization problems may also be subject to constraints:

$$\begin{array}{ll} g_i(\vec{x}) \geq 0 & i = 1,2,\ldots,m \\ h_i(\vec{x}) = 0 & i = 1,2,\ldots,p \end{array} \tag{8.2}$$

where $g_i(\vec{x})$ denotes inequality constraints and $h_i(\vec{x})$ denotes equality constraints. Constraints are said to be *active* when in the global optimum $g_i(\vec{x}) = 0$. By definition, all equality constraints are active.

8.2.1 Static Optimization

Reynolds et al. [17] and Chung and Reynolds [18] explored the use of cultural algorithms for global optimization with very encouraging results. Chung and Reynolds used a hybrid of evolutionary programming [19] and GENOCOP [20] in which they incorporated an interval constraint network to represent the constraints of the problem at hand.

Chung and Reynolds [18] used evolutionary programming with a mutation operator influenced by the best individual found so far, and the intervals where good solutions had been found. They called this approach *cultural algorithms with evolutionary programming* (CAEP) for global optimization. CAEP provided a very rich model in the belief space, and it produced very encouraging results.

Reynolds and Chung [21, 22] proposed a formal model of self-adaptation in cultural algorithms that supports the three main levels of self-adaptation in evolutionary algorithms (population, individual, and component levels). The royal road functions [23] were adopted as a case study in this work.

The CAEP was tested on a number of global optimization problems [18], showing its improvement when compared with the standard evolutionary programming algorithm. In this case, the belief space was divided in two parts or knowledge sources, specifically designed for real-valued problems: *situational knowledge* and *normative knowledge*. A general description of these knowledge sources, and those designed and added later, is provided in Table 8.1.

Jin and Reynolds [24] proposed an extension of CAEP for nonlinear constrained optimization. To handle constraints, an additional knowledge source, called *topographical knowledge*, was added to the belief space. It consists of a set of cells that store some characteristic of the region of the search space they represent. In this case, they store a map of the feasible region (i.e., a discrete representation of the search space) based on the points that have been explored so far.

Coello Coello and Landa Becerra [25, 26] extended Jin and Reynolds' approach, improving its computational efficiency and overcoming its scalability problems. In the original approach, the topographical knowledge was stored as an n-dimensional grid of the search space. This was

TABLE 8.1
Knowledge Sources for Real-Parameter Optimization in Cultural Algorithms

Knowledge Source	Description
Situational knowledge	Consists of the best exemplars found in the population, which represent leaders to follow. The individuals generated through this source, will tend to be closer to the leaders.
Normative knowledge	Consists of a set of intervals for each decision variable in which good solutions have been found. The individuals generated through this source are more likely to be within the intervals, so they exploit good regions.
Topographical knowledge	Consists of a set of cells that represent a region of the search space. Each cell stores a characteristic of the region it represents; for example, the feasibility of that region. The individuals generated through this source will be closer to the best cells.
History knowledge	Consists of a set of previous local optima, and its function is to extract patterns about their position. The individuals generated through this source will try to find in advance the location of the next local optimum. This knowledge source can also be used to add diversity to the algorithm, since it attempts to explore new regions.
Domain knowledge	It has no defined structure, because it depends of the problem which is to be solved. Its function is to exploit some knowledge about the problem, if available.

replaced by a spatial data structure that requires a controlled amount of memory even when the number of dimensions grows. Additionally, the authors presented an empirical study in which this approach was validated using a well-known benchmark adopted in evolutionary constrained optimization, and the results were compared with respect to constraint-handling techniques that were representative of the state-of-the-art in the area at that time. This approach was able to find competitive results, while performing only about 15% of the total number of fitness function evaluations required by the other approaches with respect to which it was compared.

Reynolds and Peng [27] provide a study of how the knowledge sources associated with cultural knowledge control the search process in a cultural algorithm used for solving engineering optimization problems. In this case, evolutionary programming is adopted for the population space. The authors observed that the meta-level interaction of the knowledge sources of a cultural algorithm allow the generation of feasible solutions from a fully infeasible initial population and also helps to speed up convergence. Although the study is indeed interesting, only one (relatively simple) engineering optimization problem was used to validate this approach.

Gao et al. [28] used a genetic algorithm with real numbers encoding for the population space of a genetic algorithm. However, in this case, the only modification made to the genetic algorithm consists of introducing the mutation scheme adopted in the regional-based sliding cultural algorithm proposed by Jin and Reynolds [24]. The validation is also very poor in this case, since this proposal is only tested in one constrained problem having only one decision variable and one nonlinear constraint.

Landa Becerra and Coello Coello [29, 30] have also adopted differential evolution for the population space of a cultural algorithm designed for constrained optimization. In this case, all the knowledge sources were adapted for their use with the differential evolution operator, providing some adaptation of its components. The approach was tested on a well-known benchmark commonly adopted to validate new constraint-handling techniques and on some engineering optimization problems, providing very competitive results.

Tang and Li [31] made an interesting proposal in which an anti-culture population was adopted with the purpose of having individuals that disobey the guidance of the knowledge provided by the cultural algorithm. Such individuals aim to avoid local minima and to speed up convergence. This approach was called *triple spaces cultural algorithm* (TSCA) and it was implemented using a genetic algorithm in the population space. Mutation operations are used to move individuals away from the bias introduced by the knowledge obtained during the search. The authors validated their TSCA using the same test problems from [29] and compared results with respect to that approach as well.

Nguyen and Yao [32] proposed a hybrid of a cultural algorithm with iterated local search for multimodal optimization. In this proposal, a shared knowledge space is adopted for integrating the knowledge produced from pre-defined multi-populations and knowledge migration is used to bias the search toward different directions. This approach was validated using the test problems from the special session on real-parameter optimization held at the 2005 IEEE Congress on Evolutionary Computation (CEC 2005) [33], and presented competitive results.

Yang et al. [34] proposed a cultural algorithm that integrates a hybrid of the so-called quantum-behaved particle swarm optimization (QPSO) [35] with differential evolution. In QPSO, each particle has a quantum behavior, and we can only learn the probability that a particle has of appearing in a certain position from a probability density function, the form of which depends on the potential field in which the particle lies. In this proposal, the population is divided into two sub-swarms: (1) common particles and (2) elite particles. *Elite* particles have better fitness values than *common* particles and are meant to represent the mainstream *culture* of the entire swarm. Elite particles evolved using differential evolution, and common particles evolved using QPSO. Common particles are influenced both by the other common particles and by the elite particles. This approach was validated using the test problems from the special session on real-parameter optimization held at the CEC 2005 [33], comparing results with respect to QPSO, a simple PSO, and a very competitive PSO called CLPSO [36]. The proposed approach was able to outperform QPSO and was competitive with respect to PSO and CLPSO.

Ali et al. [37, 38] introduced the use of a *social fabric influence function* in a cultural algorithm designed to solve nonlinear constrained optimization problems. This social fabric is seen by its authors as some sort of computational tool that influences the action and interactions of the different knowledge sources adopted by the cultural algorithm. PSO was adopted for the population memory in this case. In this proposal, at each time step, every individual is influenced by one of five possible knowledge sources. Individuals are connected using a certain topology and after the first one is influenced by a particular knowledge source, it passes the signal to adjacent individuals. Different particle swarm topologies were adopted by the authors for interconnecting the individuals in the population. Several benchmark problems taken from [39] were adopted to validate this proposal.

Awad et al. [40] proposed a cultural algorithm with an improved local search mechanism for global optimization. In this proposal, the initial solution to which the local search is applied is selected using a niching technique called *clearing* [41]. Once a solution is selected, several neighbors are generated using the five knowledge sources of the cultural algorithm, which give rise to five different sub-local search methods. A set of test problems with low dimensionality (10–30 decision variables) was adopted to validate this approach.

Ali and Reynolds [42] proposed a cultural algorithm that incorporated an embedded local tabu search [43] component for solving constrained optimization problems. Evolutionary programming is used in this case for the population space. The main goal of using tabu search is to improve diversity in the population to avoid premature convergence and stagnation. Since tabu search was originally designed for discrete search spaces, the authors had to adopt a version developed by Siarry and Berthiau [44] for continuous search spaces. The core idea of this approach is to use tabu search to encapsulate the history knowledge so that it can be adopted to select appropriate search paths. This algorithm was only tested with three engineering optimization problems.

Omran [45] proposed the intellect-masses optimizer (IMO), which is a variation of the cultural algorithm from Yang et al. [34]. In this case, the population is divided into two subpopulations: (1) the *intellects*, which are the fittest individuals and (2) the *masses* which refers to the rest of the population. The intellects are meant to learn from each other and to focus on exploitation, whereas the masses are meant to learn from the intellects and from themselves and focus on exploration. The author uses differential evolution for evolving the intellects and a modified artificial bee colony [46] to evolve the masses. This approach is validated using the test problems from the special session on real-parameter optimization held at the CEC 2005 [33], and its results are compared with respect to those from Yang et al. [34] and with respect to several other metaheuristics, including the artificial bee colony and CLPSO [36].

Ali et al. [47] proposed a hybrid of a cultural algorithm with multiple trajectory search (MTS) [48] for multimodal optimization. The core idea of MTS is to move in decision variable space based on different step sizes. Each step size is applied according to a certain local search method. The original MTS algorithm uses simulated orthogonal arrays to generate the initial solutions of the basic multiple trajectory model. In this case, the authors use the knowledge sources of the cultural algorithm for the same purpose. This approach was validated using the test problems from the special session on real-parameter optimization held at the CEC 2005 [33], and its results were compared with respect to a wide variety of global optimizers, obtaining very promising results in problems with up to 100 decision variables.

Ali et al. [49] proposed a framework for developing cultural algorithms based on differential evolution in which the main emphasis is to provide a proper balance between exploration and exploitation. This approach, called b-hCA-DE incorporates four knowledge sources (topographical, situational, normative, and temporal) and uses a population that is shared by all the knowledge sources. This approach is validated using the 28 test problems from the special session and competition held at the 2013 IEEE Congress on Evolutionary Computation [50] with 10, 30, and 50 decision variables. The results obtained by this approach were found to be competitive with respect to those of a high variety of state-of-the-art global optimizers.

Ali et al. [49] proposed restructuring the social fabric (social network) [51] of the connections that link the individuals in the population space of a cultural algorithm with the aim of enhancing its performance. For this restructuring, the authors adopt a dynamic neighborhood topology. The metaphor behind this proposal is that the knowledge sources of the cultural algorithm can weave a networked "fabric" of individuals that are performing the search. Thus, the population space consists of a set of subgroups called "tribes," which are meant to represent the building blocks of the population of search engines. Tribal subgroups can be merged with each other using different regrouping schemes. The restructuring of the social fabric is based on the success of each of the knowledge sources incorporated in the cultural algorithm. This approach was validated using the problems from the 2011 IEEE Congress on Evolutionary Computation (CEC 2011) Competition on Testing Evolutionary Algorithms on Real-World Numerical Optimization Problems [52]. The approach was compared with respect to other cultural algorithms and with respect to state-of-the-art global optimizers producing very promising results.

Awad et al. [53] proposed CADE, which hybridizes a cultural algorithm and differential evolution. The core idea in this case is to select the best individuals in the population and use them to update the knowledge sources in the belief space (the authors adopt topographical, situational, normative, and domain knowledge). Then, the knowledge sources that will influence the evolutionary process are selected. At this point, differential evolution is used to improve the exploratory capabilities of the cultural algorithm. In this approach, both the cultural algorithm and differential evolution are executed in parallel sharing the same population and a success-based quality function is used to guide the search. A set of 50 test problems having up to 100 decision variables each was adopted to validate this approach and results were compared with respect to six other metaheuristics. CADE had a very competitive performance.

Ravichandran [54] proposed a cultural algorithm based on decomposition (CA/D) to decompose a dynamic multi-objective optimization problem into several sub-problems that are then optimized using information shared by neighboring problems. This approach consists of a culturized version of MOEA/D-DP [55]. In this case, the historical knowledge is used to track the environmental changes, the situational knowledge is used to preserve the best solutions, and the normative knowledge is used to distribute solutions along the Pareto front. The author experimented with both Tchebycheff decomposition and reference points. Although this is clearly a multi-objective optimization algorithm, the author used it to solve single-objective optimization problems. In fact, the test problems from the special session on real-parameter optimization held at the CEC 2005 [33] were adopted to validate it.

8.2.2 Dynamic Optimization

Saleem and Reynolds [56] added two more knowledge sources to cultural algorithms to deal with dynamic environments, namely *history knowledge* and *domain knowledge*. The first of these sources was designed to extract patterns about the changes of position of optimal points at each environmental change. The second source was designed to exploit the known characteristics of the function generator. Even when these knowledge sources were designed for dynamic problems, they have also been used in static environments [29].

Peng and Reynolds [57] adopted PSO [58] for the population space of a cultural algorithm. In this case, the authors used all of the previously designed knowledge sources, and they investigated the role of the belief space in the different stages of a dynamic optimization process. The authors also argued that a cultural algorithm provides an additional degree of adaptability to the one provided by evolutionary algorithms. This work is extended in [59] in which evolutionary programming is adopted for the population space and a dynamic problems generator is used to simulate the changes of the fitness landscape that the cultural algorithm is meant to emulate. The authors show the emergence of swarms of solutions in both the population space and the belief space, which are able to detect the changes in the location of the optimum. Situational and domain knowledge are found to play a key role in this case.

Jiang et al. [60] proposed a cultural-based particle swarm optimizer for dynamic environments. The authors see the belief space as a knowledge repository that stores information about the environmental changes. Thus, the core idea of this proposal is to use this information to predict the location of the new optimum. For this sake, the authors need to learn to identify the types of belief knowledge structures required for finding and tracking a moving optimum. This approach is only validated with a single problem (a dynamic version of the Rastrigin function).

Daneshyari and Yen [61] proposed a cultural-based particle swarm optimizer in which five knowledge sources are adopted (situational, temporal, domain, normative, and space). These knowledge sources store information from the adopted particle swarm optimizer and such information is then used to detect changes in the environment. The knowledge sources also assist the optimizer to respond to these changes through a diversity maintenance mechanism (called repulsion) and a migration operator that acts among swarms in the population space.

Chen et al. [62] proposed a cultural algorithm for the path planning of an unmanned aerial vehicle (UAV) in real time. In this case, the belief space incorporates both situational and normative knowledge. This is a dynamic problem, since the environment is changing over time. When a change occurs, it is not necessary to regenerate the full path to avoid an obstacle. Only a portion of the path needs to be readjusted. This approach was compared with respect to the D* algorithm [63]. The cultural algorithm was found to have a better real-time performance, a lower path planning cost, and produced paths of a shorter length.

Kinnaird-Heether and Reynolds [64] analyzed different knowledge distribution mechanisms for a cultural algorithm. The knowledge distribution mechanism is used to handle the conflict resolution between the competing knowledge sources and the belief space. Researchers have adopted a variety of knowledge distribution mechanisms including voting [65], auctions [66], and game-theory [67]. In this work, the authors also proposed a new (sub-cultured) approach that allows the cultural algorithm to learn to use a combination of different mechanisms. This scheme was adopted for solving dynamic optimization problems. The authors reported that the use of their proposed sub-cultured mechanism was better than any of the mechanisms that it combined when used separately.

8.3 CULTURAL ALGORITHMS FOR MULTI-OBJECTIVE OPTIMIZATION

In multi-objective optimization, the aim is to solve problems of the type (without loss of generality, we will assume only minimization problems):

$$\text{minimize}\ \vec{f}(\vec{x}) := \left[f_1(\vec{x}), f_2(\vec{x}), \ldots, f_k(\vec{x})\right] \tag{8.3}$$

subject to:

$$g_i(\vec{x}) \leq 0 \quad i = 1, 2, \ldots, m \tag{8.4}$$

$$h_i(\vec{x}) = 0 \quad i = 1, 2, \ldots, p \tag{8.5}$$

where $\vec{x} = \left[x_1, x_2, \ldots, x_n\right]^T$ is the vector of decision variables, f_i: $\mathbb{R}^n \to \mathbb{R}$, $i = 1,\ldots,k$ are the objective functions, and h_j: $\mathbb{R}^n \to \mathbb{R}$, $i = 1,\ldots,m$, $j = 1,\ldots, p$ are the constraint functions of the problem.

A few additional definitions are required to introduce the notion of optimality used in multi-objective optimization:

Definition 1. Given two vectors $\vec{x}, \vec{y} \in \mathbb{R}^k$, we say that $\vec{x} \leq \vec{y}$ if $x_i \leq y_i$ for $i = 1,\ldots,k$, and that $\vec{x}$ **dominates** $\vec{y}$ (denoted by $\vec{x} \prec \vec{y}$) if $\vec{x} \leq \vec{y}$ and $\vec{x} \neq \vec{y}$.

Definition 2. We say that a vector of decision variables $\vec{x} \in X \subset \mathbb{R}^n$ is **nondominated** with respect to X, if there does not exist another $\vec{x}' \in X$ such that $\vec{f}(\vec{x}') \prec \vec{f}(\vec{x})$.

Definition 3. We say that a vector of decision variables $\vec{x} \in \mathcal{F} \subset \mathbb{R}^n$ ($\mathcal{F}$ is the feasible region) is **Pareto-optimal** if it is nondominated with respect to $\mathcal{F}$.

Definition 4. The **Pareto Optimal Set** P^* is defined by:

$$P^* = \left\{\vec{x} \in \mathcal{F} \middle| \vec{x}\ is\ Pareto-optimal\right\}$$

Definition 5. The **Pareto Front** PF^* is defined by:

$$PF^* = \left\{\vec{f}(\vec{x}) \in \mathbb{R}^k \middle| \vec{x} \in P^*\right\}$$

Therefore, our aim is to obtain the Pareto optimal set from the set $\mathcal{F}$ of all the decision variable vectors that satisfy Eqs. 8.4 and 8.5.

Apparently, the first attempt to use a Pareto-based approach for solving multi-objective problems adopting a cultural algorithm is the proposal of Coello Coello and Landa Becerra [68].

In this case, evolutionary programming is used for the population space and Pareto ranking is adopted for selecting nondominated solutions. This approach also uses the approximation of the dimensions of the Pareto front in the belief space, which works as a guide for the individuals to reach regions in which new nondominated solutions can be found. The belief space also includes a mechanism to produce a good distribution of solutions along the Pareto front (i.e., a density estimator [69]). An algorithm based on this approach was used by Gu and Wu [70] for solving a water resources problem.

Landa Becerra and Coello Coello [71] proposed combining the cultural algorithm based on differential evolution that they proposed in [29] with the ε-constraint method for solving complex multi-objective optimization problems. This approach is computationally expensive because a high number of single-objective optimizations need to be performed to solve the multi-objective problem. Thus, for justifying the use of this approach, they solved the most difficult instances from the Deb-Thiele-Laumanns-Zitzler (DTLZ) [72] and the Walking-Fish-Group test suites [73], which could not be properly solved by any of the multi-objective evolutionary algorithms available at that time (including the non-dominated sorting genetic algorithm [NSGA]-II).

Best [74] and Best et al. [75] proposed a framework for designing multi-objective cultural algorithms called MOCA in which five knowledge sources are considered: (1) situational knowledge, which influences individuals to produce children near to known good solutions; (2) domain knowledge, which searches in the four cardinal solutions with the aim of moving toward the true Pareto front of the problem; (3) normative knowledge, which tries to capture the regions in decision variable space that contain the fittest solutions; (4) historical knowledge, which is used to spread solutions along the Pareto front; and (5) topographical knowledge, which avoids getting trapped in a local Pareto front. The acceptance function is based on the use of Pareto ranking, but for selecting the best individual, a random scheme (from among the nondominated solutions generated so far) is adopted. The knowledge sources are selected to influence members of the population by sampling a dynamic probability distribution. The DTLZ test problems are adopted to validate this approach and, in general, the results of this proposal are not better than those generated by NSGA-II.

Daneshyari and Yen [76] proposed a cultural-based multi-objective PSO (MOPSO). The core idea in this case is to adapt the parameters of the MOPSO using the knowledge sources of the cultural algorithm. In this approach, Pareto dominance is used as the acceptance function and the authors adopt three knowledge sources: (1) situational knowledge, which is used to adapt the local acceleration of the MOPSO and to select the personal best; (2) normative knowledge, which is used to adapt the global acceleration and to find the global best; and (3) topographical knowledge, which is also used to adapt the global acceleration and to find the global best. This approach also incorporates the same sort of bounded global archive adopted in [77].

Reynolds and Liu [78] proposed an extension of MOCA [75] in which they allow all knowledge sources to contribute to the optimization process. What the authors actually do in this case is to improve the implementation of MOCA by introducing several changes to the previous version. For example, for the topographical knowledge the original had severe scalability problems because grid cells were used to represent decision variable space. In this new version, hypercubes in objective space are adopted instead. In this paper, a single problem is adopted to validate the proposal (from the old Zitzler-Deb-Thiele (ZDT) test suite [79]) and only a visual comparison of results is done with respect to old multi-objective evolutionary algorithms (e.g., the original NSGA [80]).

Stanley et al. [81] proposed CAPSO, which is a parallelized hybrid optimization system designed for solving multi-objective optimization problems. This approach combines cultural algorithms with PSO and the vector evaluated genetic algorithm (VEGA) [82]. Five knowledge sources are adopted in this case: (1) situational, (2) normative, (3) historical, (4) topographical, and (5) domain knowledge. For determining the best solutions, one objective is considered each time (as in VEGA). Although the authors indicate that VEGA was selected because it is easy to parallelize, it is unclear if a parallel implementation was actually developed. This approach is validated with very old (and rather simple) multi-objective test problems. Nevertheless, CAPSO has been used to analyze trends in a study of the impact of "El Niño" and climate change on artisanal fishermen behavior from the early 1980s in Peru [83] (CAPSO was adopted to perform nonlinear regressions).

Mao et al. [84] proposed a multi-objective cultural algorithm called MOFECO in which the five-elements-cycle-optimization algorithm [85] is used in the population space. This approach adopts normative, situational, topographical, and historical knowledge. This approach was validated using several problems from the ZDT [79], and the DTLZ [72] test suites and results were compared with respect to several multi-objective evolutionary algorithms (including MOEA/D [86]) using two performance measures: hypervolume [87] and inverted generational distance [88]).

8.4 SOME APPLICATIONS

Cultural algorithms have been adopted in a wide variety of applications in which they have been used to solve both single- and multi-objective optimization problems [89, 90]. A sample of these applications is provided next:

- *Electrical engineering:* Goudarzi et al. [91] proposed four different versions of a cultural algorithm (each one adopting a different knowledge source) for solving the combined environmental economic dispatch problem. This is actually a multi-objective problem in which the aim is to simultaneously minimize fuel cost and emission, while satisfying several power systems constraints. However, in this case, the authors combine the two objectives using a linear aggregating function. Three systems having 5, 20, and 50 generating units were adopted to validate this approach and results were compared with respect to a variety of metaheuristics showing promising results.

  Other authors have solved economic dispatch problems using cultural algorithms combined with differential evolution [92], evolutionary programming [93], QPSO [94], an artificial immune system [95], and the self-migrating algorithm [96].

  Additionally, this problem has also been tackled using multi-objective approaches. For example, Zhang et al. [97] proposed the *enhanced multi-objective cultural algorithm* (EMOCA) to solve this problem. EMOCA combines the framework of cultural algorithms with PSO and adopts two knowledge sources tailored for the specific problem to be solved. The authors also proposed a constraint-handling technique as part of their approach.

  Lu et al. [98] proposed the hybrid multi-objective cultural algorithm for short-term environmental/economic hydrothermal scheduling. This approach adopts a Pareto-based version of differential evolution for the population space. It is validated using two case studies in which its results were compared with respect to NSGA-II [99] and two other multi-objective optimization algorithms based on evolutionary programming and simulated annealing.

  Other related applications include the optimization of the operation of a hydropower station [100] and the optimization of a doubly-fed induction generator to attain an efficient and improved dynamic response of a wind energy conversion system [101].
- *Mechanical engineering:* Coelho and Mariani [102] proposed two Gaussian PSO approaches combined with a cultural algorithm, which are called cultural PSO

(PSO-CA) and GPSO-CA. Evidently, the main difference with respect to a traditional PSO lies in the use of a Gaussian distribution for computing the velocities of the particles. This is meant to improve the local exploration of the algorithm. This approach was used to solve mechanical engineering design problems. Engineering optimization problems similar to those adopted in [102] have been adopted by some other authors as well (see, for example, [30, 47, 49, 103]).

Jalili and Hosseinzadeh [104] proposed a cultural algorithm combined with evolutionary programming (following the proposal from [22]) for the optimal design of trusses. Results are compared with respect to those of several other metaheuristics using four test problems, which include a 120-bar dome truss. The cultural algorithm is able to produce competitive results while performing a lower number of objective function evaluations than the other algorithms.

- *Image processing:* Wang et al. [105] proposed an adaptive cultural algorithm with improved QPSO (ACA-IQPSO) for detecting underwater sonar images. The authors adopt situational, normative, and domain knowledge and introduce a new communication protocol that considers not only the acceptance function, but also an influence function that is used to guide the evolution of the particles with a poor performance using the knowledge stored in the belief space. This approach is found to be very effective for both floating and underwater object detection.

  Yan et al. [106] proposed a cultural algorithm with isolated niching (as a mechanism to improve diversity) for image matching problems. Tan and Yang [107] proposed a cultural algorithm to maximize the entropy function used to do multi-threshold image segmentation (for infrared images) to reduce the computational time as well as to improve the segmentation efficiency.

  Cai [108] proposed a scheme for increasing the detectability of sea-surface floating weak targets, which consists of using a cultural algorithm aided time-frequency distribution fusion strategy without any prior information. The authors adopted several sets of experimental data collected by an instrument-quality radar system to verify the accuracy and efficiency of this proposal. Six representative time-frequency distributions of experimental signals were obtained and their performances were quantitatively analyzed in terms of effective resolution and entropy. Additionally, the authors adopted the Volterra-series weighted averaging model as their fusion rule and a cultural algorithm for the optimization. The authors reported that their proposed approach was able to outperform other detectability techniques.
- *Scheduling:* Soza et al. [109] proposed a cultural algorithm for solving timetabling problems. Three knowledge sources are considered in this case: (1) situational, (2) normative, and (3) domain knowledge. These knowledge sources are combined with three specialized variation operators: interchange, sequencing, and simple mutation. Only one exploration operator is applied to each individual at a time. This approach was validated using a benchmark with 20 instances, and the results were compared with respect to an evolutionary algorithm with specialized crossover operators, a memetic algorithm, and a simulated annealing approach that won a timetabling competition organized by the metaheuristic network. The results obtained indicated that the cultural algorithm was a viable alternative for solving timetabling problems in an efficient manner.

  Mojab et al. [110] proposed a cultural algorithm for workshop scheduling in cloud computing. The authors considered the situation in which there is a deadline for the workflow and the goal is to minimize the monetary cost of running in the cloud while satisfying the given deadline. Three knowledge sources were adopted in this case: normative, situational, and domain knowledge. The proposed approach was validated using four synthetic workflow applications based on real scientific workflows and results were compared with respect to those of a random scheme, a genetic algorithm, a particle swarm optimizer, and a traditional cultural algorithm. The authors reported a better performance of their proposal.

- *Finance:* Sternberg and Reynolds [111] proposed to embed a fraud detection expert system called DETECT into a cultural algorithm. To simulate a dynamic environment in this application, the authors considered four objectives: characterizing fraudulent claims, nonfraudulent claims, false-positive claims (i.e., nonfraudulent claims predicted as fraudulent), and false-negative claims (i.e., fraudulent claims predicted as nonfraudulent). The authors reported that the use of a cultural algorithm allowed them to respond to changing objectives in an effective way.

  Ostrowski et al. [112] proposed a cultural algorithm for optimizing strategies in agent-based models and demonstrated its use in an application used to model pricing strategies. For a more effective evaluation of parameter configurations, the authors adopted white- and black-box testing, which are well-known software engineering techniques. The authors indicated that their proposed approach was able to derive a near-optimal pricing strategy in less periods than traditional evolutionary approaches.

8.5 FUTURE PERSPECTIVES

There are several paths for future research in this area including:

- *Use of other cultural paradigms:* Kuo and Lin [113] developed a *cultural evolutionary algorithm* based on Steward's sociocultural integration theory [114]. It would be interesting to see the development of new cultural algorithms based on other theories in the future.
- *Parallelism:* The use of cultural algorithms has been somehow limited due to their potentially high computational cost (depending on the particular application and the knowledge sources adopted). One way to deal with this limitation could be to use parallelism, but so far very few proposals of cultural algorithms seem to involve parallel implementations (see for example, [115–117]).
- *Multi-objective optimization:* Although there are several proposals to adopt cultural algorithms for multi-objective optimization, there is no multi-objective cultural optimizer currently available that has been validated with state-of-the-art test problems and compared with respect to state-of-the-art multi-objective evolutionary algorithms. Particularly, to the authors' best knowledge, no multi-objective cultural algorithm has been successfully applied to many-objective problems (i.e., multi-objective problems having four or more objectives) nor to inverted test problems [118].

  Furthermore, most of the current paradigms used in multi-objective evolutionary algorithms have not been culturized [119]. Particularly, no multi-objective cultural algorithm has been developed based on performance indicators [120] or on decomposition [121] (the approach reported in [54] was used for single-objective optimization). In other areas, such as dynamic multi-objective optimization [122], cultural algorithms seem to be an obvious choice, but to the authors' best knowledge, they have not been applied in such problems yet.
- *Culturizing other evolutionary algorithms:* Most of the current cultural algorithms are based on traditional evolutionary algorithms (i.e., evolutionary programming, genetic algorithms, and differential evolution). However, the use of other algorithms such as genetic programming [123] has been fairly limited (see, for example, [9]). The use of genetic programming would allow an extended range of applications of cultural algorithms to areas such as symbolic regression and classification.

8.6 CONCLUSIONS

This chapter has provided an overview of cultural algorithms and their use on optimization. The aim was not to be comprehensive, but to cover most of the areas (within optimization) in which they have been used. The aim is to provide a general overview of the field both to students and researchers who are interested on doing research in this area.

The topics covered in this chapter include static and dynamic single-objective optimization as well as multi-objective optimization. Additionally, a few application areas of cultural algorithms have also been provided. In the final section of this chapter, some research paths that are worth exploring in the future (from the authors' perspective) are also delineated.

ACKNOWLEDGMENTS

Carlos A. Coello Coello gratefully acknowledges support from CONACyT project 1920 (Fronteras de la Ciencia) and from a SEP-Cinvestav 2018 project (application no. 4).

REFERENCES

1. O. Cordón, E. Herrera-Viedma, C. López-Pujalte, M. Luque, and C. Zarco. A review on the application of evolutionary computation to information retrieval. *International Journal of Approximate Reasoning*, 34(2–3):241–264, November 2003.
2. Luis Gerardo de la Fraga and Carlos A. Coello Coello. A review of applications of evolutionary algorithms in pattern recognition. In Patrick S.P. Wang, editor, *Pattern Recognition, Machine Intelligence and Biometrics*, pages 3–28. Berlin: Springer-Verlag, 2011. ISBN 978-3-642-22406-5.
3. Adam Slowik and Halina Kwasnicka. Evolutionary algorithms and their applications to engineering problems. *Neural Computing and Applications*, 32:12363–12379, 2020.
4. Yaochu Jin, editor. *Knowledge Incorporation in Evolutionary Computation*. Studies in Fuzziness and Soft Computing, Vol. 167. New York: Springer, 2005. ISBN 3-540-22902-7.
5. Ricardo Landa-Becerra, Luis V. Santana-Quintero, and Carlos A. Coello Coello. Knowledge incorporation in multi-objective evolutionary algorithms. In Ashish Ghosh, Satchidananda Dehuri, and Susmita Ghosh, editors, *Multi-Objective Evolutionary Algorithms for Knowledge Discovery from Data Bases*, pages 23–46. Berlin: Springer, 2008. ISBN 978-3-540-77466-2.
6. A.E. Eiben and J.E. Smith. *Introduction to Evolutionary Computing*. Berlin: Springer, 2003. ISBN 3-540-40184-9.
7. David H. Wolpert and William G. Macready. No free lunch theorems for optimization. *IEEE Transactions on Evolutionary Computation*, 1(1):67–82, April 1997.
8. Robert G. Reynolds. An Introduction to Cultural Algorithms. In A. V. Sebald and L. J. Fogel, editors, *Proceedings of the Third Annual Conference on Evolutionary Programming*, pages 131–139. River Edge, NJ: World Scientific, 1994.
9. Robert G. Reynolds. Cultural algorithms: theory and applications. In David Corne, Marco Dorigo, and Fred Glover, editors, *New Ideas in Optimization*, pages 367–377. London: McGraw-Hill, 1999.
10. W. H. Durham. *Co-evolution: Genes, Culture, and Human Diversity*. Stanford, CA: Stanford University Press 1994.
11. A. C. Renfrew. Dynamic modeling in archaeology: What, when, and where? In S. E. van der Leeuw, editor, *Dynamical Modeling and the Study of Change in Archaelogy*. Edinburgh, Scotland: Edinburgh University Press, 1994.
12. Peter J. Richerson and Robert Boyd. *Not by Genes Alone: How Culture Transformed Human Evolution*. Chicago, IL: The University of Chicago Press, 2005.
13. Benjamin Franklin and Marcel Bergerman. Cultural algorithms: concepts and experiments. In *Proceedings of the 2000 IEEE Congress on Evolutionary Computation (CEC'2000)*, pages 1245–1251, Piscataway, NJ: IEEE Service Center, 2000.
14. Tom Mitchell. Version spaces: an approach to concept learning. PhD thesis, Computer Science Department, Stanford University, Stanford, California, 1978.

15. Stewart W. Wilson. Classifier systems and the animat problem. *Machine Learning*, 2(3):199–228, 1987.
16. John H. Holland. *Adaptation in Natural and Artificial Systems. An Introductory Analysis with Applications to Biology, Control and Artificial Intelligence.* Ann Arbor, MI: University of Michigan Press, 1975.
17. Robert G. Reynolds, Zbigniew Michalewicz, and M. Cavaretta. Using cultural algorithms for constraint handling in GENOCOP. In J. R. McDonnell, R. G. Reynolds, and D. B. Fogel, editors, *Proceedings of the Fourth Annual Conference on Evolutionary Programming*, pages 298–305. Cambridge, MA: MIT Press, 1995.
18. Chan-Jin Chung and Robert G. Reynolds. CAEP: an evolution-based tool for real-valued function optimization using cultural algorithms. *Journal on Artificial Intelligence Tools*, 7(3):239–292, 1998.
19. Lawrence J. Fogel. *Artificial Intelligence through Simulated Evolution. Forty Years of Evolutionary Programming.* New York: John Wiley & Sons, Inc., 1999.
20. Zbigniew Michalewicz. *Genetic Algorithms + Data Structures = Evolution Programs*, third edition, Berlin: Springer-Verlag, 1996.
21. Chan-Jin Chung. Knowledge-based approaches to self-adaptation in cultural algorithms. PhD thesis, Wayne State University, Detroit, Michigan, 1997.
22. Robert G. Reynolds and ChanJin Chung. Knowledge-based self-adaptation in evolutionary programming using cultural algorithms. In *Proceedings of 1997 IEEE International Conference on Evolutionary Computation (ICEC'97)*, pages 71–76, Indianapolis, IN: IEEE, 13–16 April 1997. ISBN 0-7803-3949-5.
23. Melanie Mitchell. *An Introduction to Genetic Algorithms.* Cambridge, MA: The MIT Press, 1996.
24. Xidong Jin and Robert G. Reynolds. Using knowledge-based evolutionary computation to solve nonlinear constraint optimization problems: a cultural algorithm approach. In *Proceedings of the 1999 IEEE Congress on Evolutionary Computation (CEC'99)*, pages 1672–1678, Washington, D.C.: IEEE Service Center, July 1999.
25. Carlos A. Coello Coello and Ricardo Landa Becerra. A cultural algorithm for constrained optimization. In Carlos A. Coello Coello, Alvaro de Albornoz, Enrique Sucar, and Osvaldo Cairó Battistutti, editors, *MICAI'2002: Advances in Artificial Intelligence*, pages 98–107. Berlin: Springer-Verlag. Lecture Notes in Artificial Intelligence Vol. 2313, April 2002.
26. Carlos A. Coello Coello and Ricardo Landa Becerra. Adding knowledge and efficient data structures to evolutionary programming: a cultural algorithm for constrained optimization. In Erick Cantú-Paz et al., editor, *Proceedings of the 2002 Genetic and Evolutionary Computation Conference (GECCO'2002)*, pages 201–209, San Francisco, CA: Morgan Kaufmann Publishers, July 2002.
27. Robert G. Reynolds and Bin Peng. Cultural algorithms: computational modeling of how cultures learn to solve problems: an engineering example. *Cybernetics and Systems*, 36(8):753–771, 2005.
28. Fang Gao, Gang Cui, and Hongwei Liu. Integration of genetic algorithm and cultural algorithms for constrained optimization. In Irwin King, Jun Wang, Lai-Wan Chan, and DeLiang Wang, editors, *Neural Information Processing, 13th International Conference, ICONIP 2006*, pages 817–825. Lecture Notes in Computer Science Vol. 4234, Hong Kong, China: Springer., October 3–6 2006. ISBN 978-3-540-46484-6.
29. Ricardo Landa Becerra and Carlos A. Coello Coello. Optimization with constraints using a cultured differential evolution approach. In Hans-Georg Beyer et al., editor, *Genetic and Evolutionary Computation Conference (GECCO'2005)*, Vol. 1, pages 27–34, Washington, DC: ACM Press, June 2005. ISBN 1-59593-010-8.
30. Ricardo Landa Becerra and Carlos A. Coello Coello. Cultured differential evolution for constrained optimization. *Computer Methods in Applied Mechanics and Engineering*, 195(33–36):4303–4322, July 1 2006.
31. Wanwan Tang and Yanda Li. Constrained optimization using triple spaces cultured genetic algorithm. In *Fourth International Conference on Natural Computation (ICNC 2008)*, Vol. 6, pages 589–593, October 18–20 2008. ISBN 978-0-7695-3304-9.
32. Trung Thanh Nguyen and Xin Yao. An experimental study of hybridizing cultural algorithms and local search. *International Journal of Neural Systems*, 18(1):1–17, February 2008.

33. P. N. Suganthan, N. Hansen, J. J. Liang, K. Deb, Y.-P. Chen, A. Auger, and S. Tiwari. Problem definitions and evaluation criteria for the CEC 2005 special session on real-parameter optimization. Technical Report, Nanyang Technological University, Singapore, May 2005.
34. Kaiqiao Yang, Kenjiro Maginu, and Hirosato Nomura. Cultural algorithm-based quantum-behaved particle swarm optimization. *International Journal of Computer Mathematics*, 87(10):2143–2157, August 2010.
35. Jun Sun, Wenbo Xu, and Bin Feng. A global search strategy of quantum-behaved particle swarm optimization. In *2004 IEEE Conference on Cybernetics and Intelligent Systems*, pages 111–116, Singapore: IEEE, 1–3 December 2004. ISBN 0-7803-8643-4.
36. J.J. Liang, A.K. Qin, P.N. Suganthan, and S. Baskar. Comprehensive learning particle swarm optimizer for global optimization of multimodal functions. *IEEE Transactions on Evolutionary Computation*, 10(3):281–295, June 2006.
37. Mostafa Ali, Robert Reynolds, Rose Ali, and Ayad Salhieh. Knowledge-based constrained function optimization using cultural algorithms with an enhanced social influence metaphor. In Kurosh Madani, Antonio Dourado Correia, Agostinho Rosa, and Joaquin Filipe, editors, *Computational Intelligence*, pages 103–119 Berlin: Springer, 2011.
38. Mostafa Z. Ali, Ayad Salhieh, Randa T. Abu Snanieh, and Robert G. Reynolds. Boosting cultural algorithms with an incongruous layered social fabric influence function. In *2011 IEEE Congress on Evolutionary Computation (CEC'2011)*, pages 1225–1232, New Orleans, LA: IEEE Service Center, 5–8 June 2011.
39. Zbigniew Michalewicz and Marc Schoenauer. Evolutionary algorithms for constrained parameter optimization problems. *Evolutionary Computation*, 4(1):1–32, 1996.
40. Noor H. Awad, Mostafa Z. Ali, and Rehab M. Duwairi. Cultural Algorithm with improved local search for optimization problems. In *2013 IEEE Congress on Evolutionary Computation (CEC'2013)*, pages 284–291, Cancún, México: IEEE, 20–23 June 2013.
41. Alain Pétrowski. A clearing procedure as a niching method for genetic algorithms. In *Proceedings of the 1996 IEEE International Conference on Evolutionary Computation (ICEC'96)*, pages 798–803, Nagoya, Japan: IEEE, 1996.
42. Mostafa Z. Ali and Robert G. Reynolds. Cultural algorithms: a Tabu search approach for the optimization of engineering design problems. *Soft Computing*, 18(8):1631–1644, 2014.
43. Fred Glover and Manuel Laguna. *Tabu Search*. Boston, MA: Kluwer Academic Publishers, 1997.
44. P. Siarry and G. Berthiau. Fitting of Tabu search to optimize functions of continuous variables. *International Journal for Numerical Methods in Engineering*, 40(13):2449–2457, July 1997.
45. Mahamed G.H. Omran. A novel cultural algorithm for real-parameter optimization. *International Journal of Computer Mathematics*, 93(9):1541–1563, 2016.
46. Bahriye Akay and Dervis Karaboga. A modified artificial bee colony algorithm for real-parameter optimization. *Information Sciences*, 192:120–142, 2012.
47. Mostafa Z. Ali, Noor H. Awad, Ponnuthurai N. Suganthan, Rehab M. Duwairi, and Robert G. Reynolds. A novel hybrid cultural algorithms framework with trajectory-based search for global numerical optimization. *Information Sciences*, 334:219–249, 2016.
48. Lin-Yu Tseng and Chun Chen. Multiple trajectory search for large scale global optimization. In *2008 IEEE Congress on Evolutionary Computation (CEC'2008)*, pages 3052–3059, Hong Kong, China: IEEE, 2008. ISBN 978-1-4244-1822-0.
49. Mostafa Z. Ali, Noor H. Awad, Ponnuthurai N. Suganthan, and Robert G. Reynolds. A modified cultural algorithm with a balanced performance for the differential evolution frameworks. *Knowledge-Based Systems*, 111:73–86, 2016.
50. Xiaodong Li, Ke Tang, Mohammad N. Omidvar, Zhenyu Yang, and Kai Qin. Benchmark functions for the CEC'2013 special session and competition on large-scale global optimization. Technical Report, Evolutionary Computation and Machine Learning Group, RMIT University, Australia, December 24, 2013.
51. Robert G. Reynolds and Mostafa Z. Ali. The social fabric approach as an approach to knowledge integration in Cultural Algorithms. In *2008 IEEE Congress on Evolutionary Computation (CEC'2008)*, pages 4200–4207, Hong Kong, China: IEEE, 1–6 June 2008. ISBN 978-1-4244-1822-0.
52. S. Das and P.N. Suganthan. Problem definitions and evaluation criteria for CEC 2011 competition on testing evolutionary algorithms on real world optimization problems. Technical report, Jadavpur University, India and Nanyang Technological University, Singapore, 2010.

53. Noor H. Awad, Mostafa Z. Ali, Ponnuthurai N. Suganthan, and Robert G. Reynolds. CADE: a hybridization of cultural algorithm and differential evolution for numerical optimization. *Information Sciences*, 378:215–241, February 1 2017.
54. Ramya Ravichandran. Cultural algorithm based on decomposition to solve optimization problems. Master's thesis, School of Computer Science, University of Windsor, Windsor, Ontario, Canada, 2019.
55. Leilei Cao, Lihong Xu, Erik D. Goodman, Shuwei Zhu, and Hui Li. A differential prediction model for evolutionary dynamic multiobjective optimization. In *2018 Genetic and Evolutionary Computation Conference (GECCO'2018)*, pages 601–608, Kyoto, Japan: ACM Press, July 15–19 2018. ISBN: 978-1-4503-5618-3.
56. Saleh Saleem and Robert Reynolds. Cultural algorithms in dynamic environments. In *Proceedings of the 2000 IEEE Congress on Evolutionary Computation (CEC'2000)*, pages 1513–1520, Piscataway, NJ: IEEE Service Center, July 2000.
57. Bin Peng, Robert G. Reynolds, and Jon Brewster. Cultural swarms. In *Proceedings of the 2003 IEEE Congress on Evolutionary Computation 2003 (CEC'2003)*, pages 1965–1971, Canberra, Australia: IEEE Service Center, 8–12 December 2003. ISBN 0-7803-7804-0.
58. James Kennedy and Russell C. Eberhart. *Swarm Intelligence.* San Francisco, CA: Morgan Kaufmann Publishers, 2001.
59. Bin Peng and Robert G. Reynolds. Cultural algorithms: knowledge learning in dynamic environments. In *Proceedings of the 2004 IEEE Congress on Evolutionary Computation (CEC'2004)*, pages 1751–1758, Portland, OR: IEEE, 19–23 June 2004. ISBN 0-7803-8515-2.
60. Yi Jiang, Wei Huang, and Li Chen. Cultural-based particle swarm optimization for dynamical environment. In *2009 International Symposium on Intelligent Ubiquitous Computing and Education*, pages 449–452, Chengdu, China: IEEE Computer Society Press, 15–16 May 2009. ISBN 978-0-7695-3619-4.
61. Moayed Daneshyari and Gary G. Yen. Dynamic optimization using cultural based PSO. In *2011 IEEE Congress of Evolutionary Computation (CEC'2011)*, pages 509–511, New Orleans, LA: IEEE, 5–8 June 2011. ISBN 978-1-4244-7834-7.
62. Hao Chen, Hua Wang, and Leqi Jiang. Path planning of UAV based on cultural algorithm in dynamic environments. In *2016 6th International Conference on Electronics Information and Emergency Communication (ICEIEC2'2016)*, pages 130–134, Beijing, China: IEEE, 17–19 June 2016. ISBN 978-1-5090-1998-4.
63. Sven Peyer, Dieter Rautenbach, and Jens Vygena. A generalization of Dijkstra's shortest path algorithm with applications to VLSI routing. *Journal of Discrete Algorithms*, 7(4):377–390, December 2009.
64. Leonard Kinnaird-Heether and Robert G. Reynolds. Deep social learning in dynamic environments using subcultures and auctions with cultural algorithms. In *2020 IEEE Congress on Evolutionary Computation (CEC'2020)*, Glasgow, UK: IEEE, 19–24 July 2020. ISBN 978-1-7281-6930-9.
65. Xiangdong Che, Mostafa Ali, and Robert G. Reynolds. Robust evolution optimization at the edge of chaos: commercialization of culture algorithms. In *2010 IEEE Congress on Evolutionary Computation (CEC'2010)*, Barcelona, Spain: IEEE, 18–23 July 2010. ISBN 978-1-4244-6909-3.
66. Robert G. Reynolds and Leonard Kinnaird-Heether. Optimization problem solving with auctions in cultural algorithms. *Memetic Computing*, 5(2):83–94, June 2013.
67. Faisal Waris and Robert G. Reynolds. Optimizing AI pipelines: a game-theoretic cultural algorithms approach. In *2018 IEEE Congress on Evolutionary Computation (CEC'2018)*, Rio de Janeiro, Brazil: IEEE, 8–13 July 2018. ISBN 978-1-5090-6018-4.
68. Carlos A. Coello Coello and Ricardo Landa Becerra. Evolutionary multiobjective optimization using a cultural algorithm. In *2003 IEEE Swarm Intelligence Symposium Proceedings*, pages 6–13, Indianapolis, IN: IEEE Service Center, April 2003.
69. Carlos A. Coello Coello, Gary B. Lamont, and David A. Van Veldhuizen. *Evolutionary Algorithms for Solving Multi-Objective Problems*, 2nd ed. New York: Springer, September 2007. ISBN 978-0-387-33254-3.
70. Wei Gu and Yonggang Wu. Application of multi-objective cultural algorithm in water resources optimization. In *2010 Asia-Pacific Power and Energy Engineering Conference*, Chengdu, China: IEEE, 28–31 March 2010. ISBN 978-1-4244-4812-8.

71. Ricardo Landa Becerra and Carlos A. Coello Coello. Solving Hard Multiobjective Optimization Problems Using ε-Constraint with Cultured Differential Evolution. In Thomas Philip Runarsson, Hans-Georg Beyer, Edmund Burke, Juan J. Merelo-Guervós, L. Darrell Whitley, and Xin Yao, editors, *Parallel Problem Solving from Nature – PPSN IX, 9th International Conference*, pages 543–552. Reykjavik, Iceland: Springer. Lecture Notes in Computer Science Vol. 4193, September 2006.
72. Kalyanmoy Deb, Lothar Thiele, Marco Laumanns, and Eckart Zitzler. Scalable test problems for evolutionary multiobjective optimization. In Ajith Abraham, Lakhmi Jain, and Robert Goldberg, editors, *Evolutionary Multiobjective Optimization. Theoretical Advances and Applications*, pages 105–145. London: Springer, 2005.
73. Simon Huband, Phil Hingston, Luigi Barone, and Lyndon While. A review of multiobjective test problems and a scalable test problem toolkit. *IEEE Transactions on Evolutionary Computation*, 10(5):477–506, October 2006.
74. Christopher Best. Multi-objective cultural algorithms. Master's thesis, Wayne State University, Detroit, Michigan, 2009.
75. Christopher Best, Xiangdong Che, Robert G. Reynolds, and Dapeng Liu. Multi-objective Cultural Algorithms. In *2010 IEEE Congress on Evolutionary Computation (CEC'2010)*, pages 3330–3338, Barcelona, Spain: IEEE Press, July 18–23 2010.
76. Moayed Daneshyari and Gary G. Yen. Cultural-based multiobjective particle swarm optimization. *IEEE Transactions on Systems, Man, and Cybernetics Part B–Cybernetics*, 41(2):553–567, April 2011.
77. Carlos A. Coello Coello, Gregorio Toscano Pulido, and Maximino Salazar Lechuga. Handling multiple objectives with particle swarm optimization. *IEEE Transactions on Evolutionary Computation*, 8(3):256–279, June 2004.
78. Robert Reynolds and Dapeng Liu. Multi-objective cultural algorithms. In *2011 IEEE Congress of Evolutionary Computation (CEC'2011)*, pages 1233–1241, New Orleans, LA: IEEE, 5–8 June, 2011. ISBN 978-1-4244-7834-7.
79. Eckart Zitzler, Kalyanmoy Deb, and Lothar Thiele. Comparison of multiobjective evolutionary algorithms: empirical results. *Evolutionary Computation*, 8(2):173–195, Summer 2000.
80. N. Srinivas and Kalyanmoy Deb. Multiobjective optimization using nondominated sorting in genetic algorithms. *Evolutionary Computation*, 2(3):221–248, Fall 1994.
81. Samuel Dustin Stanley, Khalid Kattan, and Robert Reynolds. CAPSO: a parallelized multiobjective cultural algorithm particle swarm optimizer. In *2019 IEEE Congress on Evolutionary Computation (CEC'2019)*, pages 3060–3069, Wellington, New Zealand: IEEE, 10–13 June 2019. ISBN 978-1-7281-2154-3.
82. J. David Schaffer. Multiple objective optimization with vector evaluated genetic algorithms. In *Genetic Algorithms and their Applications: Proceedings of the First International Conference on Genetic Algorithms*, pages 93–100, Hillsdale, NJ, Lawrence Erlbaum, 1985.
83. Khalid A. Kattan and Robert G. Reynolds. Using cultural algorithms to learn the impact of climate on local fishing behavior in Cerro Azul, Peru. In *2020 IEEE Congress on Evolutionary Computation (CEC'2020)*, Glasgow, UK: IEEE, 19–24 July 2020. ISBN 978-1-7281-6930-9.
84. Zhengyan Mao, Yue Xiang, Yijie Zhang, and Mandan Liu. A novel multi-objective cultural algorithm embedding five-element cycle optimization. In *2020 IEEE Congress on Evolutionary Computation (CEC'2020)*, Glasgow, UK: IEEE, 19–24 July 2020. ISBN 978-1-7281-6930-9.
85. Mandan Liu. Five-elements cycle optimization algorithm for solving continuous optimization problems. In *2017 IEEE 4th International Conference on Soft Computing & Machine Intelligence (ISCMI'2017)*, pages 75–79, Port Louis, Mauritius: IEEE, 23–24 November 2017. ISBN 978-1-5386-1315-3.
86. Qingfu Zhang and Hui Li. MOEA/D: A multiobjective evolutionary algorithm based on decomposition. *IEEE Transactions on Evolutionary Computation*, 11(6):712–731, December 2007.
87. Eckart Zitzler. Evolutionary algorithms for multiobjective optimization: methods and applications. PhD thesis, Swiss Federal Institute of Technology (ETH), Zurich, Switzerland, November 1999.
88. Carlos A. Coello Coello and Nareli Cruz Cortés. Solving multiobjective optimization problems using an artificial immune system. *Genetic Programming and Evolvable Machines*, 6(2):163–190, June 2005.

89. Mostafa A. El-Hosseini, Aboul Ella Hassanien, Ajith Abraham, and Hameed Al-Qaheri. Cultural-based genetic algorithm: design and real world applications. In *2008 Eighth International Conference on Intelligent Systems Design and Applications*, pages 488–493. IEEE Computer Society Press, 26–28 November 2008. ISBN 978-0-7695-3382-7.
90. N. Rychtyckyj, D. Ostrowski, G. Schleis, and R.G. Reynolds. Using cultural algorithms in industry. In *Proceedings of the 2003 IEEE Swarm Intelligence Symposium (SIS'03)*, pages 187–192, Indianapolis, IN: IEEE, April 2003. ISBN 0-7803-7914-4.
91. Arman Goudarzi, Afshin Ahmadi, Andrew G Swanson, and John Van Coller. Non-convex optimisation of combined environmental economic dispatch through cultural algorithm with the consideration of the physical constraints of generating units and price penalty factors. *SAIEE Africa Research Journal*, 107(3):146–166, September 2016.
92. Leandro dos Santos Coelho, Rodrigo Clemente Thom Souza, and Viviana Cocco Mariani. Improved differential evolution approach based on cultural algorithm and diversity measure applied to solve economic load dispatch problems. *Mathematics and Computers in Simulation*, 79(10):3136–3147, June 2009.
93. Bidishna Bhattacharya, Kamal Mandal, and Niladri Chakraborty. A multiobjective optimization based on cultural algorithm for economic dispatch with environmental constraints. *International Journal of Scientific & Engineering Research*, 3(6):1–9, June 2008.
94. Tianyu Liu, Licheng Jiao, Wenping Ma, Jingjing Ma, and Ronghua Shang. Cultural quantum-behaved particle swarm optimization for environmental/economic dispatch. *Applied Soft Computing*, 48:597–611, November 2016.
95. Richard Gonçalves, Carolina Almeida, Marco Goldbarg, Elizabeth Goldbarg, and Myriam Delgado. Improved cultural immune systems to solve the economic load dispatch problems. In *2013 IEEE Congress on Evolutionary Computation (CEC'2013)*, pages 621–628, Cancún, México: IEEE, 20–23 June 2013. ISBN 978-1-4799-0453-2.
96. Leandro dos Santos Coelho and Viviana Cocco Mariani. An efficient cultural self-organizing migrating strategy for economic dispatch optimization with valve-point effect. *Energy Conversion and Management*, 51(12):2580–2587, December 2010.
97. Rui Zhang, Jianzhong Zhou, Li Mo, Shuo Ouyang, and Xiang Liao. Economic environmental dispatch using an enhanced multi-objective cultural algorithm. *Electric Power Systems Research*, 99:18–29, June 2013.
98. Youlin Lu, Jianzhong Zhou, Hui Qin, Ying Wang, and Yongchuan Zhang. A hybrid multi-objective cultural algorithm for short-term environmental/economic hydrothermal scheduling. *Energy Conversion and Management*, 52(5):2121–2134, May 2011.
99. Kalyanmoy Deb, Amrit Pratap, Sameer Agarwal, and T. Meyarivan. A fast and elitist multiobjective genetic algorithm: NSGA–II. *IEEE Transactions on Evolutionary Computation*, 6(2):182–197, April 2002.
100. Xin Ma. Analysis on optimal operation of hydropower station based on cultural particle swarm optimization algorithm. In *2010 2nd International Symposium on Information Engineering and Electronic Commerce*, Ternopil, Ukraine: IEEE, 23–25 July 2010. ISBN 978-1-4244-6972-7.
101. C. Veeramani, Joseph Prabhakar Williams, and Punem Ramadevi. Evaluation of wind energy parameter optimization of A DFIG controller based on cultural algorithms. In *2018 International Conference on Communication and Signal Processing (ICCSP'2018)*, Chennai, India: IEEE, April 3–5 2018. ISBN 978-1-5386-3522-3.
102. Leandro dos Santos Coelho and Viviana Cocco Mariani. An efficient particle swarm optimization approach based on cultural algorithm applied to mechanical design. In *2006 IEEE International Conference on Evolutionary Computation (CEC'2006)*, pages 1099–1104, Vancouver, Canada: IEEE, July 16–21 2006. ISBN 0-7803-9487-9.
103. Carlos A. Coello Coello and Ricardo Landa Becerra. Efficient evolutionary optimization through the use of a cultural algorithm. *Engineering Optimization*, 36(2):219–236, April 2004.
104. Shahin Jalili and Yousef Hosseinzadeh. A cultural algorithm for optimal design of truss structures. *Latin American Journal of Solids and Structures*, 12:1721–1747, 2015.
105. Xingwei Wang, Wenqian Hao, and Qiming Li. An adaptive cultural algorithm with improved quantum-behaved particle swarm optimization for sonar image detection. *Scientific Reports*, 7: 17733, December 18 2017.

106. Xuesong Yan, Tao Song, and Qinghua Wu. An improved cultural algorithm and its application in image matching. *Multimedia Tools and Applications*, 76(13):14951–14968, July 2017.
107. Feng Tan and Shenyuan Yang. Application of cultural algorithm to infrared image segmentation. In *2008 Second International Symposium on Intelligent Information Technology Application*, pages 306–310, Shanghai, China: IEEE Computer Society Press, 20–22 December 2008.
108. Zhaohui Cai, Min Zhang, and Yujiao Liu. Sea-surface weak target detection scheme using a cultural algorithm aided time-frequency fusion strategy. *IET Radar, Sonar & Navigation*, 12(7):711–720, July 2018.
109. Carlos Soza, Ricardo Landa Becerra, María Cristina Riff, and Carlos A. Coello Coello. Solving timetabling problems using a cultural algorithm. *Applied Soft Computing*, 11(1):337–344, January 2011.
110. Seyed Ziae Mousavi Mojab, Mahdi Ebrahimi, Robert Reynolds, and Shiyong Lu. iCATS: scheduling big data workflows in the cloud using cultural algorithms. In *2019 IEEE Fifth International Conference on Big Data Computing Service and Applications (BigDataService)*, pages 99–106, Newark, CA: IEEE Computer Society Press, 4–9 April 2019. ISBN 978-1-7281-0060-9.
111. Michael Sternberg and Robert G. Reynolds. Using cultural algorithms to support re-engineering of rule-based expert systems in dynamic performance environments: a case study in fraud detection. *IEEE Transactions on Evolutionary Computation*, 1(4):225–243, November 1997.
112. David A. Ostrowski, Troy Tassier, Mark Everson, and Robert G. Reynolds. Using cultural algorithms to evolve strategies in agent-based models. In *Proceedings of the 2002 IEEE Congress on Evolutionary Computation (CEC'2002)*, pages 741–746, Honolulu, HI: IEEE, 12–17 May 2002. ISBN 0-7803-7282-4.
113. H.C. Kuo and C.H. Lin. Cultural evolution algorithm for global optimizations and its applications. *Journal of Applied Research and Technology*, 11(4):510–522, August 2013.
114. Julian H. Steward. *Theory of Culture Change: The Methodology of Multilinear Evolution*. Urbana, Illinois: University of Illinois Press, 1990. ISBN 978-0252002953.
115. Jianqiang Dong and Bo Yuan. GPU-accelerated standard and multi-population cultural algorithms. In *2013 International Conference on Service Sciences (ICSS'2013)*, pages 129–133, Shenzhen, China: IEEE, 11–13 April 2013. ISBN 978-1-4673-6258-0.
116. Amine Kechid and Habiba Drias. Cultural coalitions detection approach using GPU based on hybrid Bat and Cultural Algorithms. *Applied Soft Computing*, 93: 106368, August 2020.
117. Hui Min Ma, Chun Ming Ye, and Shuang Zhang. Research on parallel particle swarm optimization algorithm based on cultural evolution for the multi-level capacitated lot-sizing problem. In *2008 Chinese Control and Decision Conference*, pages 965–970, Yantai, Shandong, China: IEEE, 2008 (in Chinese).
118. Saúl Zapotecas-Martnez, Carlos A. Coello Coello, Hernán E. Aguirre, and Kiyoshi Tanaka. A review of features and limitations of existing scalable multi-objective test suites. *IEEE Transactions on Evolutionary Computation*, 23(1):130–142, February 2019.
119. Carlos A. Coello Coello, Silvia González Brambila, Josué Figueroa Gamboa, Ma Guadalupe Castillo Tapia, and Raquel Hernández Gómez. Evolutionary multiobjective optimization: open research areas and some challenges lying ahead. *Complex & Intelligent Systems*, 6:221–236, July 2020.
120. Jesús Guillermo Falcón-Cardona and Carlos A. Coello Coello. Indicator-based multi-objective evolutionary algorithms: a comprehensive survey. *ACM Computing Surveys*, 53(2):29, March 2020.
121. Anupam Trivedi, Dipti Srinivasan, Krishnendu Sanyal, and Abhiroop Ghosh. A survey of multi-objective evolutionary algorithms based on decomposition. *IEEE Transactions on Evolutionary Computation*, 21(3):440–462, June 2017.
122. Carlo Raquel and Xin Yao. Dynamic multi-objective optimization: a survey of the state-of-the-art. In Shengxiang Yang and Xin Yao, editors, *Evolutionary Computation for Dynamic Optimization Problems*, chapter 4, pages 85–106. Berlin: Springer-Verlag, 2013. ISBN 978-3-642-38415-8.
123. John R. Koza. *Genetic Programming. on the Programming of Computers by Means of Natural Selection*. Cambridge, MA: The MIT Press, 1992.

9 Application of Teaching-Learning-Based Optimization on Solving of Time Cost Optimization Problems

Vedat Toğan, Tayfun Dede, and Hasan Basri Başağa

CONTENTS

9.1 INTRODUCTION

Technological development in various fields increases the demand for the associated resources. Since these resources are often limited in nature, the efficient use of these resources has long been a question of great interest in a wide range of fields. To satisfy the demand for resource efficiency, the decisions are to be made among the various alternatives, influenced by several factors. To assist this decision-making procedure, the trial-and-error process was initially adopted. However, it was very time-consuming and proved to be cumbersome when dealing with large size system repetitions. Although early computers have formerly offered a relative solution, a satisfactory answer for the boring trial-and-error process was not found until recent developments in computer devices, together with associated advances in numerical analysis and computer-based methods. On the adoption of these advances, various decision-making models were introduced to tackle the trial-and-error-related drawbacks. Supporting the decision-making process, in the present study the mathematical, heuristic and metaheuristic-based methods are presented in chronological order. One seeks the best alternative obeying the problem's constraints through these models (optimization problem). Typically, in minimization problems, the smaller the problem's objective, the lesser the amount of resource will be.

Opposed to mathematical and heuristic models, metaheuristic-based methods have been increasingly designed by many scholars depending on the development of computational environments for over three decades. The growing interest in this area of research is due to the intrinsic ability of metaheuristic-based methods in simulation of natural phenomena, as well as its ability in achieving the problem's optimal or near-optimal solution (the best alternative), using less computational effort (Arora 2004). On the other hand, both mathematical approaches and heuristic methods, though offering good quality results, have weakness expressed, respectively, such as requiring a high computational cost (Chassiakos and Sakellaropoulos 2005) and not guaranteeing the detection of the optimal solution (Shannon and Lucko 2012). Due to the imitation of many events and according to the feature of the mimicked event, metaheuristic-based methods can be divided into various sub-classes called bio-inspired (genetic algorithm (GA; Holland 1975), evolution strategy (ES; Hansen et al. 2003), differential evolution (DE; Storn and Price 1997)), physics and chemistry based (gravitational search algorithm (GSA; Rashedi et al. 2009), charged system search (CSS; Kaveh and Talatahari 2010)), socio-inspired (particle swarm optimization (PSO; Eberhart and Kennedy 1995), ant colony optimization (ACO; Dorigo et al. 2006), cuckoo search (CS; Gandomi et al. 2013), gray wolf optimizer (GWO; Mirjalili et al. 2014), imperialist competitive algorithm (ICA; Atashpaz-Gargari and Lucas 2007)), and social group optimization (SGO; Satapathy and Naik 2016). In addition, hybrid methods are combining some advantages of each of two or more previously proposed metaheuristic-based methods in a purposed model (Cheng et al. 2016; Mlakar et al. 2016; Mortazavi et al. 2018; Mortazavi 2019). It is possible to find different forms of the classifications presented earlier in the paragraph in the literature. For example, Mohamed et al. (2020) introduced the relevant classification as evolutionary, swarm intelligence, physics-based, and human-related techniques. It is seen that the difference between these two classifications is the main classification titles of the mentioned methods.

Teaching-learning-based optimization (TLBO) was proposed as a novel metaheuristic method for solving constrained mechanical design optimization problems (Rao et al. 2011). TLBO is inspired by an interaction realized between teacher and student in a class. In light of this interaction, the teacher, as the most knowledgeable person, conveys the information associated with any subject to students, who lack the corresponding subject knowledge. This is achieved using some teaching arguments, such as discussing, solving, giving example(s), or figure(s) presentation (Figure 9.1). In addition, students can increase their knowledge level through communication among themselves or interaction with people other than their teacher. Moreover, they can use other information sources such as books, libraries, and magazines to increase their knowledge. Thence, Rao et al. (2011) imitated these processes under two different phases called teaching and

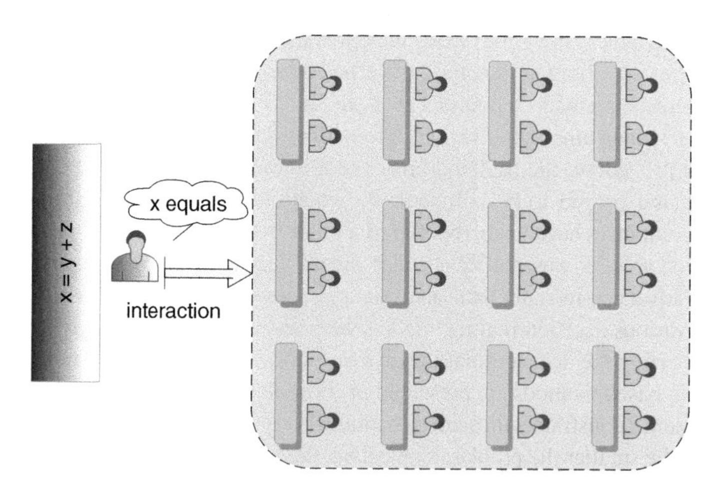

FIGURE 9.1 Teacher conveys the information.

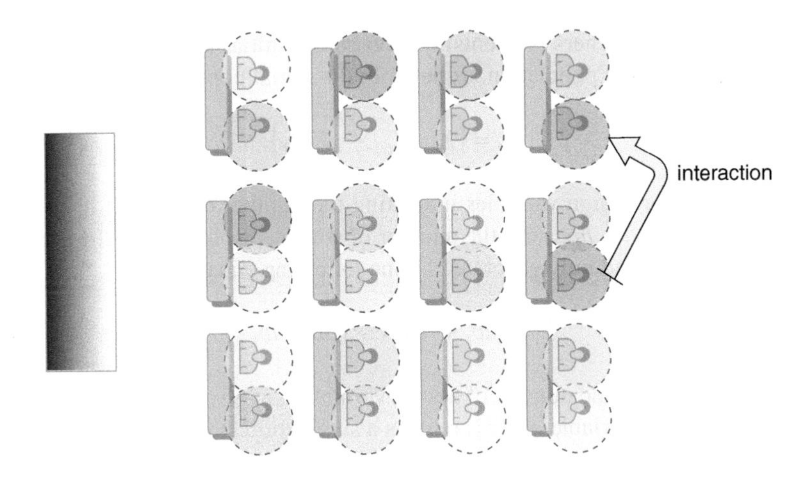

FIGURE 9.2 Interaction (randomly) between the students.

learning phases. They simulated the effect of a teacher on learners in the first phase, while only interaction between the students was imitated at the second stage (Figure 9.2).

Rao et al. (2011) explored TLBO performance on solving design optimization problems, which includes several mechanical design benchmark problems along with real-world applications. On the introduction of TLBO as a new efficient metaheuristic method, it was used for the optimization of two-dimensional (2D) and three-dimensional (3D) structural systems of trusses (Değertekin and Hayalioğlu 2013; Toğan 2013; Camp and Farshchin 2014; Dede 2014; Baghlani et al. 2017) and frames (Toğan 2012; Daloğlu et al. 2018; Shallan et al. 2018). TLBO was also utilized to optimize the structure's geometry and size (Baghlani and Makiabadi 2013; Haghpanah and Foroughi 2017). Besides, structural systems were optimized by using TLBO under natural frequencies constraints to prevent or reduce the destructive effects of dynamic loads on structures (Dede and Toğan 2015; Farshchin et al. 2016; Savsani et al. 2016; Grzywinski et al. 2019: Mishra et al. 2020). The mentioned studies executed TLBO on literature problems further solved generally for comparison and verification of the achieved results. Since there is no metaheuristic-based method performing well for all types of optimization problems, its efficiency needs to be demonstrated on various optimization problems. Therefore, in literature TLBO was frequently applied to demonstrate its efficiency and applicability within different kinds of optimization problems, i.e., optimization of retaining walls (Temur and Bekdaş 2016; Öztürk et al. 2020), optimization of grillage systems (Dede 2013), optimization of laminated plates (Topal et al. 2017; Das et al. 2018; Vosoughi et al. 2018), optimization of water demand (Kumar and Yadav 2018; Chen et al. 2019), electricity (Uzlu et al. 2014; Li et al. 2018), and optimization of time cost in the field of construction management and execution (Eirgash et al. 2019; Toğan and Eirgash 2019a, b).

This chapter discusses TLBO with detailed elaboration on its theoretical development, while illustrating TLBO through numerical examples associated with minimization of time and cost of construction projects, under the category of a multi-objective optimization (MOO) problem.

9.2 TECHNICAL BACKGROUND

9.2.1 Teaching-Learning-Based Optimization (TLBO)

Since TLBO is a population-based optimization algorithm, similar to other metaheuristic-based methods, its initialization begins with a random generation of group of learners (students). Thus, the associated group of learners (students) composes the initial population of the optimization process. So, a student in the population, as being in the initial population, is represented as:

$$\mathbf{X}_i = \{X_{i,1}, X_{i,2}, \cdots, X_{i,nd}\} \tag{9.1}$$

where $\mathbf{X}_i$ is a vector for design variables indicating the number of subjects to be taught to students. Likewise, $i = 1, 2, \ldots, np$, np is the number of students (solutions) in the population. Each design variable in $\mathbf{X}_i$ is randomly generated by the following equation.

$$X_{i,k} = lb_k + rand() \times (ub_k - lb_k) \tag{9.2}$$

where $k = 1,2, \ldots, nd$. nd is the number of design variables and lb and ub are the lower and upper bounds of the kth design variable. *rand ()* returns a single uniformly distributed random number in the interval (0,1). After evaluating the solutions, both teacher and learner phases of TLBO are initiated, respectively.

In the teacher phase of TLBO, the aim is to improve the knowledge of a group of learners. To this end, the main step of this phase is to detect the teacher among the solutions in the population. So, the solution with the best objective function value $F(\mathbf{X}_i)$ is accepted as a teacher. $F(\mathbf{X}_i)$ refers to the minimum value for the minimization problem. Depending on the teacher and the mean value of a particular subject, the updating equation for a design variable in a solution is described as given in equation below.

$$X_{i,k}^{new} = X_{i,k} + rand() \times (X_{teacher,k} - TF \times X_{mean,k}) \tag{9.3}$$

where $X_{teacher,k}$ is the kth the particular subject value for teacher in the population and *rand ()* is a random number in the range [0, 1]. Also, TF $(= round[1 + rand(0,1)])$ is a teaching factor. $X_{mean,k} = (1/np)(\Sigma X_{i,k})$ is the mean value of the particular subject for the solutions in the population. Repetition of the above updating concept for all design variables produces a new design variable vector X_i^{new} for the solution i. If $F(\mathbf{X}_i^{new}) < F(\mathbf{X}_i)$ replace $F(\mathbf{X}_i)$ with $F(\mathbf{X}_i^{new})$; otherwise, preserve $F(\mathbf{X}_i)$. The teacher phase is executed for each solution ($i = 1, 2, \ldots, np$) in the population to upgrade the knowledge through the same manner, as expressed above.

In the learner phase of TLBO, solutions in the population are updated by application of TLBO's teacher phase in a random manner, interacting with the other solution of the current population to enhance the knowledge through mutual interaction between them. Solution $\mathbf{X}_i$ is compared with solution $\mathbf{X}_j$ ($i \neq j$), and a new solution for $\mathbf{X}_i$ is updated according to the equations below:

$$\begin{gathered} if\ \left(F(\mathbf{X}_i) < F(\mathbf{X}_j)\right) \\ \mathbf{X}_i^{new} = \mathbf{X}_i + rand() \times (\mathbf{X}_i - \mathbf{X}_j) \\ else \\ \mathbf{X}_i^{new} = \mathbf{X}_i + rand() \times (\mathbf{X}_j - \mathbf{X}_i) \end{gathered} \tag{9.4}$$

where $F(\mathbf{X}_i)$ and $F(\mathbf{X}_j)$ are objective function values of solutions i and j, respectively. Afterward, a comparison is made between $F(\mathbf{X}_i^{new})$ and $F(\mathbf{X}_i)$, and then $\mathbf{X}_i^{new}$ is accepted instead of $\mathbf{X}_i$ as a solution to be added in the current population, if $F(\mathbf{X}_i^{new}) < F(\mathbf{X}_i)$. Again, this phase is continued for other solutions of i (=1, 2, …, np) to acquire knowledge communicating with another solution. Completing the learner phase of TLBO, a cycle (iteration) for TLBO process terminates. On starting the teacher phase, all steps detailed above are repeated for population updated through two phases of TLBO, until reaching a predefined number denoted as the maximum number of iterations.

Before presenting the pseudocode of TLBO (see Algorithm 9.1), to increase understandability, terminology matching will be done. Common terminologies used in an optimization problem are population, solution(s) or individual(s), the best solution, design variable(s), and objective function. However, in the language of the TLBO procedure, the terminologies are class, teacher, learner(s) or student(s), subject(s), and grade. Hence, the matching is also as follows: population = class, solution(s) = student(s), best solution = teacher, design variable(s) = subject(s), and objective function = grade.

ALGORITHM 9.1 PSEUDOCODE FOR TEACHING-LEARNING-BASED OPTIMIZATION

Begin

Set the population number as *np*, design variables as *nd*, lower and upper boundaries of design variables as *lb* and *ub*, and maximum number of iteration as it_{max}

Randomly initialize the population which is a matrix of dimension $np \times nd$ and evaluate the objective function value for each solution $\mathbf{X}_i$

Estimate $\mathbf{X}_{mean} = \sum_{i=1}^{np} \mathbf{X}_i / np$

Determine the best solution as called teacher as $\mathbf{X}_{teacher}$ from the whole population

While ($it < it_{max}$)

$it = it + 1;$

Vector_Of_Difference = ($\mathbf{X}_{teacher} - TF \times \mathbf{X}_{mean}$)

For $i = 1$ to np

For $k = 1$ to nd

$X_{i,k}^{new} = X_{i,k}^{old} + rand() \times \text{Vector\_Of\_Difference}(k)$

End for

Evaluate the objective function value of updated solution $\mathbf{X}_i^{new}$

if $F(\mathbf{X}_i^{new}) < F(\mathbf{X}_i^{old})$

$\mathbf{X}_i^{old} = \mathbf{X}_i^{new}$

End if

End for

For $i = 1$ to np

Randomly select the other solution $\mathbf{X}_j$ where ($i \neq j$) from the whole population

If $F(\mathbf{X}_i^{new}) < F(\mathbf{X}_i^{old})$

$\mathbf{X}_i^{new} = \mathbf{X}_i^{old} + rand() \times (\mathbf{X}_i - \mathbf{X}_j)$

Else

$\mathbf{X}_i^{new} = \mathbf{X}_i^{old} + rand() \times (\mathbf{X}_j - \mathbf{X}_i)$

End if

Evaluate the objective function value of updated solution $\mathbf{X}_i^{new}$

if $F(\mathbf{X}_i^{new}) < F(\mathbf{X}_i^{old})$

$\mathbf{X}_i^{old} = \mathbf{X}_i^{new}$

End if

End for

Determine the best solution in the whole population

End while

End

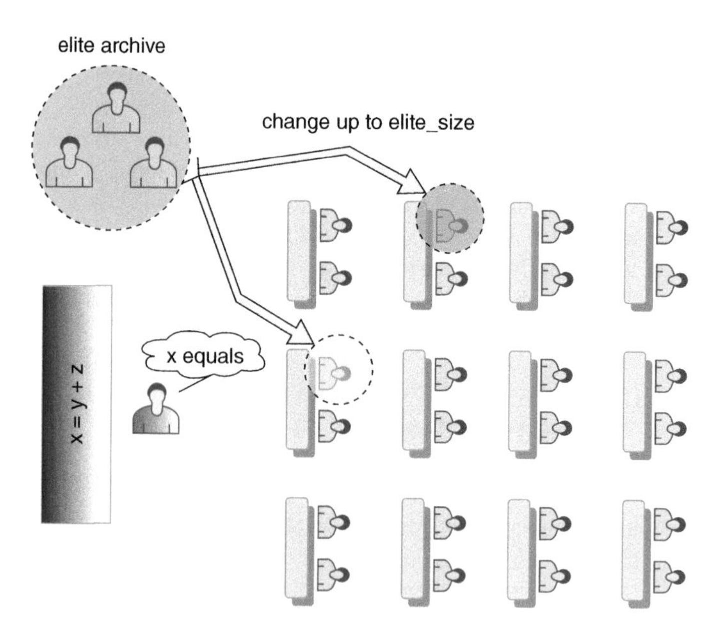

FIGURE 9.3 Elitism in Teaching-Learning-Based Optimization.

9.2.2 Elitist Teaching-Learning-Based Optimization (eTLBO)

Some minor or major improvements for the proposed metaheuristic methods were generally performed to enhance the method's exploration and exploitation capabilities. Elitism is such an improvement method, and it can be used in metaheuristics, i.e., GA (Goldberg 1989), wherein an iteration of the elite solutions (means good solutions) is replaced by the worst solutions (Figure 9.3). To empower the TLBO process, Rao and Patel (2012) implemented the concept of elitism in TLBO. After introduction of elitism, on detection of the identical solutions, the duplicate elimination step is to be executed using mutation to vary these solutions from each other. So, any design variables randomly selected in duplicate solutions are to be modified through the mutation of remaining within the bound of the variable's limits. Rao and Patel (2012) highlight the capability of the elitism strategy by enhancing the obtained results compared with its non-existence scenario. To control the effect of elitism, consideration of TLBO elite size, *elit_size,* is also to be adopted. This refers to the number of worst solutions to be updated in the current population. The pseudocode of the elitist TLBO (eTLBO) algorithm is illustrated in Algorithm 9.2.

ALGORITHM 9.2 PSEUDOCODE FOR eTLBO

Begin

Set the population number as *np*, design variables as *nd*, lower and upper boundaries of design variables as *lb* and *ub*, elite size number as *elite_size*, and maximum number of iteration as it_{max}

Randomly initialize the population which is a matrix of dimension $np \times nd$ and evaluate the objective function value for each solution $\mathbf{X}_i$

Estimate $\mathbf{X}_{mean} = \sum_{i=1}^{np} \mathbf{X}_i / np$

Determine *elite_size* best solutions and assign the first place solution as teacher $\mathbf{X}_{teacher}$ among the whole elite solutions
While ($it < it_{max}$)
carry out procedures given in Algorithm 9.1
For $i = 1$ to *elite_size*
replace the worst solutions with elitist ones
End for
For $i= 1$ to *np*
If $\mathbf{X}_i = \mathbf{X}_j$
select a random variable position for $\mathbf{X}_j$, and change it with a new one to be created within the
boundaries of that variable position
End if
End for
End while
End

9.2.3 Modified Teaching-Learning-Based Optimization (mTLBO)

In conventional TLBO, the teacher attempts to raise the student's knowledge, while considering two extreme situations using *TF*. These are known as the cause of a slower convergence rate for the optimization problems. Thus, Rao and Patel (2013) proposed refinement procedures, improving the search process as well as the convergence rate of traditional TLBO. To this end, they introduced teacher numbers and adaptive *TF* as new modifications of the basic TLBO. So multi-teachers are used to reach stability between the teacher and the students for knowledge transferring purposes. In the absence of such a modification, there might not be any results of teacher effort due to the presence of below average students. The students are grouped according to their knowledge levels and they are assigned to individual teachers who are compatible with their knowledge level (Figure 9.4). The implementation step of this modification is called the number of teachers in Rao and Patel (2013), which is given as follows:

$$\mathbf{X}_{1,teacher} = \left\{\mathbf{X}_i \mid min\ F\left(\mathbf{X}_i\right)\right\} \tag{9.5}$$

where $\mathbf{X}_i$ is the best solution vector in the entire population and $\mathbf{X}_{1,teacher}$ is the first teacher (chief teacher) to be assigned to the solutions depending on those objectives function values. Based on the chief teacher, the other teachers (*tn*) are determined as:

$$\mathbf{X}_{s,teacher} = \mathbf{X}_{1,teacher} - rand() \times \mathbf{X}_{1,teacher} \tag{9.6}$$

where $s = 1, 2, \ldots, tn$. *tn* is the teacher numbers adopted in the modified TLBO (mTLBO) process. It is sorting teachers in the ascending order of their objective function $F(.)$ values. According to the $F(.)$ values of the solutions, deliver the solutions to the teachers using:

$$If\ F\left(\mathbf{X}_{s,teacher}\right) \le F\left(\mathbf{X}_i\right) < F\left(\mathbf{X}_{s+1,teacher}\right)$$

$$\mathbf{X}_i \xrightarrow{assign\ to} teacher\ 's' \tag{9.7}$$

$$Else$$

$$\mathbf{X}_i \xrightarrow{assign\ to} teacher\ 's+1'$$

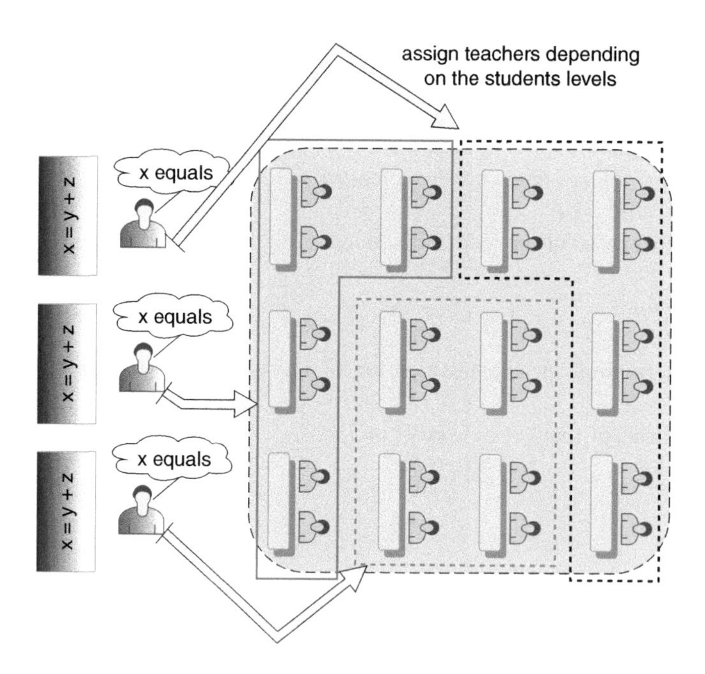

FIGURE 9.4 Multi-teacher in TLBO.

where $i = 1, 2, \ldots, np$. $s = 1, 2, \ldots, tn$-1.

Adaptive *TF* intends to give a chance of learning in any proportion from the teacher to the students, in contrast to *TF* in the classical TLBO, in which students either learn nothing or learn everything from the teacher (Rao and Patel 2013). The automatic verification of *TF* in the mTLBO process is presented below:

$$\mathbf{TF}_{s,k} = X_{s,mean,k} / X\_new_{s,k} \tag{9.8}$$

where $s = 1, 2, \ldots, tn$. $X_{s,\,mean,k} = (1/snp)(\Sigma X_{i,k})$ is the mean value of the particular subject for the solutions in the corresponding subgroup. Here, *snp* is the number of solutions in that "*s*" subgroup. $X\_new_{s,k}(= X_{s,teacher,k})$ is equal to the particular subject value of the teacher of that group. Contrary *TF* in the basic TLBO, where $\mathbf{TF}_s$ is a vector for teaching factor in mTLBO. Accordingly, the pseudocode of mTLBO is illustrated in Algorithm 9.3

ALGORITHM 9.3 PSEUDOCODE FOR MODIFIED TEACHING-LEARNING-BASED OPTIMIZATION

Begin

Set the population number as *np*, design variables as *nd*, lower and upper boundaries of design variables as *lb* and *ub*, number of teachers as *tn* and maximum number of iteration as it_{max}

Randomly initialize the population which is a matrix of dimension $np \times nd$ and evaluate the objective function value for each solution $\mathbf{X}_i$

While ($it < it_{max}$)

$it = it + 1$;

Determine the best solution as called teacher as $\mathbf{X}_{teacher}$ from the whole population and assign it as $\mathbf{X}_{1,teacher}$

For $s = 2$ to tn

$\mathbf{X}_{s,teacher} = \mathbf{X}_{1,teacher} - \text{rand}()\mathbf{X}_{1,teacher}$

Evaluate the objective function value of $\mathbf{X}_{s,teacher}$

End for

Sort the teachers in the ascending order of $F(\mathbf{X}_{s,teacher})$

For $i = 1$ to np

if $F(\mathbf{X}_{1,teacher}) \leq F(\mathbf{X}_i) < F(\mathbf{X}_{2,teacher})$

Assign the solution $\mathbf{X}_i$ to teacher 1

Else if $F(\mathbf{X}_{2,teacher}) \leq F(\mathbf{X}_i) < F(\mathbf{X}_{3,teacher})$

Assign the solution $\mathbf{X}_i$ to teacher 2

⋮

Else if $F(\mathbf{X}_{s-1,teacher}) \leq F(\mathbf{X}_i) < F(\mathbf{X}_{s,teacher})$

Assign the solution $\mathbf{X}_i$ to teacher s-1

Else

Assign the solution $\mathbf{X}_i$ to teacher s

End if

End for

Calculate mean of each group of solutions $\mathbf{X}_{s,mean} = \sum_{i=1}^{snp} \mathbf{X}_i \,/\, snp$

Assign $\mathbf{X\_new}_{s,teacher} = \mathbf{X}_{s,teacher}$ for each group of solutions

$\mathbf{TF}_s = \mathbf{X}_{s,mean} \,/\, \mathbf{X\_new}_{s,teacher}$

Vector_Of_Difference$_s$ = $(\mathbf{X\_new}_{s,teacher} - TF_s \times \mathbf{X}_{s,mean})$

Carry out the teacher and learning phases presented in Algorithm 9.1, respectively

End while

End

9.3 TIME COST OPTIMIZATION PROBLEM

As discussed earlier, this chapter discusses the application of TLBO on solving the time cost optimization (TCO) problem encountered in the field of construction management and execution. Since a TCO problem demands simultaneous time and cost minimizations, it is denoted as a MOO problem. For the solution of a MOO problem, there are many solution options, none of which are preferred over the other. In other words, there is no single solution that can be referred as the best solution for the MOO problem. So, each of these solutions is a trade-off between the measured conflicting objectives. Therefore, these solutions are forming an optimal set of solutions, whereby the related set is called a Pareto set when dealing with MOO problems. Hence, the first objective is to minimize the duration of a construction project, which tends to be implemented using different kinds of resources, such as material, man, and technology, as given by Eq. 9.9.

$$Z_1 = min.\ PD\left(\sum_{r=1}^{na} T_r^{op}\right) \tag{9.9}$$

where T_r^{op} is the duration of activity (r) on the critical path, carried out with execution option (op), and na is the number of activities in the undertaken project. The critical path method (CPM; Shaffer et al. 1965) is applied to find the critical path in the project network and the activity time.

The second objective in the TCO problem is to minimize the total cost of the project handled as given by Eq. 9.10.

$$Z_2 = min.\ PC\left(\sum_{r=1}^{na}\left[DC_r^{op} + \left(T_r^{op} + CR_r^{op}\right)\right]\right) \tag{9.10}$$

where DC_r^{op} is the direct cost of the activity (r) using resource utilization (op) and CR_r^{op} is the daily cost rate in \$/day using resource utilization.

The above expressed single objectives are together handled for the MOO problem with the help of an approach proposed previously by Zheng et al. (2004). The fitness formula for MOO is presented by the following equation (Zheng et al. 2004):

$$min.\ Z = w_1\left(\frac{Z_{1,max} - Z_1 + \varepsilon}{Z_{1,max} - Z_{1,min} + \varepsilon}\right) + w_2\left(\frac{Z_{2,max} - Z_2 + \varepsilon}{Z_{2,max} - Z_{2,min} + \varepsilon}\right) \tag{9.11}$$

where $Z_{1,max}$, $Z_{2,max}$, and $Z_{1,min}$, $Z_{2,min}$ are, respectively, maximum and minimum values of objective functions Z_1 and Z_2. Also, ε is a positive small number adopted to avoid a zero-division error. Additionally, Z is the fitness value of the ith solution in the current population. Weights w_1 and w_2 are formulated in an adaptive manner by the following equations (Zheng et al. 2004).

$$if\ Z_{1,max} \neq Z_{1,min}\ and\ Z_{2,max} \neq Z_{2,min} \tag{9.12}$$

$$\vartheta_1 = Z_{1,min} / \left(Z_{1,max} - Z_{1,min}\right)$$

$$\vartheta_2 = Z_{2,min} / \left(Z_{2,max} - Z_{2,min}\right)$$

$$w_1 = \vartheta_1 / \left(\vartheta_1 + \vartheta_2\right);\ w_2 = \vartheta_2 / \left(\vartheta_1 + \vartheta_2\right)$$

$$Else\ if\ Z_{1,max} = Z_{1,min}\ and\ Z_{2,max} = Z_{2,min} \tag{9.13}$$

$$w_1 = w_2 = 0.5$$

$$Else\ if\ Z_{1,max} = Z_{1,min}\ and\ Z_{2,max} \neq Z_{2,min} \tag{9.14}$$

$$w_1 = 0.9;\ w_2 = 0.1$$

$$Else\ if\ Z_{1,max} \neq Z_{1,min}\ and\ Z_{2,max} = Z_{2,min} \tag{9.15}$$

$$w_1 = 0.1;\ w_2 = 0.9$$

where ϑ_1 and ϑ_2 are the values of first and second objective criterion, respectively, and w_1 and w_2 are the adaptive weights for the first and second objectives of the optimization problem, respectively.

9.4 EXAMPLES

Two computational experiments are illustrated, demonstrating the application of TLBO and its variants on TCO type problems. Both tests were conducted with a computer having Intel Core i5-2.70 GHz CPU 8GB RAM. Also, for this study's cases, 10 independent runs were performed in accordance with the stochastic nature of TLBO and its variants.

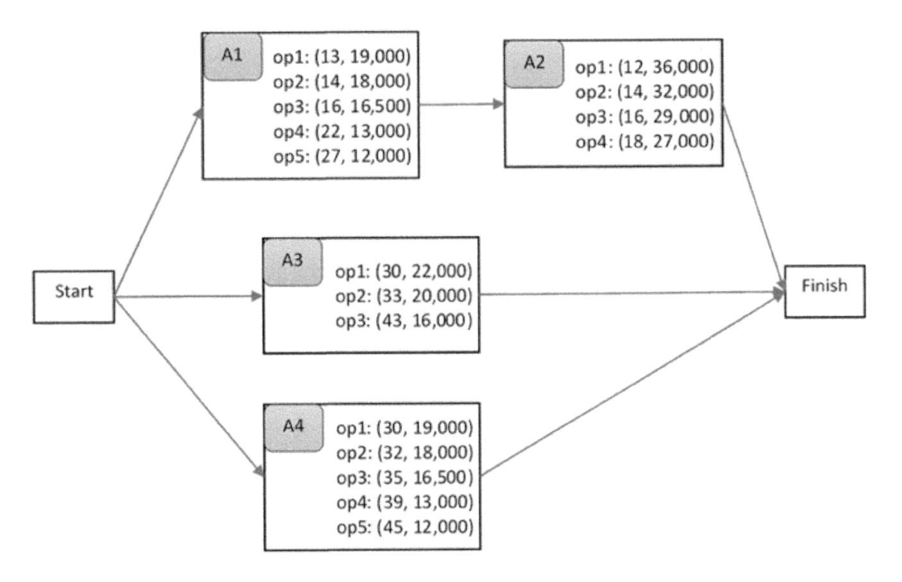

FIGURE 9.5 Case example (4-activity project).

9.4.1 Example 1: A Simple Illustrative Sample

To demonstrate the execution of TLBO on the TCO problem, a stepwise procedure for the implementation of TLBO algorithms (TLBO, eTLBO, and mTLBO) for one generation is presented in the appendix section. To this end, a case example presented in Figure 9.5 is handled to demonstrate the progress of TLBO in a stepwise manner. In Figure 9.5, A1, A2, A3, and A4 represent the activities for a simple construction project. In addition, *op1*, *op2*, …, *op5* are the resource utilization (execution) options for any of these activities. In other words, they are time cost alternatives (day, $) for the project activities needs to be constructed. Furthermore, Figure 9.5 illustrates the activity on a node diagram for the instance with a finish-to-start (FS) relationship. Here, the indirect cost is $100/day.

Table 9.1 displays the optimal solutions for the presented case example, along with a resource utilization option associated with each solution. In the optimization process conducted with

TABLE 9.1
Results for 4-Activity Project

					Project Performance	
Algorithm	**Resource Utilization Option**				**Time (days)**	**Cost ($)**
TLBO	3	3	2	1	33	91,800
	3	4	1	3	35	88,000
eTLBO	2	3	1	2	32	93,200
	3	3	2	1	33	91,800
		4	2	3	35	86,000
mTLBO	2	4	1	2	32	91,200
	2	4	2	2	33	89,300
	3	3	2	3	35	88,000
	4	3	1	4	39	84,900

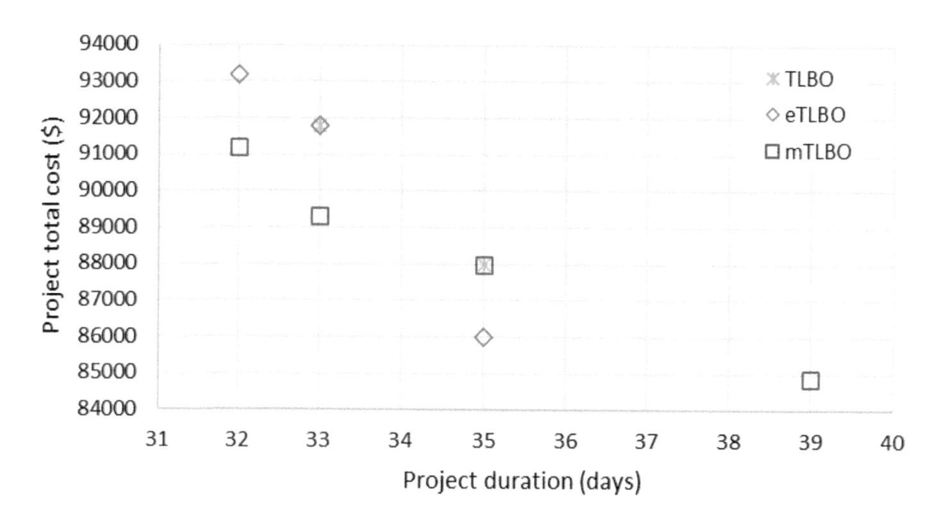

FIGURE 9.6 Optimal solutions obtained using TLBO, eTLBO, and mTLBO for 4-activity project.

TLBO and its variants, the population size and the maximum number of iterations were adjusted both as 6 in the function of optimization parameters for all optimization algorithms, whereas for eTLBO, the elite size and teacher number were both set as 2 for mTLBO. Furthermore, Figure 9.6 demonstrates these optimal solutions in 2D space. Figure 9.6 shows that results generated by mTLBO provide more compromised solutions between the two given objectives. On the other hand, the usage of optimization algorithms provides distinct characteristics, which might result in obtaining various optimal solutions.

9.4.2 Example 2: 29-Activity Project

The presented example in this study is associated with the improvement of a highway project. The project consists of 29 activities, in which 6 of them are owning single options, the other 6 with double options, and the rest of them have triplet resource utilization options. The initial test case was originally introduced by Sakellaropoulos and Chassiakos (2004), with additional restrictions on some of the start and finish times of the activities. Here, the restrictions are ignored for simplicity. Table 9.2 shows activity logical precedence relationships, activity durations, and costs for

TABLE 9.2
Data for 29-Activity Project

			Resource Utilization Options					
			Option 1		Option 2		Option 3	
Activity Number	Activity Description	Precedence Relations	Time (Days)	Cost ($)	Time (Days)	Cost ($)	Time (Days)	Cost ($)
1	Rock excavation	-	5	2030	4	2300	–	–
2	Embankment construction	1	8	1020	7	1280	6	1.510
3	Subbase and base layers	1, 2	8	1700	7	1,850	6	2090
4	Asphalt layer	3	4	590	3	730	–	–
5	Temporary marking and signing	4	2	90	–	–	–	–
6	Earth and semi-rock excavation	1	4	910	3	1100	–	–

TABLE 9.2 (*Continued*)
Data for 29-Activity Project

			Resource Utilization Options					
			Option 1		Option 2		Option 3	
Activity Number	Activity Description	Precedence Relations	Time (Days)	Cost ($)	Time (Days)	Cost ($)	Time (Days)	Cost ($)
7	Embankment construction	2, 6	2	250	–	–	–	–
8	Subbase and base layers	7, 3	7	1490	6	1650	5	1,830
9	Asphalt layer	4, 8	4	520	3	750	–	–
10	Temporary marking and signing	5, 9	2	90	–	–	–	–
11	Traffic diversion	5, 10	1	50	–	–	–	–
12	Rock excavation	11	8	3260	7	3580	6	3710
13	Earth and semi-rock excavation-existing pavement removal	12	5	1140	4	1400	3	1720
14	Subgrade stabilization, retaining wall/culvert construction	13	4	300	3	450	–	–
15	Embankment construction	12, 14	8	1020	6	1300	5	1430
16	Drainage pipe construction	15	9	790	8	900	6	1180
17	Drainage layer	15	13	3340	12	3,750	11	4060
18	Planting at roadway verges	15	9	470	8	650	7	830
19	Electrical installations at roadway verges	15	6	460	5	600	4	810
20	Ditches	17	6	1280	5	1430	–	–
21	Subbase layer	20	14	1090	12	1320	10	1560
22	Base layer	21	14	900	11	1140	9	1400
23	Median island (New Jersey)	22	14	2220	12	2510	11	2690
24	Electrical installations in median island	23	3	230	–	–	–	–
25	Asphalt layer #1	23	6	1590	5	1790	4	1990
26	Asphalt layer #2	25	10	2630	9	2930	8	3240
27	Friction course overlay	26	8	2060	7	2450	6	2660
28	Final marking and signing	27	10	320	9	440	8	610
29	Traffic restoration	28	1	50	–	–	–	–

each execution option of the project. Logical precedence relationships are accepted as FS, while the indirect cost is $150/day.

The details of MOO models presented in this chapter resulted in finding the optimal solutions for the discussed instance and are presented in Table 9.3 and Figure 9.7. This was obtained after the generation of 4040 solutions (population size = 40, max. iterations = 50). Other parameters were adjusted as elite size = 4 and teacher number = 5 for eTLBo and mTLBO, respectively. The results of the MOO models revealed that all of the optimal solutions, obtained by TLBO, eTLBO, and mTLBO, for the 29-activity project, were different due to the exploration and exploitation capacity of those algorithms. Also, the renovations employed on TLBO affected the obtained optimal solutions. It seems that delivering the students to different teachers depending on their knowledge level along with using the adaptive teaching factor yields more compromised solutions between the two objectives.

TABLE 9.3
Results for 29-Activity Project

Algorithm	Resource Utilization Option																													Project Performance	
																														Time (Days)	Cost ($)
TLBO	1	1	1	2	1	2	1	1	1	1	1	2	3	2	3	2	3	2	3	2	3	1	3	1	1	1	3	3	1	135	57,270
	1	2	2	2	1	1	1	2	2	1	1	1	2	1	1	2	3	1	2	1	1	1	1	1	1	3	3	2	1	144	56,990
	1	2	2	1	1	1	1	3	2	1	1	1	2	2	1	2	1	3	2	1	1	1	2	1	1	1	3	1	1	145	56,530
	1	1	3	2	1	1	1	1	1	1	1	1	2	1	1	1	2	2	1	2	1	1	1	1	1	1	2	3	1	148	56,300
eTLBO	2	3	3	1	1	2	1	1	2	1	1	1	3	1	2	2	3	2	3	1	3	1	3	1	2	2	3	2	1	132	57,640
	2	1	1	1	1	1	1	3	1	1	1	1	3	1	1	2	2	3	3	1	3	1	3	1	1	1	1	1	1	143	56,700
	2	2	1	1	1	2	1	1	1	1	1	1	1	1	3	1	1	2	2	1	2	1	1	1	1	1	2	1	1	145	56,460
	1	1	1	1	1	1	1	1	1	1	1	2	1	1	1	2	1	2	1	1	1	3	1	1	1	2	1	1	1	149	55,650
mTLBO	1	1	2	1	1	1	1	3	1	1	1	3	1	2	3	1	1	1	1	2	3	3	2	1	2	1	1	2	1	133	55,070
	1	1	2	1	1	1	1	3	1	1	1	1	1	1	2	1	1	1	1	1	3	3	2	1	1	1	1	3	1	138	54,910
	1	1	2	1	1	1	1	1	1	1	1	1	1	1	3	1	1	1	1	2	3	3	1	1	1	1	1	2	1	141	54,840

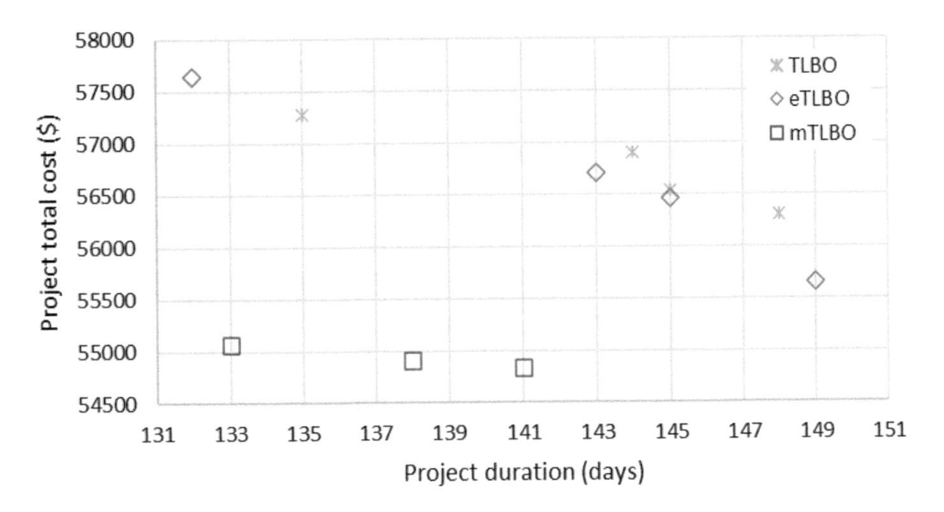

FIGURE 9.7 Convergence histories of TLBO, eTLBO and mTLBO for 29-activity project.

9.5 CONCLUSIONS

This study aims to illustrate the application of TLBO for solving the TCO problem of construction projects, classified as MOO type problems. Toward this end, basic TLBO along with its modified versions (eTLBO and mTLBO) are introduced, while their applications on the MOO problem are presented in detail using a simple illustrative example. Also, a modified version of the highway upgrading project given in Sakellaropoulos and Chassiakos (2004) is solved as an additional test problem to demonstrate the performances of TLBO and its variants. The proposed models by this study performed well, displaying their capabilities in generating the optimal solutions that correspond to the minimum project duration and cost. This study lays the groundwork for future research works on the application of MOO problems in various fields. This study also facilitates understanding the working procedure of TLBO and its variants.

APPENDIX 9.1 IMPLEMENTATION OF TLBO

Step 1: Initialize the optimization parameters for the corresponding optimization problem.

The following parameters are adopted for TCO problem given in Figure 9.1.

Population size (np) = 6, number of design variables (nd) = 4, maximum number of iteration (it_{max}) = 1, lower boundaries of design variables (lb) = [1 1 1 1], and upper boundaries of design variables (ub) = [5 4 3 5].

Step 2: Initialize the population.

Generate randomly population obeying the boundaries of design variables (see Eq. 9.2).

$$\text{Population } (np \times nd) = \begin{bmatrix} 5 & 1 & 2 & 2 \\ 2 & 2 & 2 & 3 \\ 5 & 3 & 3 & 1 \\ 2 & 2 & 2 & 4 \\ 5 & 2 & 3 & 3 \\ 2 & 4 & 3 & 4 \end{bmatrix}.$$ Any solution in the population

$\mathbf{X}_i = \{X_{i,1}, X_{i,2}, \cdots, X_{i,nd}\}$ is a candidate solution for the problem that represents any position within the search space. For the case example, any design variable (activity) in a solution shows the related activity execution option. For example, for the first solution the population is $\mathbf{X}_1 = \{X_{1,1}, X_{1,2}, X_{1,3}, X_{1,4}\} = \begin{bmatrix} 5 & 1 & 2 & 2 \end{bmatrix}$. The first design variable value is 5, and it refers to the fifth option of the first activity. From Figure 9.1, it can be inferred that the fifth execution option offers resource utilization as 27 days, and $12,000. Following the same way, options values are assigned to other design variables, and then the CPM is implemented to detect the project duration. After that,

$$\text{values of objective functions } Z_1\ (\text{days}) = \begin{bmatrix} 39 \\ 35 \\ 43 \\ 39 \\ 43 \\ 43 \end{bmatrix},\ Z_2\ (\$) = \begin{bmatrix} 92,900 \\ 92,500 \\ 84,300 \\ 90,900 \\ 83,300 \\ 82,300 \end{bmatrix}.$$

$$\text{weight factor } w_1 = 0.360;\ w_2 = 0.640$$

$$\text{and corresponding value of fitness } Z = \begin{bmatrix} 0.180 \\ 0.385 \\ 0.520 \\ 0.301 \\ 0.580 \\ 0.640 \end{bmatrix}$$

Step 3: Selection of teacher.

Based on the initial population's results, the solution with the minimum value of Z is assigned as the teacher for the iteration. $\mathbf{X}_{teacher} = Z_{min}$

$$= \begin{bmatrix} 5 & 1 & 2 & 2 \end{bmatrix} \text{ and the corresponding value of fitness function } Z = [0.180]$$

Step 4: Teacher phase.

Calculate the mean of the solutions in column wise manner.

$$\mathbf{X}_{mean} = \begin{bmatrix} 3.5 & 2.333 & 2.333 & 2.333 \end{bmatrix}$$

Update the solutions in the current population according to Eq. 9.3

$$\mathbf{X}^{new} = \begin{bmatrix} 4.27 & 0.90 & 1.71 & 2.76 \\ 3.32 & 3.91 & 1.33 & 2.77 \\ 2.65 & 4.42 & 3.66 & 4.45 \\ 3.90 & 1.41 & 1.84 & 4.03 \\ 2.25 & 3.63 & 1.90 & 3.08 \\ 3.28 & 0.90 & 2.71 & 3.16 \end{bmatrix}$$ since the design variables are discrete $\mathbf{X}^{new}$ is converted to integer one by applying an extra operation, which rounds each element of $\mathbf{X}^{new}$ to the nearest integer.

$$\mathbf{X}^{new} = \begin{bmatrix} 4 & 1 & 2 & 3 \\ 3 & 4 & 1 & 3 \\ 3 & 4 & 3 & 4 \\ 4 & 1 & 2 & 4 \\ 2 & 4 & 2 & 3 \\ 3 & 1 & 3 & 3 \end{bmatrix}$$ after this operation, boundary control is conducted for the elements exceeding the lower and upper boundaries. If any element of $\mathbf{X}^{new}$ is lower than its lower boundary value, then the related element sets to that value. For the upper boundary,

the mentioned process is opposite. Namely, the element adopts the value of its upper boundary if the element value is higher than that. Finally,

$$\mathbf{X}^{new} = \begin{bmatrix} 4 & 1 & 2 & 3 \\ 3 & 4 & 1 & 3 \\ 3 & 4 & 3 & 4 \\ 4 & 1 & 2 & 4 \\ 2 & 4 & 2 & 3 \\ 3 & 1 & 3 & 3 \end{bmatrix}, \text{ values of objective functions } Z_1 \ (\text{days}) = \begin{bmatrix} 35 \\ 35 \\ 43 \\ 39 \\ 35 \\ 43 \end{bmatrix}, \ Z_2 \ (\$) = \begin{bmatrix} 91{,}500 \\ 88{,}000 \\ 80{,}800 \\ 89{,}900 \\ 87{,}500 \\ 91{,}800 \end{bmatrix}$$

$$\text{and the corresponding value of fitness function } Z = \begin{bmatrix} 0.390 \\ 0.590 \\ 0.627 \\ 0.295 \\ 0.618 \\ 0.0004 \end{bmatrix}$$

Accept $\mathbf{X}^{new}$ if it gives better function value, then add it into the population.

$$\text{Population} = \begin{bmatrix} 5 & 1 & 2 & 2 \\ 2 & 2 & 2 & 3 \\ 5 & 3 & 3 & 1 \\ 4 & 1 & 2 & 4 \\ 5 & 2 & 3 & 3 \\ 3 & 1 & 3 & 3 \end{bmatrix}, \text{ corresponding values of objective functions } Z_1 \ (\text{days}) = \begin{bmatrix} 39 \\ 35 \\ 43 \\ 39 \\ 43 \\ 43 \end{bmatrix},$$

$$Z_2 \ (\$) = \begin{bmatrix} 92{,}900 \\ 92{,}500 \\ 84{,}300 \\ 89{,}900 \\ 83{,}300 \\ 91{,}800 \end{bmatrix}, \text{ and the corresponding value of fitness function } Z = \begin{bmatrix} 0.168 \\ 0.363 \\ 0.596 \\ 0.377 \\ 0.665 \\ 0.077 \end{bmatrix}$$

Step 5: Learner phase.

The mathematical expression is explained under Section 9.2.1. For mutual interaction, a random solution is selected from the population ($i \neq j$).

The randomly obtained solutions= $\begin{bmatrix} 3 \\ 5 \\ 1 \\ 6 \\ 2 \\ 4 \end{bmatrix}$. It means that the first solution in the population is mutually interacted with the third solution in that population (see Eq. 9.4). As explained in the teacher phase, after implementing Eq. 9.4, boundary exceeding control is performed. For example, the implementation of Eq. 9.4 between solution 1 and 3 produces $\mathbf{X}_1^{new} = \begin{bmatrix} 5.0 & -0.919 & 1.041 & 2.960 \end{bmatrix}$. Round them to the nearest integer

$\mathbf{X}_1^{new} = \begin{bmatrix} 5 & -1 & 1 & 3 \end{bmatrix}$ then control the boundaries and obtain $\mathbf{X}_1^{new} = \begin{bmatrix} 5 & 1 & 1 & 3 \end{bmatrix}$. Repeat this procedure for other solutions and obtain the population.

$$\mathbf{X}^{new} = \begin{bmatrix} 5 & 1 & 1 & 3 \\ 1 & 2 & 1 & 3 \\ 5 & 3 & 3 & 1 \\ 3 & 1 & 3 & 3 \\ 2 & 2 & 2 & 3 \\ 2 & 1 & 3 & 2 \end{bmatrix} \text{ corresponding values of objective functions } Z_1\left(\text{days}\right) = \begin{bmatrix} 39 \\ 35 \\ 43 \\ 43 \\ 43 \\ 43 \end{bmatrix},$$

$$Z_2\left(\$\right) = \begin{bmatrix} 92,900 \\ 95,500 \\ 84,300 \\ 91,800 \\ 86,300 \\ 95,300 \end{bmatrix}, \text{ and the corresponding value of fitness function } Z = \begin{bmatrix} 0.331 \\ 0.368 \\ 0.633 \\ 0.209 \\ 0.520 \\ 0.012 \end{bmatrix}$$

Accept $\mathbf{X}^{new}$ if it gives better function value then $\mathbf{X}_i$, then add it into the population instead of $\mathbf{X}_i$. Otherwise, preserve $\mathbf{X}_i$.

$$\text{Population} = \begin{bmatrix} 5 & 1 & 2 & 2 \\ 2 & 2 & 2 & 3 \\ 5 & 3 & 3 & 1 \\ 3 & 1 & 3 & 3 \\ 2 & 2 & 2 & 3 \\ 2 & 1 & 3 & 2 \end{bmatrix}$$

Step 6: Termination criterion.

Stop if the maximum number of iterations is achieved; otherwise, repeat TLBO procedure starting from *Step 3* with the population obtained at the end of *Step 5*

APPENDIX 9.2 IMPLEMENTATION OF eTLBO

The first five steps detailed in Appendix 9.1 for TLBO are the same when working with eTLBO. However, in *Step 1*, elite size (*elite_size*) can be set as 2. Then, in *Step 2*, the best solutions up to *elite_size* from the current population are saved into an external archive before going to the teacher phase.

For simplicity, the same population in Appendix 9.1 is used.

Sort the population with ascending order according to Z values

$$\text{Population} = \begin{bmatrix} 5 & 1 & 2 & 2 \\ 2 & 2 & 2 & 4 \\ 2 & 2 & 2 & 3 \\ 5 & 3 & 3 & 1 \\ 5 & 2 & 3 & 3 \\ 2 & 4 & 3 & 4 \end{bmatrix}, \text{ and the corresponding value of fitness } Z = \begin{bmatrix} 0.180 \\ 0.301 \\ 0.385 \\ 0.520 \\ 0.580 \\ 0.640 \end{bmatrix}$$

Depending on this result, save the first *elite_size* solution into external achieve elite $= \begin{bmatrix} 5 & 1 & 2 & 2 \\ 2 & 2 & 2 & 4 \end{bmatrix}$, and

$$\text{eliteZ} = \begin{bmatrix} 0.180 \\ 0.301 \end{bmatrix}$$

Step 3: Same as for TLBO.
Step 4: Same as for TLBO.
Step 5: Same as for TLBO.
Step 6: Elitism process.

For simplicity, the population obtained at the end of *Step 5* of TLBO is used.

$$\text{Population} = \begin{bmatrix} 5 & 1 & 2 & 2 \\ 2 & 2 & 2 & 3 \\ 5 & 3 & 3 & 1 \\ 3 & 1 & 3 & 3 \\ 2 & 2 & 2 & 3 \\ 2 & 1 & 3 & 2 \end{bmatrix}, \text{ and the corresponding value of fitness function } Z = \begin{bmatrix} 0.321 \\ 0.523 \\ 0.637 \\ 0.203 \\ 0.523 \\ 0.0004 \end{bmatrix}$$

Change the worst solutions in the population with the elitist ones stored in an external archive, up to *elite_size*

$$\text{Population} = \begin{bmatrix} 5 & 1 & 2 & 2 \\ 5 & 1 & 2 & 2 \\ 2 & 2 & 2 & 4 \\ 3 & 1 & 3 & 3 \\ 2 & 2 & 2 & 3 \\ 2 & 1 & 3 & 2 \end{bmatrix}.$$

Step 7: Remove duplicate.

If the population has the same solution, to remove it select a random position in that solution and change it with the new one randomly generated within the lower and upper boundaries.

As seen from the population, the first two solutions are same. The position to be changed is randomly generated as 4. The new position is 5, and the final population is $\begin{bmatrix} 5 & 1 & 2 & 2 \\ 5 & 1 & 2 & 5 \\ 2 & 2 & 2 & 4 \\ 3 & 1 & 3 & 3 \\ 2 & 2 & 2 & 3 \\ 2 & 1 & 3 & 2 \end{bmatrix}$.

Step 8: Renew the external achieve.

Sort the population with ascending order according to Z values, and save the first *elite_size* solutions into external achieve elite$= \begin{bmatrix} 2 & 1 & 3 & 2 \\ 5 & 1 & 2 & 2 \end{bmatrix}$, and eliteZ$= \begin{bmatrix} 0.048 \\ 0.377 \end{bmatrix}$

Step 9: Termination criterion.

Stop if the maximum iteration number is achieved; otherwise, repeat eTLBO procedure starting from *Step 3* with the population obtained at the end of *Step 7*

APPENDIX 9.3 IMPLEMENTATION OF mTLBO

Step 1: Initialize the optimization parameters.

In mTLBO, as well as the optimization parameters initialized in *Step 1* of TLBO, number of teachers $(tn) = 2$ is assigned as an additional optimization parameter.

Step 2: Similar to TLBO.

Step 3: Selection of teachers.

For simplicity, the population in Appendix 9.1 is used. $\mathbf{X}_{1,teacher} = Z_{min} = \begin{bmatrix} 5 & 1 & 2 & 2 \end{bmatrix}$ and corresponding value of fitness function $Z = [0.180]$ were adopted. The second teacher is obtained using the chief teacher while applying the mathematical expression explained in Section 9.2.3. Assign the closer or equal solution to this result in the population as second teacher. $\mathbf{X}_{2,teacher} = \begin{bmatrix} 2 & 2 & 2 & 3 \end{bmatrix}$ with the fitness value $= [0.385]$

Step 4: Place the solutions to the teachers according to their fitness value.

$$\text{Group } 1 = \begin{bmatrix} 5 & 1 & 2 & 2 \\ 2 & 2 & 2 & 4 \end{bmatrix}, \text{ and the corresponding value of fitness function} = \begin{bmatrix} 0.180 \\ 0.301 \end{bmatrix}$$

$$\text{Group } 2 = \begin{bmatrix} 2 & 2 & 2 & 3 \\ 5 & 3 & 3 & 1 \\ 5 & 2 & 3 & 3 \\ 2 & 4 & 3 & 4 \end{bmatrix}, \text{ and the corresponding value of fitness function} = \begin{bmatrix} 0.385 \\ 0.520 \\ 0.580 \\ 0.640 \end{bmatrix}$$

Step 5: Teacher phase.

For each group calculate the mean of the learners in a column-wise manner.

$$\mathbf{X}_{1,\,mean} = \begin{bmatrix} 4 & 2 & 2 & 3 \end{bmatrix} \text{ for group 1 and}$$

$$\mathbf{X}_{2,\,mean} = \begin{bmatrix} 4 & 3 & 3 & 3 \end{bmatrix} \text{ for group 2}$$

New mean for each group

$$\mathbf{X}\_new_1 = \mathbf{X}_{1,teacher} = \begin{bmatrix} 5 & 1 & 2 & 2 \end{bmatrix}$$

$$\mathbf{X}\_new_2 = \mathbf{X}_{2,teacher} = \begin{bmatrix} 2 & 2 & 2 & 3 \end{bmatrix}$$

Calculate the adaptive teaching factor for each group as below:

$$\mathbf{TF}_1 = \begin{bmatrix} 0.8 & 2.0 & 1.0 & 1.5 \end{bmatrix}$$

$$\mathbf{TF}_2 = \begin{bmatrix} 2.0 & 1.5 & 1.5 & 1.0 \end{bmatrix}$$

Obtain a difference between two means is expressed as

$$\mathbf{Difference\_Mean}_s = rand(\ ) \times \left(\mathbf{X}\_new_s - \mathbf{TF}_s \times \mathbf{X}_{s,\,mean}\right) \quad s = 1, \ldots, tn$$

The obtained difference is added to the current solution to update its values as

$$\mathbf{X}^{new} = \mathbf{X}_i + \mathbf{Difference\_Mean}_s$$

$$\mathbf{X}^{new} = \begin{bmatrix} 5 & 1 & 2 & 1 \\ 2 & 1 & 2 & 3 \\ 1 & 1 & 1 & 3 \\ 1 & 2 & 1 & 1 \\ 1 & 1 & 1 & 3 \\ 1 & 3 & 1 & 4 \end{bmatrix}, \text{ and the corresponding value of fitness function } Z = \begin{bmatrix} 0.407 \\ 0.372 \\ 0.108 \\ 0.284 \\ 0.108 \\ 0.761 \end{bmatrix}$$

Accept $\mathbf{X}^{new}$ if it gives better function value.

$$\text{Population} = \begin{bmatrix} 5 & 1 & 2 & 2 \\ 2 & 2 & 2 & 4 \\ 1 & 1 & 1 & 3 \\ 1 & 2 & 1 & 1 \\ 1 & 1 & 1 & 3 \\ 2 & 4 & 3 & 4 \end{bmatrix}, \text{ and the corresponding value of fitness function } Z = \begin{bmatrix} 0.360 \\ 0.438 \\ 0.200 \\ 0.698 \\ 0.200 \\ 0.675 \end{bmatrix}$$

Step 6: Learner phase.

It is known that group 1 contains the first two solutions of the population updated with the help of the teacher phase of mTLBO, while the remaining solutions of the related population are inside to group 2. Therefore, the solutions inside the corresponding group are mutually interacted to increase their knowledge. For mutual interaction, a random solution is selected from the population ($i \neq j$).

The randomly obtained solutions= $\begin{bmatrix} 2 \\ 1 \\ 4 \\ 5 \\ 6 \\ 3 \end{bmatrix}$. The new solution obtained after the learner phase

$$\mathbf{X}^{new} = \begin{bmatrix} 5 & 1 & 1 & 3 \\ 1 & 2 & 1 & 3 \\ 5 & 3 & 3 & 1 \\ 3 & 1 & 3 & 3 \\ 2 & 2 & 2 & 3 \\ 2 & 1 & 3 & 2 \end{bmatrix}, \text{ and the corresponding value of fitness function } Z = \begin{bmatrix} 0.330 \\ 0.504 \\ 0.081 \\ 0.280 \\ 0.497 \\ 0.696 \end{bmatrix}$$

Update population if $\mathbf{X}^{new}$ gives better results than the teacher phase. Obtain population after the learner phase

$$\text{Population} = \begin{bmatrix} 5 & 1 & 1 & 3 \\ 2 & 2 & 2 & 4 \\ 5 & 3 & 3 & 1 \\ 3 & 1 & 3 & 3 \\ 1 & 1 & 1 & 3 \\ 2 & 4 & 3 & 4 \end{bmatrix}$$

Step 7: Termination criterion.

Stop if the maximum iteration number is reached; otherwise, repeat from *Step 3* with the population obtained at the end of *Step 6*.

REFERENCES

Arora, J.S. 2004. *Introduction to optimum design*. Elsevier: California.

Atashpaz-Gargari, E., and C. Lucas. 2007. *Imperialist competitive algorithm: an algorithm for optimization inspired by imperialistic competition*. IEEE Congress on Evolutionary Computation, Singapore.

Baghlani, A., and M.H. Makiabadi. 2013. Teaching-learning-based optimization algorithm for shape and size optimization of truss structures with dynamic frequency constraints. *Iranian Journal of Science and Technology-Transactions of Civil Engineering* 37(C): 409–421.

Baghlani, A., Makiabadi, M.H., and M.R. Maheri. 2017. Sizing optimization of truss structures by an efficient constraint-handling strategy in TLBO. *Journal of Computing in Civil Engineering* 31(4): 04017004.

Camp, C.V., and M. Farshchin. 2014. Design of space trusses using modified teaching-learning based optimization. *Engineering Structures* 62–63: 87–97.

Chassiakos, A.P., and S.P. Sakellaropoulos. 2005. Time-cost optimization of construction projects with generalized activity constraints. *Journal of Construction Engineering and Management* 131(10): 1115–1124.

Chen, W., M. Panahi, and K. Khosravi, et al. 2019. Spatial prediction of groundwater potentiality using ANFIS ensembled with teaching-learning-based and biogeography-based optimization. *Journal of Hydrology* 572: 435–448.

Cheng, M.-Y., D. Prayogo, Y.-W. Wu, and M.M. Lukito. 2016. A hybrid harmony search algorithm for discrete sizing optimization of truss structure. *Automation in Construction* 69: 21–33.

Daloğlu, A.T., M. Artar, and K. Ozgan, et al. 2018. Optimum design of braced steel space frames including soil-structure interaction via Teaching-Learning-Based Optimization and Harmony Search Algorithms. *Advances in Civil Engineering* 2018: 3854620.

Das, A., C.K. Hirwani, and S.K. Panda, et al. 2018. Prediction and analysis of optimal frequency of layered composite structure using higher-order FEM and soft computing techniques. *Steel and Composite Structures* 29(6): 749.

Dede, T. 2013. Optimum design of grillage structures to LRFD-AISC with teaching-learning based optimization. *Structural and Multidisciplinary Optimization* 48: 955–964.

Dede, T. 2014. Application of Teaching-Learning-Based-Optimization algorithm for the discrete optimization of truss structures. *KSCE Journal of Civil Engineering* 18: 1759–1767.

Dede, T., and V. Toğan. 2015. A teaching learning-based optimization for truss structures with frequency constraints. *Structural Engineering and Mechanics* 53: 833–845.

Değertekin, S.O., and M.S. Hayalioğlu. 2013. Sizing truss structures using teaching-learning-based optimization. *Computers & Structures* 119: 177–188.

Dorigo, M., M. Birattari, and T. Stutzle. 2006. Ant colony optimization. *IEEE Computational Intelligence Magazine* 1(4): 28–39.

Eberhart, R., and J. Kennedy. 1995. A new optimizer using particle swarm theory. Proceedings of the Sixth International Symposium on Micro Machine and Human Science, Nagoya, Japan.

Eirgash M.A., V. Toğan, and T. Dede. 2019. A multi-objective decision making model based on TLBO for the time - cost trade-off problems. *Structural Engineering and Mechanics* 71: 139–151.

Farshchin, M., C.V. Camp, and M. Maniat. 2016. Multi-class teaching-learning-based optimization for truss design with frequency constraints. *Engineering Structures* 106: 355–369.

Gandomi, A.H., X.-S. Yang, and A.H. Alavi. 2013. Cuckoo search algorithm: a metaheuristic approach to solve structural optimization problems. *Engineering with Computers* 29: 17–35.

Goldberg, D.E. 1989. *Genetic algorithms in search, optimization, and machine learning*. Addison Wesley, Reading MA.

Grzywinski, M., J. Selejdak, and T. Dede. 2019. Shape and size optimization of trusses with dynamic constraints using a metaheuristic algorithm. *Steel and Composite Structures* 33(5): 747–753.

Haghpanah, F., and H. Foroughi. 2017. Size and shape optimization of space trusses considering geometrical imperfection-sensitivity in buckling constraints. *Civil Engineering Journal-Tehran* 3(12): 1314–1326.

Hansen, N., S.D. Müller, and P. Koumoutsakos. 2003. Reducing the time complexity of the derandomized evolution strategy with covariance matrix adaptation (CMA-ES). *Evolutionary Computation* 11: 1–18.

Holland, J.H. 1975. *Adaptation in natural and artificial systems: an introductory analysis with applications to biology, control, and artificial intelligence*. University of Michigan Press, Ann Arbor, MI.

Kaveh, A., and S. Talatahari. 2010. A novel heuristic optimization method: charged system search. *Acta Mechanica* 213: 267–289.

Kumar, V., and S.M. Yadav. 2018. Optimization of reservoir operation with a new approach in evolutionary computation using TLBO algorithm and Jaya algorithm. *Water Resources Management* 32(13): 4375–4391.

Li, K., X. Xie, W. Xue, et al. 2018. A hybrid teaching-learning artificial neural network for building electrical energy consumption prediction. *Energy and Buildings* 174: 232–334.

Mirjalili, S., S. M. Mirjalili, and A. Lewis. 2014. Grey wolf optimizer. *Advances in Engineering Software* 69: 46–61.

Mishra, M., S. K. Barman, D. Maity, et al. 2020. Performance studies of 10 metaheuristic techniques in determination of damages for large-scale spatial trusses from changes in vibration responses. *Journal of Computing in Civil Engineering* 34(2): 04019052.

Mlakar, U., I. Fister, and I. Fister. 2016. Hybrid self-adaptive cuckoo search for global optimization. *Swarm and Evolutionary Computation* 29: 47–72.

Mohamed, A.W., A.A. Hadi, and A.K. Mohamed. 2020. Gaining-sharing knowledge-based algorithm for solving optimization problems: a novel nature-inspired algorithm. *International Journal of Machine Learning and Cybernetics* 11: 1501–1529

Mortazavi, A. 2019. Interactive fuzzy search algorithm: a new self-adaptive hybrid optimization algorithm. *Engineering Applications of Artificial Intelligence* 81: 270–282.

Mortazavi, A., V. Toğan and A. Nuhoğlu. 2018. Interactive search algorithm: a new hybrid metaheuristic optimization algorithm. *Engineering Applications of Artificial Intelligence* 71: 275–292.

Öztürk, H.T., T. Dede, and E. Türker. 2020. Optimum design of reinforced concrete counterfort retaining walls using TLBO, Jaya algorithm. *Structures* 25: 285–296.

Rao, R.V., and V. Patel. 2012. An elitist teaching-learning-based optimization algorithm for solving complex constrained optimization problems. *International Journal of Industrial Engineering Computations* 3(4): 535–560.

Rao, R.V., and V. Patel. 2013. Multi-objective optimization of heat exchangers using a modified teaching-learning-based optimization algorithm. *Applied Mathematical Modelling* 37: 1147–1162.

Rao, R.V., V.J. Savsani, and D.P. Vakharia. 2011. Teaching–learning-based optimization: A novel method for constrained mechanical design optimization problems. *Computer-Aided Design* 43: 303–315.

Rashedi, E., H. Nezamabadi-pour, and S. Saryazdi. 2009. GSA: A gravitational search algorithm. *Information Sciences* 179: 2232–2248.

Sakellaropoulos, S., and A.P. Chassiakos. 2004. Project time–cost analysis under generalised precedence relations. *Advances in Engineering Software* 35: 715–724.

Satapathy, S., and A. Naik. 2016. Social group optimization (SGO): a new population evolutionary optimization technique. *Complex and Intelligent Systems* 2(3): 173–203.

Savsani, V.J., G.G. Tejani, and V.K. Patel. 2016. Truss topology optimization with static and dynamic constraints using modified subpopulation teaching-learning-based optimization *Engineering Optimization* 48(1): 1990–2006.

Shaffer, L.R., J.B. Ritter and W.L. Meyer. 1965. *The critical-path method*. McGraw-Hill, New York.

Shallan, O., H.M. Maaly, and O. Hamdy. 2018. A developed design optimization model for semi-rigid steel frames using teaching-learning-based optimization and genetic algorithms. *Structural Engineering and Mechanics* 66(2): 173–183.

Shannon, B.T., and G. Lucko. 2012. Algorithm for time-cost tradeoff analysis in construction projects by aggregating activity-level singularity functions. Proceedings of the 2012 Construction Research Congress, Indiana, United States.

Storn, R., and K. Price. 1997. Differential evolution – a simple and efficient heuristic for global optimization over continuous spaces. *Journal of Global Optimization* 11: 341–359.

Temur, R., and G. Bekdaş. 2016. Teaching learning-based optimization for design of cantilever retaining walls. *Structural Engineering and Mechanics* 57(4): 763–783.

Toğan, V. 2012. Design of planar steel frames using Teaching-Learning Based Optimization. *Engineering Structures* 34: 225–232.

Toğan, V. 2013. Design of pin jointed structures using teaching-learning based optimization. *Structural Engineering and Mechanics* 47: 209–225.

Toğan, V., and M.A. Eirgash. 2019a. Time-cost trade-off optimization of construction projects using teaching learning based optimization. *KSCE Journal of Civil Engineering* 23: 10–20.

Toğan, V., and M.A. Eirgash. 2019b. Time-cost trade-off optimization with a new initial population approach. *Technical Journal* 30: 9561–9580.

Topal, U., T. Dede, and H.T. Öztürk. 2017. Stacking sequence optimization for maximum fundamental frequency of simply supported antisymmetric laminated composite plates using teaching-learning-based optimization. *KSCE Journal of Civil Engineering* 21: 2281–2288.

Uzlu, E., M. Kankal, A. Akpinar, and T. Dede. 2014. Estimates of energy consumption in Turkey using neural networks with the teaching-learning-based optimization algorithm. *Energy* 75: 295–303.

Vosoughi, A.R., P. Malekzadeh, U. Topal, et al. 2018. A hybrid DQ-TLBO technique for maximizing first frequency of laminated composite skew plates. *Steel and Composite Structures* 28(4): 509–516.

Zheng, D.X.M., S.T., Ng, and M.M., Kumaraswamy. 2004. Applying a genetic algorithm-based multi-objective approach for time-cost optimization. *Journal of Construction Engineering and Management* 130(2): 168–176.

10 Social Learning Optimization

Yue-Jiao Gong

CONTENTS

Evolutionary computation has seen great success in optimizing nondeterministic polynomial complex problems by simulating Darwinian evolution or the swarm intelligence of animals. Following this paradigm, the social intelligence of humans can also be implemented on computers and be utilized in computational algorithms for solving the optimization problems. This chapter introduces a novel optimization approach that mimics the social learning process of humans. The motivation comes from the social learning behavior of humans, which exhibits the highest level of intelligence in nature, as indicated in Bandura's *Social Learning Theory*. Emulating the observational learning and reinforcement behaviors described in the social learning context, the algorithm maintains a virtual human society to seek the strongest behavioral patterns with the best outcome in the environment. This corresponds to searching for the optimum in the given optimization landscape. Experimental studies validate the powerfulness of this human intelligence-inspired approach named *social learning optimization (SLO)*.

10.1 INTRODUCTION

Optimization problems reside in a variety of scientific and engineering fields, such as industrial electronics [1], resource management [2], intelligent transportation [3, 4], and cybersecurity [5]. Typically, they involve a structure and a set of variables, each taking values (continuous or discrete) within a certain range, and aim at finding the best solution that maximizes or minimizes the objectives or "costs." Many of the problems are difficult to solve by traditional deterministic approaches even with the fastest machines.

In recent years, inspired by the intelligence from nature, a set of population-based nondeterministic algorithms, commonly known as evolutionary computation (EC), has been widely employed to deal with these hard optimization problems [6–8]. Evolutionary algorithms, such as the well-known genetic algorithms (GAs) [9], genetic programming (GP) [10], and differential evolution (DE) [11–13], draw inspiration from Darwinian evolution. On the other hand, emulating the social behaviors of animals, the swarm intelligence algorithms such as ant colony optimization (ACO) [14], particle swarm optimization (PSO) [15, 16], and artificial bee colony (ABC) [17] have attracted increasing attention. However, despite that EC algorithms have undergone rapid development in the last decade, but their performance such as global search ability and convergence efficiency still requires further improvement.

This chapter describes a novel algorithm from the perspective of mimicking the social learning process of humans, which was first developed by Gong et al. [18] in 2014. The considerations for such a method are as follows. First, compared with the natural evolution process that improves the fitness of individuals via the evolution of genotypes, the self-improvement through learning is more direct and rapid [19]. In this sense, the developed algorithm has potential to improve the efficiency of the GAs mimicking natural evolution. Second, compared with the interaction and learning behaviors in animal groups, the social learning process of human beings exhibits a higher level of intelligence. By emulating human learning behaviors, it may be possible to arrive at more effective optimizers than existing swarm intelligence algorithms.

The proposed algorithm is named a *Social Learning Optimization (SLO)* algorithm because it is derived from the Bandura's *Social Learning Theory* [20]. It is a widely accepted theory in the fields of psychology and praxeology that the human society learns most behaviors via observation rather than self-experience because, as illustrated in Figure 10.1, the cost of learning via direct experience is too high. Instead, learning from a social model (e.g., parent, teacher, friend, or even passerby) is much safer and more efficient.

SLO maintains a human society, in which the behavioral patterns of humans correspond to a set of candidate solutions for the optimization problem. The optimization objective plays as

FIGURE 10.1 Different ways of human learning.

the role of the environment for assessing the human behaviors. According to the social learning theory, the observational learning process consists of four basic steps: attention, retention, reproduction (reinforcement), and motivation. The four steps are embraced in our proposed SLO as basic operators for seeking the strongest behavioral pattern in the current environment, namely, the global optimum of the problem. This background makes the algorithm conceptually simple to follow. When considering the implementations, the operation of SLO is based on numerical calculation, which is lightweight and hence consistent with the principle of EC. The experiments and results in this chapter will validate the good performance of SLO, in terms of global search ability, local exploitation, and efficiency.

The success of SLO has encouraged the development of a series of optimization algorithms that are based on the Bandura's theory of human behaviors, such as the social group optimization (SGO) of Satapathy and Naik [21], the specific SLO variant (S-SLO) of Liu et al. [22] for cloud service composition, as well as the socio-evolution and learning optimization (SELO) algorithm of Kumar et al. [23]. This chapter introduces the vanilla SLO of Gong et al. [18].

10.2 SOCIAL LEARNING THEORY REVISIT

Social learning theory describes the cognitive learning process of humans in a social context [20, 24–28]. The adjustment of human behaviors can take place either from personal experience or by observational learning from the other persons. Although learning through self-experience of the environmental consequences is effective, the price could be high. For example, it is costly for a person to learn crossing the road, driving a car, or cooking a meal through trial and error. Most behavioral patterns displayed by humans are results of observing and imitating the behavior of others, either in a deliberate or inadvertent way. Hence, a good model can act as a better teacher than the consequences of unguided actions in most cases.

According to Bandura et al. [20], the observational learning of human is a cognitive and behavioral process that includes the following four key procedures:

1. *Attention:* Learning starts with an attention process. In a social group, members with good performance are likely to attract more attention of the observers. At the same time, informative function determines which attributes of the models capture attention and which will be ignored.
2. *Retention:* In this procedure, observers remember the details of their exemplary behavior with a symbolic system.
3. *Reproduction:* Observers organize or re-organize the behavior patterns in accordance with the remembered features of models, for the purpose of reproduction.
4. *Motivation:* After the reproduction, observers decide whether or not to exhibit the reproduced behavior according to motivations or expectations. When incentives are provided, the learned behavior, which previously remained unexpressed, will be performed by the observers.

In addition to the above four basic elements, the beauty of such human intelligence in social learning comes from the following two aspects.

First, reinforcement plays an important role in observational learning such that it distinguishes learning from simply imitating the others. In social learning theory, behavior is regulated by external reinforcement, vicarious reinforcement, and self-reinforcement, among which vicarious reinforcement has a crucial role [20]. Vicarious reinforcement is defined as the adaptation in the behavior of observers when they notice the consequences of the models. Generally, it consists of vicarious positive reinforcement and vicarious punishment. On the one hand, the observers will further enhance the learned behavior when they notice that the models get positive consequences of the behavior. On the other hand, if negative consequences are observed, the vicarious punishment prevents the observers to behave similarly as the models. By the effect of vicarious reinforcement, observers may perform even better than the models.

Second, observational learning is the primary source of innovation in a social group. This is because observers neither concentrate on a single model, nor absorb all attributes of one preferred model, but they abstract common features of diverse models to form a behavior rule or combine different attributes of the models to develop distinct personalities. The more diverse the models are, the more likely the observers exhibit creative and innovative behavioral patterns, e.g., an artist creates unique art via combining, discarding, and recombining the attributes of plenty of other works. In contrast, if the social group is small and isolated, the possible combinations of behavioral patterns are limited. After a long period of observational learning, the members in the group have a tendency to behave in similar ways and have trouble in producing innovation.

10.3 THE OPTIMIZATION METHODOLOGY

It can be noticed from the above description that, in a social group, people adapt to the environment or seek acceptance from the society via observing models with influence. The effects of the observational learning include not only imitation, but also reinforcement and innovation. As a result, the entire social group makes progress in pursuing goals and overcoming challenges as time goes by [29]. In this sense, the social learning process of humans indicates a high-order form of intelligence in nature. Based on this observation, a novel optimization technique termed the *Social Learning Optimization (SLO)* has been developed, which emulates the social intelligence of humans in computers.

SLO maintains a social group of M people. Each member in the group is assigned with a behavioral pattern vector $\boldsymbol{x}_i = [x_{i,1}, x_{i,2}, \cdots, x_{i,D}]$ and a consequence score S_i denoting the response from the environment ($i = 1, 2, \cdots, M$). Note that $\boldsymbol{x}_i$ is multi-dimensional, which consists of the member's behavioral patterns in different aspects. For example, when cooking steak, the behavioral patterns involve the use of a frying pan, the cut of steak, salt timing, frying time, and so forth, whereas the consequence score is the taste of the steak.

In the optimization context, $\boldsymbol{x}_i$ represents a candidate solution in the problem space, i.e., a D-dimensional numerical vector where D stands for the number of variables. The score of the behavior vector is evaluated as

$$S_i = f(\boldsymbol{x}_i) = f(x_{i,1}, x_{i,2}, \cdots, x_{i,D}) \tag{10.1}$$

where f denotes the objective of the problem. Note that f can have different forms, such as a numerical function, or a black-box system. SLO is a general-purpose optimizer whose implementation does not rely on the formulation of objective.

At the beginning of the algorithm, i.e., the initialization, all the behavior vectors are randomly generated in the problem space with the consequence scores evaluated by f. Then, the members perform observational learning in the society to improve their behavioral pattern vectors, so as to receive higher and higher scores. In this way, better and better solutions are found, until the global optimum is achieved.

The optimization process of SLO is based on a simplified social learning model. As illustrated in Figure 10.2, after the initialization, SLO performs an iteration process in which the members conduct "attention," "reproduction," "reinforcement," and "motivation" operators repeatedly. The four procedures are in accordance with those in the observational learning process of the social learning theory. Note that the "retention" step is omitted since it is easy for computers to store

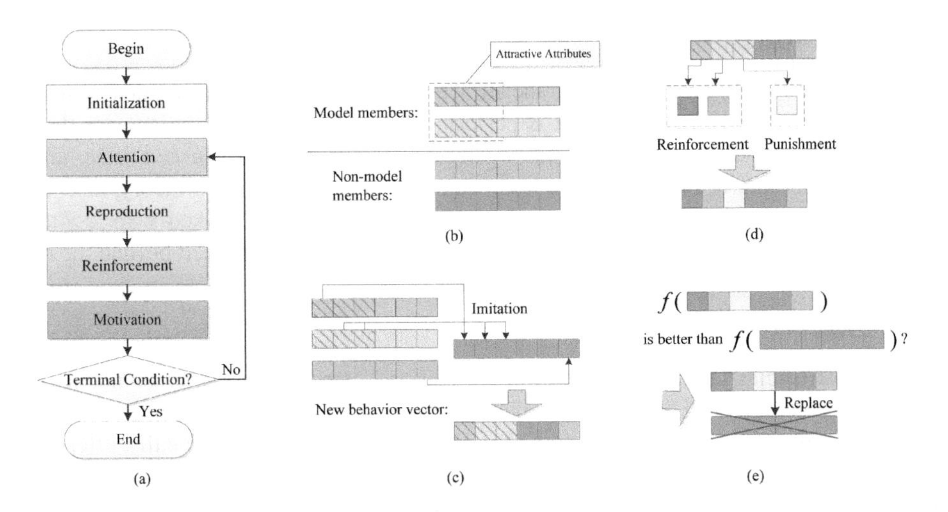

FIGURE 10.2 Schematic representation of the social learning optimization (SLO). (a) Flowchart of SLO. Starting with random initialization, the algorithm enters a loop of four operators: attention, reproduction, reinforcement, and motivation. (b) Attention captures model members and their attractive attributes according to the scores and significance test results. (c) Reproduction builds new behavior vectors for all members by imitation. (d) Reinforcement further improves the learned behavior with positive reward or negative punishment. (e) Motivation activates the new behavior vectors with incentives.

("remember") the models and exemplary attributes. In the following section, the method and operation of these four basic operators are generally described. More implementation details of the operators will be provided in Section 10.4.

10.3.1 Attention

This operator determines whose and which attributes capture attention in the social group. In SLO, a number of members that currently possess the highest scores in the society are selected as models. Then, the society is divided into two groups, one contains model members, and the other consists of non-model members. For each dimension of the behavioral pattern vectors, the statistical hypothesis test [30] is applied to determine whether the values of model members at this dimension are significantly different from those of the non-model members. If the difference is significant, the models' values on the dimension capture attention, which are named "attractive attributes." More specifically, mark the dimension with a symbol '>>' or a symbol '<<' if the attractive attributes are significantly larger or smaller than the other allelic values, respectively. In other cases, the dimensions are marked with '$\approx$' if the differences are not significant.

10.3.2 Reproduction

After attention, the reproduction process is conducted, by which the member generates a new behavior vector $\hat{\boldsymbol{x}}_i$. For each dimension that the models capture attention, $\hat{\boldsymbol{x}}_i$ imitates (copies) the attractive attribute of one model. The observer is able to combine different attributes of several models, which results in innovative behavioral patterns, as discussed in the social learning theory. For the other dimensions marked with '$\approx$', the attributes of models are not so attractive. In these cases, $\hat{\boldsymbol{x}}_i$ maintains the values of $\boldsymbol{x}_i$ or explores the entire society to enhance diversity. This way, the reproduction operator exploits good models' strengths and explores the dimensions without models.

10.3.3 Reinforcement

Members in the society not only imitate the attractive attributes, but they also make reinforcement on them. When positive reinforcement is performed, the observers further increase the learned value of their behavioral patterns, whereas, by negative punishment, the learned values are further decreased. According to the vicarious reinforcement in the social learning theory, SLO performs positive reinforcement and negative punishment on the dimensions marked with '>>' and '<<', respectively. The procedure is derived from the human's ambition to proceed beyond the models.

10.3.4 Motivation

In the natural social learning system, the motivation determines whether or not the learned behavior will be expressed by a person. Likewise, the motivation step in SLO decides whether or not to replace the current behavioral pattern vector $\boldsymbol{x}_i$ by the new $\hat{\boldsymbol{x}}_i$. Specifically, after reproduction and reinforcement, the leaned behavior vector $\hat{\boldsymbol{x}}_i$ of each member is evaluated. Only when $\hat{\boldsymbol{x}}_i$ obtains a higher score $\hat{S}_i$ than S_i of the current behavior $\boldsymbol{x}_i$ does the member update his behavior vector, i.e., replaces $\boldsymbol{x}_i$ with $\hat{\boldsymbol{x}}_i$. This step is similar to the selection in evolutionary algorithms [31].

10.4 IMPLEMENTATIONS

This section presents the detailed implementations of the attention, reproduction, reinforcement, and motivation operators in SLO.

10.4.1 Implementation of the Attention Operator

The society maintained by SLO is classified into two groups, $\boldsymbol{U}$ and $\boldsymbol{L}$, such that members of $\boldsymbol{U}$ outperform members of $\boldsymbol{L}$ in terms of the consequence scores. More specifically, the top M_{N} members are assigned into $\boldsymbol{U}$, whereas the remaining ones are assigned into $\boldsymbol{L}$. M_{N} is a control parameter of SLO, which determines the quantity and hence quality of model members. As there are M members in the society in total, we have $|\boldsymbol{U}| = M_{\mathrm{N}}$ and $|\boldsymbol{L}| = M - M_{\mathrm{N}}$. Then, for each dimension $d \in \{1,2,\cdots,D\}$ of the society, the students' t-test is performed to compare the values of $\boldsymbol{U}$ and $\boldsymbol{L}$ at this dimension:

$$\begin{aligned} \boldsymbol{U}_d &= [u_{1,d}, u_{2,d}, \cdots, u_{M_{\mathrm{N}},d}], \\ \boldsymbol{L}_d &= [l_{1,d}, l_{2,d}, \cdots, l_{(M-M_{\mathrm{N}}),d}]. \end{aligned} \tag{10.2}$$

The statistical t-value at this dimension is calculated as follows [30]. Suppose the mean values of $\boldsymbol{U}_d$ and $\boldsymbol{L}_d$ are $\bar{U}_d$ and $\bar{L}_d$, respectively. Calculate the unbiased estimators of the variances of $\boldsymbol{U}_d$ and $\boldsymbol{L}_d$ as

$$\begin{aligned} s_U^2 &= \frac{\sum_{i=1}^{M_{\mathrm{N}}} (u_{i,d} - \bar{U}_d)^2}{M_{\mathrm{N}} - 1}, \\ s_L^2 &= \frac{\sum_{i=1}^{M-M_{\mathrm{N}}} (l_{i,d} - \bar{L}_d)^2}{M - M_{\mathrm{N}} - 1}. \end{aligned} \tag{10.3}$$

Then the estimator of the common standard deviation, S_{UL}, of the two sample sets is defined as

$$S_{UL} = \sqrt{\frac{(M_{\mathrm{N}} - 1)s_U^2 + (M - M_{\mathrm{N}} - 1)s_L^2}{D.F.}} \tag{10.4}$$

where $D.F. = M - 2$ is named the degree of freedom in the comparison. Finally, the t statistical value is calculated as

$$t(d) = \frac{\bar{U}_d - \bar{L}_d}{S_{UL}\sqrt{\frac{1}{M_N} + \frac{1}{M - M_N}}} \tag{10.5}$$

which determines the significance of difference between the two sample sets.

Traditionally, the t-value is used to look up the t-distribution table for the p-value. Then, the p-value is compared with the constant threshold of the selected significance level to decide whether or not to accept the hypothesis of difference. To improve the search diversity of SLO, instead of using the constant threshold, SLO self-adjusts the threshold during the optimization process, which also releases the table look-up operation.

Define an attention threshold ξ as the absolute value of the t-value at a random dimension, i.e.:

$$\xi = |t(r)| \tag{10.6}$$

where $r \in \{1, 2, \cdots, D\}$ is a randomly generated dimension index. Based on the threshold, each dimension of the society is marked with an attention symbol $\Gamma(d)$ as

$$\Gamma(d) = \begin{cases} \text{'}\gg\text{'}, & \text{if } t(d) \geq \xi \\ \text{'}\approx\text{'}, & \text{if } -\xi < t(d) < \xi \\ \text{'}\ll\text{'}, & \text{if } t(d) \leq -\xi \end{cases} \tag{10.7}$$

where '$\gg$' and '$\ll$' denote that values of $\boldsymbol{U}_d$ are significantly larger and smaller than those of $\boldsymbol{L}_d$, respectively, whereas '$\approx$' represents that their difference is not significant. The attention symbol will be used to guild the subsequent reproduction and reinforcement procedures.

10.4.2 Implementation of the Reproduction Operator

In the reproduction, for each member $i \in \{1, 2, \cdots, M\}$ in the society, a new behavior vector $\hat{\boldsymbol{x}}_i = [\hat{x}_{i,1}, \hat{x}_{i,2}, \cdots, \hat{x}_{i,D}]$ is going to be generated. Considering the attention symbol of each dimension, if $\Gamma(d) =$ '$\gg$' and '$\ll$' (i.e., the attributes of model members are attractive), $\hat{x}_{i,d}$ imitates a modeling attribute in $\boldsymbol{U}_d$ as

$$\hat{x}_{i,d} = u_{r_i,d} \tag{10.8}$$

where r_i is a random index of members in $\boldsymbol{U}_d$. Otherwise, if the modeling attributes are not so appealing at this dimension, $\hat{x}_{i,d}$ is updated in the following way:

$$\hat{x}_{i,d} = \begin{cases} x_{e_i,d}, & \text{if } \text{rand}(0,1) < PI \\ \text{rand}(lb_d, ub_d), & \text{else if } \text{rand}(0,1) < PR \\ x_{i,d}, & \text{otherwise} \end{cases} \tag{10.9}$$

In the equation, *PI* stands for the probability of imitation. Under the probability, the observer imitates a member $x_{e_i,d}$ that is randomly selected from the entire society. Otherwise, under the probability of randomization, *PR*, the observer reproduces $\hat{x}_{i,d}$ in a completely random way, where lb_d and ub_d denote the lower and upper bounds of the variable at the dth dimension. If neither of these two conditions is satisfied, the value is kept unchanged.

10.4.3 Implementation of the Reinforcement Operator

Further, positive reinforcement and negative punishment are performed on dimensions marked with '≫' and '≪', respectively, as

$$\hat{x}_{i,d} = \begin{cases} u_{r_i,d} + \Delta_{i,d}, & \text{if } \Gamma(d) = \text{'}\gg\text{'} \\ u_{r_i,d} - \Delta_{i,d}, & \text{if } \Gamma(d) = \text{'}\ll\text{'} \end{cases} \tag{10.10}$$

where

$$\Delta_{i,d} = rand(0,1) \cdot \left|(u_{r_i,d} - x_{i,d})\right| \tag{10.11}$$

is the reinforcement step length. This way, for the dimensions that the model members possess significantly larger value than the non-model members, the observers further increase the learned value. In contrast, the learned value is further decreased when the attractive attributes are much smaller.

By the vicarious reinforcement, there is a big chance for observers to overtake their models. This mechanism brings a competition effect among the social members since the previous models that are overtaken by the others would strive to win back their leading positions. The ultimate effect here is the progress (or evolution) of the entire society.

10.4.4 Implementation of the Motivation Operator

Finally, the motivation step is performed at the end of each iteration. After evaluating the score $\hat{S}_i = f(\hat{\boldsymbol{x}}_i)$, member i will express its new behavioral pattern $\hat{\boldsymbol{x}}_i$ if and only if the evaluation score is improved.

$$\hat{\boldsymbol{x}}_i \leftarrow \begin{cases} \hat{\boldsymbol{x}}_i, & \text{if } f(\hat{\boldsymbol{x}}_i) \text{ is better than } f(\boldsymbol{x}_i) \\ \boldsymbol{x}_i, & \text{otherwise} \end{cases} \tag{10.12}$$

SLO performs the above four operators iteratively until the terminal condition is satisfied. The terminal condition could be a maximum number of iterations or a predefined acceptable value of the result. At the end of optimization, the best behavior vector with the highest score is output as the result of the problem.

Algorithm 10.1 presents the pseudocode of SLO, where Eqs. (10.8) and (10.10) are integrated together for simplification. It can be seen that the implementations of the algorithm are relatively simple. For public use, the source code of SLO is available online [32].

10.5 BEHAVIORAL ANALYSIS

The behavioral analysis of SLO provides the following two perspectives. First, it shows how SLO obtains the global optimal solution in solving an optimization problem. Second, as SLO can be considered as a simulation of the natural social learning system, the optimization process of SLO in some ways confirms or reveals the effect of humans' observational learning in a social group by computer simulation.

To investigate the search behavior of SLO, two typical benchmark problems are tested, namely the unimodal Sphere problem and the multimodal Schwefel problem [33, 34]. The landscapes of the two problems are depicted in Figures 10.3a and b, respectively. It should be emphasized that SLO is in the absence of any prior information of the problem landscapes to be optimized, neither does

ALGORITHM 10.1 SOCIAL LEARNING OPTIMIZATION

1: Initialize a society of M members randomly;
2: Evaluate all members by the objective f;
3: **while** (*stopping criterion is not satisfied*) **do**
4: Divide the society into $\boldsymbol{U}$ and $\boldsymbol{L}$;
5: **for** $d = 1$ **to** D **do**
6: Perform t-test between $\boldsymbol{U}_d$ and $\boldsymbol{L}_d$ to calculate $t(d)$;
7: **end for**
8: Generate a dimension index r and set $\xi = |t(r)|$;
9: **for** $i = 1$ **to** M **do**
10: **for** $d = 1$ **to** D **do**
11: Select an attribute $u_{r_i,d}$ in $\boldsymbol{U}_d$ randomly;
12: $\Delta_{i,d} = rand(0,1) \cdot |(u_{r_i,d} - x_{i,d})|$;
13: **if** $t(d) \geq \xi$ **then**
14: $\hat{x}_{i,d} = u_{r_i,d} + \Delta_{i,d}$;
15: **else if** $t(d) \leq \xi$ **then**
16: $\hat{x}_{i,d} = u_{r_i,d} - \Delta_{i,d}$;
17: **else**
18: Set $\hat{x}_{i,d}$ according to Eq. (10.9);
19: **end if**
20: Restrict $\hat{x}_{i,d}$ in the variable range;
21: **end for**
22: Evaluate $\hat{\boldsymbol{x}}_i$;
23: Update $\boldsymbol{x}_i$ according to Eq. (10.12);
24: **end for**
25: **end while**
26: Output the best $\boldsymbol{x}_i^*$ and S_i^* as the final solution.

it require the problem to satisfy some mathematical properties such as continuity, differentiability, and convexity. In the initialization, all members' behavior vectors are distributed in a uniformly random way in the problem space. Afterward, SLO conducts its problem-independent operation (except the evaluation of consequence scores) to search for the optimum.

10.5.1 Observations in Natural Conditions

In searching the Sphere problem space involving a single minimum, within the iterations of SLO (denoted as T), the behavior vectors gradually descend to the minimum, as can be observed in Figure 10.3a. This demonstrates the good exploitation ability and local convergence of SLO.

Further, the global exploration ability can be verified by testing on the Schwefel problem. As shown in Figure 10.3b, the landscape of the Schwefel problem involves a number of local optima. Some deep local optima look similar to the global optimum, but they are located very far from it. These local optima are named deceptive basins. Traditionally, it is easy for an optimization algorithm to get trapped into the deceptive basins. Considering the iterations of SLO, it can be observed in Figure 10.3b that, in the early stage of the optimization, the algorithm conducts full-on exploration in the problem space. Although the members are distributed around some deceptive optima at the 20th and 50th iteration, they can finally converge to the global optimum in 200 iterations. Afterward, SLO performs local exploitation as solving unimodal problems.

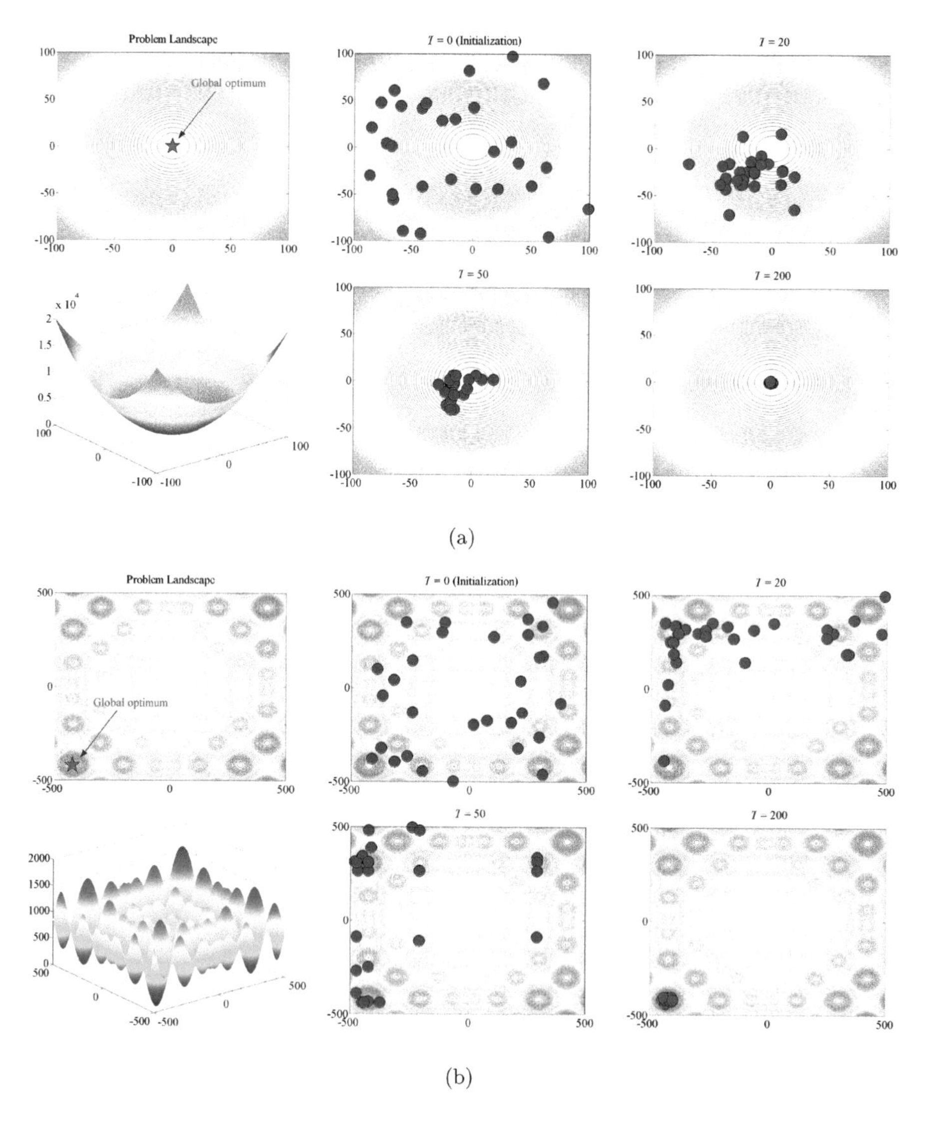

FIGURE 10.3 Search behavior of SLO in unimodal and multimodal landscapes. T stands for the number of iterations. (a) Optimization process for the Sphere problem. The landscape contains a single minimum. Members are randomly initialized and gradually descend to the minimum with iterations. (b) Optimization process for the Schwefel problem. The landscape contains many local minima. At $T = 0$, members are randomly initialized. At $T = 20$, many members are converging to a deceptive local minimum. At $T = 50$, more members find the global minimum. At $T = 200$, all members converge to the global optimum.

10.5.2 Observations in Severe Conditions

In addition, to investigate the search behavior of SLO in severe conditions, the worst-case initialization is applied to replace the uniform initialization. As depicted in Figures 10.4a and b, members in the society of SLO are initialized within a small area located far away from the global optimum. Figure 10.4a reveals an interesting phenomenon: even though the initial society is short of diversity, after some iterations of social learning, the diversity of the society can be restored. Then, by the motivation operation, inferior patterns are discarded and the entire group converges to the minimum.

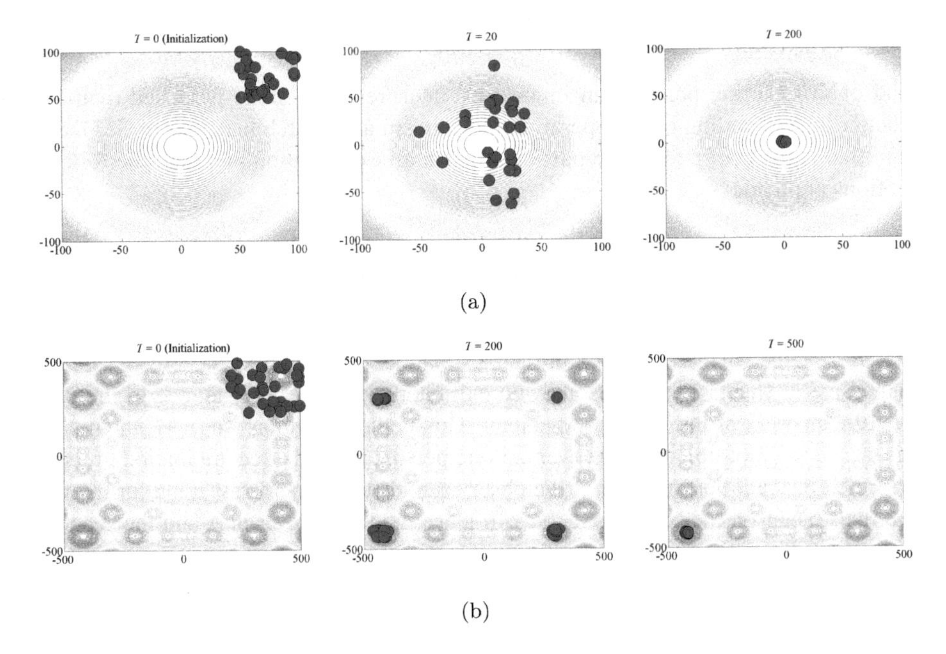

FIGURE 10.4 Search behavior of SLO in unimodal and multimodal landscapes (severe conditioned). T stands for the number of iterations. (a) Optimization process for the Sphere problem. At $T = 0$, members are initialized in a small domain so that their diversity is poor. At $T = 20$, members explore the landscape, resulting in more diversed society distribution. At $T = 200$, members converge to the optimum. (b) Optimization process for the Schwefel problem. At $T = 0$, members are initialized near a deceptive local optimum. At $T = 200$, members explore four promising basins simultaneously. At $T = 500$, all members reach the global optimum.

In Figure 10.4b, by the worst-case initialization, the entire society is generated near the deceptive local optimum located the farthest from the global optimum. Nevertheless, the society is not trapped in the local optimum. At the 200th iteration, after an exploration process, members successfully find four promising basins in the landscape and conduct local exploitation on them to determine which one is the best. Finally, at the 500th iteration, members successfully converge to the global optimum. The above behavioral analysis shows the powerfulness of SLO in optimization, which possesses both strong global exploration and local exploitation abilities even with a severe-conditioned initial environment.

10.5.3 Inspiration to the Human Society

From the perspective of psychology, SLO emulates the social learning system so that it in some ways reveals the outcomes of observational learning in a social society. The search process of SLO shows that, by observing and learning through influential models, members in the society do not simply copy the actions of others; instead, they make further improvements on the learned behavioral patterns, resulting in producing new actions better than any existing ones. Therefore, the observational learning in human society is not purely behavioral, but it is a cognitive and reciprocal process. As can be observed in the above simulation of SLO, the society makes innovation and progress constantly in both natural and severe conditions. The society exhibits a level of intelligence that is greater than the simple collection of all personal intelligence in the group. Because of such social intelligence, our human society can defeat adverse circumstances and achieve unprecedented success in the history.

10.5.4 An Interesting Interpretation of SLO

The end of this section presents an interesting interpretation of the proposed method, which helps better understanding of the operation mechanism and search behavior of SLO. Suppose a situation that a group of people are participating in an examination in which the following two assumptions are made:

- All participants are aggressive that they strive to achieve the highest score in the test.
- Communication is allowed in the examination so that everybody can refer to the answers of the others.

Now, let us imagine what will happen in such a situation. First, for reference, the participants will watch the answer sheets of the persons who exhibited excellent performance in the past, while ignoring the answers of the mediocre persons. This is the "attention" step. Meanwhile, for each participant, he will not imitate all answers of one person; instead a recombination procedure will be performed instead. For example, if the observer finds that *A* performs good for the first query while *B* does better on the second one, he will learn the first answer of *A* and the second answer of *B* together. This is the "reproduction" step, by which the observer can surpass both *A* and *B*. Then, as all participants possess their own knowledge and option for the queries, they will further reinforce the answers. The "reinforcement" step brings an effect of mutual improvement, because the reinforced answer will be referred by the others for the purpose of further enhancement, and so forth. If the examination has a long duration of time, every participant will submit the same and best answer sheet of the test (a.k.a., "motivation" and convergence).

As the ultimate effect of observational learning, the group of participants cooperates to produce an answer sheet with the highest score. This final solution cannot be achieved by any single person, even the most intelligent and knowledgeable one in the group. The example well explains the social intelligence of human society as well as the core method and procedure of SLO.

10.6 EXPERIMENTS AND COMPARISONS

10.6.1 Experimental Setup

In the experiments undertaken, eight continuous benchmark optimization problems with different characteristics are used to test the performance of SLO. The formulations and two-dimensional (2D) fitness landscapes of the benchmarks are presented in Table 10.1. P1 and P2 are unimodal problems, where P1 is separable and P2 is highly non-separable. P3–P8 are multimodal problems. P3 is the Rosenbrock problem with strong reliance between successive variables. P4 is the Schwefel problem 2.26 with lots of deceptive local optima that are similar to the global optimum. P5 and P6 contain a large number of local optima, where P5 is separable and P6 is non-separable. P7 and P8 have composite objective functions, which are extremely difficult to solve even by state-of-the-art EC algorithms.

The proposed SLO algorithm is further compared with classical EC algorithms for continuous optimization, including the GA [9], PSO [15], and two DE variants: DE/rand/1 and DE/best/1 [31]. Note that only the baseline EC algorithms are compared, because the SLO described in this chapter is also a baseline. The improvements made for the traditional algorithms, such as parameter adaptation, multi-swarm technique, and local search, can all be applied to enhance the SLO.

Considering the parameter settings, the society size is set as $M = 30$, among which half members are selected as models (i.e., $M_N = 15$). The two probability parameters, *PI* and *PR*, are empirically set to 0.7 and 0.2, respectively. Also, the parameters of the other compared algorithms are set according to the values recommended in the literature [9, 15, 31]. In GA, the probability of

TABLE 10.1
Benchmark Problems for Optimization

Type	Problem	Objective Function	Domain	Optimum	Name of Problem	Landscape in 2D
Unimodal	P1	$f_1(\boldsymbol{x})=\sum_{i=1}^{D}x_i^2$	$[-100, 100]^D$	0	Sphere	P1
	P2	$f_2(\boldsymbol{x})=\sum_{i=1}^{D}(\sum_{j=1}^{i}x_j)^2$	$[-100, 100]^D$	0	Schwefel 1.2	P2
Multimodal	P3	$f_3(\boldsymbol{x})=\sum_{i=1}^{D-1}[100(x_{i+1}-x_i^2)^2+(x_i-1)^2]$	$[-30, 30]^D$	0	Rosenbrock	P3 (The problem is multimodal when $D>2$)
	P4	$f_4(\boldsymbol{x})=\sum_{i=1}^{D}-x_i\sin(\sqrt{x_i})$	$[-500, 500]^D$	−12569.5	Schwefel 2.26	P4
	P5	$f_5(\boldsymbol{x})=\sum_{i=1}^{D}[z_i^2-10\cos(2\pi z_i)+10]-330,\ \boldsymbol{z}=\boldsymbol{x}-\boldsymbol{o}$	$[-5.12, 5.12]^D$	−330	Rastrigin	P5
	P6	$f_6(\boldsymbol{x})=\frac{\pi}{D}\{10\sin^2(\pi y_1)+\sum_{i=1}^{D-1}(y_i-1)^2[1+10\sin^2(\pi y_{i+1})]+(y_D-1)^2\}+\sum_{i=1}^{D}u(x_i,a,k,m)$	$[-50, 50]^D$	0	Penalize	P6
	P7	$f_7(\boldsymbol{x})=\sum_{i=1}^{D}(A_i-B_i(\boldsymbol{x}))^2-460$	$[-\pi, \pi]^D$	−460	Schwefel 2.13	P7
	P8	$f_8(\boldsymbol{x})=f_{gw}(f_{rb}(z_1,z_2))+f_{gw}(f_{rb}(z_2,z_3))+\ldots+f_{gw}(f_{rb}(z_{D-1},z_D))-130,\ \boldsymbol{z}=\boldsymbol{x}-\boldsymbol{o}+1$	$[-3, 1]^D$	−130	Griewank plus Rosenbrock	P8

In f_6, $y_i=1+\frac{1}{4}(x_i+1)$, $u(x_i,a,k,m)=\begin{cases}k(x_i-a)^m, & x_i>a\\ 0, & -a\le x_i\le a\\ k(-x_i-a)^m, & x_i<-a\end{cases}$, $a=10$, $k=100$, $m=4$

In f_7, $A_i=\sum_{j=1}^{D}(a_{ij}\sin\alpha_j+b_{ij}\cos\alpha_j)$, $B_i(\boldsymbol{x})=\sum_{j=1}^{D}(a_{ij}\sin x_j+b_{ij}\cos x_j)$, a_{ij} and b_{ij} are random integers in $[-100, 100]$, and α_{ij} is random numbers in $[-\pi, \pi]$

In f_8, $f_{gw}(\boldsymbol{x})=1/4000\sum_{i=1}^{D}x_i^2-\prod_{i=1}^{D}\cos(x_i/\sqrt{i})+1$, $f_{rb}(\boldsymbol{x})=\sum_{i=1}^{D-1}[100(x_{i+1}-x_i^2)^2+(x_i-1)^2]$

crossover and mutation are set as $PC = 0.7$ and $PM = 0.07$. In PSO, the inertia weight ω linearly decreases from 0.9 to 0.4 while the accelerating coefficients are set as $c_1 = c_2 = 2.0$. For the two DE variants, the scalar factor F and crossover rate CR are set to 0.5 and 0.1, respectively.

The algorithms are tested on the benchmarks with the dimensionality $D = 30$. For a fair comparison, all algorithms are allowed to conduct 300,000 problem evaluations. Each algorithm runs 30 times independently with the statistical result recorded.

10.6.2 Experimental Results

Table 10.2 reports the numerical results achieved by the five algorithms, where the best results are marked in bold. The table shows that the SLO algorithm outperforms the other algorithms on more instances. To present and compare the algorithmic performance more clearly, Figure 10.5

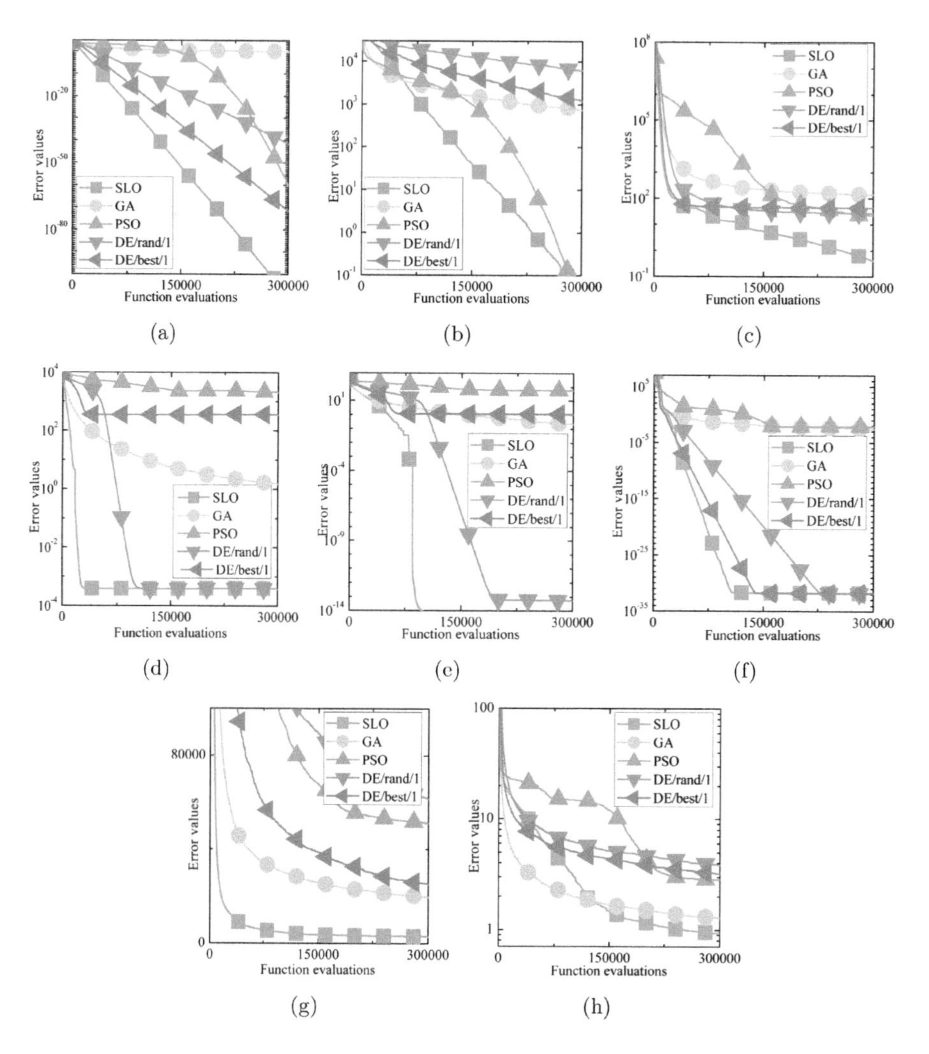

FIGURE 10.5 Convergence curves of the SLO, GA, PSO, DE/rand/1, and DE/best/1. The horizontal axis is the number of function evaluations, whereas the vertical axis indicates the obtained error values of different algorithms. (a) P1. (b) P2. (c) P3. (d) P4. (e) P5. (f) P6. (g) P7. (h) P8.

TABLE 10.2
Statistic Error Values Achieved by the Five Algorithms on the Problems

	SLO	GA	PSO	DE/rand/1	DE/best/1
Problem	**Mean ± Std**	**Mean ± Std**	**Mean ± Std**	**Mean ± Std**	**Mean ± Std**
P1	$\mathbf{4.57 \times 10^{-110} \pm 1.03 \times 10^{-109}}$	0.523104 ± 0.222082	$3.82 \times 10^{-61} \pm 7.88 \times 10^{-61}$	$2.12 \times 10^{-42} \pm 1.14 \times 10^{-42}$	$2.04 \times 10^{-72} \pm 1.91 \times 10^{-72}$
P2	$3.05 \times 10^{-2} \pm 4.22 \times 10^{-2}$	732.068 ± 330.48	$\mathbf{1.43 \times 10^{-2} \pm 1.73 \times 10^{-2}}$	6165.22 ± 1315.63	1193.15 ± 339.632
P3	**0.396473 ± 0.747604**	141.466 ± 69.8255	22.3881 ± 21.3592	27.1916 ± 2.03984	44.2551 ± 30.1971
P4	$\mathbf{3.82 \times 10^{-4} \pm 0}$	1.51579 ± 0.624245	2299.74 ± 395.116	$\mathbf{3.82 \times 10^{-4}} \pm 6.89 \times 10^{-13}$	342.13 ± 238.36
P5	**0 ± 0**	$0.225764 \pm 7.71 \times 10^{-2}$	55.9332 ± 21.1445	$5.31 \times 10^{-14} \pm 1.44 \times 10^{-14}$	1.05927 ± 1.12121
P6	$1.86 \times 10^{-32} \pm 7.13 \times 10^{-33}$	$2.40 \times 10^{-3} \pm 3.01 \times 10^{-3}$	$6.91 \times 10^{-3} \pm 2.63 \times 10^{-2}$	$\mathbf{1.57 \times 10^{-32} \pm 2.78 \times 10^{-48}}$	$\mathbf{1.57 \times 10^{-32} \pm 2.78 \times 10^{-48}}$
P7	**2996.58 ± 3169.85**	19713.3 ± 11625.8	51061 ± 55767.4	61576.6 ± 8052.4	25350.4 ± 8282.55
P8	**0.932328 ± 0.338443**	1.27586 ± 0.223778	2.84368 ± 0.508335	3.8632 ± 0.283541	3.21254 ± 0.388018
w / t / l	– / – / –	8 / 0 / 0	7 / 0 / 1	6 / 1 / 1	7 / 0 / 1

Note: *w / t / l* stands for the number of problems SLA wins / ties / loses the other algorithms.
The bold text indicates the best result obtained on each problem instance.

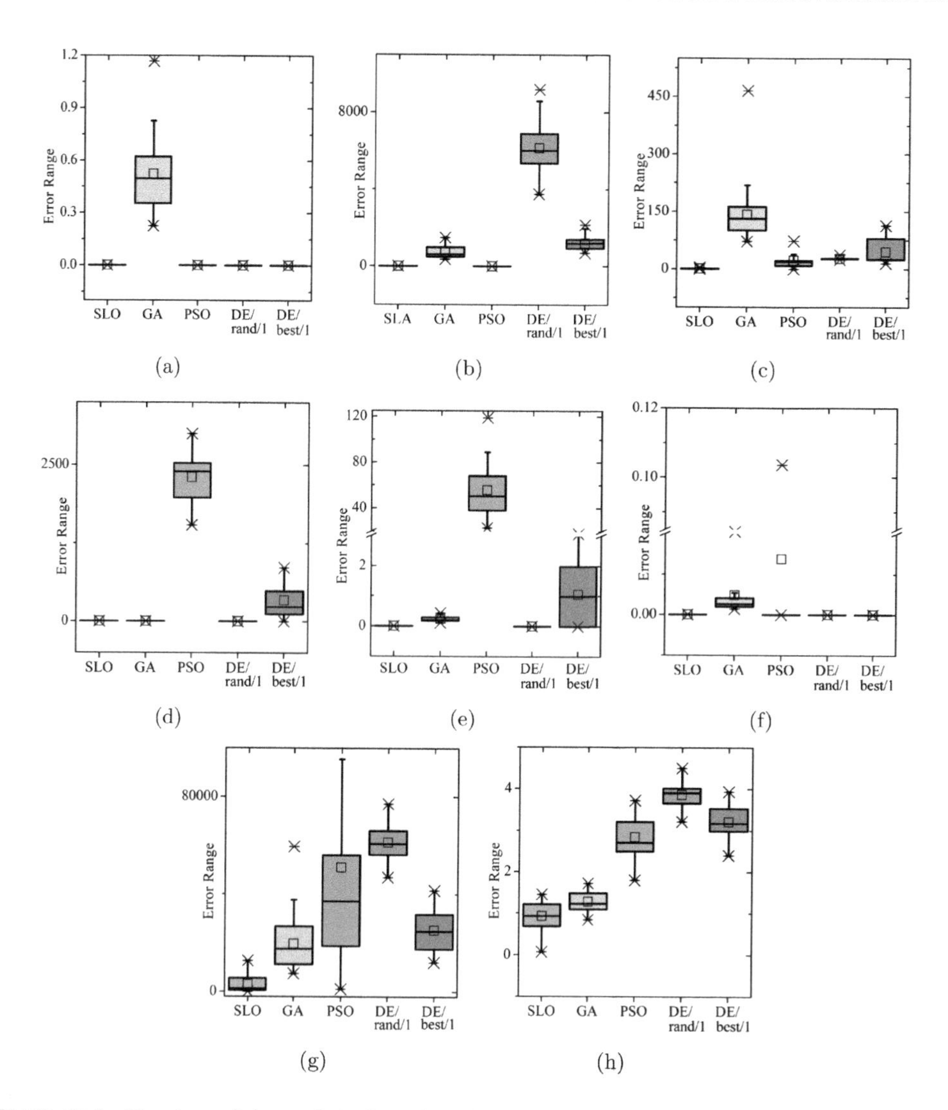

FIGURE 10.6 Boxplots of the statistical results achieved by the SLO, GA, PSO, DE/rand/1, and DE/best/1. The maximum, minimum, upper quartile, lower quartile, mean, medium, and outliers of the obtained results are depicted. (a) P1. (b) P2. (c) P3. (d) P4. (e) P5. (f) P6. (g) P7. (h) P8.

plots the convergence curves of different algorithms on the eight problems, whereas Figure 10.6 uses boxplots to show the statistical results.

In optimizing unimodal problems P1 and P2, SLO exhibits the highest convergence rate and achieves the highest solution accuracy, followed by the PSO, as can be observed in Figures 10.5a and b. DE/rand/1 and DE/best/1 performed well on the separable P1, but they reached inferior results on P2, which is a highly non-separable problem. Also, GA failed to provide high solution accuracy on the two problems, which means that the local exploitation ability of GA is not as good as the others.

In optimizing multimodal problems, as presented in Figures 10.5c–h, PSO and DE/best/1 are easily trapped in local optima. The two algorithms both employ excessively greedy information to guide the search so that their global exploration ability is reduced. Although GA is not so easy to get trapped, it still cannot obtain satisfactory results because of the weak local exploitation ability. In contrast, SLO and DE/rand/1 possess both good exploration and exploitation capabilities and hence outperform the other algorithms. However, the performance of DE/rand/1 decreases in dealing with more complex instances, such as the P7 and P8. SLO outperforms DE/rand/1 in the optimization of these two problems. Moreover, the proposed algorithm converges faster than DE/rand/1 so that it can obtain the global optimum more quickly.

Considering the boxplots in Figure 10.6, SLO provides promising statistical results on P1–P6. In solving P7 and P8, although the efficiency of the SLO is reduced, it is still the best among the algorithms compared. The solution accuracy achieved by the GA is not high for almost all the problems, as anticipated. The PSO is very efficient, but it may get trapped in poor local optima when solving multimodal problems. The two DE variants are also very promising on a number of problems, especially the DE/rand/1, but the algorithms offer inferior results on P2, P7, and P8.

10.6.3 Parameter Investigation

It is to be noticed that SLO involves four control parameters, namely, the society size *M*, the number M_N for selecting model members, and the probabilities *PI* and *PR* of imitation and randomization. This subsection is dedicated to investigate the sensitivity of the SLO algorithm to these four parameters.

As the model group is a subset of the entire society, the setting of M_N is dependent on the setting of *M*. To keep the superiority of the model group, it is suggested to use $M_N \le M/2$. In the investigation, the algorithm is tested with six different pairs of settings: $(M, M_N) = (30,10)$, $(30,15)$, $(50,10)$, $(50,25)$, $(100,10)$, and $(100,50)$, while *PI* and *PR* are fixed at 0.7 and 0.2. The convergence curves on the unimodal problem P1 and multimodal problem P6 are depicted in Figures 10.7a and b, respectively. Two observations can be obtained in the comparison. First, the smaller the society size is, the faster the convergence of the algorithm exhibits, as can be expected. Second, by setting $M_N = M/2$, the performance of the algorithm is more stable because M_N determines the sample sets joining the statistical hypothesis test. The more samples used, the more reliable are the significance test results.

According to the social leaning theory that most human behaviors are learned from imitating others, it is suggested to use a relatively large probability of imitation in the algorithm. On the other side, to avoid destroying the current solution structures, the probability of randomization should be set to small values. In the investigation, six different pairs of parameter settings are used: $(PI, PR) = (0.6,0.1), (0.6,0.2), (0.6,0.3), (0.7,0.1), (0.7,0.2)$, and $(0.7,0.3)$, while *M* and M_N are set as 30 and 15. The results are shown in Figures 10.7c and d. First, it can be seen that, all the parameter settings can contribute to satisfactory results of SLO on the two problems. Second, under the same settings of *PR*, by using $PI = 0.7$, the algorithm performs more imitation and hence converges faster than those of $PI = 0.6$. Third, with the same *PI*, adopting smaller *PR*, such as $PR = 0.1$, helps faster convergence, whereas a relatively larger *PR* value increases the search diversity.

10.7 CONCLUSIONS

Many scientific and engineering applications involve hard optimization problems. To deal with these problems, SLO is developed based on the social learning theory. Specifically, SLO imitates a high form of intelligence in nature, the social intelligence of humans, to seek the global best solution. The main process includes four basic operators, i.e., attention, reproduction, reinforcement, and motivation, based on which the society converges to the optimum of the problem at hand. Experimental results have demonstrated the effectiveness and efficiency of this new

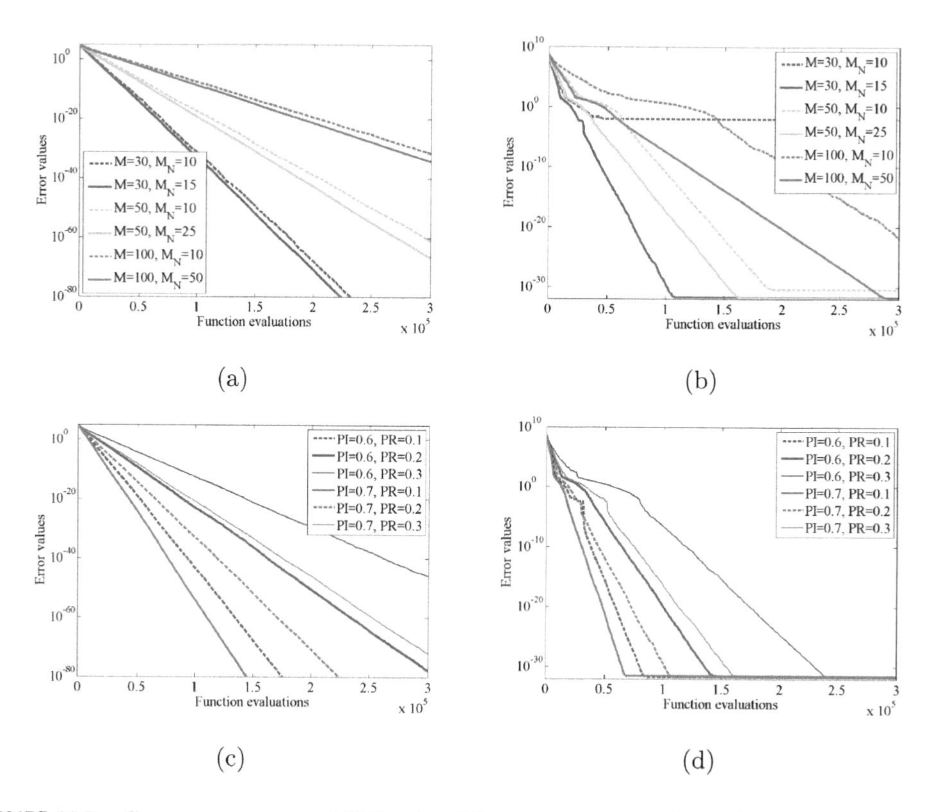

FIGURE 10.7 Convergence curves of SLO using different parameter settings. (a) Investigating (M, M_N) settings on the unimodal problem P1. (b) Investigating (M, M_N) settings on the multimodal problem P6. (c) Investigating (PI, PR) settings on the unimodal problem P1. (d) Investigating (PI, PR) settings on the multimodal problem P6.

algorithm, which has in turn also verified, through computer simulations, the outcomes of the social learning behavior in human society.

SLO provides a simple, yet powerful tool for global optimization. As a new optimization paradigm, SLO can stimulate further studies, such as the adjustment or adaptation of parameters, fine-tuning of operators, hybridization with other techniques, discretization, and parallelization. It would be appealing to extend the SLO algorithm for multi-objective, constrained, multi-solution, or dynamic optimization scenarios. The promising results of SLO also encourage a wide range of real-world applications of the algorithm, ranging from parameter optimization to system control. The methodological and performance characteristics of SLO are likely to set a new trend in nature-inspired computing.

REFERENCES

1. Mohammed R AlRashidi and Mohamed E El-Hawary. A survey of particle swarm optimization applications in electric power systems. *IEEE Transactions on Evolutionary Computation*, 13(4):913–918, 2009.
2. Mateusz Guzek, Pascal Bouvry, and El-Ghazali Talbi. A survey of evolutionary computation for resource management of processing in cloud computing. *IEEE Computational Intelligence Magazine*, 10(2):53–67, 2015.

3. Gitae Kim, Yew-Soon Ong, Chen Kim Heng, Puay Siew Tan, and Nengsheng Allan Zhang. City vehicle routing problem (city vrp): A review. *IEEE Transactions on Intelligent Transportation Systems*, 16(4):1654–1666, 2015.
4. Yi-Ning Ma, Yue-Jiao Gong, Chu-Feng Xiao, Ying Gao, and Jun Zhang. Path planning for autonomous underwater vehicles: An ant colony algorithm incorporating alarm pheromone. *IEEE Transactions on Vehicular Technology*, 68(1):141–154, 2018.
5. Janusz Kusyk, M Umit Uyar, and Cem Safak Sahin. Survey on evolutionary computation methods for cybersecurity of mobile ad hoc networks. *Evolutionary Intelligence*, 10(3–4):95–117, 2018.
6. Eric Bonabeau, Marco Dorigo, and Guy Theraulaz. Inspiration for optimization from social insect behaviour. *Nature*, 406(6791):39–42, 2000.
7. David B Fogel. *Evolutionary Computation: Toward a New Philosophy of Machine Intelligence*, volume 1. John Wiley & Sons, New York, 2006.
8. Yue-Jiao Gong, Wei-Neng Chen, Zhi-Hui Zhan, Jun Zhang, Yun Li, Qingfu Zhang, and Jing-Jing Li. Distributed evolutionary algorithms and their models: A survey of the state-of-the-art. *Applied Soft Computing*, 34:286–300, 2015.
9. Stephanie Forrest. Genetic algorithms: principles of natural selection applied to computation. *Science*, 261(5123):872–878, 1993.
10. John R Koza, Forrest H Bennett III, and Oscar Stiffelman. Genetic programming as a Darwinian invention machine. In *Genetic Programming*, pages 93–108. Springer, Berlin, 1999.
11. Swagatam Das, Sankha Subhra Mullick, and PN Suganthan. Recent advances in differential evolution–an updated survey. *Swarm and Evolutionary Computation*, 27:1–30, 2016.
12. Ferrante Neri and Ville Tirronen. Recent advances in differential evolution: a survey and experimental analysis. *Artificial Intelligence Review*, 33(1–2):61–106, 2010.
13. Rainer Storn and Kenneth Price. Differential evolution–a simple and efficient heuristic for global optimization over continuous spaces. *Journal of Global Optimization*, 11(4):341–359, 1997.
14. Marco Dorigo, Mauro Birattari, and Thomas Stützle. Ant colony optimization. *IEEE Computational Intelligence Magazine*, 1(4):28–39, 2006.
15. James Kennedy. Particle swarm optimization. In *Encyclopedia of Machine Learning*, pages 760–766. Springer, Boston, MA, 2011.
16. Yue-Jiao Gong, Jing-Jing Li, Yicong Zhou, Yun Li, Henry Shu-Hung Chung, Yu-Hui Shi, and Jun Zhang. Genetic learning particle swarm optimization. *IEEE Transactions on Cybernetics*, 46:2277–2290, 2016.
17. Dervis Karaboga and Bahriye Basturk. A powerful and efficient algorithm for numerical function optimization: artificial bee colony (ABC) algorithm. *Journal of Global Optimization*, 39(3):459–471, 2007.
18. Yue-Jiao Gong, Jun Zhang, and Yun Li. From the social learning theory to a social learning algorithm for global optimization. In *2014 IEEE International Conference on Systems, Man and Cybernetics (SMC)*, pages 222–227. IEEE, 2014.
19. Manfred Fahle and Shimon Edelman. Long-term learning in Vernier acuity: Effects of stimulus orientation, range and of feedback. *Vision Research*, 33(3):397–412, 1993.
20. Albert Bandura and David C McClelland. *Social Learning Theory*. Prentice-Hall, Englewood Cliffs, NJ, 1977.
21. Suresh Satapathy and Anima Naik. Social group optimization (SGO): a new population evolutionary optimization technique. *Complex & Intelligent Systems*, 2(3):173–203, 2016.
22. Zhi-Zhong Liu, Dian-Hui Chu, Cheng Song, Xiao Xue, and Bao-Yun Lu. Social learning optimization (SLO) algorithm paradigm and its application in QOS-aware cloud service composition. *Information Sciences*, 326:315–333, 2016.
23. Meeta Kumar, Anand J Kulkarni, and Suresh Chandra Satapathy. Socio evolution & learning optimization algorithm: A socio-inspired optimization methodology. *Future Generation Computer Systems*, 81:252–272, 2018.
24. Jonathan R Brauer and Charles R Tittle. Social learning theory and human reinforcement. *Sociological Spectrum*, 32(2):157–177, 2012.
25. A Nicolle, M Symmonds, and RJ Dolan. Optimistic biases in observational learning of value. *Cognition*, 119(3):394–402, 2011.
26. Jeanne Ellis Ormrod and Kevin M Davis. *Human Learning*. Merrill, London, 2004.

27. Luke Rendell, Robert Boyd, Daniel Cownden, Marquist Enquist, Kimmo Eriksson, Marc W Feldman, Laurel Fogarty, Stefano Ghirlanda, Timothy Lillicrap, and Kevin N Laland. Why copy others? Insights from the social learning strategies tournament. *Science*, 328(5975):208–213, 2010.
28. Henry P Sims Jr and Charles C Manz. Social learning theory: the role of modeling in the exercise of leadership. *Journal of Organizational Behavior Management*, 3(4):55–63, 1982.
29. Esther Herrmann, Josep Call, Mara Victoria Hernández-Lloreda, Brian Hare, and Michael Tomasello. Humans have evolved specialized skills of social cognition: the cultural intelligence hypothesis. *Science*, 317(5843):1360–1366, 2007.
30. John Rice. *Mathematical Statistics and Data Analysis*. Nelson Education, Toronto, ON, Canada, 2006.
31. Swagatam Das and Ponnuthurai Nagaratnam Suganthan. Differential evolution: a survey of the state-of-the-art. *IEEE Transactions on Evolutionary Computation*, 15(1):4–31, 2011.
32. Yue-Jiao Gong. *SLO source code*, 2013 (accessed February 25, 2016). https://github.com/yuejiaogong/social_learning.
33. Tim Blackwell. A study of collapse in bare bones particle swarm optimization. *IEEE Transactions on Evolutionary Computation*, 16(3):354–372, 2012.
34. Xin Yao, Yong Liu, and Guangming Lin. Evolutionary programming made faster. *IEEE Transactions on Evolutionary Computation*, 3(2):82–102, 1999.

11 Constraint Handling in Multi-Cohort Intelligence Algorithm

Apoorva S. Shastri and Anand J. Kulkarni

CONTENTS

11.1 INTRODUCTION

There have been several approximation-based optimization algorithms developed so far. the evolutionary algorithms such as the genetic algorithm (GA) (Goldberg and Holland, 1988) and differential evolution (DE) (Storn and Price, 1997), swarm-based algorithms such as ant colony optimization (ACO) (Dorigo, 1992) and particle swarm optimization (PSO) (Eberhart and Kennedy, 1995), physics-based algorithms such as simulated annealing (SA) (Kirkpatrick and Vecchi, 1983), socio-inspired algorithms such as cohort intelligence (CI) (Kulkarni et al., 2013), and league championship algorithm (LCA) (Kashan, 2014) are some of the examples. These algorithms perform well when solving unconstrained problems; however, their performance is severely affected when applied for solving problems with constraints. The penalty-based methods, feasibility-based approaches, solution repair techniques, and heuristic-based approaches are some of the approaches that have been developed so far to deal with constraints. It is important to mention that these approaches necessitate additional preliminary trials of the algorithm, information about the problems, and so forth. Moreover, these approaches may become tedious when hard constraints are involved as well as increase in number of constraints. So, it is necessary to explore the inherent constraint handling ability of the algorithms. The algorithm of CI is characterized by the inbuilt probability-based approach of handing constraints (Kulkarni and Shabir, 2016; Kulkarni et al., 2016). In 2018, Shastri and Kulkarni proposed multi-cohort intelligence (Multi-CI) in which intra-group learning and inter-group learning mechanisms were implemented. The work presented in this chapter is to further test and validate inbuilt probability-based constraint handling approach for the Multi-CI algorithm. The constrained version of the Multi-CI algorithm is tested on four test problems and two real-world problems. The solutions have been compared with the contemporary algorithms.

This chapter is organized as follows: Section 11.2 describes in detail the constrained Multi-CI. It includes the detailed procedure and flowchart of the algorithm. Section 11.3 provides the solutions and convergence plots for constrained test and engineering design problems. It is followed by the discussion on the results and comparison with several algorithms. The conclusions and a comment on future direction is provided at the end.

11.2 CONSTRAINED MULTI-COHORT INTELLIGENCE

In society several cohorts exist that interact and compete with one another, which is referred to as inter-group learning. This makes the candidates learn from the candidates within the cohort as well as the candidates from other cohorts. In the Multi-CI approach, intra-group learning and inter-group learning mechanisms were implemented unlike CI in which only intra-group learning is incorporated. Due to this, exploration is weak in the CI algorithm. In the intra-group learning mechanism, every candidate based on the roulette wheel approach chooses a behavior from within its own cohort. Then it samples certain behaviors from within the close neighborhood of the chosen behavior. In the inter-group learning mechanism, every candidate based on a roulette wheel approach chooses its behavior from within a pool of best behaviors associated with every cohort. Then it chooses the best behavior by sampling a certain number of behaviors from within the close neighborhood of both behaviors chosen using the intra-group learning and inter-group learning mechanisms. For the Multi-CI algorithm, the inherent probability-based constraint-handling mechanism is tested and validated in this chapter. The flowchart for the constrained Multi-CI algorithm is given in Figure 11.1. The detailed stepwise procedure along with probability distribution used is given below. For the detailed stepwise mathematical formulation and illustration of Multi-CI algorithm, refer to Shastri and Kulkarni (2018).

Step 1: Every candidate in each cohort generates qualities from within its associated sampling interval.

Step 2 (evaluation of behaviors): The pool of objective functions/behaviors of every associated candidate within every cohort is evaluated.

Step 3 (poolZ formation): The best behavior (objective functions with minimum value) candidate in each cohort is chosen and kept in separated PoolZ.

Step 4: Every candidate in each cohort evaluates associated constraint value.

Step 5: For every constraint, a probability distribution is developed (refer to Figure 11.2) with predefined lower and upper limits for probability calculation as $l_{lo,j}^{c}$ *and* $l_{up,j}^{c}$, respectively. In addition, any constraint value if lower than associated $l_{lo,j}^{c}$ or exceeding $l_{up,j}^{c}$ is assigned a very small probability value.

Kulkarni and Shabir (2016) proposed a modified approach to the CI method for solving knapsack problems. This approach makes use of probability distributions for handling constraints. This approach is adopted in constrained Multi-CI. For inequality constraints, the probability distribution is developed (refer to Figure 11.2) and the probability is calculated based on the following rules:

1. If $l_{lo,j}^{c} \leq g_j^c\left(\mathbf{X}\right) \leq 0$, then based on the probability distribution presented in Figure 11.2

$$p = 1 - \left(slope_{1,l_{lo,j}^{c}} \times g_j^c\left(\mathbf{X}\right)\right) \tag{11.1}$$

2. If $0 < g_j^c\left(\mathbf{X}\right) \leq l_{up,j}^{c}$, then based on the probability distribution presented in Figure 11.2

$$p = 1 - \left(slope_{2,l_{up,j}^{c}} \times g_j^c\left(\mathbf{X}\right)\right) \tag{11.2}$$

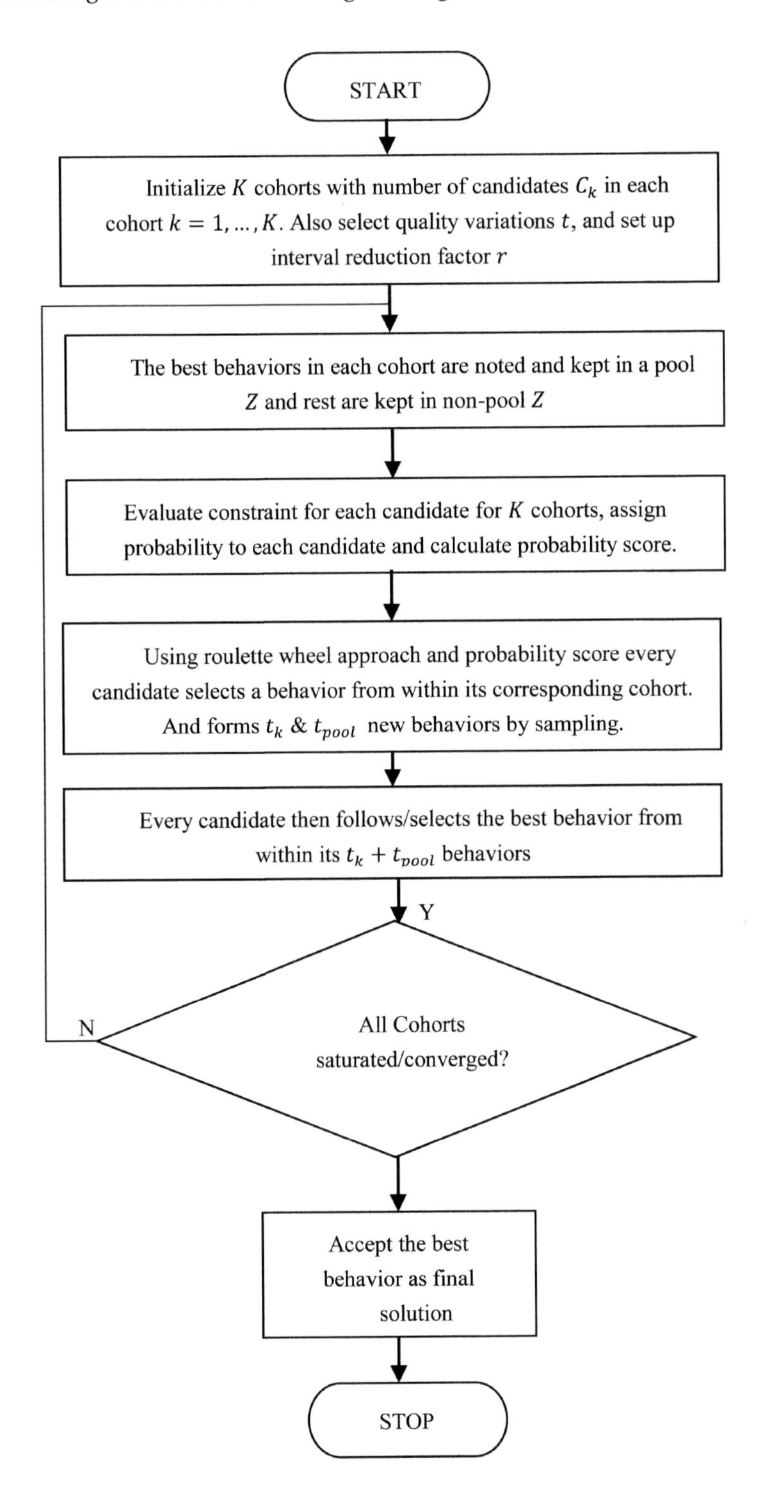

FIGURE 11.1 Flowchart of constrained multi-CI algorithm.

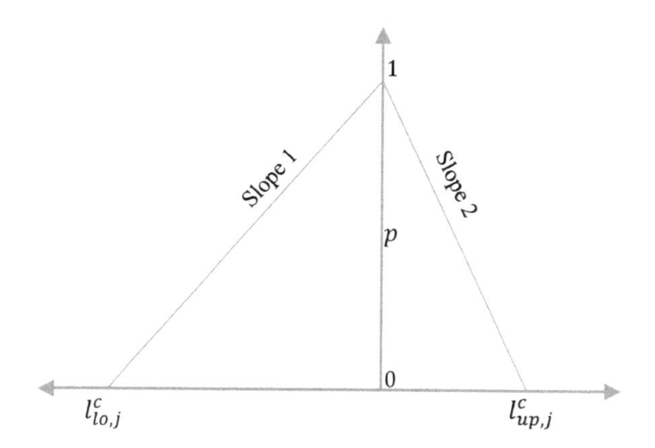

FIGURE 11.2 Probability distribution for constrained multi-CI.

3. If $l^c_{lo,j} \left\langle g^c_j(\mathbf{X}) \right\rangle l^c_{up,j}$, then $p = 0.00001$
where p is probability value, g is constraint value, and c is every individual candidate.
These three rules are applied for every candidate of each cohort.

Step 6: After probability for the individual candidate is calculated, the probability score of each candidate for every cohort is evaluated.

Step 7 (formation of non-pool behaviors): Using the roulette wheel approach and probability score, every candidate selects a behavior from within its corresponding cohort (except poolZ candidates) and form new behaviors by sampling in the close neighborhood of the associated qualities.

Step 8 (formation of poolZ behaviors): Using the roulette wheel approach and probability score every candidate selects a behavior from within poolZ and forms new behaviors by sampling in a close neighborhood of the associated qualities.

Step 9 (selection): Every candidate associated with every cohort selects the best behavior, i.e., minimum objective function value from within its behavior choices.

Step 10 (interval reduction): Every candidate shrinks its range in the close neighborhood of followed candidate using sampling interval reduction factor r. The reduction factor r is chosen based on preliminary trials of the algorithm.

Step 11 (convergence): The algorithm is assumed to have converged, if all of the convergence conditions are satisfied for successive considerable numbers of learning attempts and accept any of the current behaviors as the final solution from within cohorts. Else go to Step 2.

11.3 SOLUTION TO CONSTRAINED TEST AND ENGINEERING DESIGN PROBLEMS

The Multi-CI algorithm with the probability-based method was coded in MATLAB (R2019a) and simulations were taken on a Windows platform using an Intel Core 2 duo, 3 GHz processor speed with 2GB RAM. The Multi-CI was applied for solving four constrained test problems, viz. G02, G04, G06, and G08 and two real-world engineering design problems such as welded beam design and pressure vessel design. The solutions were compared with several other constrained optimization techniques such as GA with a penalty function approach (Deb, 1997; 2000), homomorphous mapping (Koziel and Michalewicz,1999), cultural differential evolution (CDE) (Becerra and Coello Coello, 2006), gradient repair method (Chootinan and Chen, 2006), evolution strategies (ESs) (Montes and Lopez-Ramirez, 2007), modified differential evolution (COMDE) (Mohamed and Sabry, 2012), cooperative coevolutionary differential evolution with improved augmented Lagrangian (CCiALF) (Ghasemishabankareh et al., 2016), CI (Shastri et al., 2016; Shastri et al., 2019), and CI with Panoptic Learning (CI-PL) (Krishnasamy et al., 2021).

11.3.1 Constrained Test Problems

The continuous constrained test problems G02, G04, G06, and G08 have been solved. Every problem was solved 30 times. The best, average and worst solutions for each test problem and comparison with other methods is given in Table 11.1. The associated standard deviation (SD),

TABLE 11.1
Solutions of the Constrained Test Problems

	Problem	G02 Best Avg. Worst	G04 Best Avg. Worst	G06 Best Avg. Worst	G08 Best Avg. Worst
	Best Known Solutions	0.803	–30665.539	–6961.813	0.095
Methods	Koziel and Michalewicz (1999)	0.795	–30664.500	–6952.1	0.096
		0.787	–30655.3	–6342.6	0.089
		0.783	–30645.900	–5473.900	0.029
	Montes et al. (2010)	–	–30665.539	–6961.814	0.096
			–30665.539	–6961.284	0.096
			–30665.539	–6961.482	0.096
	Becerra and Coello Coello (2006).	0.804	–30665.539	–6961.814	0.096
		0.725	–30665.539	–6961.814	0.096
		0.590	–30665.539	–6961.814	0.096
	Chootinan and Chen (2006)	0.801	–30665.539	–6961.814	0.096
		0.785	–30665.539	–6961.814	0.096
		0.745	–30665.539	–6961.814	0.096
	Deb (2000)	–	–30614.814	–	–
			–30196.404		
			–29606.596		
	Ghasemishabankareh et al. (2016)	–0.804	–30665.539	–6961.814	–0.096
		–0.793	–30665.539	–6961.814	–0.096
		–0.762	–30665.539	–6961.814	–0.096
	Mohamed and Sabry (2012)	–0.804	–30665.539	–6961.814	–0.096
		–0.804	–30665.539	–6961.814	–0.096
		N.A.	N.A.	N.A.	N.A.
	Shastri et al. (2016)	0.795	–30664.500	–6952.100	0.096
		0.787	–30655.300	–6342.600	0.089
		0.783	–30645.900	–5473.900	0.029
	Krishnasamy et al. (2021)	0.799	–30665.538	–6959.088	0.096
		0.788	–30665.533	–6903.170	0.096
		0.783	–30665.525	–6637.471	0.096
	Multi-CI	**0.806**	**–30665.410**	**–6961.848**	**0.096**
		0.824	**–30664.288**	**–6961.789**	**0.097**
		0.797	**–30664.090**	**–6961.685**	**0.094**

TABLE 11.2
Detailed Performance of Constrained Multi-CI

Problem	FE	SD	Time (sec)	Parameters (*K*, *r*)
G02	7725	4.00E-02	1.74	3, 0.989
G04	2250	2.00E-03	1.80	3, 0.941
G06	1500	5.00E-03	1.67	3, 0.989
G08	750	2.30E-02	0.80	3, 0.350

number of Function Evaluations (FE), reduction factor r, and numbers of candidates are listed in Table 11.2.

Using a Multi-CI algorithm with a probability-based approach few constrained test problems (Becerra and Coello Coello, 2006) are solved. There are two maximization problems, G02 and G08, and two minimization problems, G04 and G06. The details of the solutions achieved by Multi-CI for these problems are provided here. In problem G02, the best solution obtained was $f(\boldsymbol{X}) = 0.806$ with $\boldsymbol{X}$ = {2.604, 2.714, 2.359, 2.322, 2.549, 2.744, 2.454, 2.779, 2.673, 2.619, 2.678, 2.244, 2.565, 2.536, 2.311, 2.428, 2.360, 2.243, 2.286, 2.639} and the values of $g_i(\boldsymbol{X})$ are {–0.258, –0.000} for the constraints. In problem G08, the best result obtained was $f(\boldsymbol{X}) =$ 0.095 with $\boldsymbol{X}$ = {1.227, 4.245} and the values of $g_i(\boldsymbol{X})$ are {–1.737, –0.167} for the constraints. In problem G04, the best result obtained was $f(\boldsymbol{X}) = -30665.41$ with $\boldsymbol{X}$ = {78.000, 33.000, 30.000, 45.000, 36.754} and the values of $g_i(\boldsymbol{X})$ are {–0.003, –91.997, –11.163, –8.8362, –5.001, 0.001} for the constraints. In problem G06, the best result obtained was $f(\boldsymbol{X}) = -6961.8481$ with $\boldsymbol{X}$ = {14.1100, 0.8429} and the values of $g_i(\boldsymbol{X})$ are {–0.0828, 0.0738} for the constraints. The convergence plots showing best solutions in each cohort for G02, G04, G06, and G08 are shown in Figures 11.3a–d, respectively.

11.3.2 Engineering Design Problems

Two well-studied engineering design problems were used to evaluate the performances of constrained Multi-CI in the real-world domain. For each problem, 30 independent Multi-CI runs were carried out. The solutions were compared with several well-known optimization algorithms mentioned at the beginning of Section 11.3.

11.3.2.1 Welded Beam Design Problem

The diagram for the welded beam design problem with design variables is presented in Figure 11.4. The design variables to be optimized are the thickness of the welds (x_1), the length of the welds (x_2), the height of the beam (x_3), and the width of the beam (x_4).

The mathematical formulation of this problem taken from (Shastri et al., 2019) is given below.

$$\text{Minimize: } f(\boldsymbol{X}) = 1.10471 x_2 x_1^2 + 0.04811 x_3 x_4 (14 - x_2)$$

Subject to:

$$g_1(\boldsymbol{X}) = \tau(x) - 13000 \le 0$$

$$g_2(\boldsymbol{X}) = \sigma(\boldsymbol{X}) - 30000 \le 0$$

$$g_3(\boldsymbol{X}) = x_1 - x_4 \le 0$$

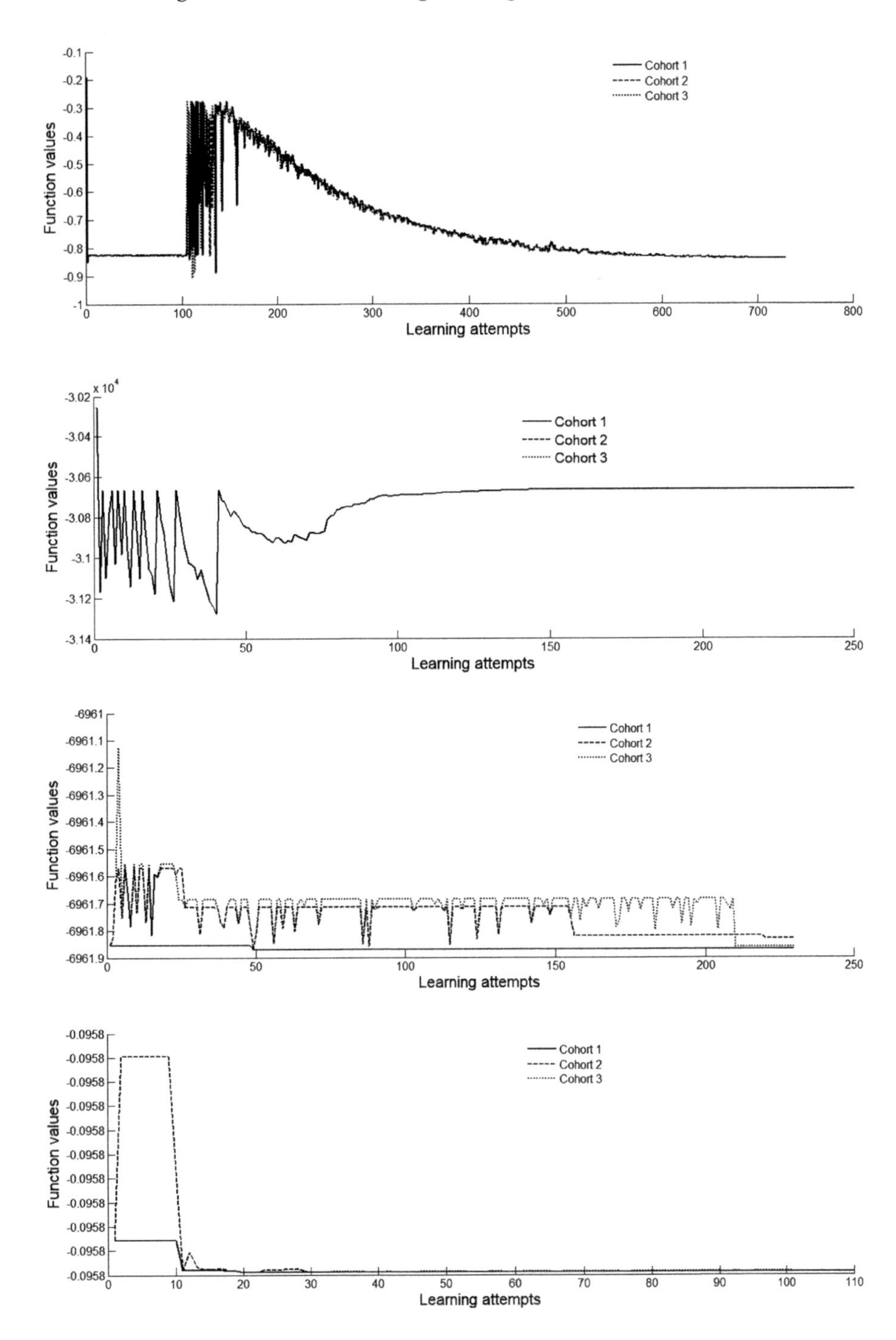

FIGURE 11.3 Convergence plots for constrained test problems.

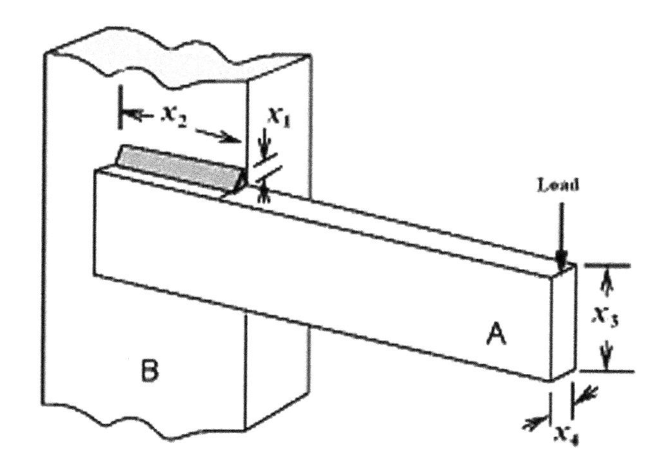

FIGURE 11.4 Welded beam design problems.

$$g_4(X) = 1.10471x_1^2 + 0.04811x_3x_4(14 + x_2) - 5 \le 0$$

$$g_5(X) = 0.125 - x_1 \le 0$$

$$g_6(X) = \delta(X) - 0.25 \le 0$$

$$g_7(X) = 6000 - Pc(X) \le 0$$

where

$$\tau(\vec{x}) = \sqrt{(\tau')^2 + (2\tau'\tau'')\frac{y_2}{2R} + (\tau'')^2}$$

$$\tau' = \frac{6000}{\sqrt{2}y_1y_2}, \ \tau'' = \frac{MR}{J}$$

$$M = 6000\left(14 + \frac{x_2}{2}\right), \ R = \sqrt{\left(\frac{x_2^2}{4}\right) + \left(\frac{x_1 + x}{2}\right)^2}, \ J = 2\left\{x_1x_2\sqrt{2}\left[\left(\frac{x_2^2}{12}\right) + \left(\frac{x_1 + x_3}{2}\right)^2\right]\right\}$$

$$\sigma(X) = \frac{504000}{x_4x_3^3}, \ \delta(X) = \frac{65856000}{(30 \times 10^6)x_4x_3^3},$$

$$Pc(X) = \frac{4.013(30 \times 10^6)\sqrt{\frac{x_3x_4^6}{36}}}{196}\left(1 - \frac{x_3\sqrt{\frac{30\times10^6}{4(12\times10^6)}}}{28}\right)$$

where $0.1 \le x_1 \le 2$, $0.1 \le x_2 \le 10$, $0.1 \le x_3 \le 10$, $0.1 \le x_4 \le 2$

Table 11.3 presents the performance comparison of variables and constraints with different optimization algorithms for solving the welded beam design problem. The simulation results, viz. the best, mean, worst, and SD are summarized in Table 11.4. Number of FEs, reduction factor r,

TABLE 11.3
Variables and Constraint Value Comparison

	Methods								
Details	**Coello Coello and Montes (2002)**	**He and Wang (2006)**	**He and Wang (2007)**	**Deb (2000)**	**Ghasemishabankareh et al. (2016)**	**Mohamed and Sabry (2012)**	**Shastri et al. (2019)**	**Krishnasamy et al. 2021**	**Multi-CI**
x_1	0.206	0.202	0.206	0.249	0.206	NA	0.206	0.206	**0.206**
x_2	3.471	3.544	3.470	6.173	3.470	NA	3.470	3.471	**3.471**
x_3	9.020	9.048	9.037	8.179	9.037	NA	9.037	9.037	**9.042**
x_4	0.206	0.206	0.206	–0.253	0.206	NA	0.206	0.206	**0.205**
$g_1(\boldsymbol{X})$	–0.074	–12.839	0.000	–5758.604	7.24E-10	–1.819E+12	0.044	–0.765	**0**
$g_2(\boldsymbol{X})$	–0.266	–1.247	–0.027	–255.577	2.00E-08	0	0.092	–1.197	**0**
$g_3(\boldsymbol{X})$	0.000	–0.001	0.000	–0.004	3.32E-13	0.00E+00	–1.89E-12	–5.74E-06	**0**
$g_4(\boldsymbol{X})$	–3.430	–3.429	–3.433	–2.983	3.433	–3.433	–3.433	–3.433	**0.000**
$g_5(\boldsymbol{X})$	–0.081	–0.079	–0.081	–0.124	0.081	–0.081	–0.807	–0.807	**0.000**
$g_6(\boldsymbol{X})$	–0.236	–0.235	–0.236	–0.234	0.236	0.236	–0.236	–0.236	**0.000**
$g_7(\boldsymbol{X})$	–58.666	–11.681	–0.030	–4465.271	0.000	3.638E+12	0.056	–0.163	**–1.892**
$f(\boldsymbol{X})$	1.728	1.728	1.725	2.433	1.725	1.725	1.725	1.725	**1.722**

TABLE 11.4
Statistical Solutions of Various Algorithms

Methods	Best	Mean	Worst	SD
Coello Coello and Montes (2002)	1.728	1.792	1.993	0.074
He and Wang (2006)	1.728	1.748	1.782	0.012
He and Wang (2007)	1.724	1.749	1.814	0.040
Deb (2000)	2.381	2.382	2.383	N.A.
Ghasemishabankareh et al. (2016)	1.724	1.724	1.724	5.11E-07
Mohamed and Sabry (2012)	1.724	1.724	1.724	1.60E-12
Shastri et al. (2019)	1.724	1.724	1.720	3.612E-11
Krishnasamy et al. (2021)	1.724	1.724	1.724	8.5554e-06
Multi-CI	**1.721**	**1.723**	**1.724**	**8.00E-05**

and numbers of candidates are listed in Table 11.5. From Table 11.3, it can be observed that the Multi-CI algorithm is able to achieve a comparable solution against other algorithms. In addition, the algorithm exhibited the robust performance (refer to Table 11.4). The convergence plot with best solutions in each cohort for welded beam design problem is shown in Figure 11.5a.

11.3.2.2 Pressure Vessel Design Problem

The diagram for the pressure vessel design problem with design variables is presented in Figure 11.6. The design variables to be optimized are the thickness of the spherical head (x_1), the thickness of the cylindrical skin (x_2), the inner radius (x_3), and the length of the cylindrical segment (x_4).

The mathematical formulation of this problem taken from (Shastri et al., 2019) is as shown below:

$$\text{Minimize: } f(X) = 0.6624x_1x_3x_4 + 1.7781x_2x_3^2 + 3.1661x_1^2x_4 + 19.84x_1^2x_3$$

Subject to:

$$g_1(X) = -x_1 + 0.0193x_3 \leq 0$$

$$g_2(X) = -x_2 + 0.00954x_1x_3 \leq 0$$

$$g_3(X) = -\pi x_3^2x_4^2 - \frac{4}{3}\pi x_3^2 + 1296000 \leq 0$$

$$g_4(X) = x_4 - 240 \leq 0$$

TABLE 11.5
Detailed Performance of Constrained Multi-CI for Engineering Design Problems

Problem	FE	Time (sec)	Parameters (K, r)
Welded Beam Design Problem	4500	1.42	3, 0.965
Pressure Vessel Design Problem	3000	1.35	3, 0.395

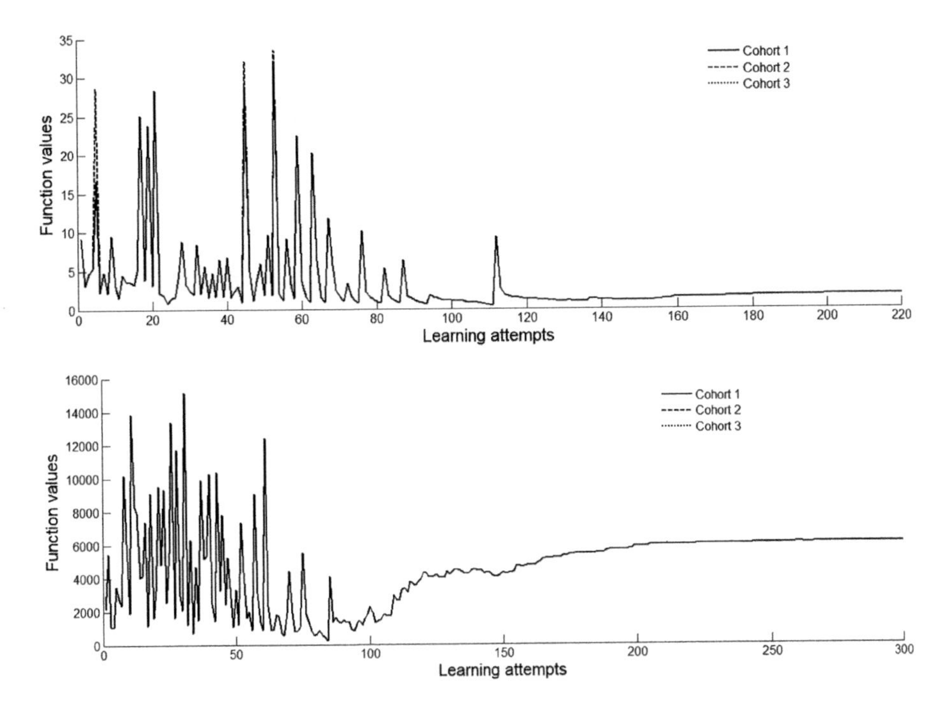

FIGURE 11.5 Convergence plots for constrained engineering design problems.

where $1 \times 0.0625 \leq x_1 \leq 99 \times 0.0625$, $1 \times 0.0625 \leq x_2 \leq 99 \times 0.0625$, $10 \leq x_3 \leq 200$ and $10 \leq x_4 \leq 200$.

The variables x_1 and x_2 are discrete values, which are integer multiples of 0.0625. Hence, the upper and lower bounds of the ranges of x_1 and x_2 are multiplies of 0.0625.

Table 11.6 presents the performance comparison for variables and constraints with different optimization algorithms for the pressure vessel design problem. The simulation results, viz. the best, mean, worst, and SD are summarized in Table 11.7. Number of FEs, reduction factor r, and numbers of candidates are listed in Table 11.5. From Table 11.6, it can be observed that the Multi-CI algorithm is able to achieve a comparable solution against other algorithms. In addition, the algorithm exhibited the robust performance (Table 11.7). The convergence plot with best solutions in each cohort for pressure vessel design problem is shown in Figure 11.5b.

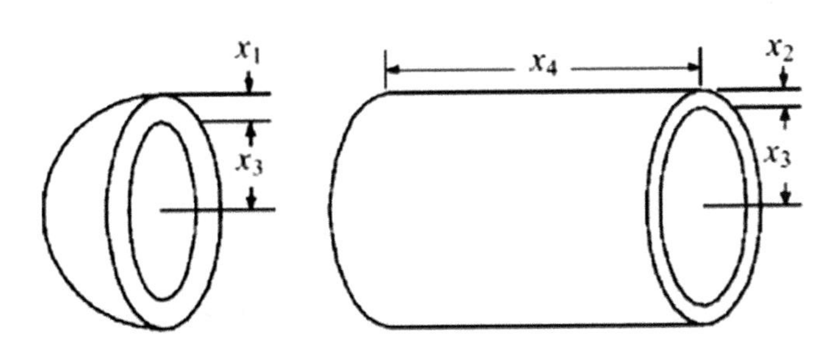

FIGURE 11.6 Pressure vessel design problems.

TABLE 11.6
Variables and Constraint Value Comparison

	Methods								
Details	**Deb (1997)**	**Coello Coello and Montes (2002)**	**He and Wang (2006)**	**He and Wang (2007)**	**Ghasemishabankareh et al. (2016)**	**Mohamed and Sabry (2012)**	**Shastri et al. (2019)**	**Krishnasamy et al. 2021**	**Multi-CI**
x_1	0.938	0.813	0.813	0.813	0.813	NA	0.182	0.813	**0.813**
x_2	0.500	0.438	0.438	0.438	0.438	NA	0.438	0.438	**0.437**
x_3	48.329	42.097	42.091	42.098	42.098	NA	42.098	41.928	**41.928**
x_4	112.679	176.654	176.747	176.637	176.637	NA	177.002	178.756	**177.995**
$g_1(X)$	–0.005	0.000	0.000	-8.8×10^{-7}	0.000	0.000	0.000	–0.003	**0.000**
$g_2(X)$	–0.039	–0.036	–0.036	–0.036	0.036	–0.036	–0.036	–0.038	**0.000**
$g_3(X)$	–3652.877	27.886	–116.383	3.123	0.000	–8.615E + 09	–0.036	0.000	**0.000**
$g_4(X)$	–127.321	–63.346	–63.254	63.363	63.360	–63.363	–63.363	–61.244	**0.000**
$f(X)$	6410.381	6059.946	6061.078	6059.714	6059.714	6053.714	6059.714	6080.536	**6056.493**

TABLE 11.7
Statistical Solutions of Various Algorithms

Methods	Best	Mean	Worst	SD
Deb (1997)	6410.381	N.A.	N.A.	N.A.
Coello Coello and Montes (2002)	6059.946	6177.253	6469.322	130.929
He and Wang (2006)	6061.077	6147.133	6363.804	86.454
He and Wang (2007)	6059.714	6099.932	6288.677	86.202
Ghasemishabankareh et al. (2016)	6059.714	6059.714	6059.714	1.01E-11
Mohamed and Sabry (2012)	6059.714	6059.714	6059.714	3.62E-10
Shastri et al. (2019)	6059.7144	6059.714	6059.714	5.023E-08
Krishnasamy et al. (2021)	6080.535	6089.599	6093.551	2.070
Multi-CI	**6056.493**	**6057.795**	**6059.089**	**8.04E-02**

11.4 DISCUSSION

The Multi-CI with probability-based constraint handling approach is validated by solving four test problems and two real-world problems. The solutions have been compared with the contemporary algorithms. For the maximization problems G02, G08 Multi-CI yielded solutions that are equal to the best known solutions. The solutions obtained for problem G08 are comparable with all the methods given in Table 11.1; however, solutions are marginally better in terms of solution quality in case of G02. The solutions obtained by Becerra and Coello Coello (2006) and Chootinan and Chen (2006) for G02 also show a high SD; therefore it is clear that Multi-CI is more robust than these methods (Table 11.2). The convergence plots for G02 and G08 are presented in Figures 11.3a and d, respectively. In Figure 11.3a solution converged to optimum solution with fluctuating pattern. Such pattern shows the solution lies in the infeasible region as constraints are violating. Even though the solution lies in an infeasible region, the probability-based approach forces the solution toward feasibility. In case of G08, candidates have shown early convergence as presented in Figure 11.3d. For the minimization problems G04, G06 Multi-CI solutions are equal to the best known solutions. The solutions obtained for problem G04 are comparable with all the methods given in Table 11.1. However, solutions obtained in case of G06 are marginally better compared with Koziel and Michalewicz (1999), Montes et al. (2010), Shastri et al. (2016), and Krishnasamy et al. (2021) in terms of solution quality and SD. The Multi-CI solution plots solving G04 and G06 problems are presented in Figures 11.3b and c, respectively. It is an important observation that for problem G04 candidates are trapped into local minima highlighted in Figure 11.3b; however, the efforts of the candidates in jumping out of the local minima are evident. In Multi-CI every candidate competes with its own local best behavior as well as the best behavior chosen from the other cohorts, and this self-supervised behavior is exhibited in this plots. It is evident from Table 11.2 that function evaluations required by Multi-CI are on the higher side as multiple cohorts exist. For engineering design domain two real-world problems are solved, viz. the welded beam design problem and the pressure vessel design problem. For these problems solutions obtained using Multi-CI are satisfactory in terms of solution quality and robustness as given in Tables 11.3 and 11.6, respectively. The welded beam design problem convergence plot is presented in Figure 11.5a. It is clear as highlighted in the plot that solution achieved in between this period was minimum; however, constraints were violating. Therefore, that solution was not accepted as an optimum solution as a mechanism of constrained Multi-CI allows one to accept the solution as an optimum solution if and only if all the constraints are satisfied. The pressure vessel design problem convergence plot is presented in Figure 11.5b. In Multi-CI exploration and exploitation the mechanism is strong as every candidate searches for the best solution from its own cohort as well as other cohorts. Because of this there are less chances of getting trapped in local minima, which results in the faster convergence as evident from Figure 11.5b. Furthermore, it is evident from the results shown in Tables 11.4 and 11.7 that the Multi-CI was robust compared with Coello Coello and Montes (2002) and He and Wang (2006, 2007) in terms of SD.

11.5 CONCLUSION AND FUTURE DIRECTIONS

This chapter presented an inbuilt probability-based constraint handling approach for Multi-CI algorithms by solving well-known constrained and engineering design problems. The performance is validated by solving G02, G04, G06, G08 and welded beam and pressure vessel design problems. The solutions were compared with several contemporary algorithms. The performance was either comparable or better because Multi-CI is characterized by improved exploration and exploitation capabilities as every candidate competes with its individual local best solution or best solution among other cohorts. In the near future efforts could be made to incorporate the variation constrained handling methods in Multi-CI to solve constrained real-world applications from some variegated domains such as supply-chain management and transportation. In addition, the inbuilt constraint handling technique may be further improved/developed using a multi-criteria optimization approach.

REFERENCES

Becerra, R., Coello Coello, C. (2006). Cultured differential evolution for constrained optimization. Computer Methods in Applied Mechanics and Engineering. 195 (33–36), 4303–4322.

Chootinan, P., Chen, A. (2006). Constraint handling in genetic algorithms using a gradient-based repair method. Computers and Operations Research. 33 (8), 2263–2281.

Coello Coello, C., Montes, E. (2002). Constraint-handling in genetic algorithms through the use of dominance-based tournament selection. Advanced Engineering Informatics. 16 (3), 193–203.

Deb, K. (1997). GeneAS: a robust optimal design technique for mechanical component design. In D. Dasgupta, and Z. Michalewicz, (Eds), Evolutionary Algorithms in Engineering Applications (pp. 497–51). Springer: Berlin, Heidelberg.

Deb, K. (2000). An efficient constraint handling method for genetic algorithms. Computer Methods in Applied Mechanics in Engineering. 186, 311–338.

Eberhart, R. and Kennedy, J. (1995). A new optimizer using particle swarm theory. In MHS'95. Proceedings of the Sixth International Symposium on Micro Machine and Human Science (pp. 39–43). IEEE.

Ghasemishabankareh, B., Li, X., Ozlen, M. (2016). Cooperative coevolutionary differential evolution with improved augmented Lagrangian to solve constrained optimisation problems, Information Sciences. 369(10), 441–456.

Goldberg, D.E., Holland, J.H. (1988). Genetic algorithms and machine learning. Machine Learning 3 (2), 95–99.

He, Q., Wang, L. (2006). An effective co-evolutionary particle swarm optimization for constrained engineering design problems. Engineering Applications of Artificial Intelligence. 20 (1), 89–99.

He, Q., Wang, L. (2007). A hybrid particle swarm optimization with a feasibility-based rule for constrained optimization. Applied Mathematics and Computation. 186 (2), 1407–1422.

Kashan, A. H. (2014). League championship algorithm (LCA): an algorithm for global optimization inspired by sport championships. Applied Soft Computing. 16, 171–200.

Kirkpatrick, S., Vecchi, M.P. (1983). Optimization by simulated annealing. Science. 220 (4598), 671–680.

Koziel, S., Michalewicz, Z. (1999). Evolutionary algorithms, homomorphous mapping, and constrained parameter optimization. Evolutionary Computation. 7 (1), 19–44.

Krishnasamy, G., Kulkarni, A.J., Shastri, A.S. (2021). An improved cohort intelligence with panoptic learning behavior for solving constrained problems, in A.J. Kulkarni, E. Mezura-Montes, Y. Wong, A.H. Gandomi, and G. Krishnasamy, (Eds), Constraint Handling in Metaheuristics and Applications. Springer: Singapore.

Kulkarni, A.J., Baki, M.F., Chaouch, B.A., (2016). Application of the cohort-intelligence optimization method to three selected combinatorial optimization problems. European Journal of Operational Research. 250 (2), 427–447.

Kulkarni, A.J., Durugkar, I.P., Kumar, M. (2013). Cohort intelligence: a self-supervised learning behavior. In: 2013 IEEE International Conference on Systems, Man, and Cybernetics (SMC), pp. 1396–1400.

Kulkarni, A.J., Shabir, H. (2016). Solving 0–1 knapsack problem using cohort intelligence algorithm. International Journal of Machine Learning and Cybernetics. 7 (3), 1–15.

M. Dorigo. (1992). *Optimization, Learning and Natural Algorithms*, PhD thesis, Politecnico di Milano, Italy.

Mohamed, A.W, Sabry H.Z. (2012). Constrained optimization based on modified differential evolution algorithm. Information Sciences. 194, 171–208.

Montes, E., Lopez-Ramirez, B. (2007). Comparing bio-inspired algorithms in constrained optimization problems. IEEE Congress on Evolutionary Computation, (pp. 662–669). Singapore.

Montes, E., Varela, M., Caemen, R., Ramon, G. (2010). Differential evolution in constrained numerical optimization. A empirical study. Information Sciences. 180 (22), 4223–4262.

Shastri, A.S., Jadhav, P.S., Kulkarni, A.J., Abraham, A. (2016). Solutions to constrained test problems using cohort intelligence algorithm. In: V. Snacel, A. Abraham, P. Kromer, M. Pant, A.K. Muda, (Eds.), Advances in Intelligent and Soft Computing 424: Innovations in Bio-Inspired Computing and Applications (pp. 427–435). Springer: Cham.

Shastri, A.S., Kulkarni, A.J. (2018). Multi-cohort intelligence algorithm: an intra-and inter-group learning behavior based socio-inspired optimisation methodology. International Journal of Parallel, Emergent and Distributed Systems. 33 (6), 675–715.

Shastri, A.S., Thorat, E.V., Kulkarni, A.J., Jadhav, P.S. (2019). Optimization of constrained engineering design problems using cohort intelligence method. In Proceedings of the 2nd International Conference on Data Engineering and Communication Technology (pp. 1–11). Springer: Singapore.

Storn, R., Price, K. (1997). Differential evolution–a simple and efficient heuristic for global optimization over continuous spaces. Journal of Global Optimization. 11 (4), 341–359.

Section IV

Swarm-Based Methods

12 Bee Colony Optimization and Its Applications

Dušan Teodorović, Tatjana Davidović, Milica Šelmić, and Miloš Nikolić

CONTENTS

12.1 INTRODUCTION

There is a significant class of metaheuristic algorithms that encompasses swarm intelligence (SI) methods (Beni and Wang 1989), such as cellular robotic systems (Beni 1988, Beni and Wang 1993), particle swarm optimization (PSO) (Eberhart and Kennedy 1995, Kennedy and Eberhart 1995), artificial immune systems (Hunt and Cooke 1996), and various algorithms based on social insects behavior (Bonabeau et al. 1999). A recent book (Yang 2020) summarizes some of the SI methods, including ant colony optimization (ACO), firefly algorithm, cuckoo search, and bat algorithm (BA). SI belongs to the area of artificial intelligence that studies behavior of individuals in various decentralized systems. Main characteristics of this behavior are autonomy, distributed functioning, and self-organizing. Swarm behavior characterizes different animal communities, such as fish schools, flocks of birds, herds of land animals, herds of elephants, colonies of social insects, (bees, wasps, ants, termites, fireflies, moths, butterflies), and so forth. Recently, several excellent papers presenting some of the novel SI methods were published. Li et al. (2020) presented an algorithm that simulates the clan updating and separation behavior of elephants. Among different evolutionary algorithms, the third version of non-dominated sorting genetic algorithm (NSGA-III), a new method capable of solving large-scale optimization problems, is used in Yi et al. (2018b). The improved version of the NSGA-III algorithm is presented in Yi et al. (2018a). Wang et al. (2015) proposed the monarch butterfly optimization. Moth behavior was also the inspiration for the new method named moth search algorithm (Wang 2015). Wang et al. (2014) proposed the algorithm based on stud krill herd behavior. Among the algorithms belonging to the SI class, numerous methods inspired by the behavior of honey bees can be distinguished: bee system (Lučić and Teodorović 2001, Sato and Hagiwara 1997), marriage in honey bees optimization (MBO) (Abbass 2001), beehive (Wedde et al. 2004), honey bees (HB) (Nakrani and Tovey 2004), artificial bee colony (ABC) (Karaboga 2005, Karaboga and Basturk 2007, Karaboga et al. 2014, Liao et al. 2013), bee system optimization (BSO) (Drias et al. 2005), virtual bee algorithm (VBA) (Yang 2005), bees algorithm (Pham et al. 2006a, b), honey bee marriage optimization (HBMO) (Afshar et al. 2007), fast marriage in honey bees optimization (MHBO) (Yang et al. 2007), opt bees (Maia et al. 2012, 2013), directed bee colony optimization algorithm (DBC) (Kumar 2014), and so forth.

The main topic in this chapter is the bee colony optimization (BCO) algorithm. The initial version of BCO, named bee system, was proposed for the first time at the beginning of the century for dealing with the well-known hard combinatorial optimization problems, such as the traveling salesman problem (TSP) (Lučić and Teodorović 2001, 2002, 2003a) and vehicle routing (VRP) (Lučić and Teodorović 2003b). The name bee colony optimization was introduced in Teodorović and Dell'Orco (2005), and that paper initiated more intensive usage of this optimization method. BCO is a metaheuristic method inspired by the foraging behavior of honey bees. It is a stochastic, random-search population-based technique. BCO is founded on the analogy between the natural behavior of bees, searching for food, and the behavior of optimization algorithms searching for the optimal solutions of combinatorial optimization problems (Teodorović 2009). The main idea is to build a multi-agent system (a colony of artificial bees) able to efficiently solve hard optimization problems. Artificial bees investigate through the search space looking for feasible solutions. To increase the quality of produced solutions, autonomous artificial bees collaborate and exchange information. By using the collective knowledge and sharing the available information between themselves, artificial bees enable directing the search toward regions containing high-quality solutions. Step by step, artificial bees collectively produce new and potentially better solutions. BCO performs its search in iterations until some predefined stopping criterion is satisfied.

The objectives in this chapter are twofold, to promote BCO as a simple and effective metaheuristic method and to try to cover all BCO applications that exist in the relevant literature since 2015. Previous papers are covered by the comprehensive review papers (Teodorović 2009, Davidović 2015, Davidović et al. 2015, Teodorović et al. 2015). BCO belongs to the class of early metaheuristics inspired by the foraging of honey bees. However, it is still not widely used by the optimization community due to lack of the proper advertisement. In addition, its description varies from paper to paper, usually because of the modifications that various authors introduce during the implementation.

This chapter is an extension of recently published survey papers (Teodorović 2009, Davidović 2015, Davidović et al. 2015, Teodorović et al. 2015). It contains the description of the BCO algorithm, and classification and analysis of its applications since 2015. The majority of the applications are related to combinatorial problems in transportation, location, and scheduling fields. However, some papers include continuous and mixed optimization problems, and extend applications to medicine, information science, chemistry, and many other domains. The recent literature is surveyed in more detail. BCO is used to successfully model complex science and engineering optimization problems. Most of the applications were reported by Teodorović and co-authors (Lučić and Teodorović 2001, 2002, 2003a, b, Teodorović and Dell'Orco 2005, 2008, Marković et al. 2007, Teodorović and Šelmić 2007, Šelmić et al. 2008, 2010, Davidović et al. 2009, 2011, 2012, Dimitrijević et al. 2011, Nikolić and Teodorović 2013a, b, Teodorović et al. 2013, Nikolić and Teodorović 2014, Nikolić and Teodorović 2015, Nikolić et al. 2015, Jovanović et al. 2017, 2019), however, since 2006 (Chong et al. 2006) other researchers from all over the world have also been developing BCO (Wong et al. 2008, 2009, 2010, Nedeljković et al. 2009, Levanova and Tkachuk 2011, Pertiwi and Suyanto 2011, Mousavinasab et al. 2011, Sohi et al. 2011, Kovač 2013, Sa'idi et al. 2013, Caraveo and Castillo 2014, Kumar 2014, Arun and Kumar 2015,

Stojanović et al. 2015, Kumar and Arun 2016). After the first PhD thesis related to the bee system (Lučić 2002), at Serbian Universities, several young researchers defended their dissertations considering the development and/or applications of BCO (Šelmić 2011, Nikolić 2015, Stojanović 2016, Jakšić Krüger 2017, Jovanović 2017).

This chapter is organized as follows. Section 12.2 provides the biological background that served as an inspiration to develop BCO described in Section 12.3. Section 12.4 is devoted to implementation insights related to the existing applications of BCO to various optimization problems. The last section contains some concluding remarks and directions for further exploration of BCO.

12.2 BIOLOGICAL BACKGROUND

When designing SI models, researchers use some principles of the swarm behavior from nature. The development of artificial systems usually does not involve the complete imitation of natural systems but involves their exploration and adaptation while searching for ideas and models.

Foraging bees in nature look for food by exploring the fields in the neighborhood of their hive (Camazine and Sneyd 1991). They collect and accumulate food to be later used by all bees. Typically, in the initial step, some scout bees search the region. Completing the search, scouts return to the hive and inform their hive mates (uncommitted bees) about the locations (direction and distance), quantity, and quality of the available food sources in the examined areas. The bees have their specific symbolic language, called a "waggle dance," for communication related to the foraging process (Figure 12.1).

By performing this figure-of-eight dance, the scout bees can inform other members of the colony about the direction and distance to the food sources. The meaning of the waggle dance was discovered by Karl von Frisch (1967) (Figure. 12.2). The direction (angle) of the waggle dance with respect to the sun is closely related to the direction of the food source being advertised by the scout bee. The distance between the hive and the food source is encoded by the number of waves in the waggle dance. Quantity of the discovered nectar is indicated by the rapidness of the waggle dance, while the quality is represented by the nectar samples brought by the scout bee. Waggle dancing is adjusted during the day with respect to the angle of the sun (as it is changing while the sun is traveling through the sky). Therefore, followers are always correctly directed to the food source.

As several bees may be dancing and attempting to recruit their hive mates on the dance floor area at the same time, it is unclear how an uncommitted bee decides which scout to follow. The only obvious fact is that "the loyalty and recruitment among bees are always a function of the quantity and quality of the food source" (Camazine and Sneyd 1991). The described process continues repeatedly, while the bees accumulate nectar in a hive and explore new areas with potential food sources.

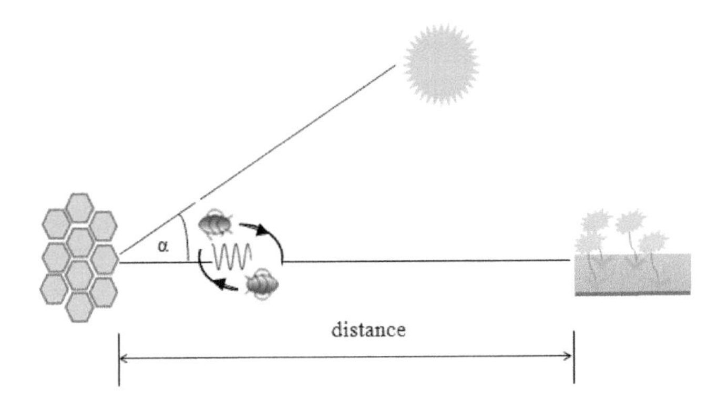

FIGURE 12.1 Illustration of honey bees waggle dance.

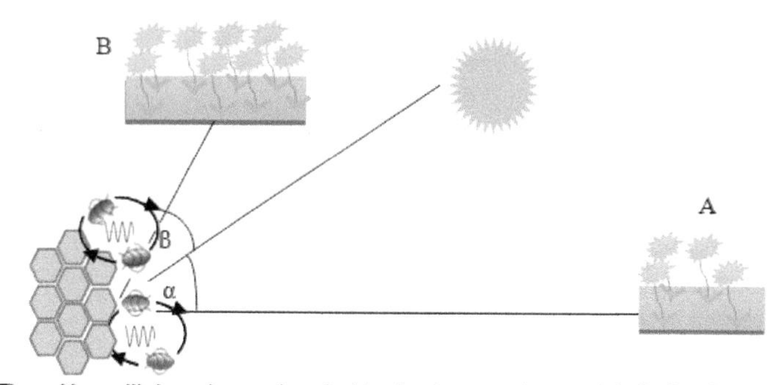

The red bee will dance longer than the blue bee because the area A is further than the area B.

The blue bee will have more waggies than the red bee because the area B has more flowers than the area A.

FIGURE 12.2 Decoding of symbolic language for communication between honey bees.

If a bee decides to leave the hive and collect the nectar (to become a foraging bee), it follows the directions given by the selected scout bee to the previously discovered food source. On arrival, the foraging bee takes a load of nectar and returns to the hive. Several scenarios are then possible for a foraging bee: (1) it can relinquish the nectar to the food store, abandon the food location, and return to the role of an uncommitted bee; (2) on relinquishing the nectar to a food store, it can continue with the foraging behavior at the discovered nectar source, without recruiting the rest of the colony; and (3) it can try to recruit its hive mates by the dance ritual before returning to the discovered food location. Each bee opts for one of the above possibilities as long as the nectar collection process is running. A bee that chooses the second or third scenario is referred to as loyal bee. Specifically, the third scenario defines the so-called recruiters.

This is just a small segment of the communication mechanism between foraging honey bees, and it is the one that is relevant for development of the BCO method.

12.3 THE BCO METAHEURISTIC

The previously described natural process inspired the development of a metaheuristic framework named BCO. It is a simplification of its predecessor bee system, the first algorithm developed by Dušan Teodorović (adviser) and Panta Lučić (PhD candidate) while doing research at Virginia Tech. The main idea behind BCO is to build the multi-agent system (a colony of artificial bees) able to efficiently solve hard optimization problems. Each artificial bee is responsible for a single solution of the considered problem. To increase the quality of produced solutions, artificial bees collaborate and exchange information in a way analogous to the bees from nature. The mechanisms of this collaboration and manipulation over solutions are described in this section.

12.3.1 History

The bee system proposed by Lučić and Teodorović (2001, 2002, 2003a,b) is a forerunner of the current BCO version and represents a constructive algorithm that has more similarities with the behavior of bees in nature than BCO. The bee system was used to deal with TSP (Lučić and Teodorović 2001, 2002, 2003a) and VRP (Lučić and Teodorović 2003b) and was described in detail in a PhD thesis (Lučić 2002). The main difference between these two versions of the algorithm is the important role of the hive location in the bee system. The hive was located in such a way to determine the beginning of the search process (for example, the initial node or the first selected component). This implied that within iteration, all bees had the same starting point and then explored the search space in the neighborhood of that point. Moreover, in each iteration of the bee system not all bees were engaged at the beginning of the search process. It was assumed that, at the beginning of iteration, the number of searching bees is zero. In the first stage, the scout bees started the search, and at each stage new bees were included. Another feature of the bee system is the exploration of the logit-based model (McFadden 1973) for calculating the probability of choosing the next component of the solution (the next node to be visited).

12.3.2 BCO Algorithm Description

BCO is a population-based method; the population of artificial bees searches for the optimal solution of a given optimization problem (Davidović et al. 2015). The algorithm runs in iterations until a stopping condition is met. The possible stopping condition could be, for example, the maximum number of iterations, the maximum number of iterations without the improvement of the objective function value, maximum allowed central processing unit (CPU) time, and so forth. Sometimes, the combination of stopping criteria is used. At the end of the BCO execution, the best found solution (the so-called global best) is reported as the final one.

During an iteration, every artificial bee is responsible for one solution of the considered problem. Its role is to make that solution as good as possible depending on the current state of the search. Each iteration contains several execution steps consisting of two alternating phases: *forward pass and backward pass* (Figure 12.3). During each forward pass, all bees are exploring the search space. Each bee applies a pre- defined number of moves, which yields a new solution. This

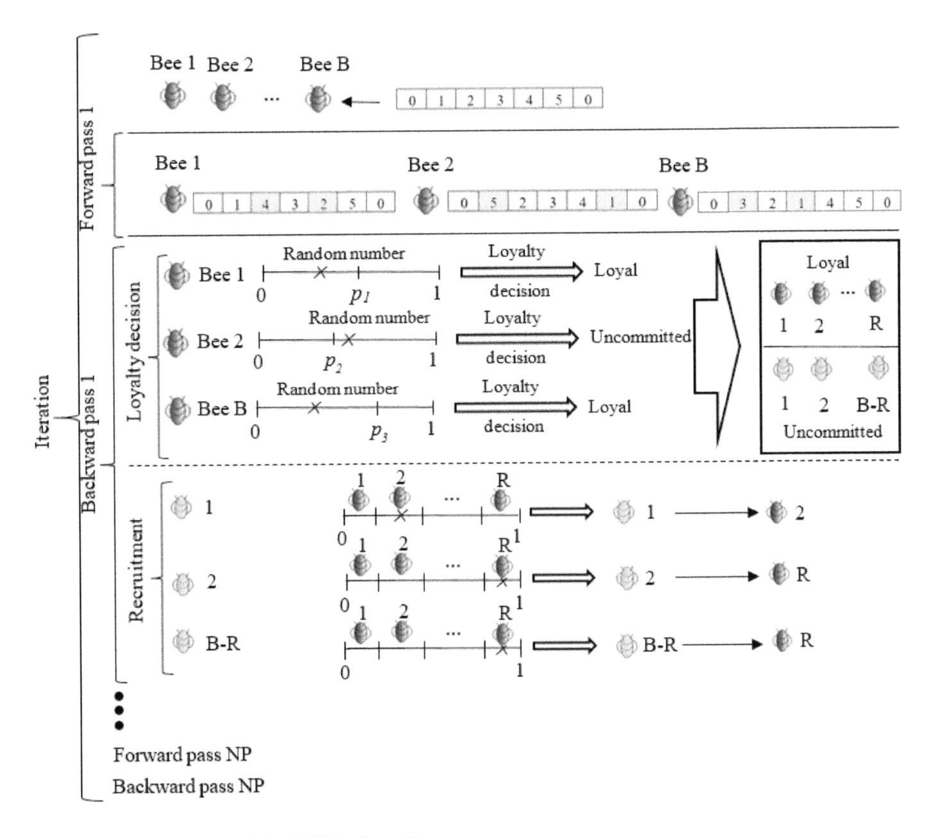

FIGURE 12.3 Main steps of the BCO algorithm.

part of the algorithm is problem dependent and should be resolved in each particular implementation of the BCO algorithm. Having obtained new solutions, the artificial bees start executing a second phase, the so-called backward pass where all bees share information about the quality of their solutions. The quality of each solution is usually defined by the current value of the objective function. When all solutions are evaluated, each bee decides with a certain probability whether it will stay loyal to its solution. Loyal bees are automatically becoming *recruiters* and advertise their solutions to other bees. It is obvious that bees with better solutions should have more chances to keep and advertise their solutions.

If a bee is not loyal, it becomes an *uncommitted follower* and has to select one of the advertised solutions. This selection is taken with a probability such that better advertised solutions have greater opportunities to be chosen for further exploration.

The differences between bees in nature and artificial bees are as follows:

- The number of artificial bees is significantly smaller than the number of honey bees in a real hive.
- All artificial bees are included in the search; the artificial hive is virtual with has no specific location.
- Communication among artificial bees is synchronous, i.e., it is performed after each forward pass and all bees are included.
- Artificial bees are divided into only two groups: loyal (recruiters) and uncommitted followers. This means that, contrary to the bees in nature, artificial bees that are loyal to their solutions are also the recruiters, i.e., their solutions are advertised and would be considered by uncommitted bees.
- Probabilities and various selection mechanisms (roulette wheel in the majority of the cases) are used to handle loyalty and recruitment decisions.

There are a lot of variants for the BCO algorithm in the recent literature. Various new parameters have been introduced to describe each particular implementation. However, for explaining the basic idea of the BCO algorithm and its pseudocode, a small number of parameters is required. There are only three parameters needed for the proper description of the BCO algorithm:

- *B:* the number of bees involved in the search
- *NP:* the number of forward/backward passes in a single BCO iteration
- *NC:* the number of changes in one forward pass

The pseudocode of the BCO algorithm can be given as follows (Nikolić and Teodorović 2019):

1. Generate initial solution.
2. **do**
3.   Set the solution to all bees.
4.   **for** *i* = 1 **to** *NP*
5.   Forward pass
6.   Backward pass
7.   **next**
8. **while** (stopping criteria is not satisfied)

The given pseudocode describes the improvement version of the BCO algorithm (referred to as BCOi). The constructive version of the algorithm (denoted here by BCOc) could be described by a similar pseudocode. Forward and backward passes are the most important steps in the both algorithms.

The initial solution should be generated at the beginning of the improved version of the BCO algorithm (step 1 in the pseudocode [this step does not exist in the constructive version of the algorithm]). At the beginning of do-while loop one complete solution should be assigned to each bee (in the constructive version of the algorithm bees start from the empty solution).

The main steps within the forward pass are as follows (Nikolić and Teodorović 2019):

1. **for** i = 1 **to** *NC*
2. **for** b = 1 **to** *B*
3. Make one modification of the bee's b solution.
4. **next**
5. Check if the new best solution has been discovered. If the new best solution has been discovered, save it.
6. **next**

Forward pass steps are problem specific and differ from implementation to implementation. Therefore, there are no directions on how to perform them. On the other hand, that gives the opportunity to maximally explore a priori knowledge about the considered problem and obtain a very powerful solution method. Within the forward pass bees modify their solution (in the constructive version of the algorithm bees add a new part to their partial solutions) and in that way try to find solutions better than the current best one.

The pseudocode of the backward pass consists of the following main steps (Nikolić and Teodorović 2019):

1. Normalization of the objective function values of the generated solutions
2. Loyalty decision for each bee
3. Recruitment (assigning every uncommitted bee to one of the loyal bees)

Depending on the optimization type, i.e., should the objective function be maximized or minimized, the following expressions for normalization of the bees' solutions are usually used:

$$O_b = \frac{T_b - T_{\min}}{T_{\max} - T_{\min}} \qquad b = 1,\ldots,B \tag{12.1}$$

for maximization, and

$$O_b = \frac{T_{\max} - T_b}{T_{\max} - T_{\min}} \qquad b = 1,\ldots,B \tag{12.2}$$

for minimization.

The loyalty decision for each bee depends on the quality of its own solution related to solutions held by other bees. In the original variant of BCO, the probability that bth bee stays loyal to its previously generated solution is expressed as follows:

$$p_b^{u+1} = e^{-\left(\frac{O_{\max} - O_b}{u}\right)}, \qquad b = 1,2,\ldots,B, \tag{12.3}$$

where:

O_b: the normalized value for the objective function (or any other fitness value) of solution created by the bth bee;

O_{max}: maximum overall normalized values of solutions to be compared; and
u: counts the forward passes, i.e., it takes values 1, 2, …, NC.

The normalization is performed in such a way that $O_b \in [0, 1]$ and larger values of O_b correspond to better solutions (Lučić and Teodorović 2003a,b, Davidović et al. 2012). In most of the BCO implementations, normalization is performed with respect to the solutions from the current forward pass. Therefore, $O_{max} = 1$ holds, and Eq. 12.1 can be simplified:

$$p_b^u = e^{-\frac{1-O_b}{u}}, \qquad b = 1,\ldots,B \tag{12.4}$$

This actually means that the bee holding current best solution always will be loyal and its solution will be advertised to the uncommitted bees.

Eq. 12.4 and a random number generator are used for each artificial bee to decide whether to stay loyal (and continue exploring its own solution) or to become an uncommitted follower (and select one among the advertised solutions for further exploration). If the generated random number from [0, 1] interval is smaller than the calculated probability p_b^u, then artificial bee b stays loyal to its own solution; otherwise, the bee becomes uncommitted. In addition to Eqs. 12.3 and 12.4, some other probability functions are examined in Jakšić Krüger (2017) and Maksimović and Davidović (2013).

For each uncommitted follower, it is decided which recruiter it will follow, taking into account the quality of all advertised solutions. The probability that the solution advertised by recruiter r would be chosen by any uncommitted bee equals:

$$p_r = \frac{O_r}{\sum_{k=1}^{R} O_k}, \qquad b = 1,\ldots,R, \tag{12.5}$$

where O_k corresponds to the kth advertised solution, and R denotes the number of recruiters. Using Eq. 12.5 and a random number generator, each uncommitted follower joins one recruiter through a roulette wheel. Three alternatives for selection criterion in the recruitment phase (tournament, rank and disruptive selection) are considered in Alzaqebah and Abdullah (2015). Another alternative, based on elitism principle, to decide on loyalty and recruitment is proposed in Todorovic and Petrovic (2013).

12.4 SURVEY OF RECENT BCO APPLICATIONS

BCO has been developing and successfully applied to various optimization problems since 2001. Our main aim in this section is to summarize as many papers related to the application of BCO from 2015 as possible. However, numerous papers appear in the recent literature, and not all of them are transparent enough to the authors; therefore, the survey may not be complete. Consequently, we have two more goals: to initiate communication between researchers that have been using BCO and to encourage others to use it in their research. According to Karaboga et al. (2014), BCO appears only in 13% publications considering algorithms based on bee swarms. The most widely used bee swarm–based method is ABC appearing in 58% of relevant publications. Another survey of algorithms based on the collective intelligence of honey bees is presented in Rajasekhar et al. (2017). It contains a part devoted to the description and applications of BCO based on 42 citations. It has to be noted that some other algorithms based on the foraging principles of honey bees are referred to as BCO in Rajasekhar et al. (2017), even ABC is sometimes renamed to BCO. However, we counted only the papers related to methods developed from

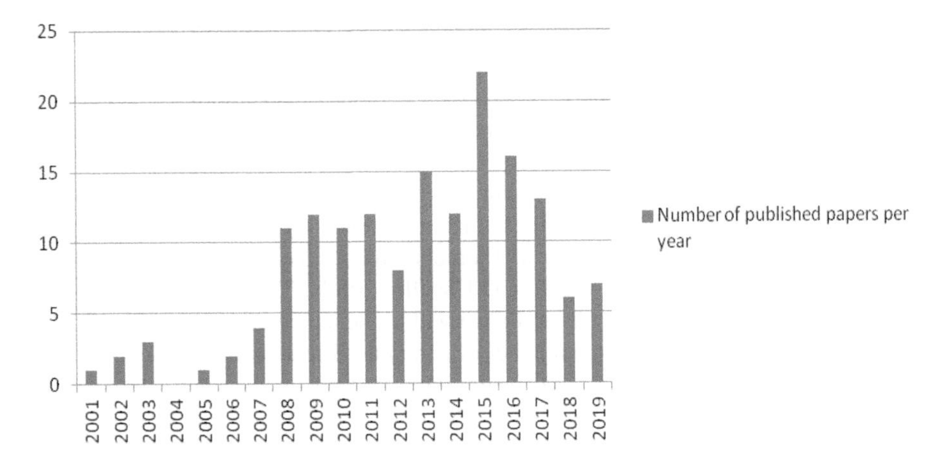

FIGURE 12.4 Number of BCO papers per year.

Lučić and Teodorović (2001, 2002, 2003a, b), and Teodorović (2009), Teodorović and Dell'Orco (2005). Taking into account only the texts (books, book chapters, PhD theses, journal, and conference papers) related to BCO developed from early works of Lučić and Teodorović (2001, 2002, 2003a, b) and later contributions by Teodorović and coauthors, mainly Teodorović and Dell'Orco (2005, 2008), Teodorović and Šelmić (2007), we made the graph representing distribution of papers over the last nineteen years (Figure 12.4). It counts all types of papers including theoretical results, application-based implementations, and survey publications. In some of the papers ABC is referred to as BCO, and we were trying to exclude these articles from this summary.

The majority of the papers proposing the application of BCO to various optimization problems up to 2015 have been already described in Teodorović et al. (2006), Teodorović (2009), Davidović (2015), and Teodorović et al. (2015). This survey focuses on the recent publications. In the following text, we have sublimated relevant journal and conference papers, PhD theses, and book chapters by years.

12.4.1 Applications of the BCO Algorithm Published in 2016

The paper by Abualhaija and Zimmermann (2016) considers automatic guessing of the meaning (sense) of the sentence on the basis of the words that appear in it. The new version of the BCOc method, called D-Bees, is proposed. In this version the role of the hive is significant; it represents the key word serving as the basis for determining meaning of the whole sentences. Hive is placed in a word that has the least number of different meanings, and it is not unambiguous. The number of senses should be within the range [2, 5]. It determines the number of bees, i.e., the number of bees is always equal to the number of meanings of the word chosen to be the hive. Therefore, the hive produces bees, while the other words represent food resources from which the bees are collecting information. Each bee generates a set of meanings of words in the sentence and calculates the value of sense in the set of considered words by randomly (with probability proportional to the frequency of the meaning) selecting a sense of other words (different from the hive). Value of sense is defined as the sum of the similarity of the meanings for the analyzed words. Two variants of D-Bees are proposed, with respect to a way of choosing the sense of a word: weighted random and completely random. It is shown that the first variant performs better. It is compared with genetic algorithm (GA), simulated annealing (SA), and ACO, and the results show that D-Bees is comparable with other methods on a standard SemEval 2007 task 7 coarse-grained English all-words corpus.

Amador-Angulo, Castillo, and coauthors (Amador-Angulo and Castillo 2014, 2015a, b, Amador-Angulo et al. 2016a, b, Caraveo et al. 2016) proposed the new variant of BCO (that will be referred to as *alternative* BCO or ACO-based BCO (BCOa). The proposed algorithm represents a modification of the original BCOc algorithm. The main modifications are related to the calculation of the selection probabilities and the implementation of waggle dance (an alternative to backward pass of BCOc). In the abovementioned papers BCOa has been applied to numerous optimization problems. In Amador-Angulo and Castillo (2016a) the authors proposed a hybrid model based on BCOa and fuzzy logic to realize the trajectory control of a mobile robot. The first modification of the BCOa algorithm is proposed in Amador-Angulo and Castillo (2016b). The authors applied fuzzy logic rules to dynamically adjust parameters of BCOa within the temperature controller. The stopping criterion for BCOa was the maximum number of iterations. The iteration counter is considered as "low" if the BCOa algorithm is close to the beginning of its work and vice versa. Possible linguistic values for the number of iterations were "low," "medium," and "high." The values for the BCOa parameters were "low," "medium low," "medium," "medium high," and "high."

In Caraveo et al. (2016) fuzzy logic is applied to dynamically change the main parameter values of the algorithm during execution. First, the traditional BCO is tested with the optimization of fuzzy controllers. Second, a modification of the original method is presented by including fuzzy logic to dynamically change the main parameter values of the algorithm during execution. Third, the proposed modification of the BCO algorithm with the fuzzy approach is used to optimize benchmark control problems. The comparison of results shows that the proposed fuzzy BCO method outperforms the traditional BCO in the optimal design of fuzzy controllers.

The job shop scheduling problem is studied in Choo et al. (2016), and the authors proposed a modification of a hybrid approach BCO + TESN5 from Wong (2012) in which BCO was combined with a local search. In his PhD thesis (Wong 2012), the author combined BCO with the two-enhancement scheme with neighborhood N5 perturbation (TESN5) local search procedure. Six neighborhood structures (N1–N6) are identified by Blazewicz et al. (1996) as the most suitable for solving the job shop scheduling problem. Among them, Wong selected N5 as the most promising neighborhood to be used within BCO framework. Neighborhood N5 is defined as the interchange of two successive operations in a block structure that belongs to a critical path. However, BCO + TESN5 has a drawback that is identified as the employment of a brute force strategy in applying the N5 neighborhood local search yielding significant processing overhead. The BCO + mTESN5 algorithm proposed in Choo et al. (2016) applies the N5 neighborhood local search only on a targeted bottleneck machine. The selection of the targeted bottleneck machine is done based on a list of bottleneck machines identified by the shifting bottleneck heuristic. Two strategies are tested to select the targeted bottleneck machine, namely the greedy selection and the linear ranking selection. The proposed algorithms are examined using a set of benchmark problems obtained from the OR-library. The results show that the proposed BCO + mTESN5 with linear ranking selection successfully solves 54% of the 82 OR-library benchmark data set producing ≤1% deviation from known optimum, and it is comparable to the BCO + TESN5 algorithm. In terms of computational time required to obtain the best makespan, the proposed BCO + mTESN5 with linear ranking selection outperforms the BCO + TESN5 algorithm by 25%.

Travel time prediction, based on a time-discrete, mesoscopic traffic flow model, is considered in Dell'Orco et al. (2016). The BCOc method adapted to the dynamic nature of the problem is applied. Sioux Falls City road network is used to compare the BCOc approach with a popular commercial software for the same problem. It is shown that BCOc predicts a more realistic travel time and average speed of the vehicle.

Jakšić Krüger et al. (2016) was the first to study the issue of the BCO convergence. This paper provides theoretical verification of the BCO algorithm by proving some convergence properties. As a result, the gap between successful practice and missing theory is reduced.

Designing a BCO method includes choosing a procedure for constructive/improvement moves, an evaluation function and setting of BCO parameters to a suitable value. The papers of Jakšić Krüger and Davidović (2016a, b) focused on detailed tuning of BCOi parameters. In their first paper the authors solve the satisfiability problem, 3-SAT, a well-known NP-hard problem. Their experimental evaluation shows that BCO can compete with a stand-alone WalkSAT heuristic, one of the state-of-the-art incomplete SAT solvers. The goal of the second paper from Jakšić Krüger and Davidović (2016b) was to address the influence of the appropriate choice of the BCOi underlying procedures, such as the choice of loyalty functions, and influence of parameter variations on algorithm performance, by means of visual inspection. Analyses are conducted for a simple variant of the scheduling problem. Also, to achieve good alternatives for reported solutions, new evaluation methods for the scheduling problem are presented.

Materialized views have been the popular mode used to achieve very fast online analytical processing operations. Selecting appropriate sets of optimal views, from among all possible views, is an NP-complete problem. In the paper from Kumar and Arun (2016), BCOc has been adapted to address the view selection problem. The experimental results show that the proposed algorithm compared with the fundamental greedy view selection algorithm is able to select comparatively better quality of views.

Leong et al. (2016) aimed to develop an optimization algorithm for optimum route selection to multiple destinations in a railway system before returning to the starting point. A variant of the BCO framework is used to generate a reliable algorithm in railway the TSP. The proposed algorithm is then verified by comparing the results with the exact solutions in 10 test cases, and a numerical case study was designed to demonstrate the application with large size sample. This algorithm can be used as a method to assist business practitioners to make better decisions in route planning.

Successful application of BCOi in telecommunications is shown in Marković (2016). Wavelength converter placement (WCP) in all-optical wavelength-division multiplexing (WDM) networks has been solved by application of metaheuristic approaches such as GAs, PSO, differential evolution (DE), and so forth. The objective is to find the best placement of a limited number of wavelength converters in a given optical network such that the overall network blocking probability is minimized. In the abovementioned paper, the author proposes the application of BCO metaheuristic as a tool. Numerous simulation experiments have been performed over some realistic optical network examples. The blocking probability performance and computational complexity are compared with optimal solutions obtained by the exhaustive search (ES) approach as well as with DE and PSO metaheuristics. It is shown that the proposed BCO-WCP algorithm is not only able to produce a high-quality (optimal) solution, but significantly outperforms the computational efficiency of other considered approaches.

Moayedikia et al. (2016) presented an algorithm based on the bee colony algorithm called REBECO to estimate worker reliability on microtask crowdsourcing platforms. This algorithm relies on the Gaussian process model to estimate reliability of workers dynamically. They compared REBECO with majority voting using two real-world data sets. The results indicate that REBECO is very competitive.

12.4.2 Applications of the BCO Algorithm Published in 2017

Amador-Angulo and Castillo (2017a) applied an interval type-2 fuzzy logic system to find the best parameters in the BCOa metaheuristic that should be used for stabilization of the trajectory in an autonomous mobile robot. Four interval type-2 fuzzy logic systems have been designed. The authors made 30 experiments and three metrics have been analyzed. Simulation results showed that the mean squared error (MSE) is one of the most important metrics.

Amador-Angulo and Castillo (2017b) used the BCOa approach when considering stabilization of the fuzzy controllers. The authors examined the proposed approach in the case of two fuzzy logic controllers: autonomous mobile robot controller and water tank controller. The authors considered several types of membership functions. The obtained results showed that the best solutions are obtained by triangular membership functions for both fuzzy logic controllers.

Amador and Castillo (2017) considered applications of the fuzzy bee colony optimization (FBCO) algorithms on two problems: filling the water tank and controlling the trajectory of an autonomous mobile robot. The authors took into consideration type-1 and type-2 fuzzy logic controllers. The quality of solutions obtained by the BCO algorithm with the adjustment of parameters was better than the results obtained by the original BCO.

The gate assignment problem is a well-known combinatorial assignment problem. The aim of this problem is to find the best flight-gate assignment to minimize the total walking distance of all passengers, or to maximize or minimize some other objective functions. When considering gate assignment problem, Dell'Orco et al. (2017) applied the FBCO algorithm. The authors used the constructive version of the BCO algorithm. In some steps in the backward pass Dell'Orco et al. (2017) used fuzzy interface systems (FISs). The proposed algorithm (FBCO) was tested on examples for two airports in Italy (Malpensa and Caselle). The obtained solutions were better than the existing solutions used for comparison. The authors also compared the proposed algorithm with the ACO metaheuristics. The results showed that the FBCO outperformed the ACO algorithm (percentage of average improvement of objective function for Malpensa airport is 7.2%, and 4.8 for Caselle airport).

Jakšić Krüger (2017) gave the comprehensive study of the BCO metaheuristic in her PhD thesis. The constructive and improvement versions (BCOc and BCOi) of the BCO metaheuristic were analyzed. The theoretical analysis of the asymptotic convergence of BCO is given. The thesis also shows various details about parallelization strategies for the BCO algorithm, as well as the methodology of the experimental study of BCOc and BCOi.

Jovanović and Teodorović (2017) considered the problem of determining cycle and phase lengths for isolated signalized intersections. The authors minimized the objective function that represents the average control delay experienced by the vehicles that arrived during the considered period. The improvement version of the bee colony optimization (BCOi) metaheuristic was proposed as the solution technique. The same BCOi algorithm was applied for undersaturated and oversaturated conditions at an isolated signalized intersection. The three test examples, or scenarios, were performed. The results obtained by BCOi were compared with the results obtained by the dynamic programming and commercial software HCS-Signals. The obtained results show that the proposed BCOi can find high-quality solutions within negligible CPU time. The authors showed that the proposed approach is competitive with the other methods in the literature.

Jovanović et al. (2017) described a new method of optimizing traffic signal settings. The area-wide urban traffic control system developed in the paper is based on the BCO technique. The authors optimized the cycle length (considered scenario assumed that the cycle length is the same for all intersections) and phase lengths for all intersections in the same time, in the way to minimize total delay experienced by all the vehicles in the network within the considered period. The considered problem was solved by the BCOi metaheuristic. The set of numerical experiments is performed on a well-known traffic benchmark network. The results obtained by the BCOi approach are compared with the results found by SA. It has been shown that the suggested BCO approach outperformed SA. Jovanović (2017) studied, in his doctoral dissertation, an isolated signalized intersection traffic control problem, as well as area-wide urban traffic control in the case of fixed time mode. By their nature, these problems fall into the area of difficult combinatorial optimization problems. Jovanović (2017) showed that the algorithms, based on the BCO technique, generate better solutions than specialized commercial software and other metaheuristic approaches.

Marković (2017a) considered the light paths provisioning problem in optical WDM networks with scarce available wavelengths under static (off-line) traffic demands. The considered problem was solved by using the BCOi metaheuristic. The objective function to be maximized was the network operator's revenue. Numerical examples show that BCOi significantly outperformed previous approaches given in the literature, such as the BCOc and DE approaches.

Marković (2017b) proposed the constructive version of the BCO algorithm (BCOc) as the solution technique for the problem of routing and spectrum allocation (RSA) in elastic optical networks (EON). The proposed BCO-RSA algorithm minimizes network spectrum utilization and the average path length criteria. The BCO-RSA algorithm was tested on different realistic size optical networks. The obtained results clearly show that proposed, the BCO-RSA, algorithm outperforms some greedy heuristics, as well as the DE metaheuristic.

Olivas et al. (2017) have made comparisons among PSO, BCO and the BA. The authors applied fuzzy versions of these algorithms (type-1 and interval type-2 fuzzy systems). These algorithms were applied to optimize the fuzzy system for controlling the trajectory of an autonomous mobile robot. Obtained results show that PSO outperforms BCO and BA algorithms.

Rajeswari et al. (2017) proposed the multi-objective directed bee colony optimization (MODBCO) algorithm for a nurse rostering problem. In the considered problem, the nurses should be assigned to shifts per day in the way to satisfy all constraints. MODBCO integrates deterministic local search, multi-agent particle system environment, and the honey bee decision-making process. The proposed approach was successfully applied to many real-world cases.

A highway detector's location problem was considered in the paper Teodorović et al. (2017). The authors tried to minimize the travel time estimated error. They applied the improvement version of the BCO algorithm (BCOi). The proposed approach was tested on a real case study for one part of the highway in the Republic of Serbia.

12.4.3 Applications of the BCO Algorithm Published in 2018

Chapter and the paper by Amador-Angulo and Castillo (2018a, b) represent a continuation of research from earlier years on the same topic. Castillo and Amador-Angulo (2018) used a generalized type-2 fuzzy logic system (GT2FLS) approach for dynamic parameter adaptation in metaheuristics and for optimal fuzzy controller design. The efficiency of the GT2FLS approach is verified by simulation. GT2FLS dynamically finds the optimal values of the heuristic parameters that are a critical part of the BCO algorithm performance. Simulation results illustrated that the implementation of the GT2FLC approach improves performances of the resulting BCO algorithm. The stability of the fuzzy controller was better.

Choong et al. (2018) proposed a modified version of the BCO algorithm with a linear dance function (denoted as the BCO-linear algorithm) developed in Wong (2012). In this algorithm, a fitter bee is allowed to dance longer, and the dance duration is determined by a linear function with a scaling parameter that requires manual tuning. Choong et al. (2018) presented BCO-fuzzy algorithm where linear function is replaced by a dynamic fuzzy-based dance mechanism to avoid the manual tuning problem. A fuzzy-based approach is applied to regulate the duration of waggle dances instead of regulating the dance duration using a linear function. The proposed BCO fuzzy algorithm comprises parameters that are dynamically controlled based on the feedback of the search process, therefore, overcoming the limitation of the manual parameter tuning of the BCO-linear algorithm. This in fact means that BCO-fuzzy algorithm, as well as the BCO-linear algorithm, are equipped with a kind of memory and are able to learn from previously discovered solutions. The BCO-fuzzy algorithm is evaluated comprehensively using a set of benchmark TSPs. The experimental results show that the performance of the BCO-fuzzy algorithm is comparable with that of the BCO-linear algorithm. Performance comparison with other nature-inspired algorithms proves the effectiveness of the proposed BCO-fuzzy algorithm.

Missing relevant data in the decision-making process is a very complex problem. It is very hard to find the method that can adequately estimate missing values in all required situations. Therefore, in the recent literature a new distance function, based on the propositional logic, which does not require determining the values of missing data, is proposed. Exploiting this distance, Davidović et al. (2018) suggested the BCOi approach for clustering incomplete data based on the p-median classification model. The efficiency of the proposed approach is demonstrated by the comparison with some recent clustering (classification) methods that include supervised learning mechanisms.

Marinelli et al. (2018) proposed a hybrid approach called biogeography-based BCO (B-BCO). The proposed approach is obtained by combining two metaheuristics: biogeography-based optimization (BBO) and the constructive BCO algorithm. The B-BCO model integrates the BBO migration operator into the bee's search behavior. It is compared with the BCOc approach to solving the flight-gate assignment problem at the Milano-Malpensa International Airport. The obtained results show better performance of the proposed B-BCO.

12.4.4 Applications of the BCO Algorithm Published in 2019

Castillo et al. (2019) presented a comparison among BCOa, DE, and harmony search (HS) algorithms developed for the optimal design of fuzzy systems for benchmark control problems, especially in fuzzy controller design. Each algorithm is present in two versions, modified and original, and then compared with each other. Simulation results provide evidence that the Fuzzy Differential Evolution (FDE) algorithm outperforms the results of the FBCO and Fuzzy Harmony Search (FHS) algorithms in the optimization of fuzzy controllers.

Assigning available buses to bus routes under conditions of bus shortages is known as the hard combinatorial optimization problem. In Nikolić and Teodorović (2019) BCOi is proposed for mitigation of bus schedule disturbances. The authors developed a model that took into account interests of the transit operator and passengers. The model reassigns available buses to bus routes and, if it is allowed, the model simultaneously changes the transportation network topology and reassigns available buses to a new set of bus routes. The proposed model is tested on the network of Rivera (Uruguay) and obtained results show that the proposed algorithm can significantly mitigate disruptions.

Jovanović et al. (2019) presented a continuation of earlier research where the same authors applied BCOc to select the optimal location of detectors in transportation networks. An increasing number of located detectors in networks provides accuracy of the obtained data, while requiring more investments and maintenance cost support. The detectors need to be placed in such a way that they can successfully sample the traffic conditions with the least possible error. On the other hand, traffic authorities tend to minimize the number of located detectors on the network to achieve investment savings. The proposed algorithm is tested on a real data collected in the Republic of Serbia.

Nikolić et al. (2019) considered a problem of estimation of freight train energy consumption. The proposed fuzzy system has three input variables and one output variable. Some fuzzy rules were generated from the numerical data using the Wang-Mendel method. The rest of the fuzzy rules were obtained according to experts' opinions. The authors proposed fine-tuning of the membership functions of fuzzy sets for input and output variables using the BCO metaheuristic. The precision of the fuzzy system was significantly increased after membership function tuning.

BCO was also applied for data mining problems (Vijayakumari and Deepa 2019). Clustering problems are essential problems in data analysis and data mining appliance. The authors adjusted the BCO algorithm for the fuzzy C-means method and compared obtained results with the PSO approach. Presented results showed that BCO gives better results when compared to PSO.

Miladić-Tešić et al. (2019), in their paper, consider the grooming capability at the optical layer in telecommunication networks. Solving the grooming problem together with the RSA, as a key issue in EON, is a highly challenging task, particularly in the case of large problem instances. The BCO metaheuristic is used to solve the traffic grooming problem with static traffic demands and is therefore suitable when designing EON. The algorithm presented in the paper aims to minimize the total spectrum usage while serving all traffic demands. Significant spectrum savings are obtained compared with the non-grooming case.

Marinelli et al. (2019) presented a novel approach consisting of a two-stage optimization system to solve the problem of the last mile distribution of goods in the context of an intelligent transportation system (ITS). They considered the vehicle routing problem with simultaneous pickup and delivery in conjunction with door-to-door operations (VRPSPD-D2D). The authors developed an algorithm that is based on two honey bee–inspired metaheuristics: ABC at the first stage and BCO at the second stage. The new instances of VRPSPD-D2D are proposed, based on the Sioux Falls City network. The results obtained by BCO are compared with two GA-based approaches.

12.5 CONCLUSION

The BCO algorithm, a metaheuristic method inspired by the foraging behavior of honey bees, belongs to the class of SI techniques. It represents a general algorithmic framework, applicable to various optimization problems in combinatorial/continuous optimization and engineering. The BCO method is based on the concept of cooperation, which increases the efficiency of artificial bees and allows them to achieve goals that could not be reached by individual actions only. Through the information exchange and recruiting process, BCO has the capability to intensify the search in the promising regions of the solution space. BCO has become a very popular algorithm due to its simplicity: it is easy to understand and has a small number of parameters (number of bees, number of forward/backward passes, and number of changes in one forward pass). This chapter presents an overview of the recent applications of the BCO algorithm to combinatorial optimization problems to promote this simple yet effective optimization method. Based on the achieved results and gained experience, new models founded on BCO principles (autonomy, distributed functioning, self-organizing, information exchange, and collaboration) are likely to significantly contribute to solving complex engineering, management, and control problems.

ACKNOWLEDGMENTS

This work was supported by the Serbian Ministry of Science, through the University of Belgrade, Faculty of Transport and Traffic Engineering and Mathematical Institute of the Serbian Academy of Science and Arts.

REFERENCES

Abbass, H. A. 2001. MBO: marriage in honey bees optimization – a haplometrosis polygynous swarming approach. In Proceedings of the Congress on Evolutionary Computation, pages 207–214, Seoul, South Korea.

Abualhaija, S., and Zimmermann K.-H. 2016. D-bees: a novel method inspired by bee colony optimization for solving word sense disambiguation. Swarm and Evolutionary Computation, 27:188–195.

Afshar, A., Bozorg Haddad, O., Mariño, M. A., and Adams, B. J. 2007. Honey-bee mating optimization (HBMO) algorithm for optimal reservoir operation. Journal of the Franklin Institute, 344(5): 452–462.

Alzaqebah, M., and Abdullah, S. 2015. Hybrid bee colony optimization for examination timetabling problems. Computers & Operations Research, 54:142–154.

Amador, L., and Castillo, O. 2017. Optimization of Type-2 Fuzzy Controllers Using the Bee Colony Algorithm. Springer, Berlin/Heidelberg, Germany.

Amador-Angulo L., and Castillo, O. 2014. Optimization of the type-1 and type-2 fuzzy controller design for the water tank using the bee colony optimization. In IEEE Conference on Norbert Wiener in the 21st Century (21CW), pages 1–8, Boston, MA.

Amador-Angulo, L., and Castillo, O. 2015a. A fuzzy bee colony optimization algorithm using an interval type-2 fuzzy logic system for trajectory control of a mobile robot. In Mexican International Conference on Artificial Intelligence, pages 460–471, Cuernavaca, Mexico. Springer.

Amador-Angulo, L., and Castillo, O. 2015b. Statistical analysis of type-1 and interval type-2 fuzzy logic in dynamic parameter adaptation of the BCO. In 2015 Conference of the International Fuzzy Systems Association and the European Society for Fuzzy Logic and Technology (IFSA- EUSFLAT-15), pages 776–783, Gijón, Asturias, Spain. Atlantis Press.

Amador-Angulo, L., and Castillo, O. 2017a. Comparative study of metrics that affect in the performance of the bee colony optimization algorithm through interval type-2 fuzzy logic systems. In North American Fuzzy Information Processing Society Annual Conference, pages 61–72. Springer, Cham, Switzerland

Amador-Angulo, L., and Castillo, O. 2017b. Comparative analysis of designing different types of membership functions using bee colony optimization in the stabilization of fuzzy controllers. In: Melin P., Castillo O., Kacprzyk J. (eds), Nature-Inspired Design of Hybrid Intelligent Systems. Studies in Computational Intelligence, vol 667. Springer, Berlin/Heidelberg, Germany.

Amador-Angulo, L., and Castillo, O. 2018a. A new fuzzy bee colony optimization with dynamic adaptation of parameters using interval type-2 fuzzy logic for tuning fuzzy controllers. Soft Computing, 22(2): 571–594.

Amador-Angulo, L., and Castillo, O. 2018b. Statistical comparison of the bee colony optimization and fuzzy BCO algorithms for fuzzy controller design using trapezoidals MFs. In: Zadeh L., Yager R., Shahbazova S., Reformat M., Kreinovich V. (eds), Recent Developments and the New Direction in Soft-Computing Foundations and Applications. Studies in Fuzziness and Soft Computing, vol 361, pages 291–306. Springer, Cham, Switzerland.

Amador-Angulo, L., Castillo, O., and Castro, J. R. 2016a. A generalized type-2 fuzzy logic system for the dynamic adaptation the parameters in a bee colony optimization algorithm applied in an autonomous mobile robot control. In IEEE International Conference on Fuzzy Systems (FUZZ-IEEE), pages 537–544, Vancouver, Canada.

Amador-Angulo, L., Mendoza, O., Castro, J., Rodrıguez-Dıaz, A., Melin, P., and Castillo O. 2016b. Fuzzy sets in dynamic adaptation of parameters of a bee colony optimization for controlling the trajectory of an autonomous mobile robot. Sensors, 16(9):1458.

Arun, B., and Kumar, T. V. V. 2015. Materialized view selection using improvement based bee colony optimization. International Journal of Software Science and Computational Intelligence (IJSSCI), 7(4): 35–61.

Beni, G. 1988. The concept of cellular robotic system. In Proceedings of the 1988 IEEE International Symposium on Intelligent Control, pages 57–62. IEEE Computer Society Press, Los Alamitos, CA.

Beni, G., and Wang, J. 1989. Swarm intelligence. In Proceedings of the Seventh Annual Meeting of the Robotics Society of Japan, pages 425–428. RSJ Press, Tokyo.

Beni, G., and Wang, J. 1993. Swarm intelligence in cellular robotic systems. In: Dario P., Sandini G., Aebischer P. (eds), Robots and Biological Systems: Towards a New Bionics?. NATO ASI Series (Series F: Computer and Systems Sciences), vol 102. Springer, Berlin, Heidelberg. https://doi.org/10.1007/978-3-642-58069-7_38.

Blazewicz, J., Domschke, W., and Pesch, E. 1996. The job shop scheduling problem: Conventional and new solution techniques. European Journal of Operational Research, 93(1):1–33.

Bonabeau, E., Dorigo, M., and Theraulaz, G. 1999. Swarm Intelligence: From Natural to Artificial Systems. Oxford University Press, Oxford.

Camazine, S., and Sneyd, J. 1991. A model of collective nectar source by honey bees: self- organization through simple rules. Journal of Theoretical Biology, 149:547–571.

Caraveo, C., and Castillo, O. 2014. Optimization of fuzzy controllers design using the bee colony algorithm. In Recent Advances on Hybrid Approaches for Designing Intelligent Systems, pages 163–175. Springer, Cham, Switzerland.

Caraveo, C., Valdez, F., and Castillo, O. 2016. Optimization of fuzzy controller design using a new bee colony algorithm with fuzzy dynamic parameter adaptation. Applied Soft Computing, 43:131–142.

Castillo, O., and Amador-Angulo L. 2018. A generalized type-2 fuzzy logic approach for dynamic parameter adaptation in bee colony optimization applied to fuzzy controller design. Information Sciences, 460: 476–496.

Castillo, O., Valdez, F., Soria, J., Amador-Angulo, L., Ochoa, P., and Peraza, C. 2019. Comparative Study in Fuzzy Controller Optimization Using Bee Colony, Differential Evolution, and Harmony Search Algorithms, Algorithms, 12(1), 9.

Chong, C. S., Low, M. Y. H., Sivakumar, A. I., and Gay, K. L. 2006. A bee colony optimization algorithm to job shop scheduling. In Proceedings of the Winter Simulation Conference, pages 1954–1961, Washington, DC.

Choo, W. M., Wong, L.P. and Khader, A. T. 2016. A modified bee colony optimization with local search approach for job shop scheduling problems relevant to bottleneck machines. International Journal of Advanced Soft Computing Applications, 8(2):52–78.

Choong, S. S., Wong L-P., and Lim C. P. 2018. A dynamic fuzzy-based dance mechanism for the bee colony optimization algorithm. Computational Intelligence, 34(4):999–1024.

Davidović, T. 2015. Bee colony optimization: Recent developments and applications. In Proceedings of the XI Balkan Conference on Operational Research, pages 225–235, BALCOR 2015, (plenary talk) Constanta, Romania.

Davidović, T., Glišović, N., and Rašković, M. 2018. Bee colony optimization for clustering incomplete data. In: S. Belim et al., ed., Proceedings of the School-Seminar on Optimization Problems and their Applications (OPTA-SCL 2018), volume 2098, pages 94–108, CEUR-WS Team, Aachen, Germany.

Davidović, T., Ramljak, D., Šelmić, M., and Teodorović, D. 2011. Bee colony optimization for the p-center problem. Computers & Operations Research, 38(10):1367–1376.

Davidović, T., Šelmić, M., and Teodorović, D. 2009. Scheduling independent tasks: bee colony optimization approach. In Proceedings of the 17th Mediterranean Conference on Control and Automation, pages 1020–1025, Makedonia Palace, Thessaloniki, Greece.

Davidović, T., Šelmić, M., Teodorović, D., and Ramljak, D. 2012. Bee colony optimization for scheduling Independent tasks to identical processors. Journal of Heuristics, 18(4):549–569.

Davidović, T., Teodorović, D., and Šelmić, M. 2015. Bee colony optimization Part I: The algorithm overview. Yugoslav Journal of Operational Research, 25(1):33–56.

Dell'Orco, M., Marinelli, M., and Altieri, M. G. 2017. Solving the gate assignment problem through the Fuzzy Bee Colony Optimization. Transportation Research Part C: Emerging Technologies, 80: 424–438.

Dell'Orco, M., Marinelli, M., and Silgu, M. A. 2016. Bee colony optimization for innovative travel time estimation, based on a mesoscopic traffic assignment model. Transportation Research Part C: Emerging Technologies, 66: 48–60.

Dimitrijević, B., Teodorović, D., Simić, V., and Šelmić, M. 2011. Bee colony optimization approach to solving the anticovering location problem. Journal of Computing in Civil Engineering, 26(6): 759–768.

Drias, H., Sadeg, S., and Yahi, S. 2005. Cooperative bees swarm for solving the maximum weighted satisfiability problem. LNCS: Computational Intelligence and Bioinspired Systems, 3512: 318–325.

Eberhart R. C., and Kennedy, J. 1995. A new optimizer using particle swarm theory. In Proceedings of the 6th International Symposium on Micromachine and Human Science (MHS '95), Nagoya, Japan.

Hunt, J. E., and Cooke, D. E. 1996. Learning using an artificial immune system. Journal of Network and Computer Applications, 19(2): 189–212.

Jakšić Krüger, T. 2017. Development, implementation, and theoretical analysis of the Bee Colony Optimization (BCO) meta-heuristic method. PhD thesis, Faculty of Techincal Science, University of Novi Sad.

Jakšić Krüger, T., and Davidović, T. 2016a. Empirical analysis of the bee colony optimization method on 3-Sat Problem. In Proceedings of the 43rd Symposium on Operations Research, SYM-OP-IS 2016, Tara, Sept. 20–23, pages 297–301.

Jakšić Krüger, T., and Davidović, T. 2016b. Sensitivity analysis of the bee colony optimization algorithm, In Proceedings of the 7th International Conference on Bioinspired Optimization Methods and their Applications, BIOMA 2016, Bled, Slovenia, May 18–20, 2016, pp. 65–78.

Jakšić Krüger, T., Davidović, T., Teodorović, D., and Šelmić, M. 2016. The bee colony optimization algorithm and its convergence. International Journal of Bio-Inspired Computation, 8(5): 340–354.

Jovanović, A. 2017. Choice of signal timing for traffic control by bee colony optimization. PhD thesis, Faculty of Traffic and Transportation, University of Belgrade.

Jovanović, A., Nikolić, M., and Teodorović, D. 2017. Area-wide urban traffic control: a bee colony optimization approach. Transportation Research Part C: Emerging Technologies, 77:329–350.

Jovanović, I., Šelmić, M., and Nikolić, M. 2019. Metaheuristic approach to optimize placement of detectors in transport networks – Case study of Serbia, Canadian journal of civil engineering, NRC Research Press, 46(3): 176–187.

Jovanović, A., and Teodorović, D. 2017. Pre-timed control for an under-saturated and over-saturated isolated intersection: a Bee colony optimization approach. Transportation Planning and Technology, 40(5): 556–576.

Karaboga, D. 2005. An idea based on honey bee swarm for numerical optimization. Technical report, Erciyes University, Engineering Faculty Computer Engineering Department Kayseri/Turkiye.

Karaboga, D., and Basturk, B. 2007. A powerful and efficient algorithm for numerical function optimization: artificial bee colony (ABC) algorithm. Journal of Global Optimization, 39: 459–471.

Karaboga, D., Gorkemli, B., Ozturk, C., and Karaboga, N. 2014. A comprehensive survey: artificial bee colony (ABC) algorithm and applications. Artificial Intelligence Review, 42: 21–57.

Kennedy, J., and Eberhart, R. 1995. Particle swarm optimization. In Proceedings of IEEE International Conference on Neural Networks IV, pages 1942–1948.

Kovač, N. 2013. Bee colony optimization algorithm for the minimum cost berth allocation problem. In XI Balcan Conference on Operational Research, pages 245–254, BALCOR 2013, Beograd- Zlatibor, Serbia.

Kovač, N. 2018. Metaheuristic approach for solving one class of optimization problems in transport. PhD thesis (in Serbian), Faculty of Mathematics, University of Belgrade.

Kumar, R. 2014. Directed bee colony optimization algorithm. Swarm and Evolutionary Computation, 17: 60–73.

Kumar, T. V. V., and Arun, B. 2016. Materialised view selection using BCO. International Journal of Business Information Systems, 22(3): 280–301.

Leong, K. H., Abdul-Rahman, H., Wang, C., Onn, C. C., and Loo, S. C. 2016. Bee inspired novel optimization algorithm and mathematical model for effective and efficient route planning in railway system. PLoS One, 11(12): e0166064.

Levanova, T. V., and Tkachuk, E. A. 2011. Development of a bee colony optimization algorithm for the capacitated plant location problem. In II International Conference, Optimization and applications (OPTIMA-2011), pages 153–156, Petrovac, Montenegro.

Li, W., Wang, G. G., and Alavi, A. H. 2020. Learning-based elephant herding optimization algorithm for solving numerical optimization problems. Knowledge-Based Systems, 195, 105675.

Liao, T., Aydın, D., and Stützle, T. 2013. Artificial bee colonies for continuous optimization: Experimental analysis and improvements. Swarm Intelligence, 7(4): 327–356.

Lučić, P. 2002. Modeling transportation problems using concepts of swarm intelligence and soft computing. PhD thesis, Virginia Polytechnic Institute, Blacksburg, VA.

Lučić, P., and Teodorović, D. 2001. Bee system: modeling combinatorial optimization transportation engineering problems by swarm intelligence. In Preprints of the TRISTAN IV Triennial Symposium on Transportation Analysis, pages 441–445. Sao Miguel, Azores Islands.

Lučić, P., and Teodorović, D. 2002. Transportation modeling: an artificial life approach. In Proceedings of the 14th IEEE International Conference on Tools with Artificial Intelligence, pages 216–223, Washington, DC.

Lučić, P., and Teodorović, D. 2003a. Computing with bees: attacking complex transportation engineering problems. International Journal on Artificial Intelligence Tools, 12(3): 375–394.

Lučić, P., and Teodorović, D. 2003b. Vehicle routing problem with uncertain demand at nodes: the bee system and fuzzy logic approach. In: J. L. Verdegay, editor, Fuzzy Sets Based Heuristics for Optimization, pages 67–82. Physica Verlag, Berlin Heidelberg.

Maia, R. D., de Castro, L. N., and Caminhas, W. M. 2012. Bee colonies as model for multimodal continuous optimization: The optbees algorithm. In IEEE Congress on Evolutionary Computation, pages 1–8.
Maia, R. D., de Castro, L. N., and Caminhas, W. M. 2013. Collective decision-making by bee colonies as model for optimization-the optbees algorithm. Applied Mathematical Sciences, 7(87): 4327–4351.
Maksimović, P., and Davidović, T. 2013. Parameter calibration in the bee colony optimization algorithm. In XI Balcan Conference on Operational Research, pages 263–272, BALCOR 2013, Beograd-Zlatibor, Serbia.
Marinelli, M., Caggiani, L., Alnajajreh, A., and Binetti, M., 2019. A two-stage metaheuristic approach for solving the vehicle routing problem with simultaneous pickup/delivery and door-to-door service. In 6th International Conference on Models and Technologies for Intelligent Transportation Systems (MT-ITS).
Marinelli, M., Palmisano, G., Dell'Orco, M., and Ottomanelli. M. 2018. Optimizing airport gate assignments through a hybrid metaheuristic. Advanced Concepts, Methodologies and Technologies for Transportation and Logistics, Book Series: Advances in Intelligence Systems and Computing, 572: 389–404.
Marković, G., Teodorović, D., and Aćimović-Raspopović, V. 2007. Routing and wavelength assignment in all-optical networks based on the bee colony optimization. AI Communications, 20(4): 273– 285.
Marković, G. 2016. Wavelength converters placement in optical networks using bee colony optimization. Advances in Electrical and Computer Engineering, 16(1): 3–10.
Marković, G. 2017a. Revenue-driven lightpaths provisioning over optical WDM networks using bee colony optimization. International Journal of Computational Intelligence Systems, 10(1): 481–494.
Marković, G. 2017b. Routing and spectrum allocation in elastic optical networks using bee colony optimization, Photonic Network Communications, 34, 356–374.
McFadden, D. 1973. Conditional Logit Analysis of Qualitative Choice Behavior, in Frontiers of Econometrics, edited by P. Zarembka, New York: Academic Press.
Miladić-Tešić, S., Marković, G., and Radojičić, V. 2019. Traffic Grooming on Designing Elastic Optical Networks, In Proceedings of ICEEST 2019, pages 35–38, Ohrid, North Macedonia.
Moayedikia, A., Ong, K. L., Boo, Y. L., and Yeoh, W. 2016. Bee colony based worker reliability estimation algorithm in microtask crowdsourcing. In 2016 15th IEEE International Conference on Machine Learning and Applications (ICMLA), Anaheim, CA, pp. 713–717.
Mousavinasab, Z., Entezari-Maleki, R., and Movaghar, A. 2011. A bee colony task scheduling algorithm in computational grids. In Digital Information Processing and Communications, Proc. Int. Conf., ICDIPC 2011, Part I, pages 200–210. Springer, Berlin.
Nakrani, S., and Tovey, C. 2004. On honey bees and dynamic server allocation in internet hosting centers. Adaptive Behavior, 12(3–4): 223–240.
Nedeljković, R., Mitrović, S., and Drenovac, D. 2009. Bee colony optimization meta-heuristic for backup allocation problem. In Proceedings of the PosTel XXVII, pages 115–122, (in Serbian), Beograd, Serbia.
Nikolić, M. 2015. Disruption management in transportation by the Bee Colony Optimization metaheuristic. PhD thesis, Faculty of Traffic and Transportation, University of Belgrade.
Nikolić, M., Ćalić, J., Šelmić, M., and Macura, D. 2019. Tuning fuzzy system for estimation of freight train energy consumption by the Bee Colony Optimization metaheuristic, In Proceedings of the SYM-OP-IS 2019, pages 672–677 (in Serbian), Kladovo, Serbia.
Nikolić, M., and Teodorović, D. 2013a. Empirical study of the bee colony optimization (BCO) algorithm. Expert Systems with Applications, 40(11): 4609–4620.
Nikolić, M., and Teodorović, D. 2013b. Transit network design by bee colony optimization. Expert Systems with Applications, 40(15): 5945–5955.
Nikolić, M., and Teodorović, D. 2014. A simultaneous transit network design and frequency setting: Computing with bees. Expert Systems with Applications, 41(16), 7200–7209.
Nikolić, M., and Teodorović, D. 2015. Vehicle rerouting in the case of unexpectedly high demand in distribution systems. Transportation Research Part C: Emerging Technologies, 55: 535–545.
Nikolić, M., and Teodorović, D. 2019. Mitigation of disruptions in public transit by Bee Colony Optimization. Transportation Planning and Technology, 42(6): 573–586.
Nikolić, M., Teodorović, D., and Vukadinović, K. 2015. Disruption management in public transit: the bee colony optimization approach. Transportation Planning and Technology, 38(2): 162–180.

Olivas, F., Amador-Angulo, L., Perez, J., Caraveo, C., Valdez, F., and Castillo, O. 2017. Comparative study of type-2 fuzzy particle swarm, bee colony and bat algorithms in optimization of fuzzy controllers. Algorithms, 10(3): 101.

Pertiwi, A. P. P., and Suyanto, P. 2011. Globally evolved dynamic bee colony optimization. In Knowledge-Based and Intelligent Information and Engineering Systems, In Proceedings of the 15th International Conference, KES 2011, Part I, pages 52–61. Springer, Berlin Heidelberg, Kaiserslautern, Germany.

Pham, D. T., Ghanbarzadeh, A., Koc, E., Otri, S., and Zaidi, M. 2006a. The bees algorithm - a novel tool for complex optimisation problems. In Proceedings of the 2nd Virtual International Conference on Intelligent Production Machines and Systems (IPROMS 2006), pages 454–459, Elsevier, Cardiff.

Pham, D. T., Soroka, A. J., Ghanbarzadeh, A., and Koc, E. 2006b. Optimising neural networks for identification of wood defects using the bees algorithm. In Proceedings of the IEEE International Conference on Industrial Informatics, pages 1346–1351. IEEE, Singapore.

Rajasekhar, A., Lynn, N., Das, S., and Suganthan, P. N. 2017. Computing with the collective intelligence of honey bees–a survey. Swarm and Evolutionary Computation, 32: 25–48.

Rajeswari, M., Amudhavel J., Pothula S., and Dhavachelvan P. 2017. Directed bee colony optimization algorithm to solve the nurse rostering problem. Computational Intelligence and Neuroscience, 2017: 1–26.

Sa'idi, M., Mostoufi, N., and Sotudeh-Gharebagh, R. 2013. Modelling and optimisation of continuous catalytic regeneration process using bee colony algorithm. The Canadian Journal of Chemical Engineering, 91(7): 1256–1269.

Sato, T., and Hagiwara, M. 1997. Bee system: finding solution by a concentrated search. In Proceedings of the IEEE International Conference on Systems, Man, and Cybernetics 'Computational Cybernetics and Simulation', pages 3954–3959, Orlando, FL.

Šelmić, M. 2011. Location problems on transport networks by computational intelligence methods. PhD thesis, Faculty of Traffic and Transportation, University of Belgrade.

Šelmić, M., Edara, P., and Teodorović, D. 2008. Bee colony optimization approach to optimize locations of traffic sensors on highways. Tehnika, 6: 9–15. (in Serbian)

Šelmić, M., Teodorović, D., and Vukadinović, K. 2010. Locating inspection facilities in traffic networks: an artificial intelligence approach. Transportation Planning and Technology, 33(6): 481–493.

Sohi, M. F., Shirdel, M., and Javidaneh, A. 2011. Applying BCO algorithm to solve the optimal dg placement and sizing problem. In Power Engineering and Optimization Conference (PE- OCO), 2011 5th International, pages 71–76.

Stojanović, T. 2016. The development and analysis of metaheuristics for satisfiability in probabilistic logics. PhD thesis, Faculty of Science, University of Kragujevac.

Stojanović, T., Davidović, T., and Ognjanović, Z. 2015. Bee-colony optimization for the satisfiability problem in probabilistic logic. Applied Soft Computing, 31: 339–347.

Teodorović, D. 2009. Bee colony optimization (BCO). In: C. P. Lim, L. C. Jain, and S. Dehuri, editors, Innovations in Swarm Intelligence, pages 39–60. Springer-Verlag, Berlin Heidelberg.

Teodorović, D., and Dell'Orco, M. 2005. Bee colony optimization – a cooperative learning approach to complex transportation problems. In Advanced OR and AI Methods in Transportation. Proceedings of the 10th Meeting of the EURO Working Group on Transportation, pages 51–60, Poznan, Poland.

Teodorović, D., and Dell'Orco, M. 2008. Mitigating traffic congestion: solving the ride-matching problem by bee colony optimization. Transportation Planning and Technology, 31(2): 135–152.

Teodorović, D., Lučić, P., Marković, G., and Dell'Orco, M. 2006. Bee colony optimization: principles and applications. In: B. Reljin and S. Stanković, editors, Proceedings of the Eight Seminar on Neural Network Applications in Electrical Engineering – NEUREL 2006, pages 151–156, University of Belgrade, Belgrade.

Teodorović, D., and Šelmić, M. 2007. The BCO algorithm for the p-median problem. In Proceedings of the XXXIV Serbian Operations Research Conference, pages 417–420, Zlatibor, Serbia (in Serbian).

Teodorović, D., Šelmić, M., and Davidović, T. 2015. Bee colony optimization Part II: the application survey. Yugoslav Journal of Operational Research, 25(2): 185–219.

Teodorović, D., Šelmić, M., and Mijatović-Teodorović, L. J. 2013. Combining case–based reasoning with bee colony optimization for dose planning in well differentiated thyroid cancer treatment. Expert Systems with Applications, 40(6): 2147–2155.

Teodorović, D., Šelmić, M., Nikolić, M., Jovanović, I., and Vidas, M. 2017. Metaheuristic approach for detector locations in transport networks, SYM-OP-IS 2017, Zlatibor, 25–28 September 2017, pp. 723–728.

Todorovic, N., and Petrovic, S. 2013. Bee colony optimization algorithm for nurse rostering. IEEE Transactions on Systems, Man, and Cybernetics: Systems, 43(2): 467–473.

Vijayakumari, K., and Deepa, V.B. 2019. Performance analysis of optimization techniques by using clustering, International Research Journal of Engineering and Technology, 6(6), 1465–1469.

Von Frisch, K. 1967. The Dance Language and Orientation of Bees. Harvard University Press, Cambridge, Massachusetts.

Wang, G. G. 2015. Moth search algorithm: a bio-inspired metaheuristic algorithm for global optimization problems. Memetic Computing, 10(2): 151–164.

Wang, G. G., Deb, S., and Cui, Z. 2015. Monarch butterfly optimization. Neural Computing Applications, 31(7): 1995–2014.

Wang, G. G., Gandomi, A. H., and Alavi, A. H. 2014. Stud krill herd algorithm. Neurocomputing, 128: 363–370.

Wedde, H. F., Farooq, M., and Zhang, Y. 2004. BeeHive: an efficient fault-tolerant routing algorithm inspired by honey bee behavior. LNCS: Ant Colony Optimization and Swarm Intelligence, 3172: 83–94.

Wong. L.-P. 2012. A generic bee colony optimization framework for combinatorial optimization problems. PhD thesis, School of Computer Engineering, Nanyang Technological University, Singapore.

Wong, L.-P., Hean Low, M. Y., and Chong, C. S. 2008. A bee colony optimization algorithm for traveling salesman problem. In 2nd Asia International Conference on Modelling & Simulation, pages 818–823.

Wong, L-P., Hean Low, M. Y., and Chong, C. S. 2009. An efficient bee colony optimization algorithm for traveling salesman problem using frequency-based pruning. In 7-th IEEE International Conference on Industrial Informatics, pages 775–782.

Wong, L.-P., Hean Low, M. Y., and Chong, C. S. 2010. Bee colony optimization with local search for traveling salesman problem. International Journal on Artificial Intelligence Tools, 19(03): 305–334.

Yang, C., Chen, J., and Tu, X. 2007. Algorithm of fast marriage in honey bees optimization and convergence analysis. In Proceedings of the IEEE International Conference on Automation and Logistics, pages 1794–1799, Jinan, China.

Yang, X.-S. 2005. Engineering optimizations via nature-inspired virtual bee algorithms. In: J. Mira and J. R. Alvarez, editors, Artificial intelligence and knowledge engineering applications: a bioinspired approach. LNCS, vol. 3562, pages 317–323. Springer-Verlag, Berlin–Heidelberg.

Yang, X. S. ed., 2020. Nature-inspired computation and swarm intelligence: algorithms. Theory and Applications. Academic Press.

Yi, J. H., Deb, S., Dong, J., Alavi, A. H., and Wang, G. G. 2018a. An improved NSGA-III Algorithm with adaptive mutation operator for big data optimization problems. Future Generation Computer Systems, 88: 571–585.

Yi, J. H., Xing, L. N, Wang, G. G., Dong, J., Vasilakos, A. V., Alavi, A. H., and Wang, L. 2018b. Behavior of crossover operators in NSGA-III for large-scale optimization problems. Information Science, 509: 470–487.

13 A Bumble Bees Mating Optimization Algorithm for the Location Routing Problem with Stochastic Demands

Magdalene Marinaki and Yannis Marinakis

CONTENTS

13.1 INTRODUCTION

The field of swarm intelligence algorithms is currently one of the most interesting fields in evolutionary computation. There is a large number of algorithms that simulate almost every living organism in nature. Indicatively, we mention some of these algorithms: the bat algorithm [1], the cuckoo search algorithm [2], the firefly algorithm [3], the krill herd algorithm [4–7], the biogeography-based optimization algorithm [8], the grey wolf optimizer [9], the moth-flame optimization algorithm [10], the ant-lion optimizer [11], the monarch butterfly optimization [12], the dragonfly algorithm [13], the moth-search algorithm [14], and the elephant herding optimization algorithm [15]. One of the insects that probably has the most algorithms devoted to it is the bee [16]. Researchers in the field proposed, in the last 20 years, a large number of bee-inspired algorithms divided in two different behaviors of bees in nature, the foraging and the mating behaviors. The main algorithms that simulate the foraging behavior are the bee colony optimization (BCO) algorithm [17], the artificial bee colony (ABC) algorithm [18, 19], the honey bee social foraging (HBSF) [20], the bees algorithm (BA) [21], the bee system (BS) [22], the wasp swarm optimization (WSO) [23, 24], the beehive algorithm [25], the artificial beehive algorithm (ABHA) [26], the OptBees algorithm [27], the bee swarm optimization (BSO) algorithm [28], the virtual bee algorithm [29], the bee colony–inspired algorithm (BCiA) [30], the simulated bee colony (SBC) [31], and the bumblebees algorithm [32]. On the other hand, there are only three algorithms that simulate the mating behavior, at least to our knowledge: the honey bees mating optimization (HBMO) algorithm [33–36], the bees life algorithm (BLA) [37], and the bumble bees mating optimization (BBMO) algorithm [38].

The BBMO algorithm simulates the mating behavior of one of the species of the bees, the bumble bees. The basis of this algorithm is the HBMO algorithm but with a number of different steps that make the algorithm a completely different and competitive algorithm with other nature-inspired algorithms. The BBMO algorithm has mainly been applied in combinatorial optimization problems (clustering, feature selection problem, capacitated vehicle routing problem (CVRP), vehicle routing problem with stochastic demands (VRPSD), open vehicle routing problem (OVRP), multicast routing problem, permutation flowshop scheduling problem, and probabilistic traveling salesman problem), but it has also been applied in the classic global optimization functions. In [39], an adaptive version of the algorithm is presented. In the present chapter, we propose a different, more effective, adaptive version of the BBMO in which the path re-linking (PR) procedure of the version of the algorithm presented in [39] has been replaced by the adaptive memory procedure similar to the one proposed in [40] in the frame of the particle swarm optimization algorithm for the solution of the vehicle routing problem with time windows. The new algorithm is used for the solution of the location routing problem with stochastic demands.

In the next section, the basic BBMO algorithm is presented and a brief explanation of the improvements of the algorithms through the years is given. In Section 13.3, the proposed adaptive memory bumble bees mating optimization (ABBMO) algorithm is presented and analyzed in detail. The location routing problem with stochastic demands is described and analyzed in Section 13.4. In Section 13.5, the computational results of the proposed algorithm are given. In the final section, conclusions and further research issues are analyzed.

13.2 BUMBLE BEES MATING OPTIMIZATION ALGORITHM

The BBMO algorithm was, initially, presented in a hybridized form [41] for solving a clustering problem. The hybridization was performed using, in the initial phase of the algorithm, the greedy randomized adaptive search procedure (GRASP) [42]. However, because we wanted to see how the algorithm performed in global unconstrained optimization problems, in the second publication of the BBMO algorithm, the algorithm was used without any hybridization for the solution of these kinds of problems. A short description of the algorithm as it was presented in [43] is the following: We have three kinds of Bumble Bees (queens, workers and drones - first difference from the Honey Bees Mating Optimization (HBMO) algorithm where only a queen and drones exist). In the initialization phase, there is a random selection of bees (solutions) that will constitute the initial population. A sorting of all these bees is performed and the best of them plays, initially, the role of the queen. The role of all other bees is to be the drones. Afterward, the queen performs the mating flights during which the most profitable drones are selected for mating based only on their fitness function (second difference from the HBMO algorithm in which speed and energy equations are used for the selection).

The queen bee begins to lay eggs when it will return to the hive and it can produce three kind of bees:

- *New queens:* Using a multiparent crossover operator where the one parent is always the queen bee and the other parents are selected between the drones using the following equation:

$$b_{ij}(t) = \begin{cases} q_j(t), & \textit{if } rand_i(0,1) \le Cr_1 \\ d_{kj}(t), & \textit{otherwise.} \end{cases} \tag{13.1}$$

  where t is the iteration number, j is the dimension of the problem, Cr_1 is a parameter that controls which elements of the broods $b_{ij}(t)$ have been selected from the solution of the queen bee, $q_j(t)$, and which elements from the selected drone k, $d_{kj}(t)$. As the value

of Cr_1 is between 0 and 1, we fix the random number generator, $rand_i(0,1)$, to produce values in the same space.
- *Workers:* Using the same procedure as in the production of the new queens, i.e., the new bees that could not be the new queens using the previous procedure are the workers.
- *Drones:* Using a random mutation operator either in an old queen or in one of the fittest workers.

The production of the new queens, drones, and workers is the third and probably the most important difference of the BBMO algorithm from the HBMO algorithm, at least in the initial version of the algorithm. The fittest of the broods are selected as new queens (another difference from the HBMO algorithm in which only one brood is a candidate for the new queen) while the rest are the workers.

In the next phase of the algorithm, the feeding procedure of the new queens is performed using Eq. 13.2. The meaning of the feeding procedure is that in the beginning, the new queens (nq_{ij}) are fed from the old queens (q_j), but as the iterations increase, the role of the old queens in the feeding of the new queens is decreasing and, at the same time, the role of the workers (w_{kj}) is increasing as the solution of the workers is improving through the iterations. Each new queen can select from the set of workers, a number M of different workers that will be used for the feeding procedure. The b_{min}, b_{max} (two parameters with values in the interval [0, 1]) and the current local search iteration (lsi) and the maximum number of local search iterations (lsi_{max}) are used for controlling the feeding procedure and if the new queens will be fed from the old queen (or queens) or the selected workers. The following equation used in the BBMO makes the BBMO algorithm suitable for continuous optimization problems and is one more different procedure compared with the HBMO algorithm:

$$nq_{ij} = nq_{ij} + \left(b_{max} - \frac{(b_{max} - b_{min}) * lsi}{lsi_{max}} \right) * (nq_{ij} - q_j) + \frac{1}{M} * \sum_{k=1}^{M} \left(b_{min} - \frac{(b_{min} - b_{max}) * lsi}{lsi_{max}} \right) * (nq_{ij} - w_{kj}) \tag{13.2}$$

The final procedure of the BBMO is the leaving of the drones from the hive in a swarm formation. This is performed as the drones (d_{ij}, d_{kj}, and d_{lj} are the solutions of drones i, k, and l, respectively) are looking for the best places in the solution space and wait there for the new queens, which have left from the hive after the feeding procedure to mate. The following equation is used (in the initial version of the BBMO algorithm, this equation is the last different procedure from the HBMO algorithm):

$$d_{ij} = d_{ij} + \alpha * (d_{kj} - d_{lj}) \tag{13.3}$$

where α is a parameter that determines the percentage that the drone i is affected by the two other drones, k and l.

In the next publication, the algorithm was used for the solution of the CVRP [38]. It was the first time that the algorithm was used for the solution of this kind of problem and some modifications were necessary to solve the problem effectively. First, a path representation of the tours was selected for each solution as the solution of a CVRP is the number of routes that are needed for the service of all the customers with the available or selected vehicles. Also, to use the feeding equation of the bees, each element of the solution was transformed into a floating point interval [0,1], the feeding equation of all bees was calculated and, then, each element of the solution was converted back into the integer domain using relative position indexing [44]. Another difference from the previous version of the algorithm is the mutation operator. The mutation operator in a CVRP usually plays the role of a local search algorithm; thus, the random mutation operator that was used for the solution of the global optimization problems was not effective for the CVRP. Instead of this operator, a very effective local search algorithm, the expanding neighborhood search [45], was used. Finally, the number of the drones per colony in each external generation was determined by:

$$number\ of\ drones\ per\ colony = \frac{number\ of\ drones}{number\ of\ queens} \tag{13.4}$$

The algorithm was tested in the two classic sets of CVRP benchmark instances and it had the second best performance among all algorithms that had been published for the solution of this problem. In [46], an improved version of the BBMO algorithm was presented for the solution of the OVRP where the equation of the movement of the drones outside the hive was replaced by the iterated local search (ILS) algorithm [47] to avoid the transformation of each solution of the drones in continuous values and vice versa. Although the algorithm without this modification had proved its effectiveness, with the addition of the ILS algorithm instead of the equation of the movement of the drones, the algorithm became faster and had better performance in less computational time. The algorithm was tested in two sets of benchmark instances and its performance ranked it among the three most effective algorithms for the solution of this problem.

In [48], a discrete version of the BBMO algorithm, the combinatorial neighborhood topology bumble bees mating optimization (CNTBBMO), was presented. The algorithm was used for the solution of the VRPSD. In this algorithm, there is no transformation at all in continuous values, thus, all the main equations of the initially proposed BBMO algorithm were replaced by suitable local search procedures. As in the previous version of the algorithm, the movement of the drones was replaced by the ILS [47] algorithm, the random mutation operator, initially, was replaced by the expanding neighborhood search algorithm and, in this version, it was replaced by the variable neighborhood search (VNS) [49] algorithm. Finally, the feeding procedure was replaced from a new, by that time, strategy denoted as combinatorial neighborhood topology (CNT) [50], which includes a PR [51] procedure. The CNT procedure replaced the feeding procedure of each new queen in which the new queen decided if it would be fed from an old queen or from a worker. Following what happens in real life, the new queen, initially, was fed from the old queen and as the local search iterations increased, it was fed more from the workers and, at the end, it was fed almost only from the workers.

In the algorithm, first a random number, different for each element of the new queen's solution, in the interval $(l,b) = (l_{bound}, u_{bound})$, is created [48]. If this value is greater than a number L_2, then either, with equal probability, no path re-linking (NOPR) is performed or a PR with a second random worker is performed (PRRW2). If the random number is between L_1 and L_2, then the new queen bee performs a path relinking (PRRW) with a random worker. Finally, if the value is less than L_1, then, a PR with a randomly selected old queen is realized (PROQ). Initially, L_2 and L_1 have large values that are decreased during the iterations to perform PR with the old queens in the beginning of the algorithm and, as the iterations are increasing, to perform the PR with larger probability with a worker. To avoid a fast convergence of the algorithm, a solution is not selected until it differs from the best queen more than 15% except if this solution improves the best solution. The values of L_1 and L_2 are as follows:

$$L_1 = (u_{bound} - l_{bound}) \times \left(w_1 - \frac{w_1}{iter_{max}} \times t \right) + l_{bound} \tag{13.5}$$

and

$$L_2 = (u_{bound} - l_{bound}) \times \left(w_2 - \frac{w_2}{2 * iter_{max}} \times t \right) + l_{bound} \tag{13.6}$$

where l_{bound} and u_{bound} are the lower and upper bounds, $iter_{max}$ is the maximum number of iterations, and t is the current iteration. As we would like to have L_1 and L_2 with large values in the beginning of the algorithm (with L_2 always greater than L_1), the w_1 and w_2 also have large values (w_2 is selected to be equal to 0.9 and w_1 equal to 0.8).

Afterward, the feeding procedure is performed using the following equation:

$$tnq_i(t) = \begin{cases} q_i(t), & \text{if } rand_i(l,b) \leq L_1 \\ wr_i(t), & \text{if } L_1 \leq rand_i(l,b) \leq L_2 \\ nq_i(t) \ or \ wr_j(t), & \text{otherwise.} \end{cases} \tag{13.7}$$

where tnq is the auxiliary solution that has been produced from the PR procedure (a possible new queen), q is the solution of the old queen bee, nq is the solution of the new queen bee, and with wr_i and wr_j are denoted the solutions of two different workers.

This version of the algorithm proved to be the most effective of all, as it was tested in the classic instances for the VRPSD. It was compared with all other versions of the BBMO and all other algorithms that have been used for the solution of this problem. The algorithm also proved to be the most effective as a number of new best solutions were produced in some instances and in all other instances the obtained solutions of the algorithm were equal to the best known solutions (BKSs) or slightly worst.

In [52], the initially proposed algorithm was used for the solution of the feature selection problem. As in the solution vector of this problem, only two values exist, 0 if the feature was not selected and 1. Otherwise, we used all the initially proposed equations and transformed suitably the solutions using the sigmoid function:

$$sig(nq_{ij}) = \frac{1}{1 + exp(-nq_{ij})} \tag{13.8}$$

and, then, the activated features are calculated by:

$$y_{ij} = \begin{cases} 1, & \text{if } rand_j(0,1) < sig(nq_{ij}) \\ 0, & \text{if } rand_j(0,1) \geq sig(nq_{ij}) \end{cases} \tag{13.9}$$

where y_{ij} is the transformed solution.

Finally, an adaptive version of the CNTBBMO algorithm was used for the solution of the permutation flowshop scheduling problem [39, 53], the probabilistic traveling salesman problem [39], and the multicast routing problem [39]. In this version of the algorithm, all the parameters are optimized during the iterations. Initially, random values are given in each one of the parameters and the only restriction is $w_1 < w_2$ and $u_{bound} > l_{bound} > 0$ to maintain the feasibility of the solutions. The parameters that were used and optimized in this procedure are the w_1, w_2, u_{bound}, and l_{bound}; the number of bees; the number of VNS iterations; and the number of iterations. The parameters were updated using three different procedures. In the first one, if for a consecutive number of iterations (this number was set equal to 20), the best solution had not been improved, the following parameters were updated:

$$w_1 = w_{1opt} + \frac{w_1 - w_{1opt}}{w_{1opt}} \tag{13.10}$$

$$w_2 = w_{2opt} + \frac{w_2 - w_{2opt}}{w_{2opt}} \tag{13.11}$$

$$u_{bound} = UB + \frac{u_{bound} - UB}{UB} \tag{13.12}$$

$$l_{bound} = LB + \frac{l_{bound} - LB}{LB} \tag{13.13}$$

$$Local\ Search\ iter = LS + \frac{Local\ Search\ iter - LS}{LS} \tag{13.14}$$

where *LS*, *LB*, *UB*, w_{2opt}, and w_{1opt} are the best performed values so far for the local search iterations, the lower bound, the upper bound, the w_2, and the w_1, respectively. In the first iteration of the algorithm, random values are given for these parameters.

In the second procedure, the number of bumble bees was increasing, if for a consecutive number of iterations the average best solutions of all bumble bees was not improving more than 2% and, in addition, no improvement of the best solution as in the first procedure had performed, using the following equation:

$$Bees = NB + \frac{Bees - NB}{NB} \tag{13.15}$$

where *NB* is the best value of the bumble bees. Finally, the decrease of the number of bumble bees was performed in the last procedure. This was performed because a bumble bee with a not very good solution probably would not give a solution better than the best queen in the next iterations. Thus, if a solution for a consecutive number of iterations would have a solution 5% worse than the solution of the best queen, it was deleted from the hive. The final outcomes of the algorithm, in addition to the best solution, were the final best values of the parameters calculated from the algorithm.

13.3 ADAPTIVE MEMORY BUMBLE BEES MATING OPTIMIZATION ALGORITHM

In the proposed version of the algorithm, we use the last version of the BBMO algorithm and we have replaced the PR procedure with an adaptive memory [54]. This modification of the CNT procedure was performed successfully in the frame of the particle swarm optimization algorithm for the solution of the vehicle routing problem with time windows [40]. All the other conditions of the CNT procedure still remain the same. We were led to this replacement because we wanted the movement from one solution to another to be realized using whole tours. Thus, two kinds of adaptive memories are created, in which the first one a number is used that records how many times a tour with a specific set of nodes (without taking into account the sequence of the nodes) participates in the solution of the queen bee. This number increases (or decreases based on some conditions) the possibilities of a tour to become a member of the new solution. In the second adaptive memory, the same procedure is followed for the most profitable workers. We do not always use the same worker, but we perform a ranking from the best to the worst and the worker that is in the first place (independently if in two iterations is a different worker) updates the first place of the adaptive memory. Respectively, the worker that is in the second place updates the second place of the adaptive memory and so on. This procedure is realized for the 10 best workers. We do not keep this memory for all workers for two reasons. First, in the BBMO algorithm the population of workers changes in each iteration and, second, if we keep a memory for each one, we will increase the memory that we need to use for the algorithm. In the CNT there is the possibility for the new queen to be fed from the old queen or from a worker or from a second worker. The addition of a number that records the consecutive iterations, in which a route has been presented in the global or in the local best solutions, is inspired by the medium and the long-term memories of tabu search [55].

Two kinds of memories are used: the one is for the old queen and it is denoted as $QueenM$ and the other memory is for the 10 most effective workers and it is denoted as $WrM(i)$, with i running from 1 to 10. In these memories, complete routes with specific nodes are recorded. Two routes with the same set of nodes but with a different sequence of the nodes are recorded as one solution in the memories, and the sequence that is kept is the one with the best cost that does not violate the constraints. In each iteration, there are two possibilities for the routes in the $QueenM$ and in the $WrM(i)$. The one possibility is that a route that already belongs to the $QueenM$ (or to the $WrM(i)$) is a route of the old queen (or of the corresponding worker); thus, the number that shows the times (iterations) in which a route is in the $QueenM$ (or in the $WrM(i)$) is increased by one. This number is denoted by $QueenMval_l$ and $WrMval_l(i)$, where l is the number of different routes in the $QueenM$ and in the $WrM(i)$, respectively. The other possibility is a new route to be constructed in the queen bee (or in the corresponding worker); thus, it has to be added in the $QueenM$ (or in the $WrM(i)$) with the counter of the routes in the $QueenM$ (or in the $WrM(i)$), l, increased by one and the $QueenMval_l$ (or the $WrMval_l$ depending on the case) to take an initial value equal to one.

In general, a new queen bee can either be fed from the old queen, from the best worker, or from another worker. In the proposed algorithm, a number of routes from the two kinds of adaptive memories are selected.

1. The routes are selected from the $QueenM$ if the new queen decides to be fed from the old queen.
2. The routes are selected from the $WrM(1)$ if the new queen decides to be fed from the first worker.
3. The routes remain the same and will not take any values from either $QueenM$ or $WrM(1)$ if the new queen decides not to be fed or the routes are selected from another worker $WrM(i)$.

Afterward, the algorithm continues as in the previous version where, initially, the values L_1 (Eq. 13.5) and L_2 (Eq. 13.6) are calculated. Then, the procedure continues with the calculation of the feeding process using Eq. 13.7 with the difference that the auxiliary vector, which probably will replace the old queen, will be calculated with the selections that are realized using the adaptive memory procedure instead of using the PR procedure.

13.4 LOCATION ROUTING PROBLEM WITH STOCHASTIC DEMANDS

In this chapter, a variant of the location routing problem is solved - the location routing problem with stochastic demands (LRPSDs) [56, 57]. It is a two-phase problem where in the one phase, the location of the facilities and the assignment of the customers are realized by solving a capacitated facility location problem (CFLP). In the other phase, the routing of the vehicles is constructed by solving a VRPSD by finding the expected cost of the a priori tour of the vehicles [56, 57]. As the demands are stochastic variables, we use a preventive restocking strategy where when the load of the vehicle becomes less than a threshold value, the vehicle returns to the depot for restocking. The formulation of the problem is the following (initially the objective function of the CFLP [first part] is presented):

$$min \sum_{i \in U} B_i y_i + \sum_{i \in U} f_{ij}(q) \tag{13.16}$$

where

$$y_i = \begin{cases} 1, & \text{if in the candidate site } i \text{ a facility is located} \\ 0, & \text{otherwise,} \end{cases} \tag{13.17}$$

$$z_{jk} = \begin{cases} 1, & \text{if vehicle } k \text{ operates out of a facility at candidate} \\ & \text{site } j \\ 0, & \text{otherwise,} \end{cases} \tag{13.18}$$

j is a customer that is assigned in the candidate facility i (U is the set of all candidate facilities).

The constraints corresponding to the first part are the classic constraints of the CFLP:

$$\sum_{k \in K} Q_k z_{ik} \leq QB_i y_i, \quad \forall i \in U \tag{13.19}$$

$$z_{ik} \leq y_i, \quad \forall i \in U, \forall k \in K \tag{13.20}$$

$$\sum_{i \in U} z_{ik} \leq 1, \quad \forall k \in K \tag{13.21}$$

The VRPSD for each candidate facility is solved using [56, 57]:

$$f_{ij}(q) = Minimum\left\{ f_{ij}^{p}(q), f_{ij}^{r}(q) \right\} \tag{13.22}$$

where

$$\begin{aligned} f_{ij}^{p}(q) &= d_{j,j+1} + \sum_{k \leq q} f_{i,j+1}(q-k) p_{j+1,k} \\ &+ \sum_{k > q} \left[2d_{j+1,0} + f_{i,j+1}(q+Q-k) \right] p_{j+1,k} \end{aligned} \tag{13.23}$$

$$f_{ij}^{r}(q) = d_{j,0} + d_{0,j+1} + \sum_{k=1}^{K} f_{i,j+1}(Q-k) p_{j+1,k} \tag{13.24}$$

with boundary condition

$$f_{in}(q) = d_{n,0}, q \in L_n \tag{13.25}$$

where the customers' demands are known only when the vehicle arrives at the current customer. The demand is a stochastic variable ($\xi_j, j = 1,...,n$) independently distributed with known distribution with upper possible value the vehicle's capacity Q and it has a discrete probability distribution $p_{jk} = Prob(\xi_j = k), k = 0,1,2,...,K \leq Q$. With QB_i and B_i, the capacity and the fixed cost of a candidate facility i are denoted, respectively. When a customer j is served from a vehicle, the remaining load of the vehicle is denoted with q. Finally, the expected costs $f_{ij}(q)$, $f_{ij}^p(q)$, and $f_{ij}^r(q)$ concern the cost from the customer j onward, the cost of the route when the vehicle goes directly to the next customer after servicing a customer, and the cost of returning the vehicle to the depot for preventive restocking, respectively.

13.5 COMPUTATIONAL RESULTS

For the implementation of the algorithm, modern Fortran was used. The algorithm was tested on 82 benchmark instances divided in three different sets (in the first set [58], there are 16 instances; in the second set [59], there are 30 instances; and in the last set [60], there are 36 instances). The main characteristics, number of customers (n), number of candidate sites (m), and vehicle capacity (Q), of each set (B1–B16 for the first set, P1–P30 for the second set, and T1–T36 for the last set) are presented in Table 13.1.

In Tables 13.2–13.4, the results of the proposed algorithm in the location routing problem with stochastic demands are presented. More precisely, in Table 13.2, the results in the Barreto instances are presented, whereas in Tables 13.3 and 13.4, the results in the other two data sets, in the Prins et al. instances and in the Tuzun and Burke instances, are presented. In these tables, the results of the best run (BS), the average results of the 10 runs, the standard deviation, and the variance in each instance are presented, respectively. Taking into account the results presented in these tables, we can conclude that the proposed algorithm gives very stable results as the standard deviations and the variances are very small. More precisely, the standard deviations in the Barreto instances vary between 0.64 and 1.12 with the average value equal to 0.90, whereas in the Prins et al. instances, they vary between 2.28 and 3.76 with the average value equal to 3.07. In the Tuzun and Burke instances, they vary between 0.66 and 1.22 with the average value equal to 0.93. On the other hand, the variances in the Barreto instances vary between 0.41 and 1.25 with the average value equal to 0.82, whereas in the Prins et al. instances, they vary between 5.22 and 14.12 with the average value equal to 9.63. In the Tuzun and Burke instances, they vary between 0.44 and 1.48 with the average value equal to 0.89.

In the literature, there are only two papers [56, 57], in which the location routing problem with stochastic demands was solved with the same formulation; thus, we compared the computational results of the proposed algorithm with the computational results of the algorithms given in these papers. More precisely, the proposed BBMO algorithm is compared with an evolutionary algorithm denoted as the RVNS [57] algorithm in which the initial solutions are created randomly and the local search algorithms that are used are the same as the ones used in this chapter, a VNS algorithm and an ILS algorithm. The other algorithms are a hybrid genetic algorithm (HGA) [57], a hybrid differential evolution (HDE) algorithm [57], a hybrid clonal selection algorithm (HCSA) [57], and the global and local combinatorial expanding neighborhood topology particle swarm optimization (GLCENTPSO) algorithm [56]. All algorithms have been executed 10 times. For more details about these algorithms, please see [56, 57].

In Tables 13.5–13.7, the comparison of the proposed ABBMO algorithm with the other algorithms from the literature is presented. As we mentioned earlier, only in two papers in the literature the problem was formulated the same way as the one used in this chapter. Thus, we compared the computational results of the proposed algorithm with the computational results

TABLE 13.1
Benchmark Instances

Barreto [58]				Prins et al. [59]				Tuzun and Burke [60]			
Instance	n	m	Q	Instance	n	m	Q	Instance	n	m	Q
B1	50	5	160	P1	20	5	70	T1	100	10	150
B2	75	10	140	P2	20	5	150	T2	100	20	150
B3	100	10	200	P3	20	5	70	T3	100	10	150
B4	88	8	9000	P4	20	5	150	T4	100	20	150
B5	150	10	8000	P5	50	5	70	T5	100	10	150
B6	21	5	6000	P6	50	5	150	T6	100	20	150
B7	22	5	4500	P7	50	5	70	T7	100	10	150
B8	29	5	4500	P8	50	5	150	T8	100	20	150
B9	32	5	8000	P9	50	5	70	T9	100	10	150
B10	32	5	11	P10	50	5	150	T10	100	20	150
B11	36	5	250	P11	50	5	70	T11	100	10	150
B12	27	5	2500	P12	50	5	150	T12	100	20	150
B13	134	8	850	P13	100	5	70	T13	150	10	150
B14	12	2	140	P14	100	5	150	T14	150	20	150
B15	55	15	120	P15	100	5	70	T15	150	10	150
B16	85	7	160	P16	100	5	150	T16	150	20	150
–	–	–	–	P17	100	5	70	T17	150	10	150
–	–	–	–	P18	100	5	150	T18	150	20	150
–	–	–	–	P19	100	10	70	T19	150	10	150
–	–	–	–	P20	100	10	150	T20	150	20	150
–	–	–	–	P21	100	10	70	T21	150	10	150
–	–	–	–	P22	100	10	150	T22	150	20	150
–	–	–	–	P23	100	10	70	T23	150	10	150
–	–	–	–	P24	100	10	150	T24	150	20	150
–	–	–	–	P25	200	10	70	T25	200	10	150
–	–	–	–	P26	200	10	150	T26	200	20	150
–	–	–	–	P27	200	10	70	T27	200	10	150
–	–	–	–	P28	200	10	150	T28	200	20	150
–	–	–	–	P29	200	10	70	T29	200	10	150
–	–	–	–	P30	200	10	150	T30	200	20	150
–	–	–	–	–	–	–	–	T31	200	10	150
–	–	–	–	–	–	–	–	T32	200	20	150
–	–	–	–	–	–	–	–	T33	200	10	150
–	–	–	–	–	–	–	–	T34	200	20	150
–	–	–	–	–	–	–	–	T35	200	10	150
–	–	–	–	–	–	–	–	T36	200	20	150

of these two papers. In all three tables, the second column presents the BKS. This solution is the best solution in each instance from the other two papers, mainly, from the particle swarm optimization algorithm, because this algorithm has the best performance of the other algorithms used in the comparisons from the literature. Then, for each algorithm the best solution (cost) and the quality of the solutions (deviation from the BKS) are presented. The quality is given from the following equation: $\omega = \frac{(cost - BKS)}{BKS}\%$.

TABLE 13.2
Results of the Proposed Algorithm in the Barreto [58] Instances

	Barreto [58]			
	BS	Average	Stdev	Var
B1	566.26	567.62	0.98	0.97
B2	852.12	852.91	0.64	0.41
B3	837.02	838.26	0.98	0.97
B4	358.87	360.33	0.87	0.76
B5	43966.75	43968.34	1.01	1.02
B6	427.34	428.50	0.94	0.89
B7	585.77	586.81	0.70	0.49
B8	514.25	515.47	0.80	0.64
B9	566.38	567.53	0.94	0.89
B10	509.91	511.19	0.98	0.96
B11	466.14	467.42	0.99	0.98
B12	3069.49	3070.36	0.74	0.55
B13	5716.21	5717.22	1.12	1.25
B14	204.11	205.46	0.77	0.59
B15	1122.78	1123.89	0.94	0.88
B16	1657.47	1658.75	0.97	0.95

TABLE 13.3
Results of the Proposed Algorithm in the Prins et al. [59] Instances

	Prins et al. [59]					Prins et al. [59]			
	BS	Average	Stdev	Var		BS	Average	Stdev	Var
P1	54808	54811.9	2.58	6.67	P16	157368	157373.8	3.32	11.04
P2	39109	39112.2	2.65	7.03	P17	200396	200399.8	3.00	9.03
P3	48925	48929.5	3.06	9.34	P18	152622	152626.9	3.54	12.50
P4	37562	37564.8	2.28	5.22	P19	294369	294373.1	3.15	9.93
P5	90137	90141.2	3.20	10.25	P20	236581	236586.9	3.40	11.56
P6	63257	63262.0	3.31	10.97	P21	243842	243845.4	2.83	8.03
P7	88351	88355.1	2.47	6.09	P22	203988	203992.2	2.30	5.27
P8	67346	67351.8	3.02	9.12	P23	253890	253893.6	3.28	10.78
P9	84063	84066.4	2.95	8.70	P24	205126	205130.6	2.62	6.85
P10	51849	51852.9	3.70	13.69	P25	479663	479668.6	3.36	11.27
P11	86238	86242.4	2.90	8.39	P26	379547	379549.4	2.30	5.31
P12	61830	61835.0	3.00	9.01	P27	450743	450746.8	3.69	13.62
P13	276524	276529.5	2.60	6.78	P28	376153	376157.9	3.16	9.99
P14	215021	215026.5	3.36	11.28	P29	475588	475593.9	3.73	13.94
P15	194325	194330.4	3.76	14.12	P30	363313	363317.4	3.63	13.15

TABLE 13.4
Results of the Proposed Algorithm in the Tuzun and Burke [60] Instances

	Tuzun and Burke [60]					Tuzun and Burke [60]			
	BS	Average	Stdev	Var		BS	Average	Stdev	Var
T1	1474.39	1475.66	1.16	1.35	T19	1212.21	1213.64	0.86	0.73
T2	1450.91	1452.35	1.10	1.20	T20	948.31	949.64	1.04	1.08
T3	1413.71	1415.02	0.94	0.89	T21	1726.84	1728.20	1.00	0.99
T4	1440.54	1441.95	0.95	0.90	T22	1404.28	1405.71	0.90	0.81
T5	1172.68	1174.33	1.05	1.09	T23	1217.71	1218.61	0.91	0.84
T6	1105.96	1107.33	0.76	0.57	T24	1156.02	1157.01	0.66	0.44
T7	793.22	794.75	1.08	1.17	T25	2299.22	2300.68	0.83	0.70
T8	736.83	738.05	0.76	0.58	T26	2213.00	2214.13	0.88	0.78
T9	1249.80	1250.95	1.17	1.37	T27	2256.02	2257.09	0.85	0.73
T10	1265.90	1267.70	0.98	0.96	T28	2271.03	2272.66	0.97	0.94
T11	902.63	904.15	1.22	1.48	T29	2100.21	2101.58	1.16	1.35
T12	1036.14	1037.63	0.98	0.96	T30	1730.95	1732.00	0.92	0.85
T13	1940.39	1941.83	1.02	1.05	T31	1469.18	1470.20	0.97	0.95
T14	1858.16	1859.50	0.93	0.87	T32	1087.58	1088.45	0.84	0.71
T15	1997.47	1998.91	0.92	0.85	T33	2003.99	2005.45	0.84	0.70
T16	1815.01	1816.48	0.85	0.72	T34	1975.14	1976.60	0.80	0.64
T17	1449.07	1450.19	0.85	0.72	T35	1786.46	1787.98	0.81	0.66
T18	1458.86	1459.99	0.80	0.64	T36	1431.40	1432.62	0.78	0.61

TABLE 13.5
Comparison of the Proposed Algorithm with other Algorithms in Barreto [58] Instances

		RVNS		HGA		HDE		HCSA		GLCENTPSO		ABBMO	
	BKS	Cost	ω	Cost	ω	Cost	ω	Cost	ω	Cost	ω	Cost	ω
B1	566.59	591.17	4.34	582.02	2.72	579.09	2.21	567.18	0.10	566.59	0	566.26	–0.06
B2	854.57	872.04	2.04	864.76	1.19	861.65	0.83	854.78	0.02	854.57	0	852.12	–0.29
B3	839.93	852.07	1.45	848.07	0.97	843.74	0.45	840.36	0.05	839.93	0	837.02	–0.35
B4	360.72	373.95	3.67	372.95	3.39	364.75	1.12	361.57	0.24	360.72	0	358.87	–0.51
B5	43967.82	43986.84	0.04	43985.80	0.04	43981.95	0.03	43967.86	0.00	43967.82	0	43966.75	0.00
B6	427.91	449.33	5.00	442.32	3.37	435.29	1.72	428.25	0.08	427.91	0	427.34	–0.13
B7	587.65	606.47	3.20	604.37	2.85	600.84	2.24	588.35	0.12	587.65	0	585.77	–0.32
B8	514.28	542.21	5.43	536.70	4.36	529.50	2.96	515.10	0.16	514.28	0	514.25	–0.01
B9	566.45	582.46	2.83	576.72	1.81	575.78	1.65	566.66	0.04	566.45	0	566.38	–0.01
B10	511.41	530.18	3.67	525.65	2.78	525.63	2.78	512.40	0.19	511.41	0	509.91	–0.29
B11	468.68	482.31	2.91	480.02	2.42	477.03	1.78	469.22	0.12	468.68	0	466.14	–0.54
B12	3069.84	3087.29	0.57	3084.08	0.46	3077.68	0.26	3070.53	0.02	3069.84	0	3069.49	–0.01
B13	5719.18	5738.81	0.34	5730.62	0.20	5726.78	0.13	5720.01	0.01	5719.18	0	5716.21	–0.05
B14	204.49	223.56	9.32	215.79	5.53	209.72	2.56	205.34	0.41	204.49	0	204.11	–0.19
B15	1123.89	1134.67	0.96	1130.85	0.62	1126.76	0.26	1124.87	0.09	1123.89	0	1122.78	–0.10
B16	1659.41	1688.43	1.75	1681.07	1.31	1679.80	1.23	1660.34	0.06	1659.41	0	1657.47	–0.12

TABLE 13.6
Comparison of the Proposed Algorithm with other Algorithms in Prins et al. [59] Instances

		RVNS		HGA		HDE		HCSA		GLCENTPSO		ABBMO	
	BKS	**Cost**	ω	**Cost**	ω	**Cost**	ω	**Cost**	ω	**Cost**	ω	**Cost**	ω
P1	54855	55179	0.59	55130	0.50	54923	0.12	54863	0.01	54855	0.00	54808	−0.09
P2	39184	39683	1.27	39584	1.02	39270	0.22	39186	0.01	39184	0.00	39109	−0.19
P3	48945	49273	0.67	49149	0.42	49075	0.27	48948	0.01	48945	0.00	48925	−0.04
P4	37630	38238	1.62	38187	1.48	37937	0.82	37637	0.02	37630	0.00	37562	−0.18
P5	90218	90537	0.35	90518	0.33	90417	0.22	90226	0.01	90218	0.00	90137	−0.09
P6	63271	63753	0.76	63591	0.51	63338	0.11	63274	0.00	63271	0.00	63257	−0.02
P7	88400	88599	0.23	88554	0.17	88544	0.16	88407	0.01	88400	0.00	88351	−0.06
P8	67347	68123	1.15	67898	0.82	67612	0.39	67355	0.01	67347	0.00	67346	0.00
P9	84076	84681	0.72	84664	0.70	84325	0.30	84084	0.01	84076	0.00	84063	−0.02
P10	51882	52176	0.57	52134	0.49	52124	0.47	51886	0.01	51882	0.00	51849	−0.06
P11	86324	86669	0.40	86482	0.18	86451	0.15	86326	0.00	86324	0.00	86238	−0.10
P12	61885	62585	1.13	62363	0.77	62030	0.23	61894	0.01	61885	0.00	61830	−0.09
P13	276594	279878	1.19	278886	0.83	277652	0.38	276594	0.00	276594	0.00	276524	−0.03
P14	215088	217242	1.00	216906	0.85	215560	0.22	215090	0.00	215088	0.00	215021	−0.03
P15	194336	196623	1.18	196500	1.11	195499	0.60	194336	0.00	194336	0.00	194325	−0.01
P16	157397	159851	1.56	159124	1.10	158145	0.48	157403	0.00	157397	0.00	157368	−0.02
P17	200446	202610	1.08	201883	0.72	200748	0.15	200452	0.00	200446	0.00	200396	−0.02
P18	152720	153699	0.64	153552	0.54	153281	0.37	152721	0.00	152720	0.00	152622	−0.06
P19	294373	296081	0.58	295877	0.51	294849	0.16	294382	0.00	294373	0.00	294369	0.00
P20	236636	238849	0.94	238765	0.90	237613	0.41	236639	0.00	236636	0.00	236581	−0.02
P21	243906	247125	1.32	245995	0.86	244972	0.44	243910	0.00	243906	0.00	243842	−0.03
P22	204015	206460	1.20	206454	1.20	205334	0.65	204021	0.00	204015	0.00	203988	−0.01
P23	253895	256071	0.86	254952	0.42	254682	0.31	253900	0.00	253895	0.00	253890	0.00
P24	205221	208267	1.48	207516	1.12	206493	0.62	205228	0.00	205221	0.00	205126	−0.05
P25	479725	480588	0.18	480537	0.17	479946	0.05	479732	0.00	479725	0.00	479663	−0.01
P26	379631	381601	0.52	380607	0.26	380264	0.17	379632	0.00	379631	0.00	379547	−0.02
P27	450795	453989	0.71	452790	0.44	452142	0.30	450798	0.00	450795	0.00	450743	−0.01
P28	376235	378431	0.58	377941	0.45	377491	0.33	376240	0.00	376235	0.00	376153	−0.02
P29	475615	477531	0.40	477101	0.31	475809	0.04	475619	0.00	475615	0.00	475588	−0.01
P30	363360	365481	0.58	365100	0.48	364213	0.23	363368	0.00	363360	0.00	363313	−0.01

The worst performing algorithm is the RVNS, as expected from the structure of the algorithm. The algorithm uses a random way to produce solutions, and there is no connection between them, but all these solutions use the same VNS and ILS procedures as the ones used in the proposed method. Afterward, the second worst algorithm is the HGA; its computational results are better only from the computational results of the RVNS algorithm. Then, with almost the same performance, follows the HDE algorithm. The three most effective algorithms are the HCSA, the GLCENTPSO algorithm, and the proposed ABBMO algorithm. The differences in the results in the last three algorithms are very small, especially, in the first two sets of instances (Tables 13.5 and 13.6). However, in all three sets of instances, the computational results of the proposed algorithm outperform the computational results of the other two algorithms. In the third set of instances, the proposed algorithm in some instances improves significantly the cost of the solution compared with the other two algorithms.

TABLE 13.7
Comparison of the Proposed Algorithm with other Algorithms in Tuzun and Burke [60] Instances

		RVNS		HGA		HDE		HCSA		GLCENTPSO		ABBMO	
	BKS	**cost**	**ω**	**cost**	**ω**	**cost**	**ω**	**cost**	**ω**	**cost**	**ω**	**cost**	**ω**
T1	1480.71	1501.82	1.43	1497.38	1.13	1488.72	0.54	1481.32	0.04	1480.71	0.00	1474.39	–0.43
T2	1452.48	1462.81	0.71	1461.02	0.59	1454.53	0.14	1452.93	0.03	1452.48	0.00	1450.91	–0.11
T3	1422.36	1432.90	0.74	1429.77	0.52	1428.64	0.44	1422.90	0.04	1422.36	0.00	1413.71	–0.61
T4	1440.93	1452.13	0.78	1451.80	0.75	1447.73	0.47	1440.94	0.00	1440.93	0.00	1440.54	–0.03
T5	1172.69	1188.96	1.39	1182.99	0.88	1182.81	0.86	1173.67	0.08	1172.69	0.00	1172.68	0.00
T6	1109.20	1128.82	1.77	1124.54	1.38	1119.60	0.94	1109.89	0.06	1109.20	0.00	1105.96	–0.29
T7	798.40	808.24	1.23	805.49	0.89	799.27	0.11	798.69	0.04	798.40	0.00	793.22	–0.65
T8	755.45	769.07	1.80	763.25	1.03	761.62	0.82	755.54	0.01	755.45	0.00	736.83	–2.46
T9	1250.54	1267.42	1.35	1262.33	0.94	1254.14	0.29	1251.38	0.07	1250.54	0.00	1249.80	–0.06
T10	1267.77	1278.72	0.86	1275.07	0.58	1273.21	0.43	1268.57	0.06	1267.77	0.00	1265.90	–0.15
T11	922.40	933.57	1.21	930.77	0.91	926.46	0.44	922.65	0.03	922.39	0.00	902.63	–2.14
T12	1050.80	1072.65	2.08	1068.32	1.67	1060.90	0.96	1051.68	0.08	1050.80	0.00	1036.14	–1.39
T13	1958.83	1967.29	0.43	1967.13	0.42	1965.19	0.32	1958.86	0.00	1958.83	0.00	1940.39	–0.94
T14	1861.05	1877.04	0.86	1871.18	0.54	1866.30	0.28	1861.12	0.00	1861.05	0.00	1858.16	–0.16
T15	2007.74	2023.46	0.78	2020.76	0.65	2014.68	0.35	2008.63	0.04	2007.73	0.00	1997.47	–0.51
T16	1815.54	1830.36	0.82	1827.58	0.66	1821.48	0.33	1816.10	0.03	1815.54	0.00	1815.01	–0.03
T17	1460.46	1470.57	0.69	1470.46	0.69	1463.73	0.22	1460.99	0.04	1460.46	0.00	1449.07	–0.78
T18	1478.47	1484.24	0.39	1481.43	0.20	1478.69	0.02	1478.49	0.00	1478.46	0.00	1458.86	–1.33
T19	1212.91	1229.24	1.35	1224.68	0.97	1216.47	0.29	1213.30	0.03	1212.91	0.00	1212.21	–0.06
T20	955.39	965.19	1.03	963.77	0.88	963.51	0.85	955.77	0.04	955.39	0.00	948.31	–0.74
T21	1742.69	1753.25	0.61	1752.28	0.55	1745.88	0.18	1743.03	0.02	1742.69	0.00	1726.84	–0.91
T22	1423.84	1436.47	0.89	1433.88	0.71	1432.26	0.59	1424.72	0.06	1423.84	0.00	1404.28	–1.37
T23	1235.82	1243.76	0.64	1239.89	0.33	1237.69	0.15	1236.74	0.07	1235.82	0.00	1217.71	–1.46
T24	1173.25	1191.43	1.55	1187.27	1.19	1178.12	0.42	1173.53	0.02	1173.25	0.00	1156.02	–1.47
T25	2304.94	2317.04	0.52	2311.27	0.27	2311.15	0.27	2305.84	0.04	2304.94	0.00	2299.22	–0.25
T26	2224.73	2240.00	0.69	2237.06	0.55	2232.27	0.34	2225.20	0.02	2224.73	0.00	2213.00	–0.53
T27	2260.72	2280.36	0.87	2277.29	0.73	2269.47	0.39	2261.68	0.04	2260.71	0.00	2256.02	–0.21
T28	2277.91	2282.05	0.18	2280.55	0.12	2279.76	0.08	2278.70	0.03	2277.90	0.00	2271.03	–0.30
T29	2108.25	2124.99	0.79	2119.39	0.53	2117.62	0.44	2108.48	0.01	2108.25	0.00	2100.21	–0.38
T30	1744.62	1760.29	0.90	1757.14	0.72	1749.86	0.30	1745.57	0.05	1744.62	0.00	1730.95	–0.78
T31	1479.07	1488.85	0.66	1485.17	0.41	1482.50	0.23	1479.77	0.05	1479.07	0.00	1469.18	–0.67
T32	1104.79	1121.18	1.48	1118.75	1.26	1113.20	0.76	1105.52	0.07	1104.79	0.00	1087.58	–1.56
T33	2010.08	2023.67	0.68	2019.72	0.48	2016.03	0.30	2010.47	0.02	2010.08	0.00	2003.99	–0.30
T34	1983.73	1990.64	0.35	1988.15	0.22	1985.26	0.08	1984.15	0.02	1983.73	0.00	1975.14	–0.43
T35	1790.15	1799.50	0.52	1795.38	0.29	1795.04	0.27	1790.57	0.02	1790.15	0.00	1786.46	–0.21
T36	1441.30	1445.84	0.31	1445.34	0.28	1442.43	0.08	1441.42	0.01	1441.30	0.00	1431.40	–0.69

13.6 CONCLUSIONS AND FUTURE RESEARCH

In this chapter, a new version of the BBMO algorithm, the ABBMO, is presented. The novelty of the proposed algorithm is the replacement of the PR method in the feeding process with an adaptive memory algorithm. The algorithm was applied in the location routing problem with stochastic demands, and the computational results were compared with the computational results of five other algorithms from the literature. In general, the computational results proved the effectiveness of the proposed algorithm. As in the previous publications of the BBMO algorithm, when it is applied in a routing problem, it almost always outperforms any other swarm intelligence algorithm. However, its performance in other problems has not yet been explored so extensively; thus, it is a challenge for us in the near future to see how the BBMO algorithm performs in other problems.

REFERENCES

1. Yang, X.S. (2010). A new metaheuristic bat-inspired algorithm. In *Nature Inspired Cooperative Strategies for Optimization (NISCO 2010), Studies in Computational Intelligence*, 284, Gonzalez, J.R, Pelta, D.A., Cruz, C., Terrazas, G., and Krasnogor, N. (Eds), 65–74, Springer-Verlag, Berlin, Heidelberg.
2. Yang, X.S., Deb, S. (2009). Cuckoo search via Levy flights. In *World Congress on Nature and Biologically Inspired Computing (NaBIC)*, 210–214, IEEE, India.
3. Yang, X.S. (2009). Firefly algorithms for multimodal optimization. In *Stochastic Algorithms: Foundations and Applications*, LNCS 5792, Watanabe, O., and Zeugmann, T. (Eds), 169–178, Springer-Verlag, Berlin, Heidelberg.
4. Gandomi, A.H., Alavi, A.H. (2012). Krill herd: a new bio-inspired optimization algorithm. *Communications in Nonlinear Science and Numerical Simulation*, 17(12), 4831–4845.
5. Wang, G.G., Guo, L., Gandomi, A.H., Hao, G.S., Wang, H. (2014). Chaotic krill herd algorithm. *Information Sciences*, 274, 17–34.
6. Wang, G.G., Gandomi, A.H., Alavi, A.H. (2014). An effective krill herd algorithm with migration operator in biogeography-based optimization. *Applied Mathematical Modelling*, 38(9–10), 2454–2462.
7. Wang, G.G., Gandomi, A.H., Alavi, A.H. (2014). Stud krill herd algorithm. *Neurocomputing*, 128, 363–370.
8. Simon, D. (2008). Biogeography-based optimization. *IEEE Transactions on Evolutionary Computation*, 12, 702–713.
9. Mirjalili, S., Mirjalili, S.M., Lewis, A. (2014). Grey wolf optimizer. *Advances in Engineering Software*, 69, 46–61.
10. Mirjalili, S. (2015). Moth-flame optimization algorithm: a novel nature-inspired heuristic paradigm. *Knowledge-Based Systems*, 89, 228–249.
11. Mirjalili, S. (2015). The ant lion optimizer. *Advances in Engineering Software*, 83, 80–98.
12. Wang, G.G., Deb, S., Cui, Z. (2019). Monarch butterfly optimization. *Neural Computing and Applications*, 31(7), 1995–2014.
13. Mirjalili, S. (2016). Dragonfly algorithm: a new meta-heuristic optimization technique for solving single-objective, discrete, and multi-objective problems. *Neural Computing and Applications*, 27(4), 1053–1073.
14. Wang, G.G. (2018). Moth search algorithm: a bio-inspired metaheuristic algorithm for global optimization problems. *Memetic Computing*, 10(2), 151–164.
15. Li, W., Wang, G.G., Alavi, A.H. (2020). Learning-based elephant herding optimization algorithm for solving numerical optimization problems. *Knowledge-Based Systems*, 195, 105675.
16. Xing, B., Gao, W.J. (2014). Innovative computational intelligence: a rough guide to 134 clever algorithms. In *Intelligent Systems Reference Library*, 62, Springer International Publishing, Switzerland.
17. Teodorovic, D., Dell'Orco, M. (2005). Bee colony optimization - a cooperative learning approach to complex transportation problems. In *Advanced OR and AI Methods in Transportation, Proceedings of the 16th Mini - EURO Conference and 10th Meeting of EWGT*, 51–60.
18. Karaboga, D., Basturk, B. (2007). A powerful and efficient algorithm for numerical function optimization: artificial bee colony (ABC) algorithm. *Journal of Global Optimization*, 39, 459–471.
19. Karaboga, D., Basturk, B. (2008). On the performance of artificial bee colony (ABC) algorithm. *Applied Soft Computing*, 8, 687–697.
20. Quijano, N., Passino, K.M. (2010). Honey bee social foraging algorithms for resource allocation: theory and application. *Engineering Applications of Artificial Intelligence*, 23, 845–861.

21. Pham, D.T., Ghanbarzadeh, A., Koc, E., Otri, S., Rahim, S., Zaidi, M. (2006). The bees algorithm – a novel tool for complex optimization problems. In *IPROMS 2006 Proceeding 2nd International Virtual Conference on Intelligent Production Machines and Systems*, Elsevier, Oxford.
22. Sato, T., Hagiwara, M. (1997). Bee system: finding solution by a concentrated search. In *IEEE International Conference on Systems, Man, and Cybernetics (SMC)*, 3954–3959.
23. Fan, H., Zhong, Y. (2012). A rough set approach to feature selection based on wasp swarm optimization. *Journal of Computational Information Systems*, 8, 1037–1045.
24. Theraulaz, G., Goss, S., Gervet, J., Deneubourg, J.L. (1991). Task differentiation in polistes wasps colonies: a model for self-organizing groups of robots. In *First International Conference on Simulation of Adaptive Behavior*, 346–355, MIT Press, Cambridge, MA.
25. Wedde, H.F., Farooq, M., Zhang, Y. (2004). Beehive: an efficient fault-tolerant routing algorithm inspired by honey bee behavior. In *Ant colony optimization and swarm intelligence, LNCS*, 3172, Dorigo, M. (Ed.), 83–94, Springer, Berlin.
26. Munoz, M.A., Lopez, J.A., Caicedo, E. (2009). An artificial beehive algorithm for continuous optimization. *International Journal of Intelligent Systems*, 24, 1080–1093.
27. Maia, R.D., Castro, L.N.D., Caminhas, W.M. (2012). Bee colonies as model for multimodal continuous optimization: the optbees algorithm. In IEEE World Congress on Computational Intelligence (WCCI), 1–8, Brisbane, Australia.
28. Drias, H., Sadeg, S., Yahi, S. (2005). Cooperative bees swarm for solving the maximum weighted satisfiability problem. In *Computational Intelligence and Bioinspired Systems, Lecture Notes in Computer Science*, Cabestany, J., Prieto, A., and Sandoval, F. (Eds), 3512, 318–325, Springer-Verlag, Berlin, Heidelberg.
29. Yang, X.S. (2005). Engineering optimizations via nature-inspired virtual bee algorithms. In *IWINAC 2005, LNCS*, 3562, Yang, J.M., and Alvarez, J.R. (Eds), 317–323, Springer-Verlag, Berlin, Heidelberg.
30. Hackel, S., Dippold, P. (2009). The bee colony-inspired algorithm (BCiA)-a two stage approach for solving the vehicle routing problem with time windows. In *Proceedings of GECCO09*, 25–32, Montreal, Quebec, Canada.
31. McCaffrey, J.D., Dierking, H. (2009). An empirical study of unsupervised rule set extraction of clustered categorical data using a simulated bee colony algorithm. In *RuleML 2009, LNCS*, 5858, Governatori, G., Hall, J., and Paschke, A. (Eds), 182–192, Springer-Verlag, Berlin, Heidelberg.
32. Comellas, F., Martinez-Navarro, J. (2009). Bumblebees: a multiagent combinatorial optimization algorithm inspired by social insect behavior. In *Proceedings of First ACM/SIGEVO Summit on Genetic and Evolutionary Computation (GECCO)*, 811–814, Shanghai, China.
33. Abbass, H.A. (2001). A monogenous MBO approach to satisfiability. In *International Conference on Computational Intelligence for Modeling, Control and Automation, CIMCA'2001*, Las Vegas, NV.
34. Abbass, H.A. (2001). Marriage in honey-bee optimization (MBO): a haplometrosis polygynous swarming approach. In *The Congress on Evolutionary Computation (CEC2001)*, 207–214, Seoul, Korea.
35. Marinaki, M., Marinakis, Y., Zopounidis, C. (2010). Honey bees mating optimization algorithm for financial classification problems. *Applied Soft Computing*, 10(3), 806–812.
36. Marinakis, Y., Marinaki, M., Dounias, G. (2008). Honey bees mating optimization algorithm for the vehicle routing problem. In *Nature INSPIRED COOPERATIVE STRATEGIES FOR OPTIMIZATION – NICSO 2007, Studies in Computational Intelligence*, Krasnogor, N., Nicosia, G., Pavone, M., and Pelta, D. (Eds), 129, 139–148, Springer-Verlag, Berlin, Heidelberg.
37. Bitam, S., Mellouk, A. (2013). Bee life-based multi constraints multicast routing optimization for vehicular ad hoc networks. *Journal of Network and Computer Applications*, 36, 981–991.
38. Marinakis, Y., Marinaki, M. (2011). Bumble bees mating optimization algorithm for the vehicle routing problem. In *Handbook of Swarm Intelligence – Concepts, Principles and Applications, Series on Adaptation, Learning, and Optimization*, Panigrahi, B.K., Shi, Y., and Lim, M.H. (Eds), 8, 347–369, Springer-Verlag, Berlin, Heidelberg.
39. Marinakis, Y., Marinaki, M., Migdalas, A. (2017). An adaptive bumble bees mating optimization algorithm. *Applied Soft Computing*, 55, 13–30.
40. Marinakis, Y., Marinaki, M., Migdalas, A. (2019). A multi-adaptive particle swarm optimization for the vehicle routing problem with time windows. *Information Sciences*, 481, 311–329.
41. Marinakis, Y., Marinaki, M., Matsatsinis, N., (2009). A hybrid bumble bees mating optimization-GRASP algorithm for clustering. In *HAIS 2009, LNAI, 5572*, Corchado, E., Wu, X., Oja, E., Herrero, A., and Baruque, B. (Eds), 549–556, Springer-Verlag, Berlin, Heidelberg.

42. Feo, T.A., Resende, M.G.C., (1995). Greedy randomized adaptive search procedure. *Journal of Global Optimization*, 6, 109–133.
43. Marinakis, Y., Marinaki, M., Matsatsinis, N., (2010). A Bumble bees mating optimization algorithm for global unconstrained optimization problems. In *Nature Inspired Cooperative Strategies for Optimization – NICSO 2010, Studies in Computational Intelligence*, 284, Gonzalez, J.R, Pelta, D.A., Cruz, C., Terrazas, G., and Krasnogor, N. (Eds), 305–318, Springer-Verlag, Berlin, Heidelberg.
44. Lichtblau D. (2002). Discrete optimization using Mathematica. In *World Multi-Conference on Systemics, Cybernetics and Informatics (SCI 2002), International Institute of Informatics and Systemics*, 16, 169–174.
45. Marinakis, Y., Migdalas, A., Pardalos, P.M. (2005). Expanding neighborhood GRASP for the traveling salesman problem. *Computational Optimization and Applications*, 32(3), 231–257.
46. Marinakis, Y., Marinaki, M. (2014). A bumble bees mating optimization algorithm for the open vehicle routing problem. *Swarm and Evolutionary Computation*, 15, 80–94.
47. Lourenco, H.R., Martin, O., Stützle, T. (2003). Iterated local search. In *Handbook of Metaheuristics. Vol. 57 of Operations Research and Management Science*, 321–353, Kluwer Academic Publishers, Dordrecht.
48. Marinakis, Y., Marinaki, M. (2015). Combinatorial neighborhood topology bumble bees mating optimization for the vehicle routing problem with stochastic demands. *Soft Computing*, 19, 353–373.
49. Hansen, P., Mladenovic, N. (2001). Variable neighborhood search: principles and applications. *European Journal of Operational Research*, 130, 449–467.
50. Marinakis, Y., Marinaki, M. (2013). Combinatorial neighborhood topology particle swarm optimization algorithm for the vehicle routing problem. In *EvoCOP 2013*, Middendorf, M. and Blum, C. (Eds), LNCS 7832, 133–144.
51. Glover, F., Laguna, M., Marti, R. (2003). Scatter search and path relinking: advances and applications. In *Handbook of Metaheuristics*, Glover, F., and Kochenberger, G.A. (Eds), 1–36, Kluwer Academic Publishers, Boston.
52. Marinaki, M., Marinakis, Y. (2016). A bumble bees mating optimization algorithm for the feature selection problem. *International Journal of Machine Learning and Cybernetics*, 7(4), 519–538.
53. Marinakis, Y., Marinaki, M. (2014). An adaptive bumble bees mating optimization algorithm for the hierarchical permutation flowshop scheduling problem. In *ANTS 2014, LNCS*, M. Dorigo et al. (Eds), 8667, 282–283.
54. Rochat, Y., Taillard, E.D. (1995). Probabilistic diversification and intensification in local search for vehicle routing. *Journal of Heuristics*, 1, 147–167.
55. Glover, F. (1989). Tabu sSearch I. *ORSA Journal on Computing*, 1(3), 190–206.
56. Marinakis, Y. (2015). An improved particle swarm optimization algorithm for the capacitated location routing problem and for the location routing problem with stochastic demands. *Applied Soft Computing*, 37, 680–701.
57. Marinakis, Y., Marinaki, M., Migdalas, A. (2016). A hybrid clonal selection algorithm for the location routing problem with stochastic demands. *Annals of Mathematics and Artificial Intelligence*, 76(1), 121–142.
58. Barreto, S., Ferreira, C., Paixao, J., Santos, B.S. (2007). Using clustering analysis in a capacitated location-routing problem. *European Journal of Operational Research*, 179(3), 968–977.
59. Prins, C., Prodhon, C., Wolfler Calvo, R. (2004). Nouveaux algorithmes pour le probleme de localisation et routage sous contraintes de capacite. In *Proceedings of the MOSIM 04*, Vol. 2, 1115–1122, Lavoisier, Ecole des Mines de Nantes, France.
60. Tuzun, D., Burke, L.I. (1999). A two-phase tabu search approach to the location routing problem. *European Journal of Operational Research*, 116, 87–99.

14 A Glowworm Swarm Optimization Algorithm for the Multi-Objective Energy Reduction Multi-Depot Vehicle Routing Problem

Emmanouela Rapanaki, Iraklis-Dimitrios Psychas, Magdalene Marinaki, and Yannis Marinakis

CONTENTS

14.1 INTRODUCTION

The GSO algorithm is one of the most successful swarm intelligence algorithms that simulates the flashing behavior of glowworms. More precisely, it simulates the movement of the glowworms in a swarm based on two quantities, the distance between them and the luminescent quantity called *luciferin* [1–3]. Initially, the algorithm was used for the solution of the classic multimodal functions ([1, 2, 4, 5]). Then, as the algorithm gained popularity and proved its efficiency, it was used for a number of different problems. In [6], a new version was applied in hazardous situations and, in the next year, it was used in the solution of problems in searching for higher dimensional spaces [3]. In [7], a multi-robot system was implemented. Theoretical foundations for a variation of the multi-agent problem were presented in [8]. Recently, a book was published in which the GSO is analyzed in depth and the most important publications of the algorithm are presented and analyzed in brief [9]. Since then, a number of papers that applied a variant of the GSO algorithm for the solution of an optimization problem have been published [10–22].

As it was previously presented, the GSO algorithm has been applied in a number of different fields. One of the fields that the effectiveness of the GSO has not been investigated thoroughly is the vehicle routing problem (VRP) [23, 24]. The VRP is a very interesting NP-hard optimization problem in which a number of customers must be served using a fleet of vehicles, minimizing one or more objective functions. A number of variants of the VRP have been published in recent years [23]. As the problem is NP-hard, there is a large number of algorithms that have been proposed for its solution. Although metaheuristic algorithms, like tabu search [25], variable neighborhood search [26], greedy randomized adaptive search procedure [27], and simulated annealing [25], to name a few, and evolutionary algorithms [28] have been usually applied for the solution of a VRP variant. Recently, a number of swarm intelligence algorithms have been used for the solution of the problem. Initially, the ant colony optimization algorithm was the first swarm intelligence algorithm that was used for the solution of the VRP, but, recently, many other swarm intelligence algorithms, mainly, the particle swarm optimization algorithm [24, 29], have been applied with remarkable results. At least to our knowledge, the only GSO algorithm for a solution of a variant of the VRP, the VRP with Stochastic Demands, has been published in [30].

In this chapter, a multi-objective energy reduction multi-depot vehicle routing problem (MERMDVRP) is solved using a new version of the GSO algorithm suitable for multi-objective optimization problems, and it is denoted as the parallel multi-start multi-objective glowworm swarm optimization (PMS-GSO) algorithm. The problem was initially proposed by Rapanaki et al. in [31, 32]; it is an extension of the single depot multi-objective problem proposed by Psychas et al. [33–35] and of the multi-objective open vehicle routing proposed by Psychas et al. [36]. In MERMDVRP, two different objective functions are minimized simultaneously: the first one concerns the total traveled time, whereas the second one concerns the energy consumption. We propose four different variants where in two of them, the customers have only pickup requests, and in the other two the customers have only delivery requests. Another distinction between the problems is that in two of them (one with pickup requests and one with delivery requests) we solve a symmetric problem in which only the load and the traveled distance are considered, whereas in the other two, we use real-life parameters (weather conditions, drivers behavior, and uphill and downhill routes).

To see the effectiveness of the proposed algorithm, we compared the computational results of the algorithm with the computational results of three other algorithms (parallel multi-start multi-objective firefly [PMS-FIREFLY], parallel multi-start multi-objective particle swarm optimization [PMS-NSPSO], and parallel multi-start non-dominated sorting genetic algorithm II [PMS-NSGA-II]) in a number of benchmark instances. All algorithms used in the comparisons have the same structure and a number of common strategies, the PMS procedure, and all of them are suitable for multi-objective optimization problems. The main difference between these algorithms is that they are based on a different evolutionary or swarm intelligence algorithm (firefly algorithm for PMS-FIREFLY, particle swarm optimization for PMS-NSPSO, and genetic algorithm for PMS-NSGA-II).

The structure of this chapter is as follows. Initially, a formulation of the problem is presented. Next, the proposed algorithm PMS-GSO is analyzed in depth. Then, computational results, comparisons with other algorithms, and analysis of the results are given. Finally, some concluding remarks and future directions are given.

14.2 MULTI-OBJECTIVE ENERGY REDUCTION MULTI-DEPOT VEHICLE ROUTING PROBLEM

There are a number of papers in the literature concerning variants of the VRP using multi-depots [37], or minimizing the energy consumption [38, 39], or combining these two objective functions [40]. In this chapter, we propose a new formulation of the problem that combines multi-depots, the minimization of the total traveled distance, and the minimization of the energy consumption. As we mentioned previously, different versions of the problem are developed and solved, where in some of them no real-life constraints are used, whereas in others real-life constraints based mainly on the behavior of the driver, the road, and the weather conditions are taken into account. Thus, four different versions of the problem are developed: the asymmetric and the symmetric delivery energy reduction VRP and the asymmetric and symmetric pick-up energy reduction VRP. In all four versions, the first objective function concerns the minimization of the **time** needed to travel between two customers or a customer and the depots, and it is given by:

$$\min OF1 = \sum_{i=I_1}^{I_\pi} \sum_{j=1}^{n} \sum_{\kappa=1}^{m} \left(t_{ij}^{\kappa} + s_j^{\kappa}\right) x_{ij}^{\kappa} \tag{14.1}$$

where t_{ij}^{κ} is time needed for the customer j to be visited by vehicle κ immediately after customer i; s_j^{κ} is service time of customer j by vehicle κ; x_{ij}^{κ} is a variable used to show that the vehicle κ visits customer j immediately after customer i; n is number of nodes; m is the number of homogeneous vehicles; π is the number of homogeneous depots; and $i = j = I_1, I_2, \ldots I_\pi$ is the subset of depots that belongs to the set of nodes $n\left(\{I_1, I_2, \ldots I_\pi, 2, 3, \ldots, n\}\right)$.

The second objective function is used for the minimization of the **route-based fuel consumption** (*RFC*) when only deliveries are performed by the vehicles taking into account real-life route parameters like the weather conditions or the uphills and the downhills or the driver's behavior. The objective function is given by:

$$\min OF2 = \sum_{h=I_1}^{I_\pi} \sum_{j=2}^{n} \sum_{\kappa=1}^{m} c_{hj} x_{hj}^{\kappa} \left(1 + \frac{y_{hj}^{\kappa}}{Q}\right) r_{hj} + \sum_{i=2}^{n} \sum_{j=I_1}^{I_\pi} \sum_{\kappa=1}^{m} c_{ij} x_{ij}^{\kappa} \left(1 + \frac{y_{i-1,i}^{\kappa} - D_i}{Q}\right) r_{ij} \tag{14.2}$$

where Q is the maximum capacity of each vehicle, D_i is the customer's i demand ($D_{I_1} = D_{I_2} = \ldots = D_{I_\pi} = 0$), y_{ij}^{κ} is the load of the vehicle, $y_{ij}^{\kappa} = \Sigma_{i=I_1}^{I_\pi} D_i$ for all vehicles (vehicle begins with load equal to the summation of the demands of all customers assigned in its route), c_{ij} is the distance from node i to node j, r_{ij} is the route parameter from node i to node j ($r_{ij} > 0$) $r_{ij} \neq r_{ji}$ is the asymmetric formulation of the problem, $r_{ij} < 1$ is the downhill route from i to j or the wind is back-wind or the driver drives with smooth shifting, $r_{ij} > 1$ is the uphill route from i to j or the wind is head-wind or the driver drives with aggressive shifting, and $r_{ij} = 1\ \forall\ (i, j)$ is the symmetric formulation of the problem. For an analytical presentation of r_{ij} please see [34].

The third objective function is used for the minimization of the *RFC* when the vehicle performs only pickups in its route and is calculated by the following equation:

$$\min OF3 = \sum_{h=I_1}^{I_\pi} \sum_{j=2}^{n} \sum_{\kappa=1}^{m} c_{hj} x_{hj}^{\kappa} r_{hj} + \sum_{i=2}^{n} \sum_{j=I_1}^{I_\pi} \sum_{\kappa=1}^{m} c_{ij} x_{ij}^{\kappa} \left(1 + \frac{y_{i-1,i}^{\kappa} + D_i}{Q}\right) r_{ij} \tag{14.3}$$

where $y_{ij}^{\kappa} = 0$ for all vehicles (vehicle begins with empty load) and D_i is the pickup amount of customer i.

Each vehicle always returns to the depot in which it started. The constraints of the problem are [34]:

$$\sum_{j=I_1}^{n} \sum_{\kappa=1}^{m} x_{ij}^{\kappa} = 1, i = I_1, \cdots, n \tag{14.4}$$

$$\sum_{i=I_1}^{n} \sum_{\kappa=1}^{m} x_{ij}^{\kappa} = 1, j = I_1, \cdots, n \tag{14.5}$$

$$\sum_{j=I_1}^{n} x_{ij}^{\kappa} - \sum_{j=I_1}^{n} x_{ji}^{\kappa} = 0, i = I_1, \cdots, n, \kappa = 1, \cdots, m \tag{14.6}$$

$$\sum_{j=I_1, j\neq i}^{n} y_{ji}^{\kappa} - \sum_{j=I_1, j\neq i}^{n} y_{ij}^{\kappa} = D_i, i = I_1, \cdots, n, \kappa = 1, \cdots, m, \textit{for deliveries} \tag{14.7}$$

$$\sum_{j=I_1, j\neq i}^{n} y_{ij}^{\kappa} - \sum_{j=I_1, j\neq i}^{n} y_{ji}^{\kappa} = D_i, i = I_1, \cdots, n, \kappa = 1, \cdots, m, \textit{for pick-ups} \tag{14.8}$$

$$Qx_{ij}^{\kappa} \geq y_{ij}^{\kappa}, i, j = I_1, \cdots, n, \kappa = 1, \cdots, m \tag{14.9}$$

$$x_{ij}^{\kappa} = \begin{cases} 1, & \text{if } (i,j) \text{ belongs to the route} \\ 0, & \text{otherwise} \end{cases} \tag{14.10}$$

Thus,

- *Constraints (Eq. 14.4 and Eq. 14.5):* Each customer must be visited only by one vehicle.
- *Constraints (Eq. 14.6):* Each vehicle that arrives at a node must also leave from that node.
- *Constraints (Eq. 14.7 and Eq. 14.8):* The reduced (for deliveries) or increased (for pickups) load (cargo) of the vehicle after it visits a node is equal to the demand of that node.
- *Constraints (Eq. 14.9):* limit the maximum load carried by the vehicle and force y_{ij}^{κ} to be equal to zero when $x_{ij}^{\kappa} = 0$.
- *Constraints (Eq. 14.10):* Only one vehicle will visit each customer.

The problems solved in this chapter are symmetric ($r_{ij} = 1 \; \forall \; (i,j)$) or asymmetric ($r_{ij} \neq r_{ji} \; \forall \; (i,j)$).

In Table 14.1, the objective functions and the r_{ij} parameter are presented for all the problems studied in this chapter.

TABLE 14.1
Objective Functions and r_{ij} for all the Problems

	Asymmetric Delivery ERVRP[a]	Asymmetric Pickup ERVRP
Objective Functions	OF1 and OF2	OF1 and OF3
r_{ij}	$r_{ij} \neq r_{ji}, \forall(i,j)$	$r_{ij} \neq r_{ji}, \forall(i,j)$
	Symmetric Delivery ERVRP	**Symmetric Pick-up ERVRP**
Objective Functions	OF1 and OF2	OF1 and OF3
r_{ij}	$r_{ij} = 1, \forall(i,j)$	$r_{ij} = 1, \forall(i,j)$

[a] ERVRP: Energy Reduction Vehicle Routing Problem

14.3 PARALLEL MULTI-START GLOWWORM SWARM OPTIMIZATION ALGORITHM

In this section, a presentation of the solution algorithm is given. In GSO algorithm [2, 3], q glowworms ($i = 1,...,q$) are initially randomly placed in the solution space, x_i, where each one of them decides to move based on the intensity of the signal it receives from its neighbors. All of them start with luciferin value l_0. Each glowworm has an objective function $f(x_i(t))$ for each location $x_i(t)$ and a luciferin value l_i with which it communicates with its neighbors. The set of neighbors of a glowworm i consists of those glowworms with the best luciferin value and located at a distance r_d^i, $0 < r_d^i < r_s$, where r_s is the radial range of the luciferin sensor. As each glowworm evaluates only data that come from useful neighbors, it selects its neighbor j by a probability p_{ij}, and then it moves toward this neighbor. This move allows glowworms to be in subgroups, and this is the reason why this algorithm can solve a problem that has many local optimal solutions.

The main stages of each iteration include the following:

- *Luciferin value updating:* It depends on the value of the objective function at the point where the glowworm is located. A value that is proportional to the glowworm's current position in the solution space is added to its previous quantity and a small quantity is removed to show the weakening of the glowworm with time. The equation of the luciferin value is [2, 3]:

$$l_i(t+1) = (1-\rho)l_i(t) + \gamma(f(x_i(t+1))) \tag{14.11}$$

where ρ is the luciferin decay constant, which takes values from 0 to 1, and γ is the luciferin enhancement constant.

- *Movement based on a transition rule:* Each glowworm i moves to a neighbor glowworm with higher luciferin value than its own, using the probability:

$$p_{ij}(t) = \frac{l_j(t) - l_i(t)}{\sum_{k \in N_i(t)} l_k(t) - l_i(t)} \tag{14.12}$$

where t is the iteration (time), $j \in N_i(t), N_i(t) = \{j : d_{ij}(t) < r_d^i(t), l_i(t) < l_j(t)\}$ is the set of neighbors of glowworm i, $d_{ij}(t)$ is the Euclidean distance between two glowworms i and j, and $r_d^i(t)$ represents the variable neighborhood radius associated with glowworm i. If eventually glowworm i chooses glowworm j, then, its new position is calculated by

$$x_i(t+1) = x_i(t) + s\frac{x_j(t) - x_i(t)}{\left\| x_j(t) - x_i(t) \right\|} \tag{14.13}$$

where $s > 0$ is the step where one glowworm moves to another. As we would like to increase the search abilities of the algorithm, we do not use a fixed length in the neighborhood's search of each glowworm. If r_0 ($r_d^i(0) = r_0, \forall i$) is the starting size of the radius that a glowworm searches for neighbors, then

$$r_d^i(t+1) = \min\left\{r_s, \max\left\{0, r_d^i(t) + \beta(n_t - |N_i(t)|)\right\}\right\} \tag{14.14}$$

where β is a constant number and n_t is the number of neighbors that a glowworm may have.

In the PMS-GSO algorithm, initially we place X glowworms (possible solutions) randomly in the solution space. The mapping of the solution should be given using a number of vehicles' routes. However, a suitable transformation of the initial solutions in continuous space is performed as the equations of the GSO algorithm are suitable for global optimization problems. After the calculations are performed using the previously described equations of the GSO algorithm, the solutions are transformed back in the discrete space to calculate the objective functions. In every iteration, the non-dominated solutions are found. Afterward, a new vector is produced by combining the solutions of the last two iterations (iteration it and $it+1$). Then the members of the new vector are sorted using the *rank* and the *crowding distance* as in the NSGA-II algorithm as it was modified in [33]. Each new solution is evaluated using each objective function separately and, then, a Variable neighborhood search (VNS) algorithm is applied to the solutions with both the vns_{max} and the $local_{max}$ equal to 10 [33].

The following issues are taken into account for determination of the new glowworm:

- If the previous best solution in iteration it dominates the current solution (iteration $it+1$), then the previous best solution remains the same.
- If the previous best solution in iteration it is dominated by the solution in the current iteration $it+1$, then the previous best solution is replaced by the current solution.
- Finally, if these two solutions are not dominated between them, then the previous best solution is not replaced.

The good solutions will not disappear from the population as the non-dominated solutions are not deleted from the Pareto front. In the next iterations, to insert a new solution in the Pareto front archive, there are two possibilities:

1. The new solution is non-dominated with respect to the contents of the archive.
2. It dominates any solution in the archive.

In both cases, all the dominated solutions, which already belong to the archive, have to be deleted from the archive. At the end of each iteration, from the non-dominated solutions from all populations, the total Pareto front is updated considering the non-dominated solutions of the last *population*. The algorithm terminates when a predefined maximum number of iterations is reached.

14.4 COMPUTATIONAL RESULTS

14.4.1 DATA DESCRIPTION

The algorithm was tested on a set of benchmark instances. Visual C++ was used to implement the whole algorithmic approach. The data were created as in the research of Rapanaki et al. [31]. These instances are suitable for multi-objective optimization VRPs and they are created using five instances with 100 nodes from the TSPLIB (kroA100, kroB100, kroC100, kroD100, and kroE100) [34]. Also, as we solve a VRP and, thus, each customer has a specific demand, we used the eighth instance from the Christofides set for selecting the demands of the customers. In this instance the number of nodes is equal to 100. Thus, a new set of instances is created by the combination of the five instances with the instance of Christofides. If we have an asymmetric problem, the distance from i to j is calculated using, for example, the coordinates of the instance kroA100 and the distance from j to i is calculated using the coordinates of the instance kroB100;

thus, the instance AB is created and a number of different instances are created. Finally, as we have a multi-depot problem, we use in all instances three depots, where the coordinates of the first depot are the first node in each example; the second depot has the opposite coordinates of the first depot, meaning that if the first depot has coordinates (x_1, y_1), then the second depot has $(x_2 = y_1, y_2 = x_1)$. The third depot's coordinates are between (100, 100) and (500, 500) depending on the instance. To see the effectiveness of the proposed algorithm, we compared the computational results of the algorithm with the computational results of three other algorithms in the previously described benchmark instances, i.e., the PMS-NSPSO [35], the PMS-NSGA-II [33], and the PMS-FIREFLY algorithm.

14.4.2 Evaluation Measures

To prove the efficiency of the proposed algorithm, we applied four different evaluation measures for comparison purposes, i.e., the M_k, the L, the Δ, and the C. It should be noted that all algorithms were executed for the combinations of the objective functions, OF1–OF2 or OF1–OF3, for 5 times using 10 instances [31]. In the following, a brief description of the four measures is given. The first applied measure is the M_k, which denotes the range that the front extends. It is calculated by

$$M_k = \sqrt{\sum_{i=1}^{K} max\{\| p' - q' \|\}} \tag{14.15}$$

where K is the number of the objective functions, and p', q' are values of the objective functions of two solutions that belong to the Pareto front.

The second measure used is the L measure, which gives the number of solutions of the Pareto front. The third measure is the Δ measure, which provides information concerning both the spread and the distribution of the solutions. It is given by

$$\Delta = \frac{d_f + d_l + \sum_{i=1}^{|S|-1} \left| dist_i - \overline{dist} \right|}{d_f + d_l + (|S| - 1)\overline{dist}} \tag{14.16}$$

where S is the number of the intermediate solutions between the extreme solutions; d_f, d_l are the Euclidean distances between the extreme solutions and the boundary solutions of the obtained non-dominated set; $dist_i$ is the distance from a boundary solution i to the next boundary solution, $i = 1, 2, ..., (S-1)$; and $\overline{dist}$: average value of all $dist_i$ distances.

Finally, the C measure is used, which is the coverage measure between the solutions of the Pareto fronts of the algorithms A_1 and B_1. The C measure is given by:

$$C(A_1, B_1) = \frac{\left| \{ b \in B_1 ; \exists \ \ a \in A_1 : a \leq b \} \right|}{|B_1|}. \tag{14.17}$$

We would like the measures L, M_k, and C to be as large as possible and the measure Δ to be as small as possible [33].

14.4.3 Parameters Selection

In order to select the best values for the parameters of the algorithms, we made a number of tests using different alternative values for them. We used the values of the parameters that gave the best computational results taking into account two issues, the quality of the solution and the computational time needed to achieve this solution. The function evaluations in all algorithms was selected to be the same. In all algorithms, the total number of individuals was set equal to 100, the total number of generations was set equal to 500 and the total number of initial populations was set equal to 10. For all algorithms, we have $vns_{max} = 10$, $local_{max} = 10$.

14.4.4 Analysis of the Results

Table 14.2 presents the results of the first three measures of the proposed algorithm PMS-GSO for each one of the problems in the 10 instances and for 5 executions of the algorithm. The average results and the best runs of all algorithms are given in Table 14.3.

In these tables, column average is the average of all five executions in each of the three first measures and the column best run is the best computational results of all executions in these measures. The best runs, in Table 14.3, are in parentheses. The Pareto fronts of the symmetric delivery problem using objective functions OF1–OF2 and of the symmetric pickup problem using objective functions OF1–OF3, for the instance "A-E" for all algorithms are presented in part (A) and in part (B) of Figure 14.1, respectively. The Pareto

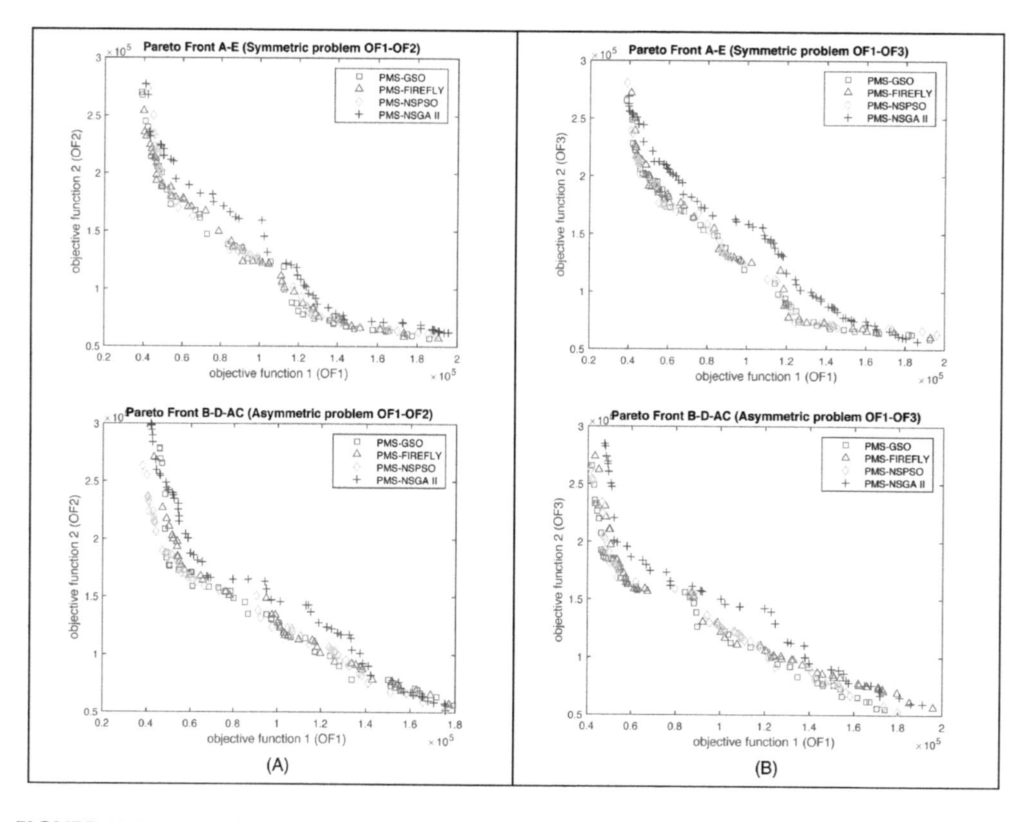

FIGURE 14.1 Pareto fronts of the four algorithms for the instances "A-E" and "B-D-AC." (A) Concerns the two figures with OF1-OF2 objective functions and (B) Concerns the two figures with OF1-OF3 objective functions.

TABLE 14.2
Results of the First Three Measures in 10 Instances for 5 Executions Using the PMS-GSO Algorithm

	Execution 1			Execution 2			Execution 3			Execution 4			Execution 5			Average			Best Run		
	L	M_k	Δ	L	M_k	Δ	L	M_k	Δ	L	M_k	Δ	L	M_k	Δ	L	M_k	Δ	L	M_k	Δ
Multi-Objective Asymmetric Delivery Route-Based Energy Reduction MDVRP																					
A-B-CD	37	593.71	0.66	49	597.87	0.66	52	593.84	0.65	46	583.15	0.69	50	610.86	0.71	46.8	595.89	0.67	49	597.87	0.66
A-C-BD	39	587.82	0.54	46	606.59	0.64	40	600.77	0.58	52	595.40	0.68	51	588.59	0.57	45.6	595.83	0.60	46	606.59	0.64
A-D-BE	48	572.45	0.58	45	605.60	0.61	44	603.86	0.64	51	595.61	0.65	55	604.44	0.62	48.6	596.39	0.62	48	572.45	0.58
A-E-BD	41	573.73	0.66	54	606.88	0.58	62	590.94	0.58	41	588.93	0.57	51	577.65	0.54	49.8	587.63	0.59	54	606.88	0.58
B-C-AD	40	590.86	0.57	52	588.06	0.71	44	595.43	0.76	43	594.61	0.66	49	606.15	0.60	45.6	595.02	0.66	49	606.15	0.60
B-D-AC	43	589.67	0.63	61	584.68	0.69	42	596.52	0.60	41	573.28	0.53	41	591.95	0.58	45.6	587.22	0.61	42	596.52	0.60
B-E-AD	53	629.19	0.59	45	604.52	0.73	50	605.01	0.63	41	568.78	0.59	44	607.94	0.69	46.6	603.09	0.64	53	629.19	0.59
C-D-AE	55	592.43	0.63	44	595.05	0.54	45	577.57	0.67	54	589.16	0.61	62	600.66	0.73	52.0	590.97	0.63	62	600.66	0.73
C-E-AB	53	591.30	0.64	47	601.40	0.67	55	566.23	0.54	43	595.38	0.56	45	599.24	0.48	48.6	590.71	0.58	47	601.40	0.67
D-E-BC	47	576.78	0.59	46	591.59	0.58	46	578.35	0.62	44	565.36	0.63	44	577.22	0.67	45.4	577.86	0.62	46	591.59	0.58
Multi-Objective Symmetric Delivery Route-Based Energy Reduction MDVRP																					
A-B	54	604.71	0.59	48	595.97	0.66	57	614.83	0.59	49	600.34	0.69	46	604.84	0.63	50.8	604.14	0.63	57	614.83	0.59
A-C	48	617.68	0.63	51	593.13	0.63	50	569.71	0.58	60	607.93	0.58	49	597.65	0.68	51.6	597.22	0.62	60	607.93	0.58
A-D	37	595.77	0.61	41	576.76	0.59	44	594.60	0.56	50	600.08	0.68	47	591.81	0.67	43.8	591.80	0.62	50	600.08	0.68
A-E	42	607.32	0.72	47	607.00	0.67	45	600.62	0.61	64	593.53	0.64	45	571.36	0.63	48.6	595.97	0.65	45	600.62	0.61
B-C	53	577.46	0.76	48	585.90	0.59	53	595.46	0.58	57	587.04	0.79	33	596.94	0.72	48.8	588.56	0.69	53	595.46	0.58
B-D	46	607.96	0.54	46	616.22	0.70	35	600.25	0.68	37	592.57	0.64	43	590.32	0.74	41.4	601.46	0.66	46	616.22	0.70
B-E	39	606.91	0.47	40	558.89	0.69	45	613.85	0.66	45	610.95	0.70	38	592.04	0.63	41.4	596.53	0.63	45	613.85	0.66
C-D	52	601.34	0.66	51	604.20	0.63	46	604.30	0.66	41	582.99	0.67	50	593.70	0.65	48.0	597.31	0.66	51	604.20	0.63
C-E	47	597.71	0.61	35	603.80	0.78	38	598.16	0.65	38	578.43	0.68	49	586.59	0.66	41.4	592.94	0.68	47	597.71	0.61
D-E	55	601.94	0.64	57	605.52	0.62	48	604.34	0.63	60	617.21	0.64	45	589.35	0.60	53.0	603.67	0.63	60	617.21	0.64

(continued)

TABLE 14.2 (*Continued*)

Results of the First Three Measures in 10 Instances for 5 Executions Using the PMS-GSO Algorithm

MDVRP	Execution 1			Execution 2			Execution 3			Execution 4			Execution 5			Average			Best Run		
	L	M_k	Δ	L	M_k	Δ	L	M_k	Δ	L	M_k	Δ	L	M_k	Δ	L	M_k	Δ	L	M_k	Δ
Multi-Objective Asymmetric Pickup Route-Based Energy Reduction MDVRP																					
A-B-CD	48	582.14	0.55	51	595.04	0.56	41	605.19	0.68	59	612.19	0.56	58	600.68	0.70	51.4	599.05	0.61	59	612.19	0.56
A-C-BD	58	586.26	0.64	44	573.01	0.61	47	600.53	0.61	49	581.76	0.61	52	598.22	0.57	50.0	587.95	0.61	52	598.22	0.57
A-D-BE	50	595.48	0.56	45	604.14	0.62	49	604.29	0.62	50	608.62	0.67	49	583.35	0.58	48.6	599.18	0.61	50	595.48	0.56
A-E-BD	45	574.53	0.60	50	602.87	0.62	39	574.71	0.55	48	607.97	0.57	47	609.38	0.63	45.8	593.89	0.60	50	602.87	0.62
B-C-AD	48	607.03	0.58	42	580.41	0.74	53	582.09	0.59	47	618.18	0.72	52	598.59	0.61	48.4	597.26	0.65	48	607.03	0.58
B-D-AC	37	581.66	0.63	42	583.58	0.73	36	572.28	0.58	42	581.25	0.59	49	585.08	0.57	41.2	580.77	0.62	49	585.08	0.57
B-E-AD	48	586.88	0.62	54	597.52	0.63	37	612.83	0.75	38	589.45	0.71	43	577.94	0.52	44.0	592.92	0.65	54	597.52	0.63
C-D-AE	45	603.63	0.76	49	585.82	0.67	41	592.85	0.55	42	582.31	0.61	49	546.11	0.62	45.2	582.15	0.64	45	603.63	0.76
C-E-AB	50	561.14	0.56	46	581.12	0.64	40	603.24	0.63	40	596.40	0.61	50	597.09	0.62	45.2	587.80	0.61	50	561.14	0.56
D-E-BC	51	572.75	0.59	47	568.93	0.61	49	575.35	0.56	47	578.77	0.58	49	594.10	0.63	48.6	577.98	0.60	49	575.35	0.56
Multi-Objective Symmetric Pickup Route-Based Energy Reduction MDVRP																					
A-B	29	621.24	0.66	42	571.01	0.62	40	608.57	0.54	47	610.85	0.70	43	626.79	0.60	40.2	607.69	0.63	43	626.79	0.60
A-C	44	624.35	0.65	45	599.47	0.70	50	586.71	0.67	49	589.51	0.62	53	589.41	0.67	48.2	597.89	0.66	49	589.51	0.62
A-D	48	588.43	0.56	57	584.65	0.62	60	608.42	0.70	53	587.39	0.72	48	590.37	0.63	53.2	591.85	0.64	60	608.42	0.70
A-E	56	599.57	0.64	47	599.55	0.69	45	608.60	0.68	54	606.16	0.68	41	602.94	0.73	48.6	603.36	0.68	56	599.57	0.64
B-C	44	541.15	0.66	42	589.07	0.72	32	579.14	0.65	48	585.69	0.78	52	603.36	0.67	43.6	579.68	0.70	52	603.36	0.67
B-D	47	600.51	0.61	49	595.76	0.65	45	587.58	0.50	40	576.10	0.66	49	593.37	0.74	46.0	590.67	0.63	47	600.51	0.61
B-E	33	550.64	0.46	48	607.85	0.68	44	598.87	0.73	45	614.05	0.58	49	603.26	0.67	43.8	594.93	0.62	45	614.05	0.58
C-D	46	584.76	0.62	47	579.83	0.63	42	602.25	0.72	47	589.67	0.64	51	595.81	0.65	46.6	590.46	0.65	51	595.81	0.65
C-E	46	591.14	0.75	59	592.96	0.67	42	595.12	0.60	52	586.67	0.57	53	617.12	0.59	50.4	596.60	0.64	53	617.12	0.59
D-E	53	601.96	0.67	48	602.80	0.66	52	597.57	0.67	47	579.87	0.65	41	595.30	0.54	48.2	595.50	0.64	48	602.80	0.66

TABLE 14.3
Average Results and Best Runs of All Algorithms Used in the Comparisons

		Asymmetric Delivery MERMDVRP			Asymmetric Pickup MERMDVRP		
	Algorithms	L	M_k	Δ	L	M_k	Δ
A-B-CD	PMS-FIREFLY	46.20(50)	593.13(585.26)	0.63(0.54)	43.60(37)	588.91(599,68)	0.64(0.59)
	PMS-GSO	46.80(49)	595.89(597.87)	0.67(0.66)	51.40(59)	599.05(612.19)	0.61(0.56)
	PMS-NSPSO	46.40(53)	598.41(592.84)	0.70(0.68)	47.00(56)	597.42(609.09)	0.66(0.62)
	PMS-NSGA-II	56.40(62)	592.33(598.84)	0.61(0.54)	59.80(63)	598.92(608.39)	0.61(0.65)
A-C-BD	PMS-FIREFLY	46.00(44)	589.25(606.89)	0.62(0.56)	46.20(52)	590.27(592.41)	0.68(0.60)
	PMS-GSO	45.60(46)	595.83(606.59)	0.60(0.64)	50.00(52)	587.95(598.22)	0.61(0.57)
	PMS-NSPSO	47.60(51)	600.33(611.96)	0.68(0.68)	47.80(53)	601.70(595.13)	0.64(0.59)
	PMS-NSGA-II	61.80(72)	594.92(602.67)	0.61(0.68)	58.80(56)	604.25(613.08)	0.59(0.64)
A-D-BE	PMS-FIREFLY	40.60(36)	595.66(598.85)	0.63(0.61)	43.80(45)	604.21(607.02)	0.64(0.55)
	PMS-GSO	48.60(48)	596.39(572.45)	0.62(0.58)	48.60(50)	599.18(595.48)	0.61(0.56)
	PMS-NSPSO	48.80(52)	592.57(606.60)	0.64(0.62)	46.00(48)	608.80(620.73)	0.66(0.65)
	PMS-NSGA-II	54.20(54)	601.74(597.52)	0.61(0.55)	54.60(63)	602.82(610.88)	0.61(0.60)
A-E-BD	PMS-FIREFLY	49.60(54)	599.17(617.97)	0.64(0.65)	44.20(47)	590.38(607.59)	0.65(0.60)
	PMS-GSO	49.80(54)	587.63(606.88)	0.59(0.58)	45.80(50)	593.89(602.87)	0.60(0.62)
	PMS-NSPSO	48.00(58)	586.44(595.07)	0.65(0.63)	49.00(60)	597.10(610.34)	0.67(0.69)
	PMS-NSGA-II	57.20(56)	595.30(604.81)	0.58(0.56)	56.40(60)	591.73(615.11)	0.58(0.53)
B-C-AD	PMS-FIREFLY	40.00(45)	595.48(590.77)	0.66(0.67)	43.80(45)	594.72(616.06)	0.66(0.55)
	PMS-GSO	45.60(49)	595.02(606.15)	0.66(0.60)	48.40(48)	597.26(607.03)	0.65(0.58)
	PMS-NSPSO	42.00(45)	587.45(587.39)	0.66(0.69)	43.20(38)	601.86(616.66)	0.67(0.57)
	PMS-NSGA-II	51.20(64)	596.11(602.55)	0.61(0.60)	53.60(60)	602.63(609.71)	0.63(0.65)
B-D-AC	PMS-FIREFLY	46.20(43)	583.95(589.69)	0.66(0.59)	44.40(53)	599.17(607.27)	0.65(0.69)
	PMS-GSO	45.60(42)	587.22(596.52)	0.61(0.60)	41.20(49)	580.77(585.08)	0.62(0.57)
	PMS-NSPSO	42.00(53)	589.78(591.38)	0.62(0.66)	42.20(50)	581.03(589.53)	0.68(0.75)
	PMS-NSGA-II	54.60(63)	593.43(618.00)	0.61(0.53)	51.40(46)	587.73(606.69)	0.60(0.51)
B-E-AD	PMS-FIREFLY	49.80(55)	600.91(598.56)	0.67(0.62)	42.20(51)	595.20(604.80)	0.68(0.73)
	PMS-GSO	46.60(53)	603.09(629.19)	0.64(0.59)	44.00(54)	592.92(597.52)	0.65(0.63)
	PMS-NSPSO	42.00(45)	584.42(543.81)	0.59(0.52)	48.00(47)	601.43(611.90)	0.65(0.62)
	PMS-NSGA-II	50.80(57)	597.52(599.73)	0.56(0.56)	50.20(55)	594.89(600.52)	0.60(0.58)
C-D-AE	PMS-FIREFLY	44.60(55)	598.24(609.38)	0.65(0.56)	44.00(52)	597.34(610.77)	0.65(0.71)
	PMS-GSO	52.00(62)	590.97(600.66)	0.63(0.73)	45.20(45)	582.15(603.63)	0.64(0.76)
	PMS-NSPSO	45.00(57)	593.92(593.63)	0.67(0.66)	44.00(36)	592.59(602.11)	0.67(0.55)
	PMS-NSGA-II	53.60(49)	597.24(589.81)	0.58(0.53)	51.80(59)	597.78(610.64)	0.63(0.59)
C-E-AB	PMS-FIREFLY	41.60(44)	585.19(601.73)	0.63(0.59)	39.60(43)	582.00(590.97)	0.62(0.62)
	PMS-GSO	48.60(47)	590.71(601.40)	0.58(0.67)	45.20(50)	587.80(561.14)	0.61(0.56)
	PMS-NSPSO	45.00(48)	586.58(577.86)	0.66(0.54)	46.80(41)	588.78(601.76)	0.63(0.53)
	PMS-NSGA-II	55.00(47)	592.95(602.86)	0.61(0.57)	51.00(65)	595.58(608.99)	0.65(0.74)
D-E-BC	PMS-FIREFLY	41.20(45)	575.61(569.15)	0.68(0.65)	46.60(56)	581.36(593.49)	0.68(0.62)
	PMS-GSO	45.40(46)	577.86(591.59)	0.62(0.58)	48.60(49)	577.98(575.35)	0.60(0.56)
	PMS-NSPSO	42.60(50)	571.71(568.50)	0.67(0.69)	39.60(44)	578.42(573.72)	0.60(0.68)
	PMS-NSGA-II	43.60(48)	526.74(503.05)	0.62(0.53)	50.80(56)	581.08(602.31)	0.66(0.69)

(continued)

TABLE 14.3 (*Continued*)
Average Results and Best Runs of All Algorithms Used in the Comparisons

		L	M_k	Δ	L	M_k	Δ
	Algorithms	**Symmetric Delivery MERMDVRP**			**Symmetric Pickup MERMDVRP**		
A-B	PMS-FIREFLY	45.00(48)	601.70(611.40)	0.71(0.64)	47.60(52)	606.74(615.63)	0.71(0.66)
	PMS-GSO	50.80(57)	604.14(614.83)	0.63(0.59)	40.20(43)	607.69(626.79)	0.63(0.60)
	PMS-NSPSO	47.80(52)	613.92(610.29)	0.68(0.62)	51.60(54)	607.75(602.86)	0.69(0.62)
	PMS-NSGA-II	56.40(61)	602.89(603.41)	0.66(0.62)	58.60(79)	596.88(605.27)	0.66(0.60)
A-C	PMS-FIREFLY	51.80(62)	608.06(602.37)	0.65(0.65)	49.00(56)	606.67(586.58)	0.66(0.61)
	PMS-GSO	51.60(60)	597.22(607.93)	0.62(0.58)	48.20(49)	597.89(589.51)	0.66(0.62)
	PMS-NSPSO	50.20(53)	604.15(615.43)	0.66(0.67)	47.40(53)	604.96(608.06)	0.68(0.72)
	PMS-NSGA-II	62.60(66)	609.36(611.80)	0.63(0.54)	56.60(63)	606.55(615.56)	0.64(0.61)
A-D	PMS-FIREFLY	44.40(52)	577.81(578.33)	0.67(0.62)	49.20(55)	581.42(574.20)	0.67(0.59)
	PMS-GSO	43.80(50)	591.80(600.08)	0.62(0.68)	53.20(60)	591.85(608.42)	0.64(0.70)
	PMS-NSPSO	48.40(54)	577.25(575.63)	0.62(0.61)	49.00(58)	586.10(594.99)	0.63(0.64)
	PMS-NSGA-II	54.40(57)	592.85(587.50)	0.67(0.63)	58.60(66)	580.64(582.86)	0.66(0.60)
A-E	PMS-FIREFLY	50.80(54)	597.18(589.47)	0.66(0.58)	42.60(46)	594.91(602.84)	0.65(0.66)
	PMS-GSO	48.60(45)	595.97(600.62)	0.65(0.61)	48.60(56)	603.36(599.57)	0.68(0.64)
	PMS-NSPSO	48.00(47)	595.33(602.24)	0.68(0.69)	41.60(46)	592.88(612.23)	0.68(0.81)
	PMS-NSGA-II	51.20(51)	598.85(608.34)	0.62(0.60)	61.00(74)	598.56(598.90)	0.65(0.59)
B-C	PMS-FIREFLY	46.20(47)	586.89(614.90)	0.64(0.60)	45.00(50)	600.21(596.60)	0.75(0.81)
	PMS-GSO	48.80(53)	588.56(595.46)	0.69(0.58)	43.60(52)	579.68(603.36)	0.70(0.67)
	PMS-NSPSO	41.00(51)	589.04(587.03)	0.68(0.62)	47.40(49)	595.11(605.91)	0.67(0.64)
	PMS-NSGA-II	58.80(55)	591.49(602.55)	0.65(0.55)	56.20(61)	597.28(590.74)	0.65(0.61)
B-D	PMS-FIREFLY	53.80(65)	600.97(605.45)	0.66(0.60)	43.80(48)	574.57(599.79)	0.67(0.60)
	PMS-GSO	41.40(46)	601.46(616.22)	0.66(0.70)	46.00(47)	590.67(600.51)	0.63(0.61)
	PMS-NSPSO	43.40(55)	594.80(598.42)	0.65(0.54)	43.20(49)	591.07(600.78)	0.64(0.54)
	PMS-NSGA-II	58.60(56)	595.91(609.20)	0.64(0.63)	55.80(54)	592.55(608.67)	0.65(0.66)
B-E	PMS-FIREFLY	43.40(50)	604.38(617.95)	0.65(0.73)	45.00(41)	598.04(609.86)	0.67(0.69)
	PMS-GSO	41.40(45)	596.53(613.85)	0.63(0.66)	43.80(45)	594.93(614.05)	0.62(0.58)
	PMS-NSPSO	49.40(52)	606.42(590.66)	0.60(0.53)	41.40(40)	604.84(618.80)	0.68(0.63)
	PMS-NSGA-II	57.60(59)	603.48(607.87)	0.61(0.60)	58.60(63)	603.78(618.11)	0.63(0.55)
C-D	PMS-FIREFLY	44.00(40)	578.14(609.98)	0.70(0.69)	47.80(50)	588.27(603.99)	0.74(0.72)
	PMS-GSO	48.00(51)	597.31(604.20)	0.66(0.63)	46.60(51)	590.46(595.81)	0.65(0.65)
	PMS-NSPSO	48.80(46)	584.51(602.85)	0.64(0.59)	46.60(56)	586.85(594.19)	0.67(0.73)
	PMS-NSGA-II	56.60(59)	587.97(604.68)	0.63(0.64)	51.20(57)	586.43(587.41)	0.59(0.60)
C-E	PMS-FIREFLY	49.20(51)	599.70(599.48)	0.71(0.65)	48.00(55)	598.85(606.21)	0.66(0.62)
	PMS-GSO	41.40(47)	592.94(597.71)	0.68(0.61)	50.40(53)	596.60(617.12)	0.64(0.59)
	PMS-NSPSO	48.00(51)	598.45(604.81)	0.64(0.65)	43.20(33)	599.71(609.57)	0.67(0.59)
	PMS-NSGA-II	60.40(63)	599.00(592.77)	0.64(0.58)	60.40(61)	607.48(613.53)	0.65(0.60)
D-E	PMS-FIREFLY	50.00(56)	599.83(616.32)	0.66(0.63)	49.00(53)	606.20(616.22)	0.65(0.66)
	PMS-GSO	53.00(60)	603.67(617.21)	0.63(0.64)	48.20(48)	595.50(602.80)	0.64(0.66)
	PMS-NSPSO	49.20(57)	606.90(620.98)	0.71(0.80)	48.00(55)	609.46(617.27)	0.69(0.70)
	PMS-NSGA-II	60.20(52)	601.63(610.06)	0.67(0.66)	59.00(53)	606.62(617.50)	0.63(0.57)

TABLE 14.4
Results of the *C* Measure for the Four Algorithms in 10 Instances When the Asymmetric Delivery Problem Is Solved Using Objective Functions OF1–OF2

OF1–OF2	Multi-Objective Asymmetric Delivery Route-Based Energy Reduction MDVRP								
A-B-CD	**NSPSO**	**NSGA-II**	**GSO**	**FIREFLY**	**B-D-AC**	**NSPSO**	**NSGA-II**	**GSO**	**FIREFLY**
NSPSO	–	0.82	0.27	0.24	NSPSO	–	0.89	0.45	0.72
NSGA-II	0.04	–	0.08	0.06	NSGA-II	0,02	–	0.21	0.07
GSO	0.58	0.89	–	0.42	GSO	0.43	0.67	–	0.58
FIREFLY	0.57	0.94	0.37	–	FIREFLY	0.23	0.65	0.29	–
A-C-BD	NSPSO	NSGA-II	GSO	FIREFLY	**B-E-AD**	NSPSO	NSGA-II	GSO	FIREFLY
NSPSO	–	0.78	0.63	0.59	NSPSO	–	0.75	0.30	0.35
NSGA-II	0.02	–	0.07	0.07	NSGA-II	0.07	–	0.08	0.05
GSO	0.45	0.68	–	0.36	GSO	0.33	0.75	–	0.31
FIREFLY	0.43	0.68	0.39	–	FIREFLY	0.40	0.84	0.47	–
A-D-BE	NSPSO	NSGA-II	GSO	FIREFLY	**C-D-AE**	NSPSO	NSGA-II	GSO	FIREFLY
NSPSO	–	0.74	0.33	0.28	NSPSO	–	0.82	0.50	0.47
NSGA-II	0.15	–	0	0	NSGA-II	0.07	–	0.03	0.13
GSO	0.71	0.91	–	0.53	GSO	0.47	0.82	–	0.45
FIREFLY	0.54	0.91	0.33	–	FIREFLY	0.39	0.82	0.39	–
A-E-BD	NSPSO	NSGA-II	GSO	FIREFLY	**C-E-AB**	NSPSO	NSGA-II	GSO	FIREFLY
NSPSO	–	0.89	0.37	0.26	NSPSO	–	0.79	0.49	0.48
NSGA-II	0.03	–	0.06	0.07	NSGA-II	0.13	–	0.17	0.23
GSO	0.40	0.84	–	0.48	GSO	0.46	0.66	–	0.45
FIREFLY	0.38	0.88	0.41	–	FIREFLY	0.42	0.68	0.34	–
B-C-AD	NSPSO	NSGA-II	GSO	FIREFLY	**D-E-BC**	NSPSO	NSGA-II	GSO	FIREFLY
NSPSO	–	0.86	0.47	0.42	NSPSO	–	0.38	0.26	0.24
NSGA-II	0.04	–	0.14	0.07	NSGA-II	0.56	–	0.59	0.47
GSO	0.42	0.83	–	0.44	GSO	0.44	0.40	–	0.27
FIREFLY	0.38	0.86	0.43	–	FIREFLY	0.62	0.44	0.59	–

fronts of the asymmetric delivery problem using objective functions OF1–OF2 and of the asymmetric pickup problem using objective functions OF1–OF3, for the instance "B-D-AC" for all algorithms, are given in part (C) and in part (D) of the same figure, respectively. In addition, in Tables 14.4–14.7 we present the results of the C measure for the four algorithms in 10 instances in all problems. In Tables 14.4–14.7, when we mention GSO, FIREFLY, NSPSO, and NSGA-II, we mean PMS-GSO, PMS-FIREFLY, PMS-NSPSO, and PMS-NSGA-II algorithms, respectively.

In Tables 14.2 and 14.3, from the comparison of the four algorithms, we can see that for the L measure, the performance of the PMS-NSGA-II algorithm is better than the other three algorithms in 72.5% of the instances, whereas the performances of the PMS-NSPSO, the PMS-FIREFLY, and the PMS-GSO compared with other algorithms are better in 10%, 7.5%, and 5%, respectively. The same results are taken from the algorithms in 5% of the instances. For the M_k measure, the performance of the PMS-NSGA-II algorithm is better than the other algorithms in 30% of the instances. The performances of the PMS-NSPSO, the PMS-GSO, and the PMS-FIREFLY algorithms compared with the other algorithms are

TABLE 14.5

Results of the *C* Measure for the Four Algorithms in 10 Instances When the Symmetric Delivery Problem Is Solved Using Objective Functions OF1–OF2

OF1–OF2	Multi-Objective Symmetric Delivery Route-Based Energy Reduction MDVRP								
A-B	**NSPSO**	**NSGA-II**	**GSO**	**FIREFLY**	**B-D**	**NSPSO**	**NSGA-II**	**GSO**	**FIREFLY**
NSPSO	–	0.97	0.61	0.56	NSPSO	–	0.96	0.28	0.42
NSGA-II	0.02	–	0.04	0.02	NSGA-II	0	–	0.07	0.03
GSO	0.35	0.95	–	0.56	GSO	0.45	0.82	–	0.46
FIREFLY	0.35	0.92	0.44	–	FIREFLY	0.45	0.93	0.35	–
A-C	NSPSO	NSGA-II	GSO	FIREFLY	B-E	NSPSO	NSGA-II	GSO	FIREFLY
NSPSO	–	0.97	0.42	0.42	NSPSO	–	0.78	0.18	0.32
NSGA-II	0	–	0	0.03	NSGA-II	0.10	–	0.02	0.04
GSO	0.53	1.00	–	0.47	GSO	0.52	0.85	–	0.50
FIREFLY	0.43	0.95	0.40	–	FIREFLY	0.50	0.88	0.38	–
A-D	NSPSO	NSGA-II	GSO	FIREFLY	C-D	NSPSO	NSGA-II	GSO	FIREFLY
NSPSO	–	0.96	0.50	0.40	NSPSO	–	0.81	0.41	0.30
NSGA-II	0	–	0.04	0.06	NSGA-II	0.07	–	0.02	0.15
GSO	0.39	0.84	–	0.27	GSO	0.41	0.95	–	0.48
FIREFLY	0.41	0.88	0.42	–	FIREFLY	0.48	0.73	0.31	–
A-E	NSPSO	NSGA-II	GSO	FIREFLY	C-E	NSPSO	NSGA-II	GSO	FIREFLY
NSPSO	–	0.92	0.18	0.31	NSPSO	–	0.81	0.34	0.45
NSGA-II	0.06	–	0	0	NSGA-II	0.20	–	0.06	0.12
GSO	0.66	0.96	–	0.59	GSO	0.59	0.86	–	0.57
FIREFLY	0.60	1.00	0.36	–	FIREFLY	0.49	0.78	0.21	–
B-C	NSPSO	NSGA-II	GSO	FIREFLY	D-E	NSPSO	NSGA-II	GSO	FIREFLY
NSPSO	–	0.78	0.34	0.15	NSPSO	–	0.92	0.40	0.38
NSGA-II	0.12	–	0.09	0.11	NSGA-II	0.02	–	0.05	0.02
GSO	0.51	0.78	–	0.36	GSO	0.63	0.96	–	0.46
FIREFLY	0.69	0.82	0.42	–	FIREFLY	0.54	0.92	0.40	–

better in 27.5%, 25%, and 17.5% of the instances, respectively. Also, for the Δ measure, the performance of the PMS-NSGA-II algorithm compared with the other algorithms is better in 40% of the instances, whereas the performances of the PMS-NSPSO, the PMS-FIREFLY and the PMS-GSO algorithms compared with the other algorithms are better in 22.5%, 15%, and 10% of the instances, respectively. The same results are taken from all algorithms in 12.5% of the instances.

For the *C* measure presented in Table 14.4, the PMS-GSO algorithm gives better results than the PMS-NSGA-II algorithm in 90% of the instances. The PMS-GSO algorithm has better performance compared with the PMS-FIREFLY algorithm in 70% of the instances. Finally, the PMS-GSO algorithm performs equally well with the PMS-NSPSO algorithm in 50% of the instances. Taking into account the results of Table 14.5, the PMS-GSO algorithm has better performance than the PMS-NSGA-II algorithm in all instances and it performs better compared with the PMS-NSPSO algorithm in 80% of the instances and compared with the PMS-FIREFLY algorithm in 70% of the instances, respectively. Finally, from Tables 14.6 and 14.7, the performance of the proposed PMS-GSO algorithm is better than the PMS-NSGA-II algorithm in all instances in both tables. The proposed algorithm compared with the PMS-NSPSO algorithm gives better results in 60% and 70% of the instances, respectively. Finally, the performance of the proposed

TABLE 14.6
Results of the *C* Measure for the Four Algorithms in 10 Instances When an Asymmetric Pickup Problem Is Solved Using Objective Functions OF1–OF3

OF1–OF3	Multi-Objective Asymmetric Pickup Route-Based Energy Reduction MDVRP								
A-B-CD	**NSPSO**	**NSGA-II**	**GSO**	**FIREFLY**	**B-D-AC**	**NSPSO**	**NSGA-II**	**GSO**	**FIREFLY**
NSPSO	–	0.67	0.51	0.30	NSPSO	–	0.96	0.18	0.75
NSGA-II	0.09	–	0.17	0.05	NSGA-II	0	–	0	0.06
GSO	0.34	0.75	–	0.14	GSO	0.66	1.00	–	0.79
FIREFLY	0.68	0.83	0.68	–	FIREFLY	0.32	0.83	0.10	–
A-C-BD	NSPSO	NSGA-II	GSO	FIREFLY	B-E-AD	NSPSO	NSGA-II	GSO	FIREFLY
NSPSO	–	0.91	0.44	0.54	NSPSO	–	0.75	0.59	0.51
NSGA-II	0.02	–	0	0.08	NSGA-II	0.11	–	0.24	0.18
GSO	0.49	0.88	–	0.67	GSO	0.28	0.65	–	0.35
FIREFLY	0.38	0.71	0.17	–	FIREFLY	0.38	0.64	0.44	–
A-D-BE	NSPSO	NSGA-II	GSO	FIREFLY	C-D-AE	NSPSO	NSGA-II	GSO	FIREFLY
NSPSO	–	0.87	0.40	0.31	NSPSO	–	0.90	0.51	0.29
NSGA-II	0.08	–	0.10	0.07	NSGA-II	0.03	–	0.09	0.02
GSO	0.48	0.89	–	0.36	GSO	0.14	0.78	–	0.31
FIREFLY	0.56	0.87	0.40	–	FIREFLY	0.36	0.83	0.44	–
A-E-BD	NSPSO	NSGA-II	GSO	FIREFLY	C-E-AB	NSPSO	NSGA-II	GSO	FIREFLY
NSPSO	–	0.80	0.48	0.47	NSPSO	–	0.85	0.36	0.33
NSGA-II	0.08	–	0.02	0.04	NSGA-II	0.10	–	0	0.02
GSO	0.48	0.80	–	0.53	GSO	0.59	0.91	–	0.56
FIREFLY	0.38	0.75	0.40	–	FIREFLY	0.46	0.91	0.42	–
B-C-AD	NSPSO	NSGA-II	GSO	FIREFLY	D-E-BC	NSPSO	NSGA-II	GSO	FIREFLY
NSPSO	–	0.90	0.63	0.69	NSPSO	–	0.77	0.29	0.63
NSGA-II	0	–	0.06	0.13	NSGA-II	0.09	–	0.08	0.16
GSO	0.24	0.82	–	0.64	GSO	0.52	0.79	–	0.71
FIREFLY	0.13	0.75	0.21	–	FIREFLY	0.20	0.71	0.16	–

algorithm compared with the PMS-FIREFLY algorithm is better in 60% of the instances in both tables. Taking into account the *C* measure, we can conclude that the proposed algorithm has a high quality of the solutions and gives very satisfactory results. It has a competitive performance compared with the other algorithms used in the comparisons. Finally, in Figure 14.1 we conclude that the Pareto front produced from the PMS-GSO algorithm is slightly better than the Pareto fronts produced from the other three algorithms.

In Table 14.8, a total comparison of all algorithms for all measures is given. In the first part of the table, a heat map of the algorithms in *C* measure is presented, whereas in the second part of the table, the average values in all runs for the three other measures are given. In the heat map, the red cells mean that the algorithm in the corresponding row performs better than the algorithm in the corresponding column in almost all instances (i.e., the proposed algorithm PMS-GSO is better in 39 of 40 instances compared with PMS-NSGA-II). The orange cells mean that the algorithm in the corresponding row performs better than the algorithm in the corresponding column in more than 50% of the instances (i.e., the proposed algorithm PMS-GSO is better in 25 of 40 instances compared with PMS-NSPSO and with PMS-FIREFLY). On the other hand, the light green cells mean that the algorithm in the corresponding row performs better than the algorithm in the corresponding column in less than 50% of the instances (i.e., the

TABLE 14.7
Results of the *C* Measure for the Four Algorithms in 10 Instances When a Symmetric Pickup Problem Is Solved Using Objective Functions OF1–OF3

OF1 – OF3	Multi-Objective Symmetric Pickup Route-Based Energy Reduction MDVRP								
A-B	**NSPSO**	**NSGA-II**	**GSO**	**FIREFLY**	**B-D**	**NSPSO**	**NSGA-II**	**GSO**	**FIREFLY**
NSPSO	–	1.00	0.35	0.48	NSPSO	–	0.94	0.40	0.42
NSGA-II	0	–	0	0	NSGA-II	0.02	–	0.02	0
GSO	0.43	0.99	–	0.42	GSO	0.37	0.89	–	0.29
FIREFLY	0.28	1.00	0.33	–	FIREFLY	0.45	0.96	0.30	–
A-C	NSPSO	NSGA-II	GSO	FIREFLY	B-E	NSPSO	NSGA-II	GSO	FIREFLY
NSPSO	–	0.84	0.29	0.43	NSPSO	–	0.68	0.42	0.34
NSGA-II	0.08	–	0	0.02	NSGA-II	0.15	–	0.04	0.12
GSO	0.70	0.95	–	0.64	GSO	0.33	0.89	–	0.39
FIREFLY	0.64	0.92	0.33	–	FIREFLY	0.58	0.83	0.62	–
A-D	NSPSO	NSGA-II	GSO	FIREFLY	C-D	NSPSO	NSGA-II	GSO	FIREFLY
NSPSO	–	0.86	0.53	0.29	NSPSO	–	0.88	0.25	0.52
NSGA-II	0.07	–	0.07	0.02	NSGA-II	0.07	–	0.02	0.10
GSO	0.40	0.89	–	0.24	GSO	0.43	0.88	–	0.58
FIREFLY	0.57	0.94	0.58	–	FIREFLY	0.29	0.79	0.22	–
A-E	NSPSO	NSGA-II	GSO	FIREFLY	C-E	NSPSO	NSGA-II	GSO	FIREFLY
NSPSO	–	0.85	0.36	0.43	NSPSO	–	0.85	0.42	0.44
NSGA-II	0.09	–	0.05	0.04	NSGA-II	0.03	–	0.02	0
GSO	0.37	0.88	–	0.59	GSO	0.52	0.95	–	0.33
FIREFLY	0.30	0.85	0.30	–	FIREFLY	0.36	0.97	0.49	–
B-C	NSPSO	NSGA-II	GSO	FIREFLY	D-E	NSPSO	NSGA-II	GSO	FIREFLY
NSPSO	–	0.95	0.38	0.50	NSPSO	–	0.89	0.29	0.30
NSGA-II	0	–	0	0	NSGA-II	0.02	–	0	0.02
GSO	0.51	0.97	–	0.50	GSO	0.55	0.89	–	0.38
FIREFLY	0.57	0.97	0.50	–	FIREFLY	0.55	0.94	0.52	–

TABLE 14.8
Comparisons of All Algorithms in all Measures

	C Measure			
	PMS-NSPSO	**PMS-NSGA-II**	**PMS-GSO**	**PMS-FIREFLY**
PMS-NSPSO		39	15	16
PMS-NSGA-II	1		1	1
PMS-GSO	25	39		25
PMS-FIREFLY	24	39	15	
	L	M_k	**Δ**	
PMS-NSPSO	45.92	594.76	0.66	
PMS-NSGA-II	55.74	595.15	0.63	
PMS-GSO	47.02	593.45	0.63	
PMS-FIREFLY	45.75	593.77	0.66	

PMS-FIREFLY is better in 15 of 40 instances compared with the proposed algorithm PMS-GSO). Finally, the green cells mean that the algorithm in the corresponding row performs better than the algorithm in the corresponding column in almost no instance (i.e., the PMS-NSGA-II is better in 1 of 40 instances compared with the proposed algorithm PMS-GSO). Thus, we can say that the proposed algorithm in C measure performs much better than all the other algorithms. From the second part of the table, we cannot make safe conclusions as the average results in all the other measures are almost the same for all algorithms, except for the number of solutions (L measure), where the PMS-NSGA-II produces around 10 solutions on average more than the other three algorithms.

14.5 CONCLUSIONS AND FUTURE RESEARCH

In this chapter, the formulation and the solution of the multi-objective route-based energy reduction multi-depot VRP are performed using the proposed PMS-GSO algorithm. Comparisons of the PMS-GSO algorithm with the PMS-FIREFLY, the PMS-NSPSO, and the PMS-NSGA-II algorithms are given in a number of benchmark instances from the literature, and the results proved the effectiveness of the PMS-GSO algorithm.

In general, in the four different problems that are solved in this chapter, the PMS-GSO algorithm performs better than the other algorithms in the C measure, except from the symmetric pickup problem, where the PMS-FIREFLY algorithm performs better than the other algorithms, whereas PMS-NSGA-II performs better in the other three measures. As expected, the behavior of the algorithms was different when different problems were solved (symmetric or asymmetric delivery problem, symmetric or asymmetric pickup problem). In the future, our research will concern mainly the use of other algorithms for the solution of the multi-objective route-based energy reduction multi-depot VRP. Also, we will use these algorithms to solve other multi-objective problems in the frame of supply chain management.

REFERENCES

1. Kaipa, K.N., Ghose, D. (2006). Glowworm swarm based optimization algorithm for multimodal functions with collective robotics applications. *Multiagent and Grid Systems: Special Issue on Recent Progress in Distributed Intelligence*, 2(3), 209–222.
2. Kaipa, K.N., Ghose, D. (2009). Glowworm swarm optimisation: A new method for optimising multi-modal functions. *International Journal of Computational Intelligence Studies*, 1(1), 93–119.
3. Kaipa, K.N., Ghose, D. (2009). Glowworm swarm optimization for searching higher dimensional spaces. *Innovations in Swarm Intelligence, SCI 248*, Chee Peng Lim, Lakhmi C. Jain, and Satchidananda Dehuri (Eds.), 61–76, Springer-Verlag, Berlin, Heidelberg.
4. Kaipa, K.N., Ghose, D. (2009). Glowworm swarm optimization for simultaneous capture of multiple local optima of multimodal functions, *Swarm Intelligence*, 3, 87–124.
5. Kaipa, K.N., Ghose, D. (2011). Glowworm swarm optimization for multimodal search spaces. *Handbook of Swarm Intelligence*, Panigrahi, B.K., Shi, Y., Lim, M.-H. (Eds.), 8, 451–467, Springer-Verlag, Berlin Heidelberg.
6. Kaipa, K.N., Ghose, D. (2008). Glowworm swarm optimization algorithm for hazard sensing in ubiquitous environments using heterogeneous agent swarms. *Soft Computing Applications in Industry*, StudFuzz, Prasad, B. (Ed.), 226, 165–187, Springer, Berlin, Heidelberg.
7. Kaipa, K.N., Ghose, D. (2009). A glowworm swarm optimization based multi-robot system for signal source localization. *Design and Control of Intelligent Robotic Systems, SCI 177*, Liu, D., Wang, L., Tan, K.C. (Eds.), 49–68, Springer-Verlag, Berlin, Heidelberg.
8. Kaipa, K.N., Ghose, D. (2008). Theoretical foundations for rendezvous of glowworm-inspired agent swarms at multiple locations. *Robotics and Autonomous Systems*, 56, 549–569.

9. Kaipa K.N., Ghose, D. (2017). *Glowworm Swarm Optimization: Theory, Algorithms, and Applications*. Studies in Computational Intelligence. Vol. 698. Springer, Switzerland.
10. Chowdhury, A., De, D. (2020). FIS-RGSO: Dynamic Fuzzy Inference System Based Reverse Glowworm Swarm Optimization of energy and coverage in green mobile wireless sensor networks. *Computer Communications*, 163, 12–34.
11. Chowdhury, A., De, D. (2020). MSLG-RGSO: Movement Score based Limited Grid-mobility approach using Reverse Glowworm Swarm Optimization Algorithm for Mobile Wireless Sensor Networks. *Ad Hoc Networks*, 106, 102191.
12. Goel, U., Varshney, S., Jain, A., Maheshwari, S., Shukla, A. (2018). Three dimensional path planning for UAVs in dynamic environment using glow-worm swarm optimization. *Procedia Computer Science*, 133, 230–239.
13. Karthikeyan, R., Alli, P. (2018). Feature selection and parameters optimization of support vector machines based on hybrid glowworm swarm optimization for classification of diabetic retinopathy. *Journal of Medical Systems*, 42(10), 195.
14. Khan, A., Aftab, F., Zhang, Z. (2019). Self-organization based clustering scheme for FANETs using Glowworm Swarm Optimization. *Physical Communication*, 36, 100769.
15. Liu, Z., Guo, S., Wang, L., Du, B., Pang, S. (2019). A multi-objective service composition recommendation method for individualized customer: hybrid MPA-GSO-DNN model. *Computers and Industrial Engineering*, 128, 122–134.
16. Prashant, P., Anupam, S., Ritu, T. (2018). Three-dimensional path planning for unmanned aerial vehicles using glowworm swarm optimization algorithm, *International Journal of System Assurance Engineering and Management*, 9 (4), 836–852.
17. Pruthi, J., Khanna, K., Arora, S. (2020). Optic Cup segmentation from retinal fundus images using Glowworm Swarm Optimization for glaucoma detection. *Biomedical Signal Processing and Control*, 60, 102004.
18. Puttamadappa, C., Parameshachari, B. D. (2019). Demand side management of small scale loads in a smart grid using glow-worm swarm optimization technique. *Microprocessors and Microsystems*, 71, 102886.
19. Sun, Y., Ma, R., Chen, J., Xu, T. (2020). Heuristic optimization for grid-interactive net-zero energy building design through the glowworm swarm algorithm. *Energy and Buildings*, 208, 109644.
20. Turgeman, A., Werner, H. (2018). Multiple Source Seeking using Glowworm Swarm Optimization and Distributed Gradient Estimation, *Annual American Control Conference (ACC)*, June 27–29, Wisconsin Center, Milwaukee, USA.
21. Wang, X., Yang, K. (2019). Economic load dispatch of renewable energy-based power systems with high penetration of large-scale hydropower station based on multi-agent glowworm swarm optimization. *Energy Strategy Reviews*, 26, 100425.
22. Xiuwu, Y., Qin, L., Yong, L., Mufang, H., Ke, Z., Renrong, X. (2019). Uneven clustering routing algorithm based on glowworm swarm optimization. *Ad Hoc Networks*, 93, 101923.
23. Marinakis, Y., Marinaki, M., Migdalas, A. (2018). Variants and Formulations of the Vehicle Routing Problem, P. Pardalos, A. Migdalas (Eds.), *Open Problems in Optimization and Data Analysis*, 91–127, Springer Nature, Switzerland.
24. Marinakis, Y., Marinaki, M., Migdalas, A. (2018). Particle Swarm Optimization for the Vehicle Routing Problem: A Survey and a Comparative Analysis, R. Marti, P. Pardalos, M. Resende (Eds.), *Handbook of Heuristics*, 1163–1196, Springer International Publishing, Switzerland.
25. Osman, I. H. (1993). Metastrategy simulated annealing and tabu search algorithms for the vehicle routing problem. *Annals of Operations Research*, 41(4), 421–451.
26. Braysy, O. (2003). A reactive variable neighborhood search for the vehicle-routing problem with time windows. *INFORMS Journal on Computing*, 15(4), 347–368.
27. Marinakis, Y. (2012). Multiple phase neighborhood search-GRASP for the capacitated vehicle routing problem. *Expert Systems with Applications*, 39(8), 6807–6815.
28. Braysy, O., Dullaert, W., Gendreau, M. (2004). Evolutionary algorithms for the vehicle routing problem with time windows. *Journal of Heuristics*, 10(6), 587–611.
29. Marinakis, Y., Marinaki, M. (2010). A hybrid genetic-Particle Swarm Optimization Algorithm for the vehicle routing problem. *Expert Systems with Applications*, 37(2), 1446–1455.
30. Marinaki, M., Marinakis, Y. (2016). A glowworm swarm optimization algorithm for the vehicle routing problem with stochastic demands. *Expert Systems with Applications*, 46, 145–163.

31. Rapanaki, E., Psychas, I.D., Marinaki, M., Marinakis, Y., Migdalas, A. (2018). A Clonal Selection Algorithm for Multiobjective Energy Reduction Multi-Depot Vehicle Routing Problem, *International Conference on Machine Learning, Optimization, and Data Science*, LNCS, Springer, 381–393.
32. Rapanaki, E., Psychas, I. D., Marinaki, M., Marinakis, Y. (2019). An Artificial Bee Colony Algorithm for the Multiobjective Energy Reduction Multi-Depot Vehicle Routing Problem. *International Conference on Learning and Intelligent Optimization*, LNCS, Springer, 208–223.
33. Psychas, I.D., Marinaki, M., Marinakis, Y., Migdalas, A. (2015). A parallel multi-start NSGA II algorithm for multiobjective energy reduction vehicle routing problem. *International Conference on Evolutionary Multi-Criterion Optimization,* LNCS, Springer, 336–350.
34. Psychas, I.D., Marinaki, M., Marinakis, Y., Migdalas, A. (2017). Non-dominated Sorting Differential Evolution Algorithm for the Minimization of Route based Fuel Consumption Multiobjective Vehicle Routing Problems, *Energy Systems*, 8(4), 785–814.
35. Psychas, I.D., Marinaki, M., Marinakis, Y., Migdalas, A. (2017). Parallel Multi-Start Non-dominated Sorting Particle Swarm Optimization Algorithms for the Minimization of the Route-Based Fuel Consumption of Multiobjective Vehicle Routing Problems, *Springer Optimization and Its Applications*, 130, 425–456.
36. Psychas, I. D., Delimpasi, E., Marinaki, M., Marinakis, Y. (2018). Influenza virus algorithm for multiobjective energy reduction open vehicle routing problem. *Shortest Path Solvers. From Software to Wetware*, 145–161, Springer, Switzerland.
37. Montoya-Torres, J. R., Franco, J. L., Isaza, S. N., Jimenez, H. F., Herazo-Padilla, N. (2015). A literature review on the vehicle routing problem with multiple depots. *Computers and Industrial Engineering*, 79, 115–129.
38. Demir, E., Bektaş, T., Laporte, G. (2014). A review of recent research on green road freight transportation. *European Journal of Operational Research*, 237(3), 775–793.
39. Lin, C., Choy, K. L., Ho, G. T. S., Chung, S. H., Lam, H. Y. (2014). Survey of green vehicle routing problem: Past and future trends. *Expert Systems with Applications*, 41(4), 1118–1138.
40. Li, J., Wang, R., Li, T., Lu, Z., Pardalos, P. (2018). Benefit analysis of shared depot resources for multi-depot vehicle routing problem with fuel consumption. *Transportation Research Part D: Transport and Environment*, 59, 417–432.

15 Monarch Butterfly Optimization

Liwen Xie and Gai-Ge Wang

CONTENTS

15.1 INTRODUCTION

In the areas of computer science, mathematics, control, and decision making, a relatively new set of algorithms, called nature-inspired algorithms, have been proposed and used to address an array of complex optimization problems. Among various nature-inspired algorithms, swarm-based algorithms and evolutionary algorithms (EAs) are two of the most representative paradigms.

Swarm-based algorithms, also called swarm intelligence (SI) methods (Cui and Gao 2012), are one of the most well-known paradigms in nature-inspired algorithms that have been widely used in various applications, such as scheduling, directing orbits of chaotic systems (Cui et al. 2013), wind generator optimization (Gao et al. 2012), and fault diagnosis (Gao et al. 2007). SI concerns the collective, emerging behavior of multiple, interacting agents who follow some simple rules (Fister Jr et al. 2013). Two of widely used SI methods are particle swarm optimization (PSO) (Kennedy and Eberhart 1995, Ram et al. 2014, Mirjalili et al. 2014b, Wang et al. 2014d, Wang et al. 2015a) and ant colony optimization (ACO) (Dorigo et al. 1996, Krynicki et al. 2014). The idea of PSO (Kennedy and Eberhart 1995) originated from the social behavior of bird flocking when searching for food. The ants in nature are well capable of keeping the past paths in mind by pheromones. Inspired by this phenomenon, the ACO algorithm by Dorigo et al. (1996). Recently, more superior SI algorithms have been proposed, such as artificial bee colony (ABC) (Karaboga and Basturk 2007, Li and Yin 2012), cuckoo search (CS) (Yang and Deb 2009, 2013, Li et al. 2013, Ouaarab et al. 2014, Wang et al. 2014b), bat algorithm (BA) (Yang 2010b, Mirjalili et al. 2013, Fister Jr et al. 2014), gray wolf optimizer (GWO) (Mirjalili et al. 2014, Saremi et al. 2014b), ant lion optimizer (ALO) (Mirjalili 2015), firefly algorithm (FA) (Yang 2010a, Gandomi et al. 2011, Guo et al. 2013, Wang et al. 2014e), chicken swarm optimization (CSO) (Meng et al. 2014), and krill herd (KH) (Gandomi and Alavi 2012, Wang et al. 2013a, Li et al. 2014a). These are inspired by the swarm behavior of honey bees, cuckoos, bats, gray wolves, chickens, and krill, respectively.

By simplifying and idealizing the genetic evolution process, different kinds of EAs have been proposed and used in a wide range of applications. Genetic algorithm (GA) (Goldberg 1998, Gao and Ovaska 2002), evolutionary programming (EP) (Bäck 1996, Zhao et al. 2007), genetic programming (GP) (Hand 1992), and evolutionary strategy (ES) (Beyer and Schwefel 2002) are four of the most classical EAs among them. With the development of the evolutionary theory, some new methods have been proposed over the last decades that significantly improved the theory and search capacities of EAs. Differential evolution (DE) (Storn and Price 1997, Wang et al. 2014c) is a very efficient search algorithm that simulates the biological mechanisms of natural selection and mutation. The best-to-survive criteria is adopted in the above algorithms on a population of solutions. Stud genetic algorithm (SGA) (Khatib and Fleming 1998, Wang et al. 2014a) is a special kind of GA that uses the best individual and the other randomly selected individuals at each generation for a crossover operator. By incorporating the sole effect of predictor variable as well as the interactions between the variables into the GP, Gandomi and Alavi (2011) proposed an improved version of the GP algorithm, called multi-stage genetic programming (MSGP), for nonlinear system modeling. Recently, motivated by the natural biogeography, Simon (2008) has provided the mathematics of biogeography and accordingly proposed a new kind of EA called biogeography-based optimization (BBO) (Li and Yin 2013, Saremi et al. 2014a, Wang et al.

2013b, 2014c). Inspired by the animal migration behavior, animal migration optimization (AMO) (Li et al. 2014b) is proposed and compared with other well-known heuristic search methods.

The goal of this chapter is twofold. First, the new optimization method called monarch butterfly optimization (MBO) is introduced. It is carried out by first studying the migration behavior of monarch butterflies and then generalizing it to formulate a general-purpose meta-heuristic method. Second, a comparative study of the performance of MBO with respect to other population-based optimization methods is done. This has been addressed by looking at the commonalities and differences from an algorithmic point of view as well as by comparing their performances on an array of benchmark functions.

The second part of this chapter mainly introduces some related work on the MBO algorithm. The third section details the application of the MBO algorithm in solving the 0-1 knapsack (KP) problem. The fourth section shows the experimental results of this application. In the fifth part, I summarized this chapter and looked forward to the future work of this algorithm.

15.2 RELATED WORK OF MONARCH BUTTERFLY OPTIMIZATION

15.2.1 Monarch Butterfly Optimization Algorithm

15.2.1.1 The Swarm Intelligence Algorithms Related to Butterfly Species

The algorithms mentioned above almost belong to the category of SI, which is the collective behavior of decentralized, self-organized systems. Examples in natural systems of SI include bird flocking, fish schooling, ant colonies, animal herding, and so forth. Due to the effectiveness and efficiency of the SI algorithm in solving complex optimization problems, this technology is becoming an important research topic and has been applied in various fields. Butterflies belong to the class of *Lepidoptera* and there are more than 18,000 species around the world. So far, there are three other important optimization techniques associated with butterflies that have been proposed along with the MBO algorithm:

1. By mimicking the food search and mating behavior of butterflies, the butterfly optimization algorithm (BOA) was proposed by Arora and Singh (2019). It is employed to solve global optimization problems and engineering problems.
2. Inspired by the mate-finding mechanism of some butterfly species, the artificial butterfly optimization algorithm (ABO) was presented by Qi et al. (2017). It provides a new effective computational framework for solving optimization problems.
3. Similarly, inspired by the mating phenomenon of butterfly species in nature, the butterfly mating optimization algorithm (BMO) was proposed by Jada et al. (2016). It is divided into four phases to implement a search process.

Among the four algorithms mentioned above, MBO becomes the most widely studied algorithm and has been applied for solving various optimization problems. The distribution of publications of the related algorithms based on butterfly species is shown in Figure 15.1. In total, 76% of publications belong to the MBO algorithm and the ABO algorithm accounts for merely 2%. However, as far as we know, there is no literature review of the MBO algorithm. This is the purpose of this study.

15.2.1.2 Studies on the Monarch Butterfly Optimization Algorithm

After the presentation of MBO by Wang *et al.* in 2015 (Wang et al. 2019a), MBO has been widely researched. As of June 23, 2021, the original paper has been cited 426 times according to Google Scholar (https://scholar.google.com). This review first presents the annual distribution of

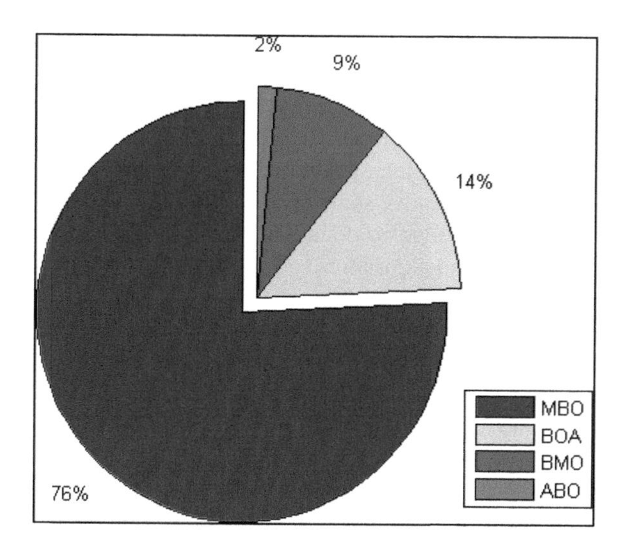

FIGURE 15.1 Percentages of publications concerning MBO, BOA, BMO, and ABO.

the related papers of MBO in Figure 15.2, which illustrates that the number of published articles shows an upward trend year by year. Because the published information has not been updated, the publication is researched the most in 2018. As far as we know, there are 50 papers in total on the improvement and application of MBO. There are more than 30 papers published in peer-reviewed journals, such as the *International Journal of Bio-Inspired Computation* (Yi et al. 2020), *Neural Computing and Applications* (Ghanem and Jantan 2018, Feng et al. 2018b), *Soft Computing* (Soltani and Hadavandi 2019), *Operational Research* (Wang et al. 2018), *Memetic Computing* (Feng et al. 2018c), *Journal of Cleaner Production* (Ehteram et al. 2017), and so forth. There are more than 15 papers published on the peer-reviewed conferences, such as the International Conference on Swarm Intelligence (ICSI), the International Conference on Hybrid Intelligent Systems (HIS),

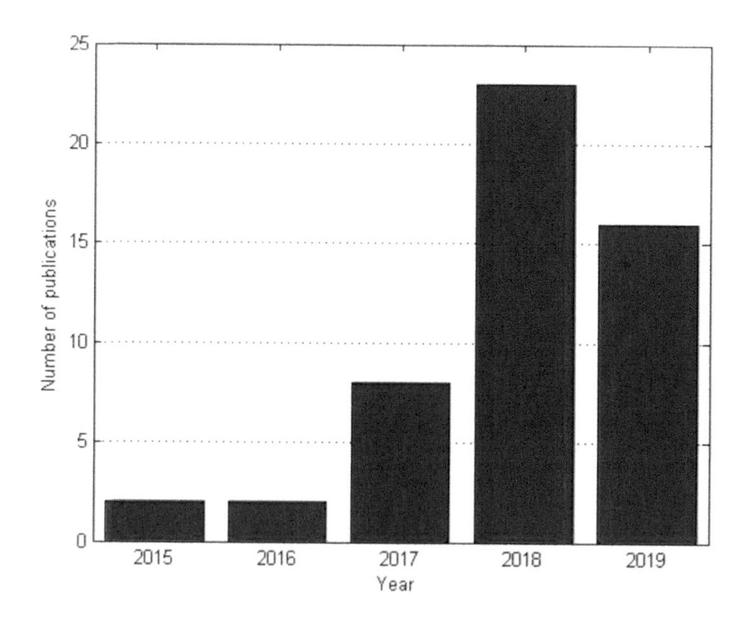

FIGURE 15.2 Number of publications per year.

the International Conference on Soft Computing & Machine Intelligence (ISCMI), and so forth. Additionally, there is one master's thesis based on the MBO algorithm.

15.2.1.3 Monarch Butterfly Optimization Algorithm

The original MBO algorithm (Wang et al. 2019a) is simple and easy to implement. In MBO, there are two equal-sized subpopulations: subpopulation1 and subpopulation2. Note that the half of the individuals with the best fitness value constitutes subpopulation1, and the rest forms subpopulation2. Accordingly, there are two strategies in MBO, migration operator, and butterfly adjusting operator. After one iteration, the global optimal information is reserved and the two subpopulations are reorganized into a population again. After that the whole population is separated into two subpopulations according to the new fitness value. The process is repeated until the termination condition is met.

15.2.1.3.1 Migration Operator

For individual i in subpopulation1, the movement is influenced by the position of other individuals in subpopulation1 or subpopulation2, which can be balanced by the adjusting ratio p. The aim of the migration operator is to enhance the exchange of information between the two subpopulations and within the subpopulation1. The movement of individual i in subpopulation1 can be mathematically expressed as follows:

$$x_{i,k}^{t+1} = \begin{cases} x_{r1,k}^{t}, & if \ r \leq p \\ x_{r2,k}^{t}, & else \end{cases} \tag{15.1}$$

where $r1$, $r2$ is an integer index randomly selected from subpopulation1 and subpopulation 2, respectively. The parameter $r = rand * peri$, in which $rand$ is a random real number in [0, 1] and $peri$ represents migration period.

15.2.1.3.2 Butterfly Adjusting Operator

For individual i in subpopulation2, the movement is formulated in terms of the current global best individual, a random individual in subpopulation2, and Lévy flight, which can be determined by the adjusting ratio p and butterfly adjusting rate BAR. The usage of the butterfly adjusting operator mainly considers three factors: (1) influence of the social model by moving to the global optimum, (2) the cognitive effects of other individuals by moving to a random individual, and (3) introduction of Lévy flight to increase population diversity and expand search scope. The new individual of subpopulation2 is generated according to the following function:

$$x_{i,k}^{t+1} = \begin{cases} x_{best,k}^{t} & if \ rand \leq p \\ x_{r3,k}^{t} & if \ rand > p \wedge rand \leq BAR \\ x_{i,k}^{t} + \alpha \times (dx_k - 0.5) & if \ rand > p \wedge rand > BAR \end{cases} \tag{15.2}$$

where $x_{best,k}^{t}$ is the kth element of the current global best optimum in generation t; the parameter $r3$ is an integer index, which is randomly selected from subpopulation2; and BAR indicates the butterfly adjusting rate. The weighting factor α and dx can be given as:

$$\alpha = S_{\max} / t^2 \tag{15.3}$$

$$dx = Levy(x_i^t) \tag{15.4}$$

where $S_{\max}$ is max walk step.

Monarch butterfly optimization algorithm

Begin

Step 1: Initialization. Randomly initialize the population of *NP* monarch butterfly individuals; Set the related parameters: migration period *peri*, the migration ratio *p*, butterfly adjusting rate *BAR*, the max walk step of Levy flights S_{max}, the maximum generation *GMax*

Step 2: While the termination criteria is not met **do**

Sort the population according to the fitness value in non-ascending order.

Construct subpopulation1 with better NP_1 individuals and subpopulation2 with the rest.

Update subpopulation1 by performing migration operator.

Update subpopulation2 by performing butterfly adjusting operator.

Recombine the newly-generated subpopulations into one whole population.

Find the current best.

end while

Step 3: Find the current best

End.

FIGURE 15.3 Monarch butterfly optimization algorithm.

15.2.1.3.3 Algorithmic Structure of MBO

According to the analysis mentioned above, the search process and the corresponding flowchart of the MBO algorithm is shown in Figures 15.3 and 15.4, respectively. Note that the source codes of the MBO can be obtained at https://github.com/ggw0122/Monarch-Butterfly-Optimization and http://www.mathworks.com/matlabcentral/fileexchange/50828-monarch-butterfly-optimization.

15.2.2 Modifications of MBO Algorithm

The position updating of the monarch butterfly individual in the MBO algorithm depends primarily on the migration operator and the butterfly adjusting operator. These two factors allow the algorithm to be greatly improved over a wide range. Hence, the different variants of the classical MBO algorithm have been proposed. A list of these modifications is shown in Table 15.1. A more detailed introduction to these variants is given in this section.

TABLE 15.1
The Modifications of MBO Algorithm

Name	Problem	Author	Year
Chaotic MBO (CMBO)	0-1 knapsack problem	Feng et al. (2018c)	2018
Opposition-based learning MBO (OMBO)	0-1 knapsack problem	Feng et al. (2018a)	2018
Opposition-based learning MBO (OPMBO)	Clustering optimization	Sun et al. (2019)	2019
Multi-strategy MBO (MMBO)	Discounted {0-1} KP	Feng et al. (2018b)	2018
Self-adaptive population MBO (SPMBO)	Numerical optimization	Hu et al. (2018)	2018
Self-adaptive crossover operator MBO (GCMBO)	Numerical optimization	Wang et al. (2018)	2015
F/T mutation MBO	Numerical optimization	Wang et al. (2017)	2017

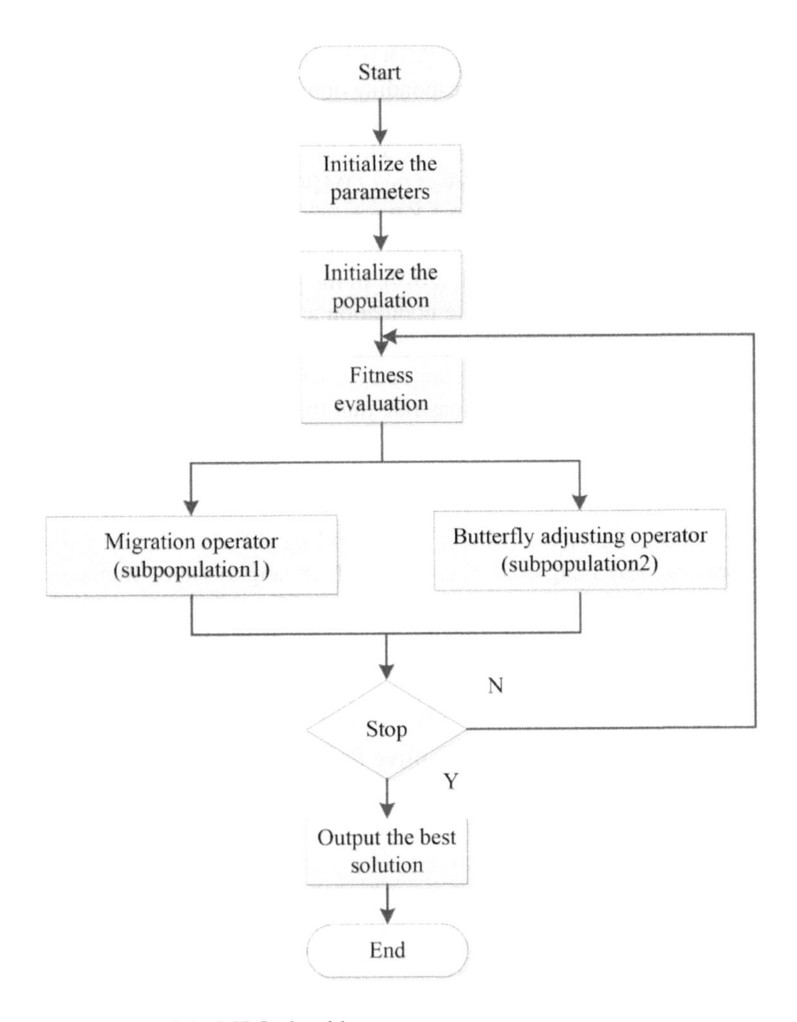

FIGURE 15.4 Flowchart of the MBO algorithm.

15.2.2.1 Chaotic-Based MBO Algorithm

With the purpose of taking advantage of important characteristics of chaotic theory, Feng (2018c) introduced chaotic maps into monarch butterfly optimization algorithm (CMBO) for solving the 0-1 KP problem. In CMBO, the two most important parameters, namely, the ratio of monarch butterflies in subpopulation1 p and butterfly adjusting rate BAR is replaced with 12 one-dimensional chaotic maps, respectively. The comparative study shows that BAR is more effective than p when replacing with chaotic maps. At the same time, the singer map or tent map can be selected as BAR to generate the chaotic MBO, and then it is compared with the other eight state-of-the-art meta-heuristic algorithms, including ABC (Karaboga and Basturk 2007), CS (Yang and Deb 2009), DE (Storn and Price 1997), GA (Goldberg and Holland 1988), FA (Yang 2010c), shuffled frog leaping algorithm (SFLA) (Eusuff and Lansey 2003b), harmony search (HS) (Geem 2001), and MBO (Wang et al. 2019a).

15.2.2.2 Opposition-Based Learning MBO Algorithm

Opposition-based learning (OBL) is a new concept and effective population generation strategy in computational intelligence. Feng et al. (2018a) proposed an improved MBO, combined with

OBL strategy and Gaussian perturbation (OMBO). In OMBO, OBL acts on half of the individuals with poor fitness to produce their corresponding opposite individuals in the later stage of evolution. Gaussian perturbation is applied to the worst part of individuals in each generation. The former accelerates the convergence of OMBO, while the latter avoids the algorithm from falling into the local optimal. The authors observed that OMBO can offer high-quality solutions when dealing with three groups of large-scale 0-1 KP.

Another version of the MBO algorithm based on the OBL strategy and random local perturbation (OPMBO) is introduced by Sun et al. (2019). In the population initialization phase, the OBL strategy is applied to generate an opposite population according to the original population. The better individuals are selected from the two populations and then pass to the next generation. When the performance of OPMBO is evaluated by a series of benchmark functions and clustering optimization, the results clearly demonstrate that the proposed OPMBO is superior to some existing algorithms in most of the test cases.

15.2.2.3 Multi-Strategy MBO Algorithm

A multi-strategy MBO algorithm (MMBO) for tackling the discounted {0-1}knapsack problem (DKP) is presented by Feng et al. (2018b), in which neighborhood mutation with crowding and Gaussian perturbation are introduced into MMBO. The authors proved that the former can strengthen the global search capability and the latter can enhance the local search ability by dealing with three types of large-scale DKP instances. The results showed that MMBO can offer a better approximate solution than the six methods.

15.2.2.4 MBO Algorithm with Self-Adaptive Population

A modified MBO algorithm with self-adaptive populations MBO algorithm (SPMBO) is proposed by Hu et al. (2018) for the global numeric optimization problem. In the migration operation of SPMBO, only the newly generated offspring individuals who are better than their parents are accepted and pass to the next generations. It is discovered that the performance of the proposed SPMBO outperforms the original MBO algorithm when 20 benchmark functions are verified.

15.2.2.5 Self-Adaptive Crossover Operator MBO Algorithm

The original MBO is an effective SI algorithm when it is verified by a series of unimodal and multimodal benchmark functions (Wang et al. 2019a). However, the performance of MBO can be further improved. Based on this, Wang et al. (2015b) proposed a new version of the MBO algorithm, incorporating a self-adaptive crossover operator (SAC) and a greedy strategy (GCMBO). In GCMBO, an additional crossover operator and greedy strategy are utilized in updated the butterfly adjusting operator to increase the population diversity, especially in the late stage of evolution. What needs to be emphasized is that the crossover rate Cr is related to the fitness of each individual in subpopulation2. Numerical experiment showed that the GCMBO significantly outperforms the original MBO.

15.2.2.6 Other Modifications

Wang et al. (2017) presented the modification of the MBO algorithm by designing a new population initialization strategy. In fact, the entire search space is evenly divided into NP (population size) parts at each dimension, the purpose of which is to distribute each individual throughout the search space, and to further develop the whole space in an all-round way. Next, the random distribution is applied to perturb each dimension of each individual. The authors select T and F distribution to mutate the population. The experimental results clearly show that the variants of MBO have superior performances to the original MBO.

15.2.3 Hybridizations of MBO Algorithm

In accordance with the no free lunch theorem (Wolpert and Macready 1997), when comparing the average performance of all possible problems, any two optimization algorithms can be regarded as equivalent. In the domain of optimization, the hybrid approach has always been a very popular and feasible method, and its idea comes from improving the performance of the algorithm by hybridizing the beneficial components from other optimization techniques. In fact, some optimization techniques can be incorporated into MBO to improve the performance for a given problem. The research has shown that the performance of MBO has been enhanced by integrating the operators of other metaheuristic algorithms. In contrast, MBO can also be embedded in other algorithms as a heuristic because of some of its characteristics, such as fast convergence and implementation simplicity. The detailed descriptions of these hybrid algorithms are presented in this section, and the basics are listed in Table 15.2.

15.2.3.1 Hybridization with Differential Evolution

It has been proved that the mutation operator of DE has a good global exploration ability (Gong et al. 2010). Yazdani and Hadavandi (2019) proposed a hybrid algorithm termed a linearized and hybrid version of MBO (LMBO-DE) for solving the numerical optimization problem. The authors utilized a new linearized migration operator and butterfly adjusting operator by embedding components of differential mutation. They investigated their algorithm with 18 benchmark functions in different dimensionality and the experimental results demonstrated that the LMBO-DE is very effective than the original MBO, one MBO variant, and three other classical algorithms.

Ibrahim and Tawhid (2019) overcame the poor exploitation ability of the MBO algorithm by combining the DE with MBO (DEMBO) when applied to solve nonlinear systems. In DEMBO, in addition to the initial population P ($P = P1 \cup P2$), an additional population V is generated based on the differential mutation. Similarly, population V is divided into subpopulation $V1$ and subpopulation $V2$. Unlike the original MBO, the new individual in DEMBO is generated by randomly selecting two individuals from $P1$ and $V2$, respectively. Based on the experiential results, it showed that the DEMBO is more efficient than other methods.

TABLE 15.2
The Hybridizations of MBO Algorithm

Name	Problem	Author	Year
Differential evolution (LMBO-DE)	Numerical optimization	Yazdani and Hadavandi (2019)	2018
Differential evolution (DEMBO)	Nonlinear system of equations	Ibrahim and Tawhid (2019)	2018
Differential evolution (DE-LSMBO)	PID tuning and FIR filter design	Cui et al. (2018)	2018
Differential evolution (DEMBO)	Discounted {0-1}knapsack problem	Feng et al. (2017a)	2018
Harmony search (HMBO)	Numerical optimization	Ghetas (2015)	2018
Artificial bee colony (HAM)	Numerical optimization	Ghanem and Jantan (2018)	2018
Artificial bee colony (HAM)	Intrusion detection	Ghanem and Jantan (2019)	2019
Artificial bee colony (HAMBO-CHLD)	Energy saving in MANET	Shukla and Goyal (2019)	2019
Firefly algorithm (MBO-FS)	Numerical optimization	Strumberger et al. (2018)	2018
Simulated annealing (SAMBO)	Numerical optimization	Wang et al. (2019b)	2019

The combinations of DE and local search strategy with the MBO algorithm (DE-LSMBO) is introduced by Cui et al. (2018) to tackle proportional-integral-derivative (PID) tuning, finite impulse response (FIR) filter design, and global numerical optimization. In DE-LSMBO, the individuals in subpopulation1 are generated by a new migration operator with local search strategy in which the information of a newly generated individual comes from a randomly selected individual, and the vector difference between the randomly selected one and the original state of the individual. Additionally, the entire population is updated with a mutation operator and crossover operator of DE as well as a greedy strategy after recombining the two newly generated subpopulations. Computational experiments proved that the local search strategy and DE can enhance the performance of MBO when verified on benchmark functions and PID tuning, FIR filter design.

Feng, Yang, et al. (2017a) solved the DKP problem with a hybrid DEMBO where seven kinds of DE mutation strategies are estimated and the best DE/best/2/bin is selected. A new migration operator is introduced by replacing the original individual update strategy with the mutation operator. When the performance of the DEMBO is evaluated by 30 typical DKP instances, the results demonstrate that the solution's quality and convergence speed have improved noticeably.

15.2.3.2 Hybridization with Harmony Search

Ghetas et al. (2015) improved the performance of the MBO algorithm by incorporating HS into the MBO algorithm (HMBO). With the purpose of increasing the population diversity and improving the solutions with poor fitness, HS is embedded into the butterfly adjusting process of HMBO as a mutation operator. The performance of the HMBO was estimated on 14 standard benchmark functions, which demonstrated that HMBO is superior to the original MBO and the other eight state-of-the-art metaheuristic algorithms.

15.2.3.3 Hybridization with Artificial Bee Colony

Through the analysis of the original MBO, Ghanem and Jantan (2018) draws the conclusion that the MBO possesses a weak exploitation ability due to the large random steps of Lévy flight. On the other hand, the ABC algorithm is proficient in local search. Therefore, a hybrid algorithm of ABC and MBO (HAM) is proposed for solving numerical optimization. The major improvements of HAM include (1) replacing the butterfly adjusting operator of MBO with the employee phase of ABC and (2) executing Lévy flight with a smaller probability. When testing the ABC, MBO, and HAM with eight benchmark functions, the simulation results showed that the HAM is more efficient than ABC and MBO.

Another hybrid method, called ABC algorithm and MBO (HAM), is presented by Ghanem and Jantan (2019) for intrusion detection. The authors enhanced the global search efficiency of the original ABC via integrating the modified butterfly adjusting operator of MBO into the employee phase. The performance of the HAM is verified on three types of intrusion detection evaluation data sets, and the simulation results showed that the hybrid method is more efficient than the other nine classification algorithms.

A novel algorithm HAMBO-CHLD is proposed by Shukla and Goyal (2019) to optimize the energy efficiency in mobile ad hoc networks, in which the ABC algorithm is incorporated with MBO (HAMBO) to form the cluster. The experimental results demonstrate that the proposed load distribution scheme increases the energy efficiency and improves the lifetime of a network.

15.2.3.4 Hybridization with Firefly Algorithm

Strumberger et al. (2018) proposed a hybrid algorithm by combining the advantages of FA and MBO (MBO-FS) to tackle the global numerical optimization problems. The author analyzes the defects of MBO algorithm, that is, monarch butterfly individuals usually move directly to the global optimal in the early search phase when applying the butterfly adjusting operator. Therefore, an improved butterfly adjusting operator is presented by utilizing the flight operator of FA. The results are promising when compared with the original MBO and five other state-of-the-art metaheuristics.

15.2.3.5 Hybridization with Simulated Annealing

Normally, the population-based SI algorithm like MBO is skillful in searching multiple regions of the search space at the same time. However, the original MBO cannot effectively exploit each region of the search space deeply. Conversely, local search techniques like the simulated annealing (SA) algorithm is adept at exploiting a single search region. Hence, Wang et al. (2019b) introduced a new MBO algorithm with SA strategy to address 14 continuous nonlinear functions (SAMBO), where the SA is embodied in the migration operator and butterfly adjusting operator. The experimental results show that the proposed SAMBO performs significantly better than the original MBO on most benchmark functions.

15.2.4 Variants of MBO Algorithm

To enhance the optimization performance of the original MBO algorithm, or to tackle the optimization problems in different fields, several typical MBO variants have been proposed by researchers and discussion of these variants is presented herein. Table 15.3 records the basic information about these variants.

15.2.4.1 Binary MBO Algorithm

Feng et al. (2017b) introduced a transfer function method to establish the mapping relationship in continuous space and binary space, which integrated into the original MBO algorithm and generated a binary version MBO algorithm (BMBO) for tackling the 0-1 KP problems. The proposed BMBO outperforms four other methods when evaluated on 28 0-1 KP instances.

Feng et al. (2019) proposed a novel MBO algorithm embedded with the global position update operator for solving large-scale 0-1 KP problems. The comprehensive comparative experiments of GMBO with seven optimization algorithms is conducted, and the results demonstrate that GMBO can obtain better solutions on almost all of the two groups of 0-1 KP instances.

15.2.4.2 Discrete MBO Algorithm

Application of the discrete MBO algorithm (DMBO) for the Chinese traveling salesman problem (TSP) (CTSP) problem is proposed by Wang et al. (2016) where the original MBO method is

TABLE 15.3
The Variants of MBO Algorithm

Name	Problem	Author	Year
Binary MBO	0-1 knapsack problem	Feng et al. (2017b)	2017
Binary MBO	0-1 knapsack problem	Feng et al. (2019)	2019
Discrete MBO	Chinese traveling salesman problem (TSP)	Wang et al. (2016)	2016
Quantum MBO	Path planning navigation problem	Yi et al. (2020)	2017
Multi-objective MBO	Multi-objective optimization	Strumberger et al. (2018)	2018

discretized via round function. Additionally, the authors make a more in-depth investigation into the influence of the butterfly adjusting rate (*BAR*). The experimental results clearly show that the DMBO method is a promising optimization technique when compared with the other three methods.

15.2.4.3 Quantum MBO Algorithm

The performance of the MBO should be enhanced if quantum computing theory is embedded. Yi et al. (2020) introduced the quantum-inspired MBO algorithm (QMBO) for uninhabited combat air vehicle (UCAV) path planning navigation problem. The utilization of the quantum operators is to update some of the worst individuals. The experimental results demonstrate that QMBO can find a much shorter UCAV path than the original MBO.

15.2.4.4 Multi-Objective MBO Algorithm

The literature shows extensive and in-depth research on the MBO algorithm for solving single-objective problems. Strumberger et al. (2018) extended MBO to tackle multi-objective optimization problems. The authors proposed a modified and a hybridized MBO variant, namely, MMBO and MBO-FS. In MMBO, note that the crossover operator of GA was utilized on the current best solutions in the whole population in later phases of evolution. In MBO-FS, an adjusting MBO-FS operator based on the FA algorithm is integrated into the butterfly adjusting operator. Four multi-objective benchmarks were employed to verify the performance of the proposed two algorithms and found that MBO-FS outperformed the other comparative algorithms.

15.3 OPPOSITION-BASED LEARNING MBO WITH GAUSSIAN PERTURBATION FOR THE 0-1 KP

15.3.1 Hybrid Encoding Scheme

0-1 KP is a discrete optimization problem based on binary value. It is natural that the solution space is composed of a number of discrete points rather than contiguous area. In MBO, monarch butterfly individuals are real vectors; moreover, the migration operator and butterfly adjusting operator are defined on the operation of the real number field. Hence, the closure of two operators on a discrete optimization problem cannot be guaranteed, that is to say, MBO is unable to directly solve a discrete problem. In this chapter, a hybrid encoding scheme (Feng et al. 2017) is employed without changing the search strategy of MBO.

In OMBO, each monarch butterfly individual is expressed as a two-tuple <X, Y>. Here, two operators of MBO still perform the search on real space composed by X, while binary vectors Y constitute solution space.

Furthermore, it is necessary that a surjective function g should realize the mapping relationship from X to the corresponding Y.

$$g(x) = \begin{cases} 1 & if\ sig(x) \geq 0.5 \\ 0 & else \end{cases} \tag{15.5}$$

where, $sig(x) = 1/(1+e^{-x})$ is the sigmoid function.

15.3.2 Individual Repair and Optimization Method Based on Greedy Strategy

When solving constraint problems with EAs, infeasible solution individual will naturally occur. There are usually four main methods to handle this kind of problem, i.e., penalty function method, repair method, purist method, and separatist method (Runarsson and Yao 2000).

Previous research work on 0-1 KP suggests that the repair method based on greedy strategy is more effective than the penalty function method, especially for large-scale 0-1 KP.

He et al. (2016) proposed a two-stage repair and optimization method based on the greedy strategy. In the first stage, all the infeasible solution individuals are repaired to feasible individuals with greedy modification operator (GMO). In the second stage, all the feasible individuals are further optimized with greedy optimization operator (GOO). The basic steps are as follows:

Step 1: Sort all items in the non-increasing order of p_i / w_i and the indexes of the items are saved in array $H[1], H[2], \ldots, H[n]$, respectively, *i.e.*, $p_{H[1]} / w_{H[1]} \geq p_{H[2]} / w_{H[2]} \geq \ldots \geq p_{H[n]} / w_{H[n]}$.

Step 2: Given an infeasible solution individual $X = \{x_1, x_2, \ldots, x_n\} \in \{0,1\}^n$, the repair process is performed with GMO.

Step 3: Given a normal encoding individual $X = \{x_1, x_2, \ldots, x_n\} \in \{0,1\}^n$, the optimization process is performed with GOO.

OMBO FOR 0-1 KP PROBLEM

Begin

Step 1: Sorting. Sort all items in the non-ascending order by p_i / w_i, $1 \leq i \leq n$ and the indexes of items are stored in array $H[1...n]$.
Step 2: Initialization. Generate NP monarch butterfly individuals randomly $\{<X_1, Y_1>, <X_2, Y_2>, \ldots, <X_{NP}, Y_{NP}>\}$. Calculate the fitness of each individual, $f(Y_i)$, $1 \leq i \leq NP$. Divide the entire population into subpopulation_1 and subpopulation_2. Set the maximum iteration number *MaxGen* and iteration number g=1. Set the generation interval of recombining two subpopulations *RG*. Set the *Ga*=0.2.
Step 3: While(stopping criterion is not satisfied)
Evaluate the fitness of the swarm and record the $<X_{gbest}, Y_{gbest}>$.
Update subpopulation_1 with migration operator.
Update subpopulation_2 with butterfly adjusting operator.
Update the worst *Ga*NP* individuals with Gaussian perturbation.
Utilize GMO to repair the infeasible solutions.
Utilize GOO to improve the feasible solutions.
Keep best solutions.
Find the current best solution (Y_{gbest}, $f(Y_{gbest})$).
g=g+1.
if g>=*MaxGen*/2
Perform OBL on the latter half of the individuals.
Utilize GMO to repair the infeasible solutions.
Utilize GOO to improve the feasible solutions.
end if
if Mod(g, *RG*)=0
Subpopulation_1 and subpopulation_2 recombine to one.
Divide the whole population into two subpopulations according to fitness
end if
Step 5: end while
Step 6: Output the best results.

End

15.3.3 The Procedure of OMBO for the 0-1 KP

In this section, the detailed procedure of OMBO for 0-1 KP is illustrated. There are four main operation processes.

In the migration process, each monarch butterfly individual in subpopulation_1 is updated by obtaining more information about the entire population. This process can be seen as exploitation. In the butterfly adjusting process, the excellent genes of the global best individual can be inherited to the next generation. Additionally, Lévy flights also come into play when exploring the search space. This process can be regarded as exploration in which new solutions may be found in the unknown regions of the search space. In the OBL process, OBL is used by the latter half individuals of the whole population in the late stage of evolution according to Eq. (15.6).

$$H'(x_i)=\left\{x_i' \middle| x_i'=r*(a_i+b_i)-x_i, x_i\in R\wedge x_i\in[a_i,b_i], i\in\{1,2,\ldots,n\}\right\} \tag{15.6}$$

where r is *a* uniform random real number in [0, 1]. It should be noted that if x_i' is out of the original space, x_i' is reassigned a new random number draw from uniform distribution within the current interval $[a_i, b_i]$. Here, a represents the lower bound on each dimension and b is the upper bound on each dimension.

In the Gaussian perturbation process, this strategy is applied to the worst *Ga*NP* individuals in each generation for the purpose of escaping the local optimum. Different from the original MBO, the newly generated two subpopulations merge into one at a certain generation rather than at each generation.

15.3.4 Simulation Experiments

15.3.4.1 Experimental Data Set

In this section, three kinds of correlation characteristics of profit and weight are shown in Table 15.4.

Here function *Rand* (*a*, *b*) returns a random integer drawn from the uniform distribution in interval [*a*, *b*]. Therefore, each group includes five large-scale 0-1 KP instances generated randomly whose dimensions are 800, 1000, 1200, 1500, and 2000, respectively. These 15 instances are represented by KP1–KP15, respectively. Additionally, the maximum capacity of the KP equals 0.75 times of the sum of the weights.

TABLE 15.4
The Correlation Characteristics of Profit and Weight

Correlation	w_i	p_i	C
Uncorrelated instances	$Rand(10,100)$	$Rand(10,100)$	$0.75*\sum_{i=1}^{n} w_i$
Weakly correlated instances	$Rand(10,100)$	$Rand(w_i-10, w_i+10)$	$0.75*\sum_{i=1}^{n} w_i$
Strongly correlated instances	$Rand(10,100)$	w_i+10	$0.75*\sum_{i=1}^{n} w_i$

15.3.4.2 A Summary of the Comparison Algorithms Involved

A comprehensive comparison study was conducted to investigate the optimization capability of OMBO, along with the convergence speed and numerical stability. Seven typical optimization methods, including ABC, CS, DE, GA, FA (Cui et al. 2017), SFLA (Eusuff and Lansey 2003a), and MBO (Wang et al. 2019), are chosen for solving 15 0-1 KP instances. In our experiments, the parameter settings are recorded in Table 15.5.

To reflect the fairness of the comparison process, the population sizes of eight methods are set to 50. Additionally, the maximum running time is 8 seconds for the instances with dimensions less than 1500, and others are set to 10 seconds. To carry out a more comprehensive investigation on each method, 50 independent runs are conducted for each approach on each instance.

We use C++ as the programming language and run the codes on a PC with Intel (R) Core(TM) i5-2415M CPU, 2.30GHz, 4GB RAM.

15.3.4.3 Comparisons of OMBO With Other Methods on Three Groups of 0-1 KP Instances

In this section, the proposed OMBO method is evaluated in conjunction with the quality of solution, convergence speed, and stability by using 15 large-scale 0-1 KP instances. The comparative results achieved by eight methods on uncorrelated, weakly correlated, and strongly correlated 0-1 KP instances are presented in Tables 15.6–15.8, respectively.

From Tables 15.6–15.8, it can be clearly seen that OMBO significantly outperforms the other seven methods on KP1–KP15 regarding the best, the mean, the worst, and the median values. Additionally, considering the stability, OMBO has excellent performance, which is reflected in a standard deviation not more than 1.0 for most problems, except KP3, KP5, KP6, and KP7. After careful observations about FA and SFLA, there is almost no difference in numerical stability if

TABLE 15.5
The Parameter Settings of Eight Methods on 0-1 KP

	Parameter	Value
ABC	*Number of food sources*	25
	Maximum search times	100
CS	P_a	0.25
DE	*Crossover probability*	0.9
	Amplification factor	0.3
GA	*Crossover probability*	0.6
	Mutation probability	0.001
FA	*Alpha*	0.2
	Beta	1.0
	Gamma	1.0
SFLA	*M*	5
	N	10
MBO	*Migration ratio*	3/12
	Migration period	1.4
	Butterfly adjusting rate	1/12
	Max step	1.0
OMBO	*The same parameters as MBO*	
	Ga 0.2	

TABLE 15.6
The Experimental Results of Eight Methods on Uncorrelated 0-1 KP Instances

		ABC	CS	DE	GA	FA	SFLA	MBO	OMBO
KP1	Best	39816	40445	39486	39190	40683	40683	40276	40686
	Mean	39639	39602	39323	38838	40683	40683	40036	40684
	Worst	39542	39411	39154	38274	40683	40683	39839	40683
	Median	39641	39527	39326	38867	40683	40683	40029	40683
	Std	55.5	218.11	80.60	196.70	0.00	0.00	100.34	0.86
KP2	Best	49374	50104	49303	48955	50590	50590	50023	50592
	Mean	49256	49211	48945	48384	50590	50590	49743	50590
	Worst	49150	49056	48696	47809	50590	50590	49411	50590
	Median	49256	49154	48940	48416	50590	50590	49734	50590
	Std	58.56	205.08	111.08	256.69	0.00	0.00	133.40	0.70
KP3	Best	60222	60490	59921	59578	61840	61840	61090	61845
	Mean	60059	59938	59645	58996	61840	61840	60732	61842
	Worst	59867	59764	59435	58106	61840	61840	60401	61840
	Median	60046	59926	59643	59061	61840	61840	60734	61843
	Std	82.28	120.76	114.17	362.53	0.00	0.00	163.76	1.82
KP4	Best	74959	75828	74671	74372	77031	77031	75405	77033
	Mean	74742	74666	74319	73584	77031	77031	75072	77031
	Worst	74571	74472	74077	72477	77031	77031	74815	77031
	Median	74723	74583	74310	73612	77031	77031	75068	77031
	Std	90.07	245.28	113.92	414.02	0.00	0.00	149.57	0.56
KP5	Best	99353	99248	98943	98828	102313	102313	99946	102316
	Mean	99035	98926	98645	97765	102313	102313	99512	102314
	Worst	99822	98706	98330	96830	102313	102313	99017	102313
	Median	99021	98922	98648	97852	102313	102313	99497	102313
	Std	130.80	124.58	154.40	480.15	0.00	0.00	187.15	1.11

TABLE 15.7
The Experimental Results of Eight Methods on Weakly Correlated 0-1 KP Instances

		ABC	CS	DE	GA	FA	SFLA	MBO	OMBO
KP6	Best	34706	34975	34629	34585	35068	35064	34850	35069
	Mean	34675	34676	34588	34476	35064	35064	34795	35067
	Worst	34650	34621	34549	34361	35064	35064	34724	35064
	Median	34672	34663	34586	34481	35064	35064	34797	35068
	Std	16.00	65.25	20.93	60.91	0.96	0.00	31.41	1.47
KP7	Best	43321	43708	43251	43172	43786	43781	43487	43786
	Mean	43275	43326	43187	43049	43781	43781	43425	43785
	Worst	43243	43215	43140	42901	43781	43781	43349	43782
	Median	43271	43266	43181	43067	43781	43781	43421	43785
	Std	18.74	143.50	23.94	74.36	1.13	0.00	31.78	1.03
KP8	Best	52061	52848	51900	51460	53552	53552	52720	53553
	Mean	51876	51838	51547	50945	53552	53552	52449	53552
	Worst	51711	51617	51289	50112	53552	53552	52185	53552
	Median	51873	51768	51538	50991	53552	53552	52455	53552
	Std	79.72	260.46	123.67	281.41	0.00	0.00	111.26	0.14

TABLE 15.7 (*Continued*)
The Experimental Results of Eight Methods on Weakly Correlated 0-1 KP Instances

		ABC	CS	DE	GA	FA	SFLA	MBO	OMBO
KP9	Best	64864	65549	64770	64769	65708	65708	65144	65710
	Mean	64806	64932	64692	64535	65708	65708	65041	65709
	Worst	64752	64749	64620	64315	65708	65708	64941	65708
	Median	64802	64844	64689	64543	65708	65708	65050	65709
	Std	25.45	245.06	35.66	85.75	0.00	0.00	48.66	0.52
KP10	Best	115305	116597	114929	114539	118200	118200	116466	118200
	Mean	114922	114879	114462	113674	118200	118200	115998	118200
	Worst	114586	114560	114199	112681	118200	118200	115273	118200
	Median	114927	114784	114421	113699	118200	118200	115982	118200
	Std	123.59	428.93	160.77	405.23	0.00	0.00	248.70	0.00

TABLE 15.8
The Experimental Results of Eight Methods on Strongly Correlated 0-1 KP Instances

		ABC	CS	DE	GA	FA	SFLA	MBO	OMBO
KP11	Best	40127	40127	40137	40069	40166	40109	40137	40167
	Mean	40116	40108	40119	40023	40149	40109	40119	40167
	Worst	40107	40096	40087	39930	40109	40109	40102	40167
	Median	40117	40107	40117	40025	40150	40109	40117	40167
	Std	4.52	6.59	10.19	31.12	14.77	0.00	7.18	0.00
KP12	Best	49390	49393	49363	49333	49443	49407	49393	49443
	Mean	49376	49364	49340	49287	49424	49407	49379	49443
	Worst	49363	49353	49323	49231	49407	49407	49363	49441
	Median	49373	49363	49339	49294	49425	49407	49383	49443
	Std	5.61	6.80	8.31	29.76	10.80	0.00	9.94	0.34
KP13	Best	60567	60559	60545	60520	60639	60556	60580	60640
	Mean	60554	60543	60518	60451	60621	60556	60562	60640
	Worst	60540	60533	60498	60391	60556	60556	60539	60640
	Median	60554	60541	60517	60449	60623	60556	60560	60640
	Std	5.54	5.39	10.16	29.87	15.83	0.00	10.77	0.00
K14	Best	74822	74837	74778	74766	74932	74873	74849	74932
	Mean	74805	74794	74759	74689	74916	74873	74822	74932
	Worst	74792	74779	74737	74606	74873	74873	74778	74931
	Median	74802	74792	74759	74694	74919	74873	74822	74932
	Std	6.85	9.19	10.23	37.20	13.73	0.00	14.70	0.14
KP15	Best	99523	99517	99501	99461	99683	99631	99573	99683
	Mean	99506	99489	99459	99382	99663	99631	99536	99683
	Worst	99490	99473	99436	99305	99631	99631	99496	99679
	Median	99503	99489	99456	99383	99665	99631	99533	99683
	Std	7.29	8.19	14.03	38.42	16.45	0.00	15.63	0.58

we only consider uncorrelated and weakly correlated KP instances. From Table 15.5, the stable performance of FA shows a slight decrease compared with OMBO and SFLA, which implicitly shows that the optimization ability of EA is affected by the type of KP instances to some degree. A similar phenomenon can be observed in other algorithms, such as ABC and MBO.

In summary, Tables 15.6–15.8 indicate that the OMBO method is overwhelmingly superior. In addition, the worst values obtained by OMBO even surpass the best values of the other seven methods.

To compare the performance of each algorithm more clearly, the rank of eight methods according to the best values, the mean values, and the worst values is illustrated in Figure 15.5. The detailed rank based on the three criteria is as follows:

$$GA < DE < ABC < CS < MBO < SFLA < FA < OMBO \tag{15.7}$$

$$GA < DE < CS < ABC < MBO < SFLA < FA < OMBO \tag{15.8}$$

$$GA < DE < CS < ABC < MBO < SFLA < FA < OMBO \tag{15.9}$$

In this chapter, two special evaluation metrics are put forward: (1) the approximate ratio of the optimal solution value *(Opt)* to the best approximate solution value *(Best)* (abbreviated as *ARB*) and (2) the approximate ratio of the optimal solution value *(Opt)* to the mean approximate solution value (*Mean*) (abbreviated as *ARM*).

$$ARB = \frac{Opt}{Best} \geq 1.0 \tag{15.10}$$

$$ARM = \frac{Opt}{Mean} \geq 1.0 \tag{15.11}$$

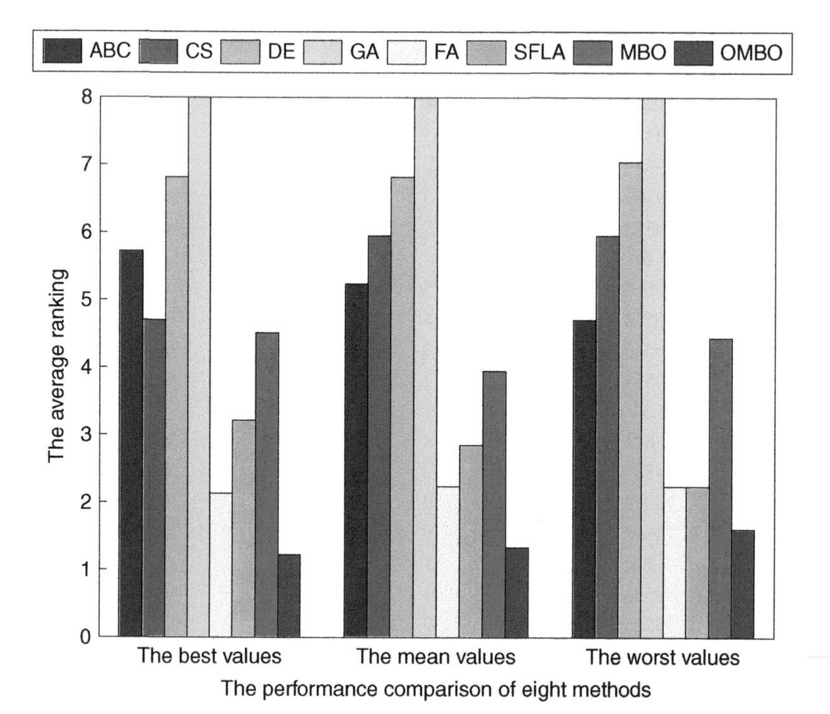

FIGURE 15.5 Average ranking on three evaluation criteria.

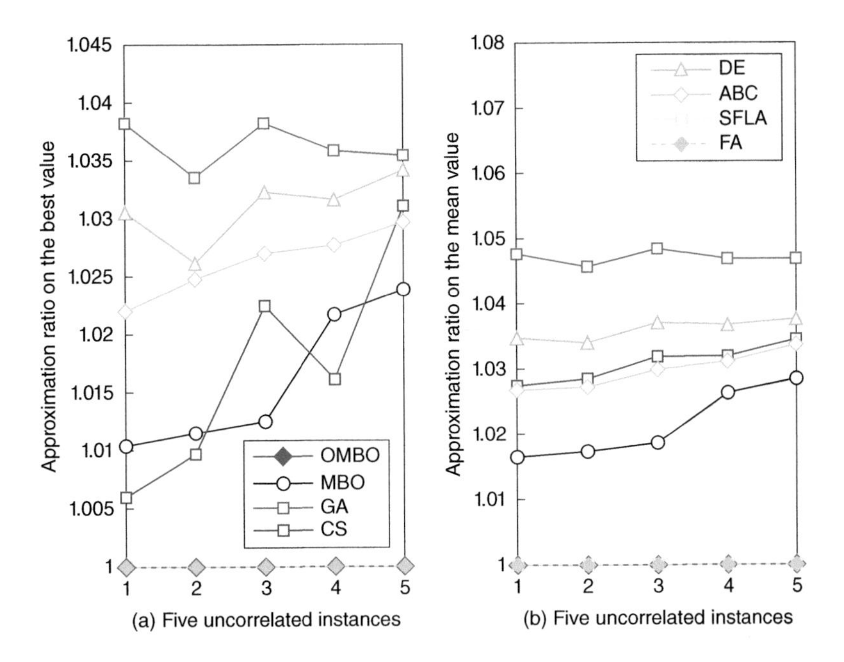

FIGURE 15.6 Comparison on KP1–KP5 of (a) ARB and (b) ARM.

Figures 15.6–15.8 illustrate the comparative results of *ARB* and *ARM* among eight methods on three instances. Figures 15.6–15.8 show that OMBO, FA, and SFLA have better *ARB* and *ARM* values, which are equal or nearly equal to 1.0 in 10 instances. GA obtains the worst results compared with the other seven methods. In particular, the *ARB* and *ARM* of FA and SFLA in Figure 15.8 are worse than that of OMBO. In addition, the curves of Figures 15.6 and 15.8 show

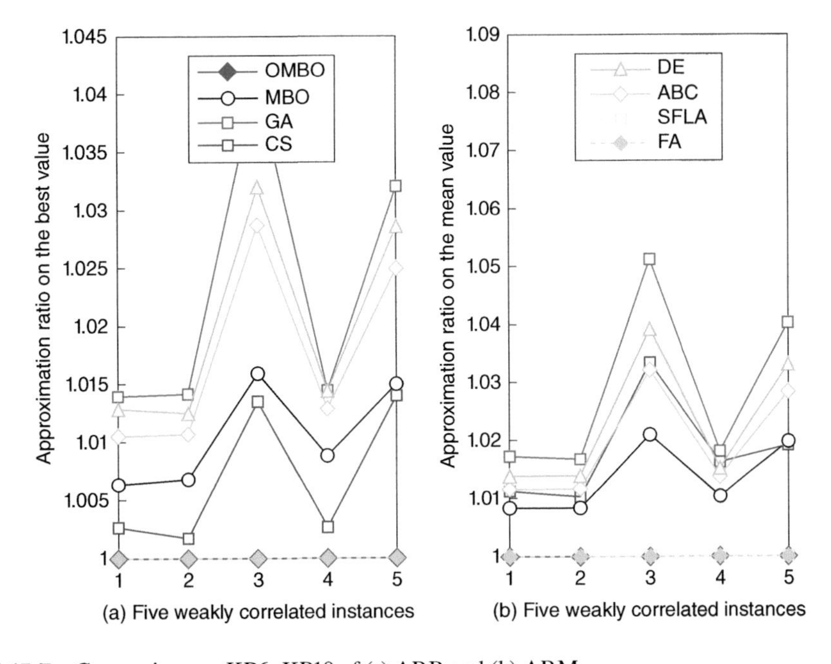

FIGURE 15.7 Comparison on KP6–KP10 of (a) ARB and (b) ARM.

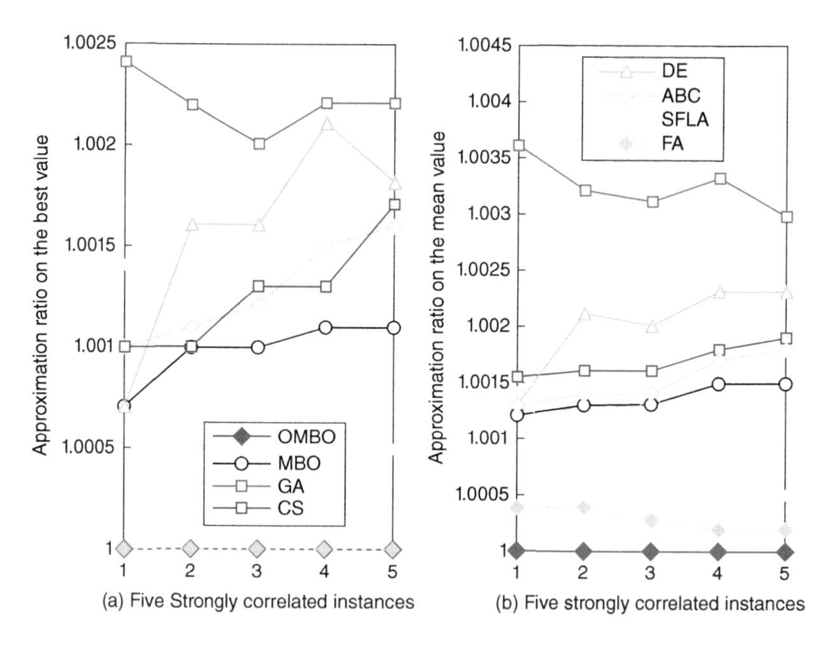

FIGURE 15.8 Comparison on KP11–KP15 of (a) ARB and (b) ARM.

a gentle upward trend along with the extending of the 0-1 KP problem, whereas this phenomenon does not appear in Figure 15.7.

Boxplots based on statistical view are shown in Figures 15.9–15.11, which are more appropriate to observe the obvious advantage of OMBO than other methods. Similarly, KP5, KP10, and KP15 are specially selected. Apparently, the boxplot of OMBO is always at the highest point in every figure and is close to a straight line, which demonstrates that OMBO has the

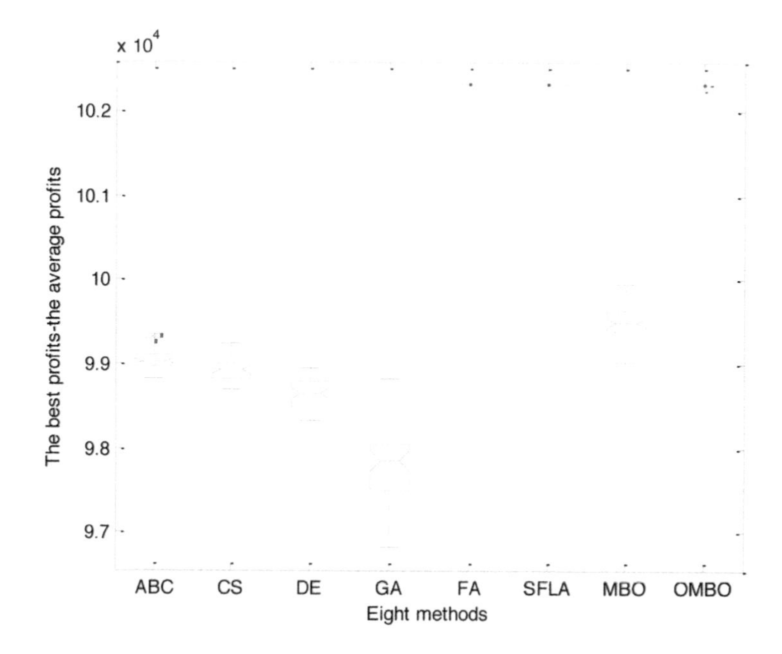

FIGURE 15.9 Boxplot of the best values on KP5 in 50 runs.

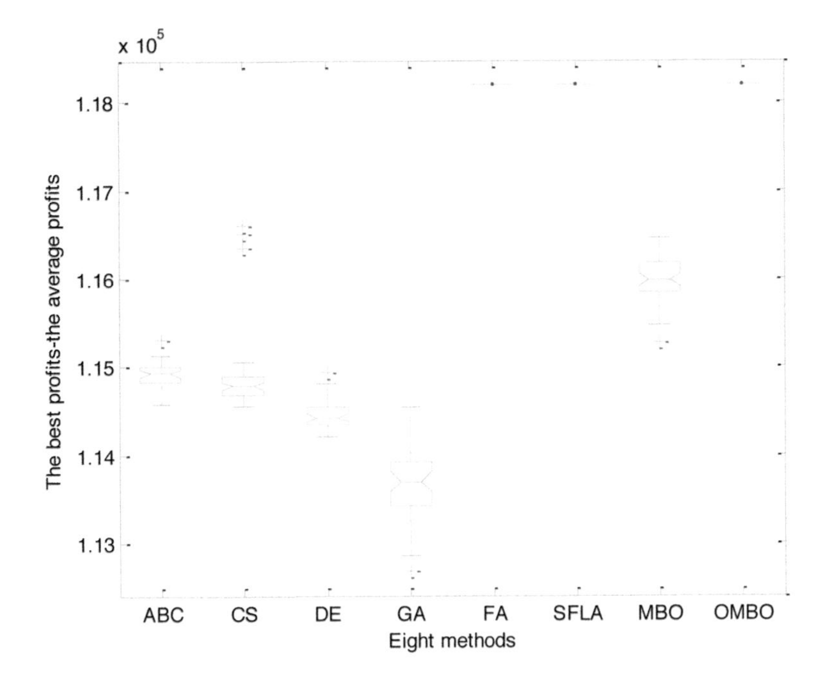

FIGURE 15.10 Boxplot of the best values on KP10 in 50 runs.

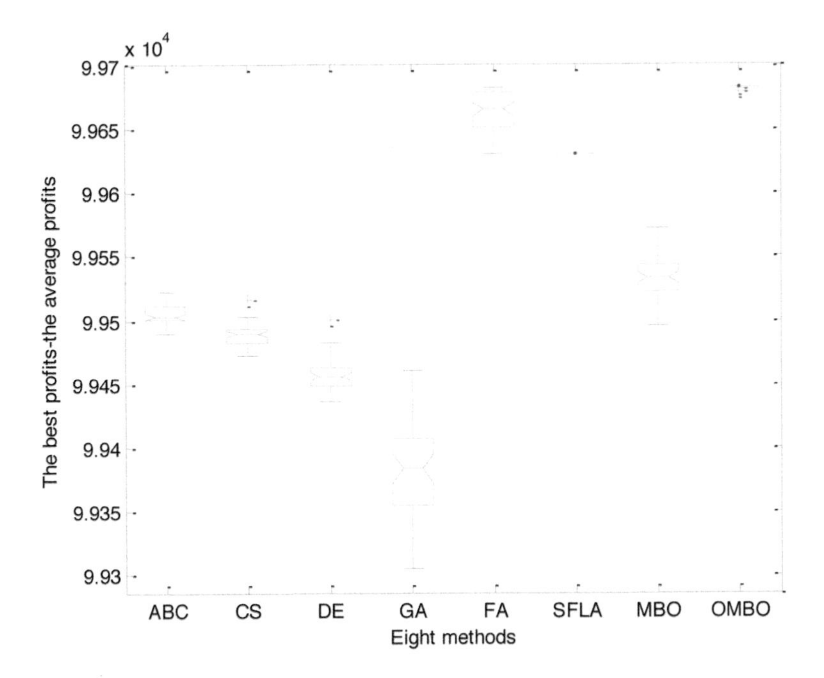

FIGURE 15.11 Boxplot of the best values on KP15 in 50 runs.

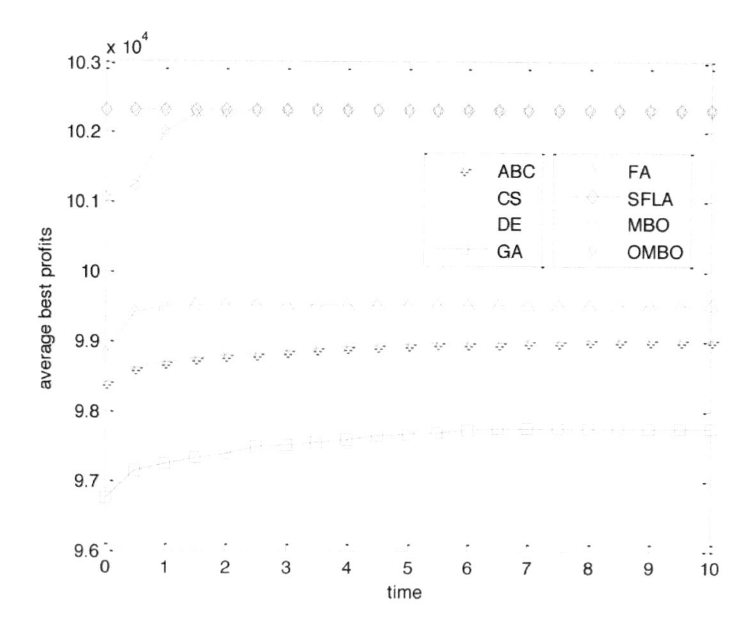

FIGURE 15.12 The convergence graph on KP5 in 10 seconds.

least difference degree of the best value, the worst value, the mean value, and the median value compared with the other seven methods. In addition, OMBO, FA, and SFLA are the three most effective methods among the eight methods in our experimental studies.

To clearly show the convergence property of OMBO, the evolution curve of the eight methods on the three typical large-scale 0-1 KP instances mentioned above are illustrated in Figures 15.12–15.14. In these figures, each curve reflects the changes of the average fitness based on 50 independent runs.

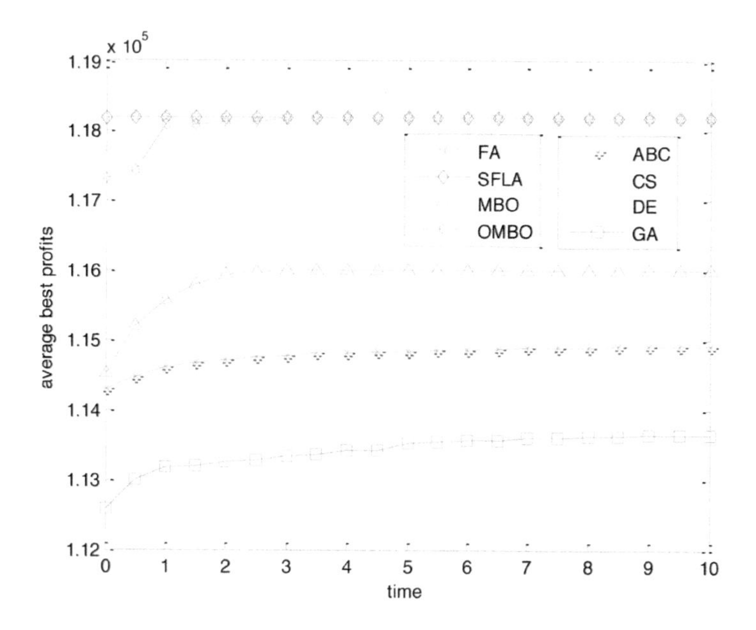

FIGURE 15.13 The convergence graph on KP10 in 10 seconds.

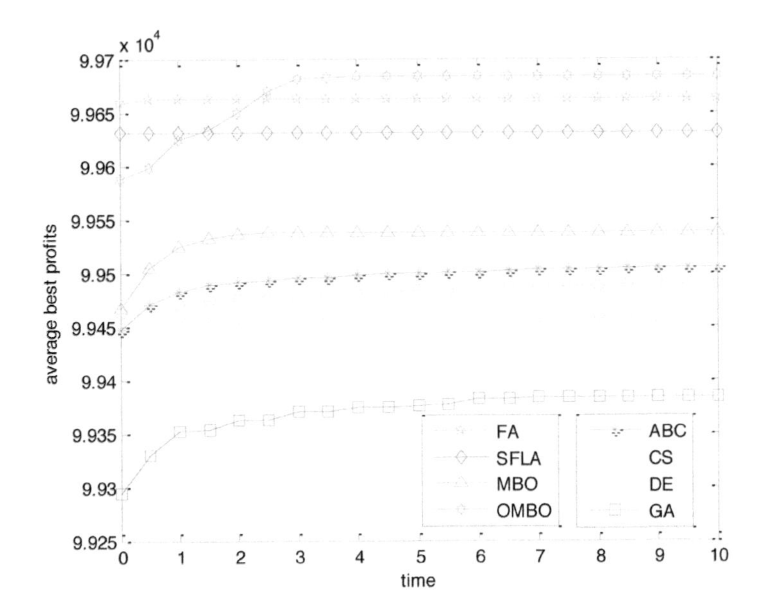

FIGURE 15.14 The convergence graph on KP15 in 10 seconds.

Figures 15.12–15.14 illustrate the similarity of these curves. All of these figures have a common feature, i.e., FA, SFLA, and OMBO are capable of finding a better solution at the beginning of the evolution. Although the initial solution obtained by OMBO is just a little worse than FA and SFLA, OMBO converges quickly to the best value in about 1 second. In short, OMBO, FA, and SFLA have the identical final value.

From Tables 15.6–15.8 and Figures 15.5–15.14, we can conclude that our proposed OMBO algorithm has some excellent characteristics of better optimization ability, numerical stability, and good convergence. OMBO is a promising variant of MBO and has a strong ability to solve the large-scale 0-1 KP problem. In other words, OBL and Gaussian perturbation have played an important role in enhancing the performance of OMBO.

15.4 CODE OF MONARCH BUTTERFLY OPTIMIZATION

In the previous chapter, we introduced the application of the MBO algorithm in solving the 0-1 KP problem. Below I intend to show the specific code of this algorithm, Note that the source codes of the MBO can be obtained at https://github.com/ggw0122/Monarch-Butterfly-Optimization and http://www.mathworks.com/matlabcentral/fileexchange/50828-monarch-butterfly-optimization.

We use MATLAB as the programming language and run the codes on a PC with Intel (R) Core(TM) i7-10750M CPU, 2.60GHz, 16GB RAM.

MBO.M

```
%% Notes: Different run may generate different solutions, this is
determined by the nature of metaheuristic algorithms.
%%
function [MinCost] = MBO(ProblemFunction, DisplayFlag, RandSeed)
```

```
% Monarch Butterfly Optimization (MBO) software for minimizing a
general function
% The fixed generation is considered as termination condition.
% INPUTS: ProblemFunction is the handle of the function that
returns
% the handles of the initialization, cost, and feasibility
functions.
% DisplayFlag = true or false, whether or not to display and plot
results.
% ProbFlag = true or false, whether or not to use probabilities to
update emigration rates.
% RandSeed = random number seed
% OUTPUTS: MinCost = array of best solution, one element for each
generation
% Hamming = final Hamming distance between solutions
%% CAVEAT: The "ClearDups" function that is called below replaces
duplicates with randomly-generated individuals, but it does not
then recalculate the cost of the replaced individuals.
%%
tic
if ~exist('ProblemFunction', 'var')
   ProblemFunction = @Ackley;
end
if ~exist('DisplayFlag', 'var')
   DisplayFlag = true;
end
if ~exist('RandSeed', 'var')
   RandSeed = round(sum(100*clock));
end
[OPTIONS, MinCost, AvgCost, InitFunction, CostFunction,
FeasibleFunction, MaxParValue, MinParValue, Population] =
Init(DisplayFlag, ProblemFunction, RandSeed);
% Initial parameter setting
   Keep = 2;
% elitism parameter: how many of the best habitats to keep from
    one generation to the next
  maxStepSize = 1.0;            %Max Step size
  partition = OPTIONS.partition;
  numButterfly1 = ceil(partition*OPTIONS.popsize);              % NP1
in paper
  numButterfly2 = OPTIONS.popsize - numButterfly1;              % NP2
    in paper
  period = 1.2;            % 12 months in a year
  Land1 = zeros(numButterfly1, OPTIONS.numVar);
  Land2 = zeros(numButterfly2, OPTIONS.numVar);
  BAR = partition; % you can change the BAR value in order to get
    much better performance
% End of Initial parameter setting
% Begin the optimization loop
for GenIndex = 1 : OPTIONS.Maxgen
% Elitism Strategy
% Save the best monarch butterflies in a temporary array.
```

```
for j = 1 : Keep
    chromKeep(j,:) = Population(j).chrom;
    costKeep(j) = Population(j).cost;
end
% End of  Elitism Strategy
% Divide the whole population into two subpopulations
%% Divide the whole population into Population1 (Land1) and
Population2 (Land2) according to their fitness. %%
% The monarch butterflies in Population1 are better than or equal
to Population2.
% Of course, we can randomly divide the whole population into
Population1 and Population2.
% We do not test the different performance between two ways.
for popindex = 1 : OPTIONS.popsize
  if popindex <= numButterfly1
     Population1(popindex).chrom = Population(popindex).chrom;
  else
     Population2(popindex-numButterfly1).chrom =
Population(popindex).chrom;
  end
end
% End of Divide the whole population into two subpopulations
% Migration operator
for k1 = 1 : numButterfly1
  for parnum1 = 1 : OPTIONS.numVar
    r1 = rand*period;
  if r1 <= partition
    r2 = round(numButterfly1 * rand + 0.5);
    Land1(k1,parnum1) = Population1(r2).chrom(parnum1);
  else
    r3 = round(numButterfly2 * rand + 0.5);
    Land1(k1,parnum1) = Population2(r3).chrom(parnum1);
  end
end
%% for parnum1
  NewPopulation1(k1).chrom =  Land1(k1,:);
end
% for k1
% End of Migration operator
% Evaluate NewPopulation1
  SavePopSize = OPTIONS.popsize;
  OPTIONS.popsize = numButterfly1;
% Make sure each individual is legal.
  NewPopulation1 = FeasibleFunction(OPTIONS, NewPopulation1);
% Calculate cost
  NewPopulation1 = CostFunction(OPTIONS, NewPopulation1);
  OPTIONS.popsize = SavePopSize;
% End of Evaluate NewPopulation1
% Butterfly adjusting operator
for k2 = 1 : numButterfly2
  scale = maxStepSize/(GenIndex^2); %Smaller step for local walk
  StepSzie = ceil(exprnd(2*OPTIONS.Maxgen,1,1));
```

```
  delataX = LevyFlight(StepSzie,OPTIONS.numVar);
for parnum2 = 1:OPTIONS.numVar,
  if (rand >= partition)
  Land2(k2,parnum2) = Population(1).chrom(parnum2);
  else
    r4 = round(numButterfly2*rand + 0.5);
    Land2(k2,parnum2) =  Population2(r4).chrom(1);
  if (rand > BAR) % Butterfly-Adjusting rate
    Land2(k2,parnum2) =  Land2(k2,parnum2) +
      scale*(delataX(parnum2)-0.5);
    end
  end
end
% for parnum2
  NewPopulation2(k2).chrom =  Land2(k2,:);
end
% for k2
% End of Butterfly adjusting operator
% Evaluate NewPopulation2
  SavePopSize = OPTIONS.popsize;
  OPTIONS.popsize = numButterfly2;
% Make sure each individual is legal.
  NewPopulation2 = FeasibleFunction(OPTIONS, NewPopulation2);
% Calculate cost
  NewPopulation2 = CostFunction(OPTIONS, NewPopulation2);
  OPTIONS.popsize = SavePopSize;
% End of Evaluate NewPopulation2
%Combine two subpopulations into one and rank monarch butterflies
% Combine Population1 with Population2 to generate a new Population
  Population = CombinePopulation(OPTIONS, NewPopulation1,
    NewPopulation2);
% Sort from best to worst
  Population = PopSort(Population);
% End of Combine two subpopulations into one and rank monarch
   butterflies
% Elitism Strategy
% Replace the worst with the previous generation's elites.
n = length(Population);
for k3 = 1 : Keep
   Population(n-k3+1).chrom = chromKeep(k3,:);
   Population(n-k3+1).cost = costKeep(k3);
   end
% end for k3
% End of  Elitism Strategy
% Precess and output the results
% Sort from best to worst
  Population = PopSort(Population);
% Compute the average cost
[AverageCost, nLegal] = ComputeAveCost(Population);
% Display info to screen
  MinCost = [MinCost Population(1).cost];
  AvgCost = [AvgCost AverageCost];
```

```
if DisplayFlag
    disp(['The best and mean of Generation # ', num2str(GenIndex),
      ' are ',...
    num2str(MinCost(end)), ' and ', num2str(AvgCost(end))]);
  end
% End of Precess and output the results
end
% end for GenIndex
  Conclude1(DisplayFlag, OPTIONS, Population, nLegal, MinCost,
    AvgCost);
toc
% End of Monarch Butterfly Optimization implementation
  function [delataX] = LevyFlight(StepSize, Dim)
% Allocate matrix for solutions
  delataX = zeros(1,Dim);
% Loop over each dimension
for i=1:Dim
% Cauchy distribution
  fx = tan(pi * rand(1,StepSize));
  delataX(i) = sum(fx);
end
```

15.5 DISCUSSION AND CONCLUSION

By simulating the migration behavior of the monarch butterflies in nature, a new kind of nature-inspired metaheuristic algorithm, called the monarch butterfly optimization is presented for continuous optimization problems in this chapter. In MBO, all the monarch butterfly individuals are idealized and only located in two lands: southern Canada and the northern United States (land 1) and Mexico (land 2). Accordingly, the positions of the monarch butterflies are updated in two ways. First, the offspring are generated (position updating) by the migration operator, which can be adjusted by the migration ratio. Second, for other butterflies, their positions are tuned by means of a butterfly adjusting operator. In other words, the search direction of the monarch butterfly individuals in the MBO algorithm are mainly determined by the migration operator and butterfly adjusting operator. With the aim of showing the performance of the MBO method, it is compared with five other metaheuristic algorithms through 38 benchmark problems. The results show that the MBO method is able to find better function values on most benchmark problems than five other metaheuristic algorithms.

In addition, the MBO algorithm is simple with no complicated calculation and operators. This makes the implementation of the MBO algorithm easy and fast.

Despite various advantages of the MBO method, the following points should be clarified and focused on in the future research.

1. It is well known that the parameters used in a metaheuristic method have great influence on its performance. In the present work, we make little effort to fine-tune the parameters used in the MBO method. The best parameter settings will be selected through theoretical analyses or empirical experiments.
2. Computational requirements are of vital importance for any metaheuristic method. It is imperative to improve the search speed by analyzing the MBO method.
3. We use only 38 benchmark functions to test our proposed MBO method. In the future, more benchmark problems, especially real-world applications, should be used for

effective implementation of the MBO method, such as image segmentation; constrained optimization; the KP problem; scheduling; dynamic optimization; antenna and microwave design problems; and water, geotechnical, and transport engineering.

4. In the current work, the characteristics of the migration behavior (essentially migration operator and butterfly adjusting operator) are idealized to form the MBO method. In the future, more characteristics, such as swarm, defense against predators, and human interactions, can be idealized and simplified to be added to the MBO method.
5. As discussed in Section 15.4, the MBO method has the absolute advantage over the other five methods on best performance. However, for the average performance, MBO is not the best one among six methods. Furthermore, bad average performance must lead to bad standard deviation. Efforts should be made to improve the average performance by updating the search process.
6. As shown in Section 15.4, the MBO method has the absolute advantage over the other five methods when dealing with high-dimensional functions. However, for the low-dimensional functions, MBO performs equally to or worse than the other five methods. We plan to investigate this and endeavor to find out the reasons by addressing this disadvantage in our future research.
7. In the current work, the performance of the MBO method is experimentally tested only using benchmark problems. The convergence of the MBO method will be analyzed theoretically by dynamic systems and the Markov chain, which can ensure stable implementation of MBO method.

REFERENCES

Arora, Sankalap, and Satvir Singh. 2019. "Butterfly optimization algorithm: a novel approach for global optimization." *Soft Computing* 23 (3):715–734.

Bäck, Thomas. 1996. *Evolutionary Algorithms in Theory and Practice: Evolution Strategies, Evolutionary Programming, Genetic Algorithms*. Oxford, UK: Oxford University Press.

Beyer, H., and H. Schwefel. 2002. *Natural Computing*. Dordrecht Netherlands: Kluwer Academic Publishers.

Cui, Xingyue, Zhe Chen, and Fuliang Yin. 2018. "Differential evolution and local search based monarch butterfly optimization algorithm with applications." *International Journal of Computational Intelligence Systems* 12 (1):149–163.

Cui, Z., B. Sun, G. Wang, Y. Xue, and J. Chen. 2017. "A novel oriented cuckoo search algorithm to improve DV-Hop performance for cyber-physical systems." *Journal of Parallel and Distributed Computing* 103: 42–52.

Cui, Zhihua, Shujing Fan, Jianchao Zeng, and Zhongzhi Shi. 2013. "APOA with parabola model for directing orbits of chaotic systems." *International Journal of Bio-Inspired Computation* 5 (1):67–72.

Cui, Zhihua, and Xiaozhi Gao. 2012. "Theory and applications of swarm intelligence." *Neural Computing and Applications* 21 (2):205–206. doi: 10.1007/s00521-011-0523-8.

Dorigo, M., V. Maniezzo, and A. Colorni. 1996. "Ant system: optimization by a colony of cooperating agents." *IEEE Transactions on Systems, Man, and Cybernetics, Part B: Cybernetics* 26 (1):29–41. doi: 10.1109/3477.484436.

Ehteram, Mohammad, Hojat Karami, Sayed-Farhad Mousavi, Saeed Farzin, and Ozgur Kisi. 2017. "Optimization of energy management and conversion in the multi-reservoir systems based on evolutionary algorithms." *Journal of cleaner production* 168:1132–1142.

Eusuff, M.M., and K. E. Lansey. 2003a. "Optimization of water distribution network design using the Shuffled Frog Leaping Algorithm." *Journal of Water Resources Planning and Management-ASCE* 129 (3):210–225.

Eusuff, Muzaffar M., and Kevin E. Lansey. 2003b. "Optimization of water distribution network design using the shuffled frog leaping algorithm." *Journal of Water Resources Planning and Management* 129 (3):210–225.

Feng, Y., G.-G. Wang, S. Deb, M. Lu, and X.-J. Zhao. 2017. "Solving 0-1 knapsack problem by a novel binary monarch butterfly optimization." *Neural Computing & Applications* 28 (7): 1619–1634.

Feng, Y., J. Yang, Y. He, and G. G. Wang. 2017. "Monarch butterfly optimization algorithm with differential evolution for the discounted {0-1} knapsack problem." *Acta Electronica Sinica* 45 (6): 1343.

Feng, Y., X. Yu, and G.-G. Wang. 2019. "A novel monarch butterfly optimization with global position updating operator for large-scale 0-1 knapsack problems." *Mathematics* 7 (11): 1056.

Feng, Yanhong, Gai-Ge Wang, Suash Deb, Mei Lu, and Xiang Jun Zhao. 2017. "Solving 0-1 knapsack problem by a novel binary monarch butterfly optimization." *Neural Computing and Applications* 28 (7):1–16.

Feng, Yanhong, Gai-Ge Wang, Junyu Dong, and Ling Wang. 2018. "Opposition-based learning monarch butterfly optimization with Gaussian perturbation for large-scale 0-1 knapsack problem." *Computers & Electrical Engineering* 67:454–468.

Feng, Yanhong, Gai-Ge Wang, Wenbin Li, and Ning Li. 2018. "Multi-strategy monarch butterfly optimization algorithm for discounted {0-1} knapsack problem." *Neural Computing and Applications* 30 (10):3019–3036.

Feng, Yanhong, Juan Yang, Congcong Wu, Mei Lu, and Xiang-Jun Zhao. 2018. "Solving 0-1 knapsack problems by chaotic monarch butterfly optimization algorithm with Gaussian mutation." *Memetic Computing* 10 (2):135–150.

Fister Jr, Iztok, Simon Fong, Janez Brest, and Iztok Fister. 2014. "Towards the self-adaptation in the bat algorithm." Proceedings of the 13th IASTED International Conference on Artificial Intelligence and Applications.

Fister Jr, Iztok, Xin-She Yang, Iztok Fister, Janez Brest, and Dušan Fister. 2013. "A brief review of nature-inspired algorithms for optimization." *arXiv preprint arXiv:1307.4186.*

Gandomi, Amir Hossein, and Amir Hossein Alavi. 2011. "Multi-stage genetic programming: A new strategy to nonlinear system modeling." *Information Sciences* 181 (23):5227–5239. doi: 10.1016/j.ins.2011.07.026.

Gandomi, Amir Hossein, and Amir Hossein Alavi. 2012. "Krill herd: a new bio-inspired optimization algorithm." *Communications in Nonlinear Science and Numerical Simulation* 17 (12):4831–4845. doi: 10.1016/j.cnsns.2012.05.010.

Gandomi, Amir Hossein, Xin-She Yang, and Amir Hossein Alavi. 2011. "Mixed variable structural optimization using firefly algorithm." *Computers & Structures* 89 (23–24):2325–2336. doi: 10.1016/j.compstruc.2011.08.002.

Gao, X. Z., and S. J. Ovaska. 2002. "Genetic algorithm training of Elman neural network in motor fault detection." *Neural Computing and Applications* 11 (1):37–44. doi: 10.1007/s005210200014.

Gao, X. Z., S. J. Ovaska, X. Wang, and M. Y. Chow. 2007. "A neural networks-based negative selection algorithm in fault diagnosis." *Neural Computing and Applications* 17 (1):91–98. doi: 10.1007/s00521-007-0092-z.

Gao, X. Z., X. Wang, T. Jokinen, S. J. Ovaska, A. Arkkio, and K. Zenger. 2012. "A hybrid PBIL-based harmony search method." *Neural Computing and Applications* 21 (5):1071–1083. doi: 10.1007/s00521-011-0675-6.

Geem, Zong Woo, Joong Hoon Kim, and Gobichettipalayam Vasudevan Loganathan. 2001. "A new heuristic optimization algorithm: harmony search." *Simulation* 76 (2):60–68.

Ghanem, Waheed A. H. M, and Aman Jantan. 2018. "Hybridizing artificial bee colony with monarch butterfly optimization for numerical optimization problems." *Neural Computing and Applications* 30 (1):163–181.

Ghanem, Waheed A. H. M, and Aman Jantan. 2019. "Training a neural network for cyberattack classification applications using hybridization of an artificial bee colony and monarch butterfly optimization." *Neural Processing Letters* 1–42.

Ghetas, Mohamed, Chan Huah Yong, and Putra Sumari. 2015. "Harmony-based monarch butterfly optimization algorithm." 2015 IEEE International Conference on Control System, Computing and Engineering (ICCSCE), Penang, Malaysia.

Goldberg, D. E. 1998. *Genetic Algorithms in Search, Optimization and Machine learning.* New York: Addison-Wesley.

Goldberg, David E., and John H. Holland. 1988. "Genetic algorithms and machine learning." *Machine Learning* 3 (2):95–99.

Gong, Wenyin, Zhihua Cai, and Charles X. Ling. 2010. "DE/BBO: a hybrid differential evolution with biogeography-based optimization for global numerical optimization." *Soft Computing* 15 (4):645–665.

Guo, Lihong, Gai-Ge Wang, Heqi Wang, and Dinan Wang. 2013. "An effective hybrid firefly algorithm with harmony search for global numerical optimization." *The Scientific World Journal* 2013:1–10. doi: 10.1155/2013/125625.

Hand, D. J. 1992. *Genetic Programming: on the Programming of Computers by Means of Natural Selection*. Cambridge, MA: MIT Press.

He, Y.C., X.L. Zhang, W.B. Li, X. Li, W.L. Wu, and S.G. Gao. 2016. "Algorithms for randomized time-varying knapsack problems." *Journal of Combinatorial Optimization* 31 (1):95–117.

Hu, Hui, Zhaoquan Cai, Song Hu, Yingxue Cai, Jia Chen, and Sibo Huang. 2018. "Improving monarch butterfly optimization algorithm with self-adaptive population." *Algorithms* 11 (5):71.

Ibrahim, Abdelmonem M., and Mohamed A. Tawhid. 2019. "A hybridization of differential evolution and monarch butterfly optimization for solving systems of nonlinear equations." *Journal of Computational Design and Engineering* 6 (3):354–367.

Jada, Chakravarthi, Anil Kumar Vadathya, Anjumara Shaik, Sowmya Charugundla, Parabhaker Reddy Ravula, and Kranthi Kumar Rachavarapu. 2016. "Butterfly mating optimization." In *Intelligent Systems Technologies and Applications*, 3–15. Cham, Switzerland: Springer.

Karaboga, Dervis, and Bahriye Basturk. 2007. "A powerful and efficient algorithm for numerical function optimization: artificial bee colony (ABC) algorithm." *Journal of Global Optimization* 39 (3):459–471. doi: 10.1007/s10898-007-9149-x.

Kennedy, J., and R. Eberhart. 1995. "Particle swarm optimization." Proceeding of the IEEE International Conference on Neural Networks, Perth, Australia, 27 November-1 December.

Khatib, W., and P. Fleming. 1998. "The stud GA: A mini revolution?" Proceedings of the 5th International Conference on Parallel Problem Solving from Nature, New York.

Krynicki, Kamil, Javier Jaen, and Jose A. Mocholí. 2014. "Ant colony optimisation for resource searching in dynamic peer–to–peer grids." *International Journal of Bio-Inspired Computation* 6 (3):153–165. doi: 10.1504/IJBIC.2014.062634.

Li, Junpeng, Yinggan Tang, Changchun Hua, and Xinping Guan. 2014. "An improved krill herd algorithm: krill herd with linear decreasing step." *Applied Mathematics and Computation* 234:356–367. doi: 10.1016/j.amc.2014.01.146.

Li, Xiangtao, Jianan Wang, and Minghao Yin. 2013. "Enhancing the performance of cuckoo search algorithm using orthogonal learning method." *Neural Computing and Applications* 24 (6):1233–1247. doi: 10.1007/s00521-013-1354-6.

Li, Xiangtao, and Minghao Yin. 2012. "Self-adaptive constrained artificial bee colony for constrained numerical optimization." *Neural Computing and Applications* 24 (3–4):723–734. doi: 10.1007/s00521-012-1285-7.

Li, Xiangtao, and Minghao Yin. 2013. "Multiobjective binary biogeography based optimization for feature selection using gene expression data." *IEEE Transactions on Nano Bioscience* 12 (4):343–353. doi: 10.1109/TNB.2013.2294716.

Li, Xiangtao, Jie Zhang, and Minghao Yin. 2014. "Animal migration optimization: an optimization algorithm inspired by animal migration behavior." *Neural Computing and Applications* 24 (7–8):1867–1877. doi: 10.1007/s00521-013-1433-8.

Meng, Xianbing, Yu Liu, Xiaozhi Gao, and Hengzhen Zhang. 2014. "A new bio-inspired algorithm: chicken swarm optimization." In *Advances in Swarm Intelligence*, Ying Tan, Yuhui Shi, and Carlos A. Coello Coello, eds., 86–94. Cham, Switzerland: Springer International Publishing.

Mirjalili, Seyedali. 2015. "The ant lion optimizer." *Advances in Engineering Software* 83:80–98. doi: 10.1016/j.advengsoft.2015.01.010.

Mirjalili, Seyedali, Seyed Mohammad Mirjalili, and Andrew Lewis. 2014. "Grey wolf optimizer." *Advances in Engineering Software* 69:46–61. doi: 10.1016/j.advengsoft.2013.12.007.

Mirjalili, Seyedali, Seyed Mohammad Mirjalili, and Xin-She Yang. 2013. "Binary bat algorithm." *Neural Computing and Applications* 25 (3–4):663–681. doi: 10.1007/s00521-013-1525-5.

Mirjalili, Seyedali, Gai-Ge Wang, and Leandro dos S. Coelho. 2014. "Binary optimization using hybrid particle swarm optimization and gravitational search algorithm." *Neural Computing and Applications* 25 (6):1423–1435. doi: 10.1007/s00521-014-1629-6.

Ouaarab, Aziz, Belaïd Ahiod, and Xin-She Yang. 2014. "Discrete cuckoo search algorithm for the travelling salesman problem." *Neural Computing and Applications* 24 (7–8):1659–1669. doi: 10.1007/s00521-013-1402-2.

Qi, Xiangbo, Yunlong Zhu, and Hao Zhang. 2017. "A new meta-heuristic butterfly-inspired algorithm." *Journal of Computational Science* 23:226–239.

Ram, Gopi, Durbadal Mandal, Rajib Kar, and Sakti Prasad Ghoshal. 2014. "Optimal design of non–uniform circular antenna arrays using PSO with wavelet mutation." *International Journal of Bio-Inspired Computation* 6 (6):424–433.

Runarsson, T.P., and X. Yao. 2000. "Stochastic ranking for constrained evolutionary optimization." *IEEE Transactions on Evolutionary Computation* 4 (3):284–294.

Saremi, Shahrzad, Seyedali Mirjalili, and Andrew Lewis. 2014. "Biogeography-based optimisation with chaos." *Neural Computing and Applications* 25 (5):1077–1097. doi: 10.1007/s00521-014-1597-x.

Saremi, Shahrzad, Seyedeh Zahra Mirjalili, and Seyed Mohammad Mirjalili. 2014. "Evolutionary population dynamics and grey wolf optimizer." *Neural Computing and Applications.* doi: 10.1007/s00521-014-1806-7.

Shukla, A., and V. Goyal. 2019. "Formation of artificial bee colony with monarch butterfly optimization and cluster head load distribution for energy saving in MANET." *Journal of Advanced Research in Dynamical and Control Systems* 11 (10):647–654. doi: 10.5373/JARDCS/V11SP10/20192854.

Simon, D. 2008. "Biogeography-based optimization." *IEEE Transactions on Evolutionary Computation* 12 (6):702–713. doi: 10.1109/TEVC.2008.919004.

Soltani, Parham, and Esmaeil Hadavandi. 2019. "A monarch butterfly optimization-based neural network simulator for prediction of siro-spun yarn tenacity." *Soft Computing* 23 (20):10521–10535.

Storn, Rainer, and Kenneth Price. 1997. "Differential evolution-a simple and efficient heuristic for global optimization over continuous spaces." *Journal of Global Optimization* 11 (4):341–359.

Strumberger, Ivana, Marko Sarac, Dusan Markovic, and Nebojsa Bacanin. 2018. "Hybridized monarch butterfly algorithm for global optimization problems." *International Journal of Computers* 3:63–68.

Strumberger, Ivana, Eva Tuba, Nebojsa Bacanin, Marko Beko, and Milan Tuba. 2018. "Modified and hybridized monarch butterfly algorithms for multi-objective optimization." International Conference on Hybrid Intelligent Systems, Porto, Portugal.

Sun, Lin, Suisui Chen, Jiucheng Xu, and Yun Tian. 2019. "Improved monarch butterfly optimization algorithm based on opposition-based learning and random local perturbation." *Complexity* 2019:4182148.

Wang, G.G., S. Deb and Z.H. Cui. 2019. "Monarch butterfly optimization." *Neural Computing & Applications* 31 (7):1995–2014.

Wang, Gai-Ge, Suash Deb, and Zhihua Cui. 2019. "Monarch butterfly optimization." *Neural Computing and Applications* 31 (7):1995–2014.

Wang, Gai-Ge, Suash Deb, Xinchao Zhao, and Zhihua Cui. 2018. "A new monarch butterfly optimization with an improved crossover operator." *Operational Research* 18 (3):731–755.

Wang, Gai-Ge, Amir H. Gandomi, and Amir H. Alavi. 2014a. "Stud krill herd algorithm." *Neurocomputing* 128:363–370. doi: 10.1016/j.neucom.2013.08.031.

Wang, Gai-Ge, Amir H. Gandomi, Amir H. Alavi, and Suash Deb. 2015. "A hybrid method based on krill herd and quantum-behaved particle swarm optimization." *Neural Computing and Applications.* doi: 10.1007/s00521-015-1914-z.

Wang, Gai-Ge, Amir H. Gandomi, Xiangjun Zhao, and Hai Cheng Eric Chu. 2014. "Hybridizing harmony search algorithm with cuckoo search for global numerical optimization." *Soft Computing.* doi: 10.1007/s00500-014-1502-7.

Wang, Gai-Ge, Amir Hossein Gandomi, and Amir Hossein Alavi. 2013. "A chaotic particle-swarm krill herd algorithm for global numerical optimization." *Kybernetes* 42 (6):962–978. doi: 10.1108/K-11-2012-0108.

Wang, Gai-Ge, Amir Hossein Gandomi, and Amir Hossein Alavi. 2014b. "An effective krill herd algorithm with migration operator in biogeography-based optimization." *Applied Mathematical Modelling* 38 (9-10):2454–2462. doi: 10.1016/j.apm.2013.10.052.

Wang, Gai-Ge, Amir Hossein Gandomi, Xin-She Yang, and Amir Hossein Alavi. 2014. "A novel improved accelerated particle swarm optimization algorithm for global numerical optimization." *Engineering Computations* 31 (7):1198–1220. doi: 10.1108/EC-10-2012-0232.

Wang, Gai-Ge, Lihong Guo, Hong Duan, and Heqi Wang. 2014. "A new improved firefly algorithm for global numerical optimization." *Journal of Computational and Theoretical Nanoscience* 11 (2):477–485. doi: 10.1166/jctn.2014.3383.

Wang, Gai-Ge, Guo-Sheng Hao, Shi Cheng, and Zhihua Cui. 2017. "An improved monarch butterfly optimization with equal partition and F/T mutation." International Conference on Swarm Intelligence, Fukuoka, Japan.

Wang, Gai-Ge, Guo-Sheng Hao, Shi Cheng, and Quande Qin. 2016. "A discrete monarch butterfly optimization for Chinese TSP problem." International Conference on Swarm Intelligence, Bali, Indonesia.

Wang, Gai-Ge, Xinchao Zhao, and Suash Deb. 2015. "A novel monarch butterfly optimization with greedy strategy and self-adaptive." 2015 Second International Conference on Soft Computing and Machine Intelligence (ISCMI), Hong Kong, China.

Wang, Gaige, Lihong Guo, Hong Duan, Heqi Wang, Luo Liu, and Mingzhen Shao. 2013. "Hybridizing harmony search with biogeography based optimization for global numerical optimization." *Journal of Computational and Theoretical Nanoscience* 10 (10):2318–2328. doi: 10.1166/jctn.2013.3207.

Wang, Xitong, Xin Tian, and Yonggang Zhang. 2019. "A new monarch butterfly optimization algorithm with SA strategy." International Conference on Knowledge Science, Engineering and Management, Athens, Greece.

Wolpert, D. H., and W. G. Macready. 1997. "No free lunch theorems for optimization." *IEEE Transactions on Evolutionary Computation* 1 (1):67–82.

Yang, X. S. 2010a. "Firefly algorithm, stochastic test functions and design optimisation." *International Journal of Bio-Inspired Computation* 2 (2):78–84. doi: 10.1504/IJBIC.2010.032124.

Yang, X. S. 2010b. *Nature-Inspired Metaheuristic Algorithms*. 2nd ed. Frome: Luniver Press.

Yang, Xin-She. 2010c. "Firefly algorithm, stochastic test functions and design optimisation." *arXiv preprint arXiv:1003.1409*.

Yang, Xin-She, and Suash Deb. 2009. "Cuckoo search via Levy flights." 2009 World Congress on Nature & Biologically Inspired Computing (NaBIC 2009), Coimbatore, India.

Yang, Xin-She, and Suash Deb. 2013. "Cuckoo search: recent advances and applications." *Neural Computing and Applications* 24 (1):169–174. doi: 10.1007/s00521-013-1367-1.

Yazdani, Samaneh, and Esmaeil Hadavandi. 2019. "LMBO-DE: a linearized monarch butterfly optimization algorithm improved with differential evolution." *Soft Computing* 23 (17):8029–8043.

Yi, Jiao-Hong, M. Lu, and X. J. Zhao. 2020. "Quantum inspired monarch butterfly optimization for UCAV path planning navigation problem." *International Journal of Bio-Inspired Computation.* 15 (2):75–89.

Zhao, Xinchao, Xiao-Shan Gao, and Ze-Chun Hu. 2007. "Evolutionary programming based on non-uniform mutation." *Applied Mathematics and Computation* 192 (1):1–11. doi: 10.1016/j.amc.2006.06.107.

Index

Note: *Italicized* page numbers refer to figures, **bold** page numbers refer to tables

A

B

C

D

E

F

G

H

I

K

L

M

N

O

P

Q

R

S

T

U

V

W

9780367755355